Engineering Materials

More information about this series at http://www.springer.com/series/4288

Alexei Nazarov · Francis Balestra
Valeriya Kilchytska · Denis Flandre
Editors

Functional Nanomaterials and Devices for Electronics, Sensors and Energy Harvesting

Springer

Editors
Alexei Nazarov
Lashkaryov Institute of Semiconductor Physics
National Academy of Science of Ukraine
Kyiv
Ukraine

Valeriya Kilchytska
Microelectronics Laboratory, ICTEAM Institute
Université catholique de Louvain
Louvain-la-Neuve
Belgium

Francis Balestra
Sinano Institute IMEP-LAHC, CNRS Grenoble
IMEP-LAHC
Grenoble
France

Denis Flandre
Electrical Engineering (ELEN), ICTEAM Institute
Université catholique de Louvain
Louvain-la-Neuve
Belgium

ISSN 1612-1317 ISSN 1868-1212 (electronic)
ISBN 978-3-319-08803-7 ISBN 978-3-319-08804-4 (eBook)
DOI 10.1007/978-3-319-08804-4

Library of Congress Control Number: 2014946761

Springer Cham Heidelberg New York Dordrecht London

Printed on acid-free paper

Springer is part of Springer Science+Business Media (www.springer.com)

Preface

This book is devoted to fast the evolving field of modern material science and nanoelectronics, and more particularly to physics and technology of functional nanomaterials and devices. The book focuses on nanodevices for electronics, sensors, and energy harvesting, considering as main device structure—the semiconductor-on-insulator (SemOI) one. The book reports the recent achievements in this field from leading companies and universities in Europe, Russia, and Ukraine. It is articulated around four main topics: (1) Nanoscale CMOS materials and devices; (2) Beyond CMOS materials, devices and their diagnostics; (3) New functional nanomaterials and nanoscaled devices for energy harvesting, light emission, optoelectronics and THz range: (4) NanoSensors and MEMS/NEMS.

Part I is focused on new SemOI materials for More Moore and More-than-Moore applications. Ultrathin silicon SOI structures are necessary for production of fully depleted SOI devices of the 22 nm technology node and beyond. The materials innovation for RF electronics, Si-based photonics, and 3D integration are presented. Device solutions for very low-energy computing, high-performed tunnel FETs, 3D nanowire RAM cells, and mechanical flexible CMOS devices on plastics are described.

Part II of the Book deals with the physics of novel "beyond CMOS" devices such as IR memory cells on basis of Si/Ge nanoheterostructures, nonvolatile memory based on graphene on ferroelectric substrate, Si spintronic devices and the AFM diagnostics for different functional nanostructurated material and devices.

In Part III, we focus on functional nanomaterials and structures regarding self-powered systems, solar energy harvesting structures, and THz electronics. Also nanocomposite dielectric materials for light-emitting materials and other optoelectronics applications are also discussed.

Part IV considers the application of SemOI nanowire structures for radiation sensors, biosensors, chemical sensors, and MEMS. The use of SemOI substrates allows a considerable increase of the sensitivity of the sensors, as well as the fabrication of MEMS compatible with CMOS technology. Additionally, Si and SiC nanodot materials are considered as fluorescent markers in different biomedical applications.

This book will be useful not only to specialists in nano, microelectronics, and functional nanomaterials but also to students and to the wider audience of readers who are interested by new directions in modern material science, electronics, and optoelectronics.

Alexei Nazarov
Francis Balestra
Valeriya Kilchytska
Denis Flandre

Contents

Part I
Nanoscale CMOS Materials and Devices

Engineered Substrates for Advanced CMOS Technology Nodes and More-Than-Moore Applications

Konstantin K. Bourdelle

Abstract Traditional planar bulk or partially depleted SOI (PDSOI) CMOS transistor architectures at present leading edge of miniaturization are plagued by limitations due to unacceptably high current leakages and variability. To cope with these intrinsic limitations there is a need to introduce innovative technologies which take advantage of the benefits of Fully Depleted (FD) devices. There are two main architectures for the undoped channel FD device: 3D FinFETs (SOI or bulk-based) and 2D FDSOI-based transistors. Both of them are being introduced in high volume manufacturing (HVM). A pioneer of SOI concept, silicon on sapphire (SOS) substrates, have recently entered a mainstream radio frequency (RF) application market. Other flavors of engineered substrates, e.g. for photonics or 3D-based applications, have moved from research to industrial development phase. In this work an overview of the recent advances in the development of the engineered substrates for More Moore and More-than-Moore applications will be presented.

1 Introduction

Over past decades, exponential gains in computational power have fuelled unprecedented progress in innovation and economical growth which has provided considerable benefits to society and has also enabled entirely new businesses such as e-commerce, social networking and mobile devices. This expected progress, however, now becomes threatened: an end to the gains in computing power and the raise of power consumption. The purely dimensional scaling era worked well down to 130 nm node. Since then such model of scaling has become less productive in improving CMOS performance due to increases short-channel effects (SCE). The local strain technologies were then introduced as mobility boosters.

K.K. Bourdelle (✉)
Soitec, Parc Technologique Des Fontaines, 38190 Bernin, France
e-mail: konstantin.bourdelle@soitec.com

A. Nazarov et al. (eds.), *Functional Nanomaterials and Devices for Electronics, Sensors and Energy Harvesting*, Engineering Materials,
DOI 10.1007/978-3-319-08804-4_1

These boosters, however, have started to run out of steam in the recent years mostly due to the bottleneck issues related to the leakage and device/circuit variability. That prompted the leading industry players to adopt FD device architectures: ultra-thin body and buried oxide (BOX) UTBB SOI (at 28 nm node) [1] and FinFET (at 22 nm node) [2]. Most important these two innovations feature new device geometry as compared to their planar bulk or PDSOI predecessors.

The limitations of Si-based CMOS, including FD devices in reaching high performance requirement are mostly due to reduction of the efficiency of uniaxial strain at 14 nm node and beyond. High mobility channel/substrate engineering is essential to boost performance especially for NMOSFETs. The corresponding innovative solutions must be: (i) Compatible with large diameter Si substrates to fully utilize existing platform therefore necessitating co-integration of Ge and III–V materials; (ii) Multigate devices will be needed to address the SCE issues; (iii) The solutions must meet defectivity as well as multi-V_T requirements. They also must assure functional system on chip (SOC) circuits with high yield in HVM and be "economically affordable" based on the criteria of a broad consumer electronics market. Using Smart Cut™ technology SOI/UTBB-like devices with undoped and high mobility channels on thin insulator film can be fabricated in the heterogeneous integration concept. The technology is capable to address lattice mismatch challenges and transfer high quality ultra-thin (15–25 nm) Ge/III–V films on Si substrates.

The definition of the substrates for More than Moore applications is rather broad and varies from source to source. Such applications would/might incorporate RF, power, photonics, imagers and micro electro mechanical systems (MEMS) devices. Soitec SOI product line to address some of these applications is depicted in Fig. 1. In addition, in the recent years 3D integration has appeared is a promising solution to overcome the limitations of size reduction and to increase circuit performances and functionality. By offering heterogeneous integration, shorter interconnection lengths and lower cost, this technology can address both More Moore and More than Moore applications.

2 Engineered Substrates for More Moore Applications

In this section the emphasis will be made on FD device architectures (2D and 3D) with Si channels. In addition the new channel materials, strained Si (sSi), Ge and III–V alloys, and corresponding substrates requirements will be reviewed in the context of the potential introduction of the nanowire (NW) devices for sub-10 nm technology nodes.

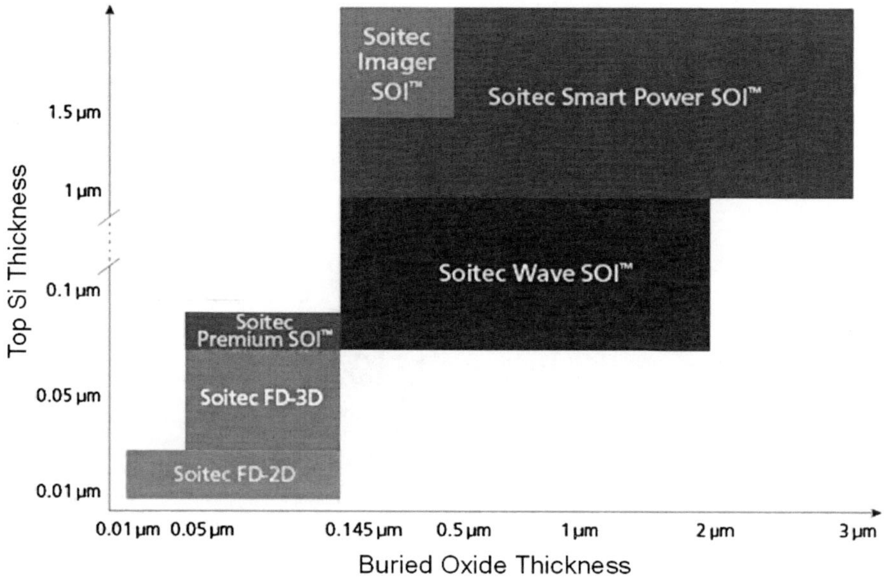

Fig. 1 SOI product line. The *top* Si layers with *thickness* larger than 1 μm are obtained with the additional epitaxial step

2.1 Planar FDSOI Technology

2D FDSOI-based devices have been extensively studied for more than 20 years. Their benefits are well documented: e.g. better immunity to SCEs and to random dopant fluctuations [1]. In addition, FDSOI substrates with thin, ≤25 nm, BOX (UTBOX) address a major challenge of undoped channel devices, i.e. fabrication of multiple threshold voltage, V_T, devices with dynamic V_T modulation. That is done by formation of different ground planes (GP) below the BOX and back biasing (V_B). Such strategy is also attractive due to the possibility to reuse the bulk forward (FBB) and reverse (RBB) biasing techniques historically developed for planar bulk CMOS technologies. Top Si layer thickness (typical thickness ≤15 nm) non-uniformity was a formidable challenge to overcome before considering industrialization of 2D FDSOI technology for mainstream SOC applications. In FDSOI devices the Si thickness is strongly coupled to the MOSFET parameters, e.g. significant thickness non uniformity results in non acceptable V_T fluctuation. Typical uniformity requirements include within-the-wafer and wafer-to-wafer uniformity. Their combination is defined as layer total thickness variation (LTTV) and determines the overall manufacturing process window. LTTV of top Si has to be in the nanometer range for all wafers and all sites in order to meet the FD specifications.

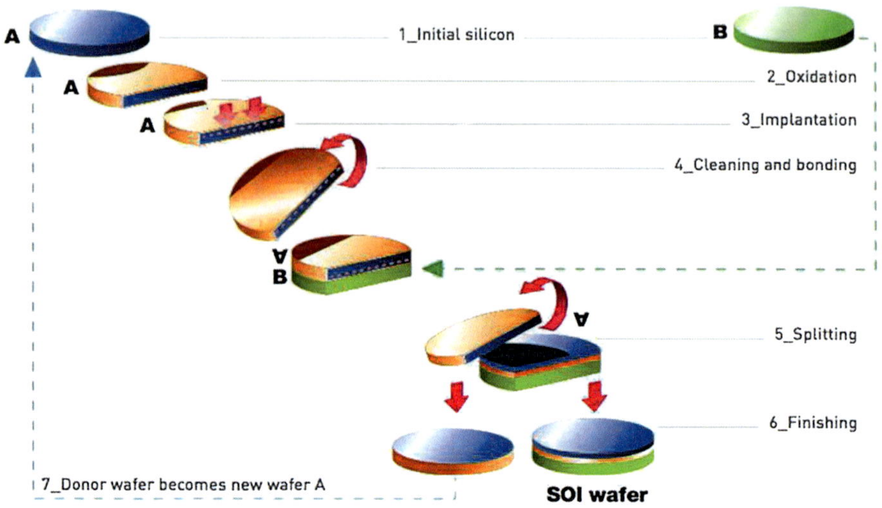

Fig. 2 Schematics of smart cut process flow used in the fabrication of UTBOX wafers [3]

Fabrication of UTBOX substrates is based on the conventional Smart Cut process depicted in Fig. 2. Uniformity of thin BOX thickness is defined by the donor wafer oxidation process step. Top Si thickness control is determined by ion implantation, splitting and finishing steps. Figure 3a shows that via thorough optimization of these steps within the wafer thickness uniformity ± 4 Å is achieved, as measured by optical ellipsometry with 41 points mapping [3]. To address wafer to wafer uniformity challenge the Advanced Process Control (APC) was implemented at post-split process module [4]. APC includes a specific tailored clean, which reduces a wafer to wafer SOI thickness variation from a standard ± 10 Å (as obtained through usual tool to tool matching) down to less than ± 2 Å. The LTTV performance for UTBOX (12 nm top Si and 25 nm BOX) substrates currently available for HVM is currently as follows: top Si thickness maximum variation of ± 5 Å (6σ value, all sites, all wafers), corresponding to an overall 1.6 % 1σ thickness control (Fig. 3b).

Device scale thickness variation is monitored by micro-roughness measurements using Atomic Force Microscopy (AFM) technique. Optimization of the thermal smoothing step in the finishing module (the step is driven by the surface diffusion of Si atoms) has lead to significant improvement in surface roughness of the fully processed UTBOX wafers. As a result the surface quality of the UTBOX family of substrates is currently very close to that of the polished bulk wafers. That is confirmed by the comparison of Power Spectral Density (PSD) curves obtained from $30 \times 30\ \mu m^2$ AFM scans (see Fig. 4). The corresponding root mean square (RMS) values have been reduced from the previously reported 2.0 Å [4] down to current 0.8 Å. Peak to valley (PV) local thickness variation has improved from 20 to 8 Å. The reported 8 Å PV measured performance is close to the theoretical minimum of 7 Å [3].

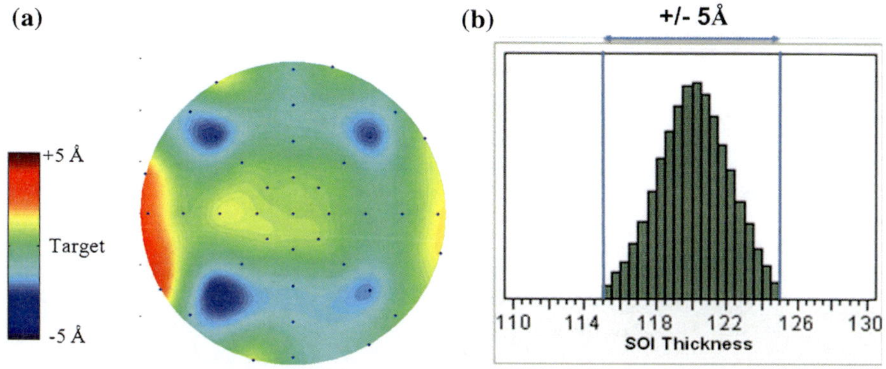

Fig. 3 **a** Ellipsometry map of within the wafer thickness uniformity; **b** UTBOX 12/25 nm all wafers all points top Si thickness distribution [3]

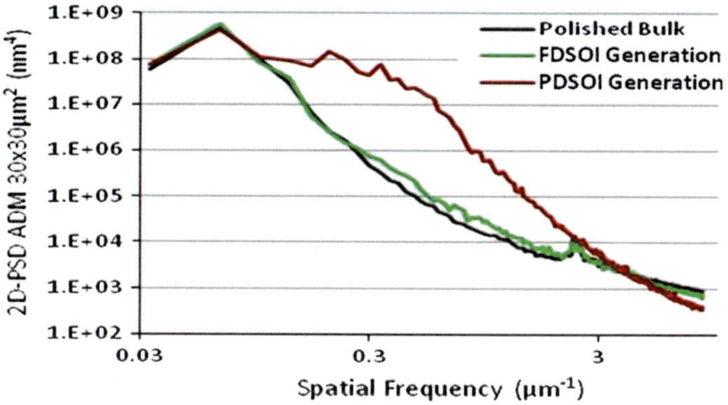

Fig. 4 Power spectral distribution from AFM 30 × 30 μm^2 scans for polished bulk, PDSOI and current generation of FDSOI/UTBOX25 substrates [3]

As mentioned above ultrathin Si films are today combined with ultrathin BOX layers to fabricate multi-*Vt* devices for low power (LP) and SOC applications [1]. Thinning of the BOX leads to increase of the electrostatic coupling with the substrate via GP contact. This results in effective dynamic *Vt* modulation of more than 100 mV for a back gate (BG) biasing in the range of ±1 V (with BOX thickness of 25 nm). However, to replace 145 nm thick BOX, industry standard PDSOI technologies, these ultrathin BOX layers must exhibit the same high quality and reliability. It is of importance therefore to verify that the presence of many critical steps within Smart Cut flow does not degrade the intrinsic reliability of ultrathin BOX. In [5] the BOX breakdown reliability of UTBOX25 substrates was investigated using simple MOS capacitors. Figure 5 shows time dependent dielectric breakdown (TDDB) distributions obtained for UTBOX25 wafers

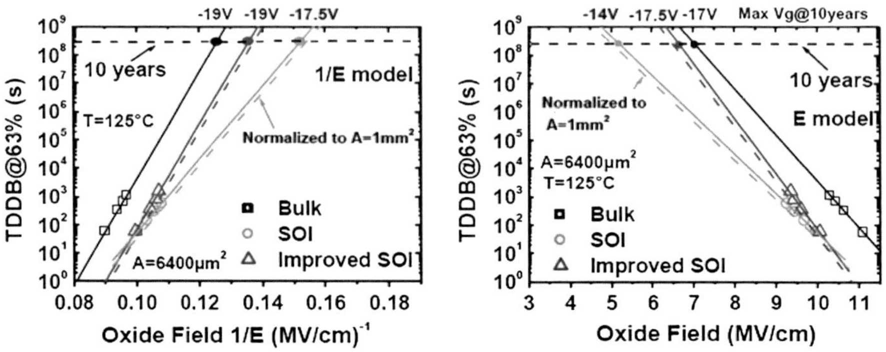

Fig. 5 (*left*) 10-years extrapolation results reported through 1/E model. (*right*) 10-years extrapolation results reported through E model [5]

fabricated with two different process flows (denoted as "SOI" and "improved SOI") compared with that for the oxide grown on bulk Si wafers. For this range of oxide thicknesses (>10 nm), TDDB is driven by the oxide electric field, E_{OX}, rather than by the maximum gate voltage V_G. Therefore authors [5] used the "E" and "1/E" models to extrapolate the TDDB lifetime at 10 years. It is seen in Fig. 5 that "improved SOI" and bulk oxide exhibit similar 10 years breakdown lifetimes. By using the "1/E" model, the maximum gate voltage is estimated to 19 V. The "E" model itself is more pessimistic than the "1/E" model consistently with literature data that lead to lower values. For the standard "SOI" oxide, the lifetime margin is reduced. Nevertheless, the extracted maximum voltages are 3–6 times larger than typical back-biases used for the 28 nm node [5].

2.2 3D Multiple Gate Devices

Device and circuit variability has become a major concern, forcing the industry to consider new transistor architectures for the 28 nm technology generation and below [1, 2]. Multiple gate devices like FinFETs or Tri-Gate can be fabricated using either bulk [2] or SOI substrates (see [6] for the recent reference). In FD architectures the electrostatic integrity and electrical characteristics are controlled by the thickness of the channel and, in the 3D case, its height and shape. The process control of these dimensions will play a critical role in the variability of the new technologies. In [7] the process-induced variability in a FinFET suitable for the 16 nm technology generation and built on SOI or bulk substrates was studied. Authors [7], using extensive numerical simulations, have compared the process sensitivity of key figures of merit including V_T, subthreshold slope (*SS*), drain induced barrier lowering (*DIBL*) and fixed overdrive drive current (I_{ODSAT}). The structures of the simulated n-channel FinFETs (FF) are schematically illustrated in Fig. 6 (left) and main device design parameters are listed in Fig. 6 (right).

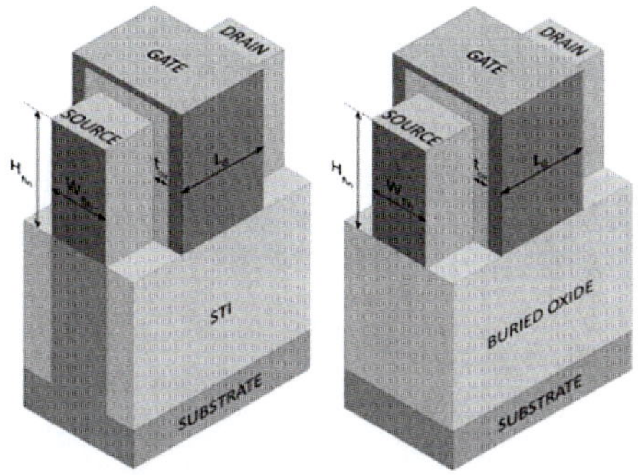

Parameter	Value
Lg	25 nm
Fin Width, W_{fin}	12 nm
Fin Height, H_{fin}	30 nm
Spacer	6 nm
Extension	No extension doping
HDD profile	2nm/dec (σ=2.1nm)
HDD doping	1×10^{20} cm^{-3}
EOT	0.8 nm
Stopper doping	5×10^{18} cm^{-3}
Channel doping	1×10^{15} cm^{-3}

Fig. 6 Schematics of the bulk (*left*) and SOI (*center*) FinFETs. Device design used in the simulations (*right*) (from [7])

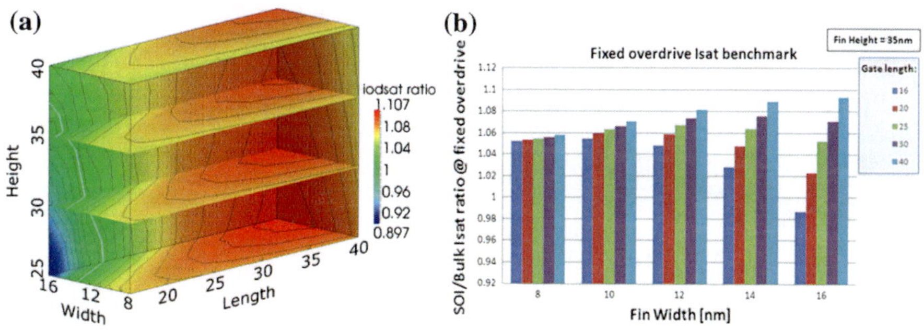

Fig. 7 **a** Colour and contour map of I_{ODSAT}(SOI)/I_{ODSAT}(bulk) in the experiment space. The *white* contour marks the unity ratio where bulk performance is equal to that of SOI. **b** Detailed comparison for $H_{fin} = 35$ nm of this I_{ODSAT} boost for SOI [7]

The map of the I_{ODSAT}(SOI)/I_{ODSAT}(bulk) ratio in the $H/W/L_G$ space is shown in Fig. 7a. It indicates a performance advantage for the SOI-FF which is typically 5 % over the parameter space. Figure 7b provides a more detailed comparison for $H_{Fin} = 35$ nm of this I_{ODSAT} boost for SOI, showing improved current for narrower Fins, especially when the gate length is short. These data are of interest when used to assess the performance corners resulting from process variability. SOI-FF will benefit from a better Fin definition which is defined by the top Si thickness uniformity of the starting wafers [8]. The process-induced variations scenarios for bulk-FF and SOI-FF are summarized in Table 1. SOI-FF does have V_T corner range improved by factor two. In addition the range of I_{SAT} variation is reduced from 28.7 % in the bulk case down to 7.3 % for the SOI. That proves that better Fin height control directly translates into reduced I_{SAT} variability [7].

Table 1 Process-induced variation assumptions and process corners for device electrostatics [7]

	H control (Å)	*W* control (Å)	V_T min (V)	V_T max (V)	V_T range (V)	*SS* min (mV/dec)	*SS* max (mV/dec)	*SS* range (mV/dec)	*SS* range (%)
Bulk fin	±5	±2	0.147	0.212	0.065	69.5	77.6	8.1	11.05
SOI fin	±2	±2	0.167	0.229	0.062	71.9	80.3	8.5	11.2
SOI fin	±1	±1.5	0.175	0.221	0.046	72.8	79.2	6.33	8.37

2.3 New Channel Materials and New Device Architectures

Carrier mobility improvement becomes necessary to enhance device performance for the advanced nodes. The combination of high mobility channel and FD devices architectures is most appealing for high density, high performance, and low leakage applications. The challenges of scaling of Si-based devices below 11 nm node (using local strain options) have prompted researchers in recent years to consider fundamental paradigm change, i.e. replacement of the conventional Si channel by materials with much higher carrier mobilities: wafer scale strain Si (sSi), Ge and III–V (see Table 2 below). That would advance CMOS beyond traditional Si (SOI or bulk), and as a consequence open a new path to performance increase while still leveraging mature CMOS technology for large diameter Si wafers.

(a) Tensile strained silicon on insulator (sSOI) substrates represent a promising option to improve the carrier mobility of NMOS devices, without significant degradation of PMOS performance (with proper process integration know-how). Technical details of fabrication of 300 mm sSOI substrates with top Si thickness of 12 nm, BOX thickness of 25 nm, and stress level of 1.3 GPa are described in [9]. Schematics of corresponding Smart Cut process flow is shown in Fig. 8. Within the wafer sSi thickness uniformity is determined by donor wafer epitaxy and finishing process steps. Figure 9a shows typical sSi thickness uniformity maps (at the thickness range levels at 12 Å and down to 9 Å, measured by optical ellipsometry technique with 41 points). Note: to get the industrial acceptance within the wafer sSi layer uniformity is required to be at the level of more mature UTBOX family of substrates (described in previous section). Wafer to wafer sSi thickness control is largely determined by the epitaxy step and is currently at par with that of UTBOX substrates due to implementation of the APC module (see Fig. 9b) [9]. Stress uniformity measurements are usually performed using well established Raman spectroscopy technique [10]. Figure 10a shows typical stress maps (with 41 measurement points) for an sSOI wafer (sSi thickness of 12 nm and BOX thickness of 25 nm). Device scale stress variation results obtained with the same technique (scan size of $80 \times 80\ \mu m^2$) are shown in Fig. 10b. The stress variation of typically 0.02 GPa RMS is observed (related to the SiGe buffer layer local relaxation conditions) [9].

The devices fabricated on such substrates showed the gain in performance of at least 20 % for NMOS as compared to the standard SOI substrate [11, 12]. The efficiency of back gate bias for V_T modulation has been confirmed as well as an additional performance booster [11]. Figure 11 illustrates main results published recently for the devices fabricated on the sSOI substrates (stress level of 1.3 GPa and BOX thickness of 25 nm).

The sSOI substrates are also considered as potential candidates for sub-10 nm nodes with NW-based devices architectures. The reader is referred, e.g., to [13, 14] for recent advances. In [13] top-down Si NWs under tensile elastic uniaxial strain up to 4.5 % were demonstrated. Authors of [14] fabricated inverters based on

Table 2 Physical properties of different semiconductor materials

Parameter	Si	Ge	GaAs	InP	$In_{0.53}Ga_{0.47}As$	$In_{0.7}Ga_{0.3}As$	InAs	GaSb	InSb
Lattice constant (Å)	5.431	5.658	5.653	5.869	5.8687	5.937	6.058	6.09593	6.4793
Band gap (eV)	1.12	0.66	1.42	1.34–1.35	0.74	0.58	0.35–0.36	0.726	0.17
Electron mobility (cm^2/Vs)	1500–1600	3900	8500–9200	4600–5400	12000	20000	33000–40000	≤3000	77000–80000
Hole mobility (cm^2/Vs)	430–450	1900	400	150–200	300	400	460–500	≤1000	850–1250
Effective density of states in conduction band, N_c (cm^{-3})	2.8×10^{19}	1.04×10^{19}	4.7×10^{17}	5.7×10^{17}	2.1×10^{17}	1.6×10^{17}	8.7×10^{16}	2.1×10^{17}	4.2×10^{16}
Effective density of states in valence band, N_v (cm^{-3})	1.04×10^{19}	6.0×10^{18}	7.0×10^{18}	1.1×10^{19}	5.5×10^{18}	5.9×10^{18}	6.6×10^{18}	1.8×10^{19}	7.3×10^{18}
Dielectric constant, k	11.9	16	13.1	12.4	13.9	NA	15.5	15.7	17.7
Melting point, T_m (°C)	1412	937	1240	1060	NA	NA	942	712	527

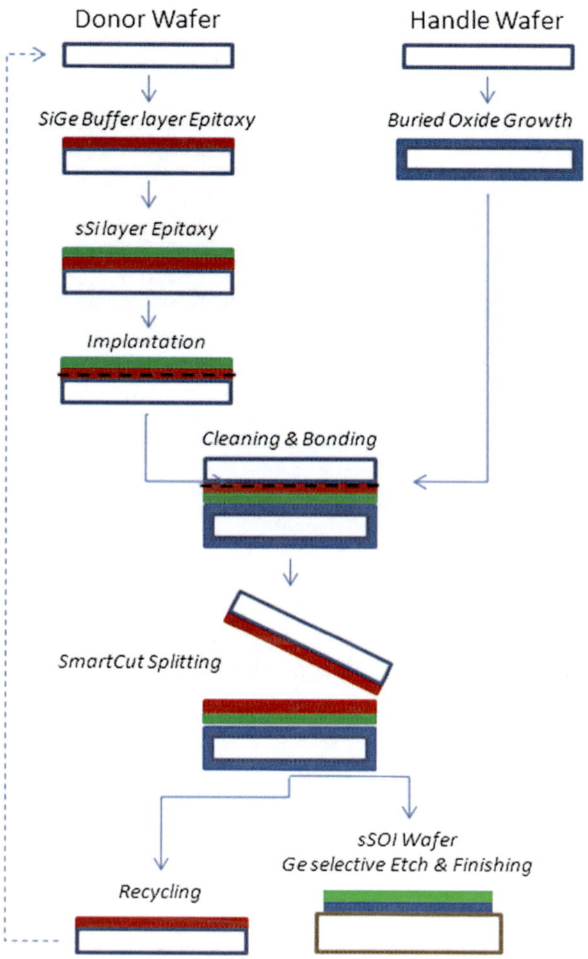

Fig. 8 Schematics of smart cut process flow used in the fabrication of sSOI wafers [9]

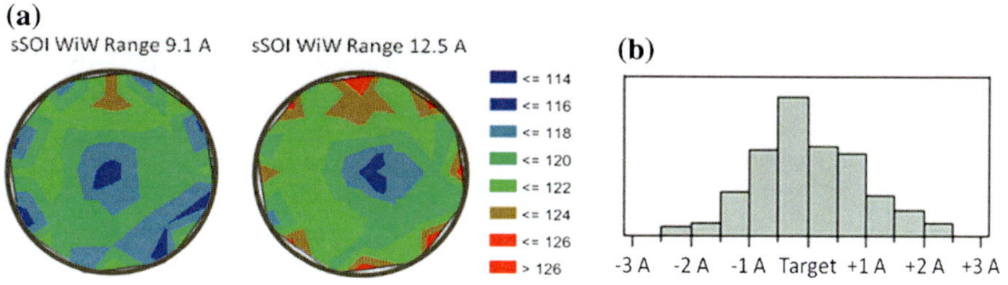

Fig. 9 **a** Typical sSOI within the wafer thickness uniformity mapping. **b** sSOI wafer to wafer mean thickness distribution [9]

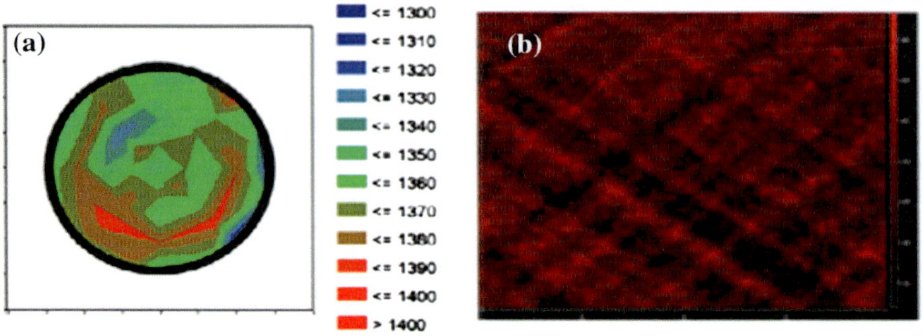

Fig. 10 **a** Within the wafer stress variation for a sSOI wafer. **b** 80 × 80 μm^2 local strain mapping [9]

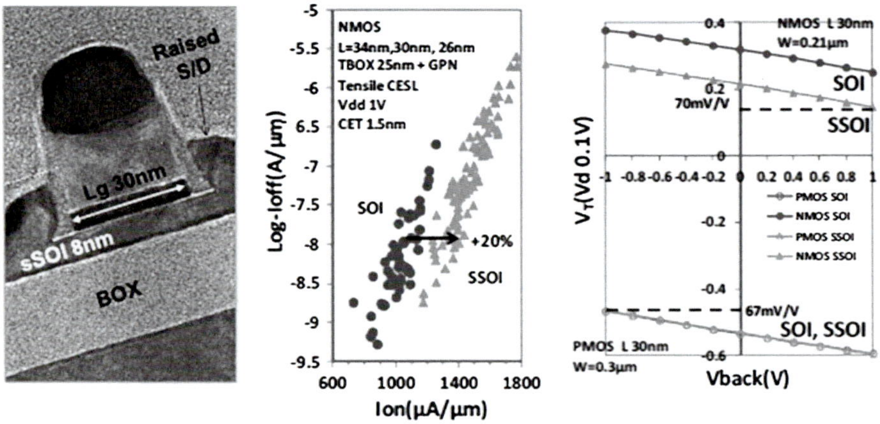

Fig. 11 XTEM micrograph of sSOI NMOS device with $L_G = 30$ nm (*left*). NMOS I_{ON}/I_{OFF} ratio for sSOI and SOI wafers (*center*). NMOS and PMOS linear V_T evolution on back bias voltage in RBB and FBB modes (*right*) (from [11])

uniaxially tensile sSi NW tunnelling field-effect transistors (TFETs). These first sSi NW complementary TFET inverters show sharp transitions and fairly high static gain even at very low $V_{DD} = 0.2$ V [14].

(b) In the last years, the semiconductor industry was re-introducing Ge to the front end process flows with a steady increase of Ge content in critical device areas. Pure Ge nowadays (despite limitations related to band-to-band tunnelling) is considered one of the promising channel materials to replace Si in future PMOS devices due to its high hole mobility. Technical details of fabrication of 200 mm GeOI substrates using Smart Cut technology are described in [15]. Figure 12a shows the picture of fully processed 200 mm GeOI wafer. Since GeOI-based devices are considered for sub-10 nm nodes such solution must be compatible with 300 mm and potentially 450 mm production environment. That defines the structure of the donor wafer, i.e. Ge layer epitaxially grown on the bulk Si

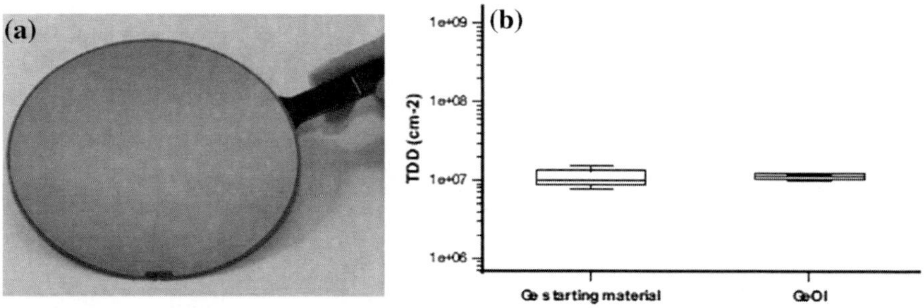

Fig. 12 **a** Picture of fully processed 200 mm GeOI wafer. **b** TDD in Ge epitaxial film before and after smart cut layer transfer [15]

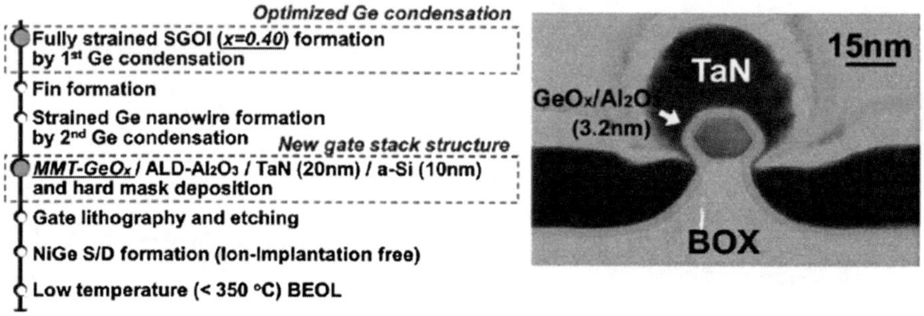

Fig. 13 (*left*) Fabrication process of strained-Ge nanowire MOSFETs with GeOx /ALD-Al_2O_3 gate stack. (*right*) Cross-sectional TEM image of the fabricated device [16]

substrate. The crystal quality of the transferred Ge layer is of importance for the device fabrication. The main figure of merit is threading dislocation density, TDD. Figure 12b shows the current level of TDD in starting material and illustrates the fact that Smart Cut technology does not degrade the crystal quality obtained in the donor substrate.

Another way to fabricate the GeOI substrates is condensation technique which uses SOI wafers as starting material (see [16] for details). Authors of [16] reported uniaxially strained-Ge nanowire channel formed by improved two-step Ge condensation technique, which induced uniaxial stress along the channel direction. To improve the gate interface characteristics, a GeOx layer was formed by using the modified magnetron typed plasma at 300 °C on the strained-Ge nanowire surface. A 3.2 nm Al_2O_3 layer was then grown by using an atomic layer deposition (ALD) system. NiGe-metal source-drain, S/D, structures were formed by salicide-like process on the unintentionally doped Ge wire. The fabrication process flow and cross-section of PMOS device are shown in Fig. 13. Authors of [16] demonstrated a record-high hole mobility ($\mu_{eff} = 1922$ cm^2/Vs) of a Ge nanowire MOSFET. High intrinsic transconductance G_{msat} of 1.21 mS/µm and low off-current of

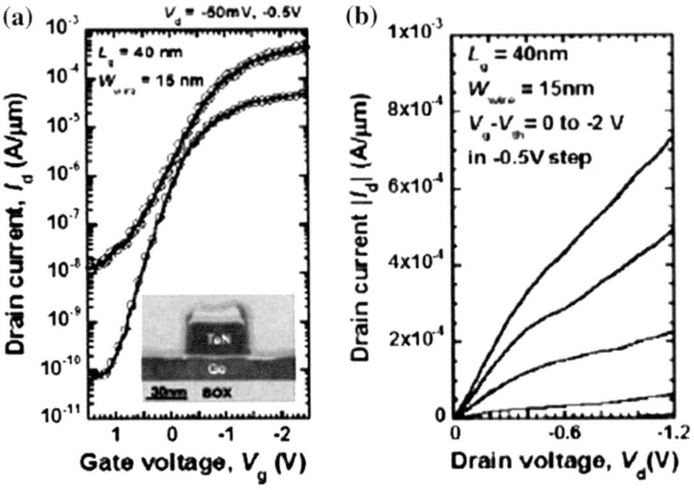

Fig. 14 **a** I_D-V_G and **b** I_D-V_D characteristics of fabricated strained Ge nanowire MOSFET with $W_{wire} = 15$ nm, $L_G = 40$ nm [16]

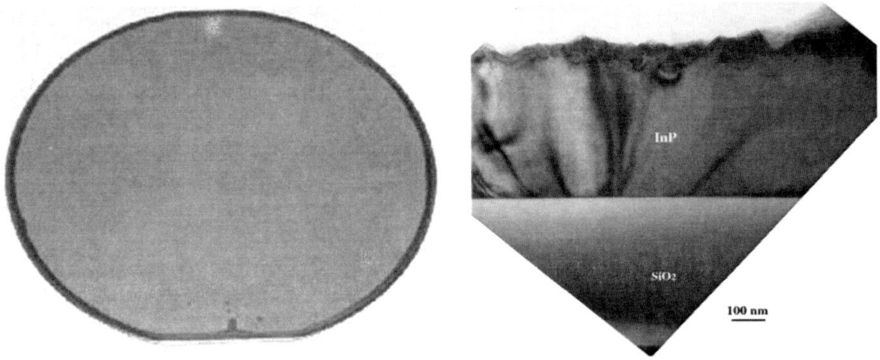

Fig. 15 (*left*) Picture of 3" GaAs thin film transferred onto Si [18]. (*right*) XTEM cross section of InP film transferred on Si wafer covered with oxide [19]

2.7×10^{-9} A/μm at $V_D = -0.5$ V were achieved for the device with the L_G of 45 nm due to the reduced interface state density. $I_D - V_G$ and $I_D - V_D$ characteristics of fabricated strained Ge nanowire MOSFETs with $L_G = 40$ nm and $W_{wire} = 15$ nm are shown in Fig. 14.

(c) III–V channel materials, particularly InGaAs alloys, are considered to address the challenges of NMOS performance boost. That would necessitate co-integration of III–V and Ge materials: proof of concept was recently reported in [17]. The transfer of GaAs and InP layer onto Si substrates using Smart Cut technology has been described in [18–20]. The pictures of GaAsOI and InPOI

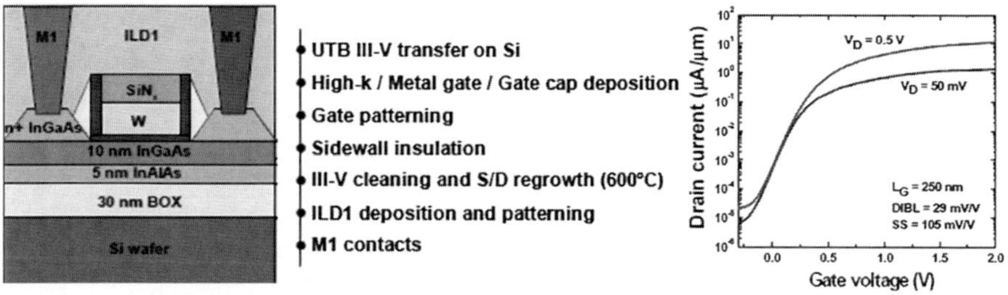

Fig. 16 (*left*) Process flow and schematic for gate-first implant-free UTB InGaAs/InAlAs MOSFETs on silicon. (*right*) I_D-V_G characteristic of a short-channel MOSFET as described in Fig. 16a with an EOT of 17.5 Å, 10 nm InGaAs channel and 5 nm InAlAs back barrier. Good electrostatic integrity and I_{ON}/I_{OFF} ratio are obtained due to the thin-body structure. The on-current is limited by the high access resistance *below* the sidewalls [21]

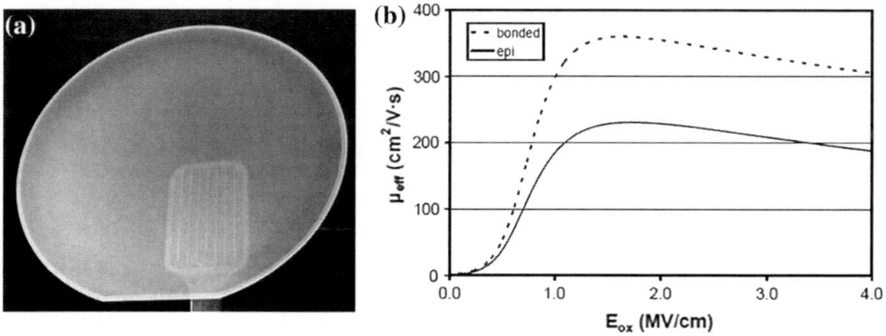

Fig. 17 **a** Picture of 6" fully processed SOS wafer. **b** Comparison of effective electron mobility for BSOS and SOS [24]

substrates are shown in Fig. 15. Recently, IBM-Zurich team has reported high quality InGaAs/InAlAs heterostructure devices (channel thickness <10 nm) fabricated on Si substrates using Smart Cut approach [21]. The thermal stability of the bonded layer enabled to integrate III–V MOSFETs at 500 nm pitch using a gate-first flow featuring raised S/D grown at 600 °C [21]. The process flow and I_D–V_G curves are show in Fig. 16. The authors of [21] have stated that the donor wafer can be recycled for a cost-effective process. This publication proves the fact that Smart Cut implants do not degrade the electrical quality of InGaAs/InAlAs layers. The fabrication of strained [22] and high performance (with $L_G = 20$ nm) [23] InGaAs-OI devices on Si substrates using bonding and etchback (BESOI) technique has been reported recently by the University of Tokyo team.

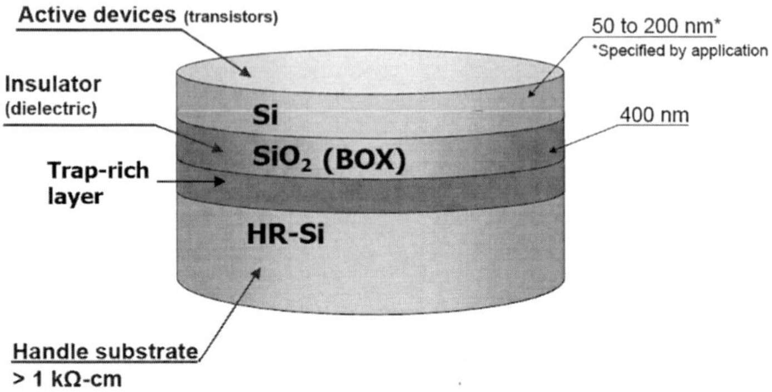

Fig. 18 Wafers commercially available from Soitec for RF and mixed-mode integration [27]

3 Engineered Substrates for More-Than-Moore Applications

As mentioned above the definition of the engineered Substrates for More-than-Moore applications is rather broad. In this section the author will focus in on the RF, photonics and 3D applications.

3.1 RF Applications: SOS and HR-SOI Substrates

Modern engineered substrates technologies are currently successfully used to fabricate silicon on sapphire (SOS) and Si layers on the high resistivity (HR) substrate (HR-SOI) to reduce parasitic capacitance and improve RF performance. SOS substrates feature base wafer resistivity above 10 kΩ-cm, when for HR-SOI this value is typically around 1–2 kΩ-cm. SOS, historically developed via heteroepitaxial Si growth on a crystalline Al_2O_3, has become a first commercial SOI-based technology. The high crystal defectivity levels inherent to the heteroepitaxial technique did limit, however, industrial adaptation of SOS substrates. An alternative, SOS wafers fabricated with BESOI technique (BSOS), has emerged recently as a substrate solution to address the growing market needs. The photograph of 6" fully processed BSOS substrate is shown in Fig. 17a. The data, published by Peregrine and Soitec team, confirm that BSOS shows >50 % higher electron mobility as compared to conventional SOS fabricated by direct epitaxy (see Fig. 17b). These technical results have helped to open new market opportunities for "first-of-a-kind" SOI-based technology.

The availability of CMOS foundry technologies on 200 mm HR-SOI wafers has made possible high volume fabrication of RF systems, including high quality passive devices and RF switches [25–27]. The achieved performance was sufficient to integrate wireless and RF multi-standard multi-band functionalities in Si-based

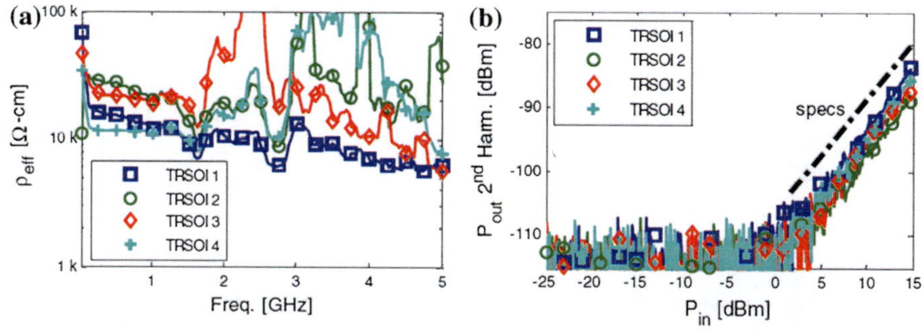

Fig. 19 **a** Effective resistivity versus frequency of CPWs on different TR HR-SOI wafers. **b** Harmonic distortion of CPWs on different trap-rich HR-SOI wafers. The specification (specs *straight line*) for the harmonic distortion corresponds to that of RF switches for GSM/EDGE transmitter modules (from [27])

SOCs. The inherent transmission line losses, due to the presence of parasitic surface conduction (PSC), however, limit RF performance of standard HR-SOI substrates [27]. The non-linear behavior of HR-SOI substrates is also increased due to PSC, and the generated harmonics at the passive components can have higher power levels than those from RF switches or other active devices fabricated on HR-SOI. The increase of BOX thickness provides better isolation, but does not reduce the parasitic effects. These effects are characterized by a threshold limit in the substrate losses reduction when increasing Si HR and high harmonic distortion levels introduced by the substrate itself [26]. Several methods were developed in the last years to reduce these parasitic effects and to enhance the HR properties of the base substrate [27]. The introduction of a trap-rich (TR) layer has proved to be an effective technique to reach the industry standards for transmission line performance. The Smart Cut technology used in the fabrication TR HR-SOI substrates has a moderate cost impact and is compatible with the thermal budgets of typical CMOS process flows [27]. Figure 18 shows schematics of TR HR-SOI wafer commercially available from Soitec. TR HR-SOI wafers manufactured with different process flavours were RF-characterized after fabrication of coplanar waveguides (CPW): for the details the reader is referred to Ref. [27]. Figure 19a depicts the effective resistivity versus frequency of CPWs. It confirms the effective resistivity higher than 4 kΩ-cm for all wafers being tested. Regarding the linear behaviour parameter, CPW harmonic distortion was always found to be lower than −81 dBm (for an input power of +15 dBm (Fig. 19b), i.e. more than 95 dBc).

3.2 Substrates for Photonics Applications

The push towards ever increasing data rates coupled with requirement for LP consumption suggests integration of optical components together with Si-electronics, i.e. development and industrialization of silicon photonics technology. Figure 20a

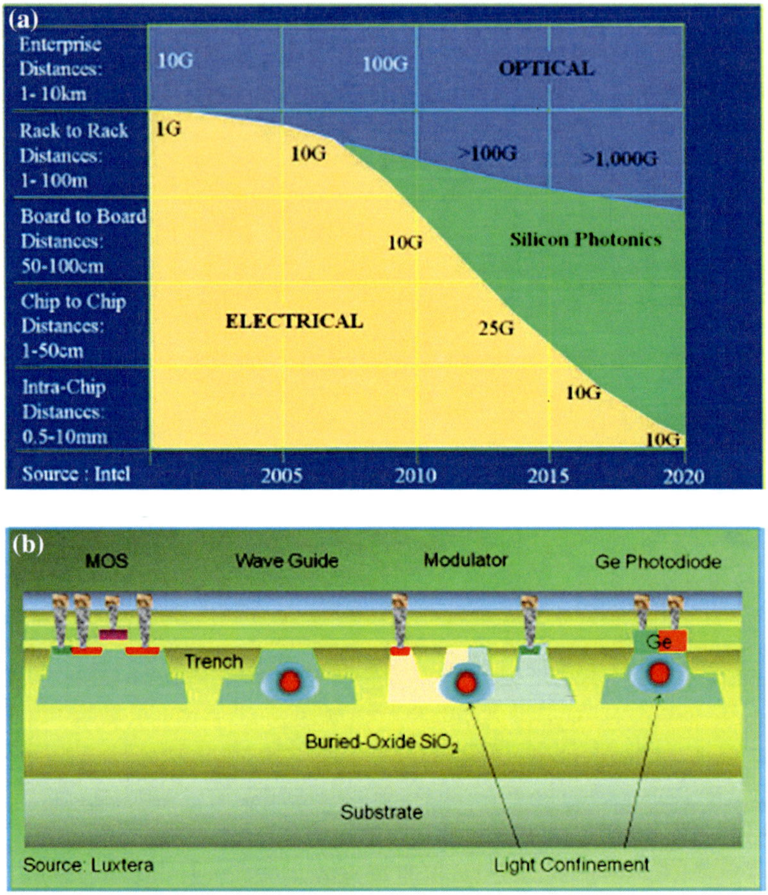

Fig. 20 **a** Electrical data rate limits for different data communication distances. **b** CMOS silicon photonics device cross-section [28]

depicts the foreseen electrical limits versus data communication distances [28]. A typical silicon photonics CMOS process flow starts from an SOI wafer (see Fig. 20b). SOI is required for the confinement of the light in the optical waveguides. The light is confined and guided into the silicon strip layer due to the difference of the two refractive indexes, n: top Si ($n = 3.48$) and BOX ($n = 1.45$ for SiO_2). The thicknesses of top Si and BOX layers are chosen accordingly to the selected range of the operating wavelengths. The top Si thickness is typically in the range of about 0.2–0.5 μm, whereas for the BOX it is >1 μm. To support today's R&D efforts top Si thickness uniformity is currently about ±10 nm. In order to gain wide-range industrial adoption the ultimate uniformity is targeted at the level of ±1 nm.

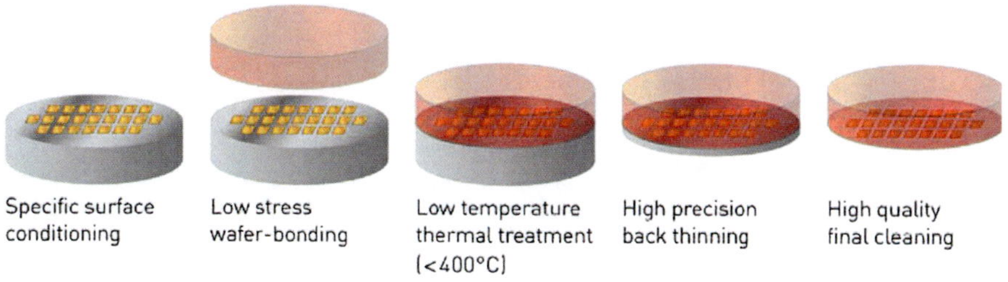

Fig. 21 Process schematics of the smart stacking technology [30]

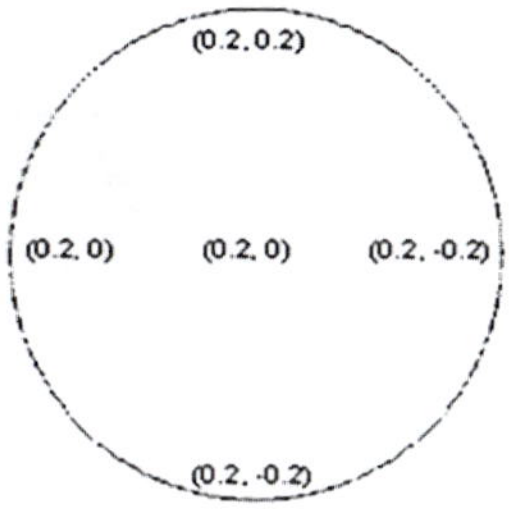

	data overview/ all runs		data 1 run (10 wafers)	
	x (μm)	y (μm)	x (μm)	y (μm)
average	-0,21	-0,24	-0,25	-0,17
std dev (sigma)	0,49	0,27	0,36	0,10
3 sigma	1,47	0,81	1,08	0,30

Fig. 22 W2W alignment accuracy across 300 mm wafer [31, 32]

Another potential application of Smart Cut technology in photonics is related to the fabrication of Bragg reflectors with top crystalline Si layer. Such reflectors provide high coupling efficiency, represent a fundamental building block for a number of devices (such as optical microcavities), and increase the quantum efficiency in resonant cavity enhanced photodetectors (while preserving high-speed operation) [29]. The proof of concept was demonstrated in [29] where conventional BOX was replaced with alternating multi-layer Si/SiO_2 stack.

3.3 Substrates for 3D Integration

The emerging field of 3D integration aims at providing highly integrated systems by vertical stacking, connecting various materials and functional components together. Due to stringent reliability standards, a fundamental challenge of direct bonding technology is to achieve a mechanical bond between the face-to-face bonded wafers without intermediate materials. To address these challenges, Soitec and CEA-LETI have developed two wafer-to-wafer (W2W) stacking technologies: Smart Stacking^{TM} and low-temperature Smart Cut^{TM}. These technologies provide reduction of process cost and sub-micron alignment capability. Smart Stacking is a W2W stacking platform for partially or fully processed wafers (see Fig. 21).

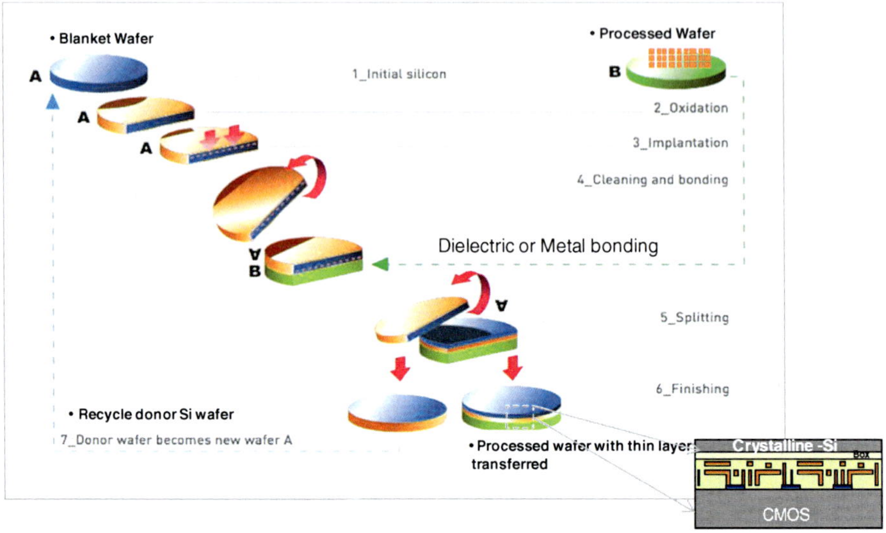

Fig. 23 Process schematics of the smart cut technology for 3D stacking [33]

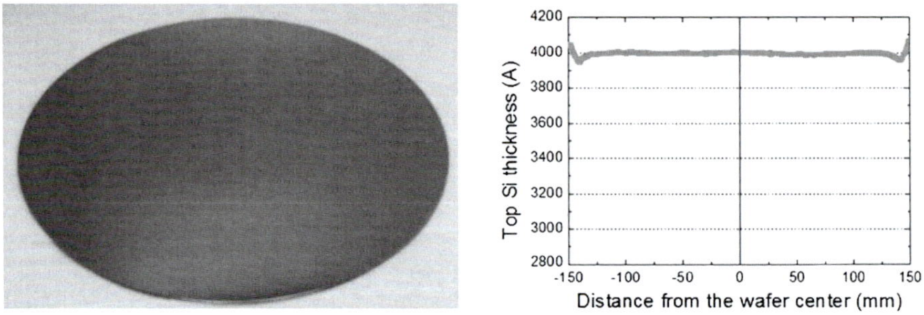

Fig. 24 Uniform thin Si film with low roughness transferred at low temperatures [33]

The technology incorporates surface conditioning, room temperature (RT) low stress bonding with precision alignment, low-temperature post-bond anneal, grinding and thinning. RT bonding ensures low thermal expansion mismatch, less stress and mechanical deformation, which are critical to achieve required level of alignment accuracy. W2W alignment accuracy of $\sim$0.2 μm across 300 mm wafers was demonstrated (see Fig. 22). To address the thermal budget constraint imposed by stacking of the backend of the line (BEOL) processed wafers (i.e. <400 °C), specific pre-bonding surface conditioning and a post-bonding thermal treatment were developed with the aim to increase the bonding strength. High bonding energies were achieved, providing compatibility with following wafer grinding and thinning processes [31, 32].

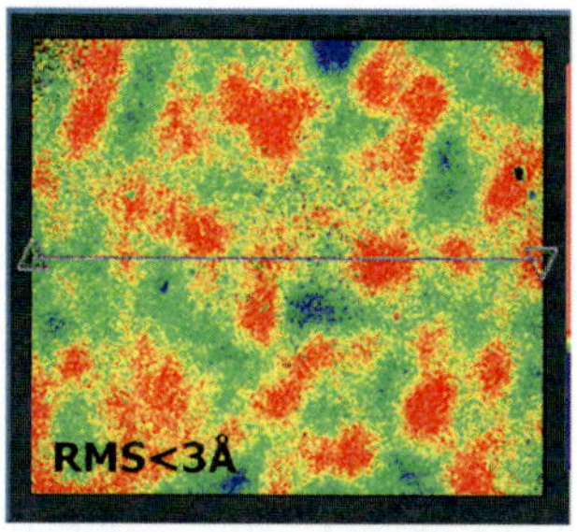

Fig. 25 Uniform thin Si film with low roughness transferred at low temperatures. The shown results are obtained with 5 × 5 μm^2 AFM scan [30, 33]

(a) Well established oxide-oxide bonding technology is a promising way for Smart Stacking 3D integration. Direct bonding of two oxide surfaces, however, requires thorough control of surface preparation. The Smart Stacking technology was demonstrated on 300 mm wafers with more than 10 layers of backend metal and a deposited oxide layer on top acting as the bonding layer. Wafer topology was controlled by a chemical-mechanical polishing process that maintains wafer edge quality and wafer micro-roughness <5 Å RMS. This process is fully compatible with common interconnect planarization technologies [32].

(b) Smart Stacking based on low-temperature patterned metal bonding enables the formation of electrical contacts during the stacking process: a promising path for high-density 3D interconnects. Non-thermo compression bonding, unlike thermo compression option, is performed RT under atmospheric pressure and is based on molecular adhesion between surfaces in contact. The bonding process does not require any additional processing steps: standard dual-damascene processes and surface treatment techniques are optimized to ensure smooth hydrophilic surfaces [31]. The bonding can be done directly after standard backend processes and can be adapted for via middle and via last integration schemes. Authors [31] investigated this bonding option for copper pad sizes varying from 5 to 500 μm with a pitch of 10 and 40 μm; excellent bonding interface defectivity was demonstrated.

(c) Smart Cut technology is a promising option for monolithic 3D integration. E.g. blanket single-crystal Si layers can be transferred on a processed wafer and "second level" of devices and circuits will be fabricated afterwards in the Si film (see Fig. 23). This approach is based on the optimization of the bonding energy with the reduction of the overall thermal budget (as done for Smart Stacking) as well as optimizing the splitting kinetics to achieve layer transfer at low temperature (<350 °C) [32]. The W2W direct bonding used in Smart Cut technology could include oxide-oxide or metal-metal bonding [33]. The later option allows the formation of electrical connections during the bonding process. Figures 24 and 25 illustrates the recent results (demonstrated on 300 mm wafers) for the transfer of the 200–400 nm Si layers (with typical uniformity range of ±15 nm) with low surface micro-roughness [33].

4 Conclusions and Perspectives

In the last 2–3 years, the application domain of the engineered substrates has gone through fundamental paradigm change. The ultra-thin-body SOI-based devices/ circuits with undoped channels are entering the advanced phase of industrialization. Such SOCs feature good performance, high circuit density, competitive overall cost with completion solutions and low power consumption. The later is of critically importance for rapidly growing market of portable consumer electronics. In addition, engineered substrates enable the applications that represent formidable technical challenge for bulk Si: RF devices with high resistivity substrates, Si-based photonics, 3D integration, as well as backside imagers and MEMS devices.

Acknowledgments Fruitful discussions and advices provided by B.-Y. Nguyen, W. Schwarzenbach, C. Aulnette, N. Daval, O. Bonnin, F. Allibert, L. Ecarnot, P. Nguyen, X. Cauchy, I. Radu, M. Sadaka, E. Desbonnets, C. Cailler, A. Rigny, and many other colleagues at Soitec, CEA-LETI, IMEP-LAHC, and STMicroelectronics are gratefully acknowledged.

References

1. Skotnicki, T., Arnaud, F., Faynot, O.: UTBB SOI: a wolf in sheep's clothing. Fut. Fab Int. **42**, 72–79 (2012)
2. Auth, C., et al.: A 22 nm high performance and low-power CMOS technology featuring fully-depleted tri-gate transistors, self-aligned contacts and high density MIM capacitors. In: Symposium on VLSI Technology. Digest of Technical Papers, pp. 131–132 (2012)
3. Schwarzenbach, W., Daval, N., Barec, V., Bonnin, O., Acosta-Alba, P., Maddalon, C., Chibko, A., Robson, T., Nguyen, B.Y., Maleville, C.: Atomic scale thickness control of SOI wafers for fully depleted applications. ECS Trans. **53**(5), 39–46 (2013)
4. Schwarzenbach, W., Cauchy, X., Boedt, F., Bonnin, O., Butaud, E., Girard, C.B.-Y., Nguyen, B.Y., Mazure, C., Maleville, C.: Excellent silicon thickness uniformity on ultra-thin SOI for controlling Vt variation of FDSOI. In: IEEE International Conference on ICICDT 2012–Paper B3 (2011)
5. Besnard, G., Garros, X., Nguyen, P., Andrieu, F., Reynaud, P., Van Den Daele, W., Bourdelle, K.K., Schwarzenbach, W., Reimbold, G., Cristoloveanu, S.: Reliability of ultrathin buried oxides for multi-V_T FDSOI technology. In: EuroSOI Conference 2013–Paper 3.3 (2013)
6. Hashemi, P., et al.: High-performance $Si_{1-x}Ge_x$ channel on insulator tri-gate PFETs featuring an implant-free process and aggressively-scaled fin and gate dimensions. In: Symposium on VLSI Technology. Digest of Technical Papers–Paper 2.2 (2013)
7. Brown, A.R., Daval, N., Bourdelle, K.K., Nguyen, B.Y., Asenov, A.: Simulation analysis of process-induced variability in nanoscale SOI and bulk FinFETs. In: IEEE International SOI Conference–Paper 11.4 (2012)
8. Dobrovolný, P., Zuber, P., Miranda, M., Garcia Bardon, M., Chiarella, T., Buchegger, P., Mercha, K., Verkest, D., Steegen, A., Horiguchi, N.: Impact of fin height variations on SRAM yield. In: Symposium on VLSI-TSA. Digest of Technical Papers–Paper T45 (2012)
9. Schwarzenbach, W., Daval, N., Kerdilès, S., Chabanne, G., Figuet, C., Guerroudj, S., Bonnin, O., Cauchy, X., Nguyen, B.Y., Maleville, C.: Strained silicon on insulator substrates for fully depleted application. In: IEEE International ICICDT Conference–Paper H3 (2012)

10. Tiberj, A., Paillard, V., Aulnette, C., Daval, D., Bourdelle, K.K., Moreau, M., Kennard, M., Cayrefourcq, I.: Stress metrology: the challenge for the next generation of engineered wafers. In: Proceedings of Materials Research Society Symposium–Paper B3.1.1, vol. 809 (2004)
11. Fenouillet-Beranger, C., Perreau, P., Weber, O., Ben-Akkez, I., Cros, A., Bajolet, A., Haendler, S., Fonteneau, P., Gouraud, P., Richard, E., Abbate, F., Barge, D., Pellissier-Tanon, D., Dumont, B., Andrieu, F., Passieux, J., Bon, R., Barral, V., Golanski, D., Petit, D., Planes, N., Bonin, O., Schwarzenbach, W., Poiroux, T., Faynot, O., Haond, M., Boeuf, F.: Enhancement of devices performance of hybrid FDSOI/Bulk technology by using UTBOX sSOI substrates. In: Symposium on VLSI Technology. Digest of Technical Papers, pp. 115–116 (2012)
12. Allibert, F., Cheng, K., Schwarzenbach, W., Khakifirooz, A., Ecarnot, L., Nguyen, B.Y., Doris, B.: Evaluation of sSOI wafers for 22 nm node and beyond. In: SOI Conference–Paper 7.2 (2012)
13. Minamisawa, R.A., Süess, M.J., Spolenak, R., Faist, J., David, C., Gobrecht, J., Bourdelle, K.K., Sigg, H.: Top-down fabricated silicon nanowires under tensile elastic strain up to 4.5 %. Nat. Commun. **3**, 1096 (2012). doi: 10.1038/ncomms2102
14. Knoll, L., Zhao, Q.T., Nichau, A., Trellenkamp, S., Richter, S., Schäfer, A., Esseni, D., Selmi, L., Bourdelle, K.K., Mantl, S.: Inverters with strained Si nanowire complementary tunnel field-effect transistors. Electron Device Lett. IEEE **34**, 813–815 (2013)
15. Daval, N., Figuet, C., Aulnette, C., Landru, D., Drazek, C., Bourdelle, K.K., Guiot, E., Letertre, F., Nguyen, B.Y., Mazure, C.: SiGe and Ge on insulator wafers. ECS Trans. **35**(5), 29–38 (2011)
16. Ikeda, K., Kamimuta, Y., Moriyama, Y., Ono, M., Koji, U., Oda, M., Irisawa, T., Furuse, K., Tezuka, T.: Enhancement of hole mobility and cut-off characteristics of strained Ge nanowire pMOSFETs by using plasma oxidized GeOx inter-layer for gate stack. In: Symposium on VLSI Technology. Digest of Technical Papers–Paper 3.3 (2013)
17. Ivana Subramanian, S., Owen, M.H.S., Tan, K.H., Loke, W.K., Wicaksono, S., Yoon, S.F., Yeo, Y.C.: N-channel InGaAs field-effect transistors formed on Germanium-on-insulator substrates. Appl. Phys. Express **5**(116502), 1–3 (2012)
18. Jalaguier, E., Aspar, B., Pocas, S., Michaud, J.F., Zussy, M., Papon, A.M., Bruel, M.: Transfer of 3in GaAs film on silicon substrate by proton implantation process. Electron. Lett. **34**, 408–409 (1998)
19. Jalaguier, E., Aspar, B., Pocas, S., Michaud, J.F., Papon, A.M., Bruel, M.: Transfer of thin InP films onto silicon substrate by proton implantation process. In: Proceedings of the 11th International Conference on Indium Phosphide and Related Materials, pp. 26–27 (1999)
20. Di Cioccio, L., Jalaguier, E., Letertre, F.: III–V layer transfer onto silicon and applications. Phys. Status Solidi A **202**, 509–515 (2005)
21. Czornomaz, L., Daix, N., Caimi, D., Sousa, M., Erni, R., Rossell, M.D., El-Kazzi, M., Rossel, C., Marchiori, C., Uccelli, E., Richter, M., Siegwart, H., Fompeyrine, J.: An integration path for gate-first UTB III–V-on-insulator MOSFETs with silicon, using direct wafer bonding and donor wafer recycling. In: Proceedings of the IEEE International Conference on Electron Devices Meeting, IEDM 2012. Technical Digest, pp. 517–520 (2012)
22. Kim, S.H., Yokoyama, M., Nakane, R., Ichikawa, O., Osada, T., Hata, M., Takenaka, M., Takagi, S.: Strained extremely-thin body $In_{0.53}Ga_{0.47}As$-on-insulator MOSFETs on Si substrates. In: Symposium on VLSI Technology. Digest of Technical Papers–Paper 5-1 (2013)
23. Kim, S.H., Yokoyama, M., Nakane, R., Ichikawa, O., Osada, T., Hata, M., Takenaka, M., Takagi, S.: High performance extremely-thin body InAs-on-insulator MOSFETs on Si with Ni-InGaAs metal S/D by contact resistance reduction technology. In: Symposium on VLSI Technology. Digest of Technical Papers–Paper 5-2 (2013)
24. Imthurn, G.P., Miscione, A.M., Landry, K., Vaufredaz, A., Barge, T., Lagahe-Blanchard, C.: Strain reduction in silicon-on-sapphire by wafer bonding. In: SOI Conference–Paper 6.2 (2011)

25. Ben Ali, K., Roda Neve, C., Gharsallah, A., Raskin, J.P.: RF SOI CMOS technology on commercial trap-rich high resistivity SOI wafer. In: SOI Conference–Paper 9.2 (2012)
26. Roda Neve, C., Raskin, J.P.: RF harmonic distortion of CPW lines on HR-Si and trap-rich HR-Si substrates. Electron. Devices IEEE Trans. **59**, 924–932 (2012)
27. Roda Neve, C., Ben Ali, K., Malaquin, C., Allibert, F., Desbonnets, E., Bertrand, I., Van Den Daele, W., Raskin, J.P.: RF and linear performance of commercial 200 mm trap-rich HR-SOI wafers for SoC applications. In: Proceedings of the 13th Topical Meeting on Silicon Monolithic Integrated Circuits in RF Systems–SiRF'13–Paper MO2B-2 (2013)
28. Zuffada, M.: The industrialization of the silicon photonics: technology road map and applications. In: 42nd European Solid-State Device Research Conference, ESSDERC 2008, pp. 7–13 (2012)
29. Akiyama, S., Grawert, F.J., Liu, J., Wada, K., Celler, G.K., Kimerling, L.C., Kaertner, F.X.: Fabrication of highly reflecting epitaxy-ready Si/SiO_2 Bragg reflectors. Photonics Technol. Lett. IEEE **17**, 1456–1458 (2005)
30. Sadaka, M., Radu, I., Lagahe-Blanchard, C., Di Cioccio, L.: Stacking^{TM} and smart cut^{TM} technologies for wafer level 3D integration. In: ICICDT Conference–Paper L4 (2012)
31. Radu, I., Landru, D., Gaudin, G., Riou, G., Tempesta, C., Letertre, F., Di Cioccio, L., Gueguen, P., Signamarcheix, T., Euvrard, C., Dechamp, J., Clavelier, L., Sadaka, M.: Recent developments of Cu-Cu non-thermo compression bonding for wafer-to-wafer 3D stacking. In: Proceedings of the IEEE International Conference on 3D Systems Integration (3DIC) 2010, pp. 1–6 (2010)
32. Gaudin, G., Riou, G., Landru, D., Tempesta, C., Radu, I., Sadaka, M., Winstel, K., Kinser, E., Hannon, R.: Low temperature direct wafer to wafer bonding for 3D integration: Direct bonding, surface preparation, wafer-to-wafer alignment. In: Proceedings of the IEEE International Conference on 3D Systems Integration (3DIC) 2010, pp. 1–4 (2010)
33. Di Cioccio, L., Radu, I., Baudin, F., Mounier, A., Lacave, T., Delaye, V., Imbert, B., Chevalier, N., Mariolle, D., Thieffry, S., Mazen, F., Gaudin, G., Signamarcheix, T.: Wafer level 3D stacking using smart cut^{TM} and metal-metal direct bonding technology. ECS Trans. **50**(7), 169–175 (2012)

Perspectives of UTBB FD SOI MOSFETs for Analog and RF Applications

Valeriya Kilchytska, Sergej Makovejev, Mohd Khairuddin Md Arshad, Jean-Pierre Raskin and Denis Flandre

Abstract Ultra-thin body and buried oxide (UTBB) fully depleted (FD) silicon-on-insulator (SOI) MOSFETs are widely recognized as a promising candidate for 20 nm technology node and beyond, due to outstanding electrostatic control of short channel effects (SCE). Introduction of a highly-doped layer underneath thin buried oxide (BOX), so called ground-plane (GP), targets suppression of detrimental parasitic substrate coupling and opens multi-threshold voltage (V_{Th}) and dynamic-V_{Th} opportunities within the same process as well as the use of back-gate control schemes [1, 2]. Electrostatics, scalability and variability issues in UTBB MOSFETs as well as their perspectives for low power digital applications are widely discussed in the literature [1–5]. At the same time assessment of UTBB FD SOI for analog and RF applications received less attention. This *chapter* will discuss Figures of Merit (FoM) of UTBB MOSFETs of interest for further analog/RF applications summarizing our original research over the last years [6–15]. Device analog/RF performance is assessed through the key parameters such as the transconductance, g_m, the output conductance, g_d, the intrinsic gain, A_v and the cut-off frequencies, f_T and $f_{\max}$. Particular attention is paid to (1) a wide-frequency band assessment, the only approach that allows fair performance prediction for analog/RF applications; (2) the effect of parasitic elements, whose impact on the device performance increases enormously in deeply downscaled devices, in which they can even dominate device performance. Whenever possible, we will compare FoM achievable in UTBB FD SOI devices with those reported for other advanced devices.

V. Kilchytska (✉) · S. Makovejev · J.-P. Raskin · D. Flandre
ICTEAM Institute, Université Catholique de Louvain, 1348 Louvain-la-Neuve, Belgium
e-mail: valeriya.kilchytska@uclouvain.be

M.K. Md Arshad
Institute of Nanoelectronic Engineering, Universiti Malaysia Perlis, 01000 Kangar, Perlis, Malaysia

A. Nazarov et al. (eds.), *Functional Nanomaterials and Devices for Electronics, Sensors and Energy Harvesting*, Engineering Materials,
DOI 10.1007/978-3-319-08804-4_2

1 Introduction

Enormous progress of the semiconductor technology during the last decade is mostly driven by the continuous demand for the increase of the operation speed and the integration density of complex digital circuits. In order to be able to continue device scaling down to 20 nm and beyond, both new materials and new device architectures are unavoidably employed. Therefore, nowadays we deal not simply with proportional shrinking of respective device dimensions, but as well as with new architectures and materials in both channel and gate oxide. Both these factors will evidently affect analog/RF device features. From the device architecture point of view, two main contenders clearly appear as able to satisfy ITRS requirements for device downscaling: planar FD SOI with ultra-thin body and ultra-thin BOX (so-called UTBB, or UTBOX, or UT2B or ETSOI) and multiple-gate devices or MuGFET.

Focus of this chapter is UTBB FD SOI MOSFETs, while some benchmarking with other devices is provided whenever possible. Main features which allow UTBB FD SOI withstanding ITRS requirements for 20 nm-node and beyond are the following:

- outstanding electrostatic integrity;
- effective suppression of fringing fields through the BOX (by BOX thinning);
- ease of heat exhaust through the thin BOX and thus attenuated self-heating (SH) is expected;
- possibility of back-gate control schemes implementation.

However, lateral coupling of source and drain through the substrate is enhanced in the case of thin-BOX devices. This calls for realization of highly-doped layer just under the BOX, or so-called Ground Plane (GP), which screens or prevents electric field lines penetration into the substrate and thus coupling through the substrate. Furthermore, GP opens a practical way for multi-V_{Th} and back-gate biasing schemes realization.

Technological aspects, electrostatics, scalability and variability issues in UTBB FD SOI MOSFETs as well as their perspectives for low power digital applications are widely discussed and shown to be excellent [1–5]. However, till now almost no attention has been paid to analog and RF performance of these devices. Similarly, UTBB MOSFET small-signal behavior in a wide frequency range is rarely discussed except in our works issued during last 3 years [6–15].

2 Devices

UTBB FD SOI MOSFETs discussed in this chapter have been processed at CEA-Leti on UNIBOND^{TM} SOI wafers with either 25 or 10 nm-thick BOX. Wafers without GP, with n- and p-type GP are considered. The Si film in the channel region is thinned down to 7–8 nm, depending on the wafer, and left undoped. Elevated source-drain structures are employed to reduce parasitic resistance. The gate stack is formed by a HfSiON dielectric with the equivalent oxide thickness of ~1.3 nm and a TiN electrode. More process details can be found e.g. in [1, 2].

The studied devices are n-channel MOSFETs with the gate length L ranging from 30 nm to 10 μm and the channel width W from 80 nm to 10 μm. Multi-fingers devices are used and embedded in coplanar waveguide access pads for performing RF characterization.

3 MOSFET's Key Parameters and Methodology Applied

It is worth firstly to list the key device parameters and FoM of interest for analog/RF applications and describe the methodological approach, which allows a fair comparison of different devices under different bias conditions.

Main key-factors of any MOSFET are: cut-off frequencies (f_T and $f_{\max}$) and intrinsic voltage gain (A_{v0}), which itself varies in a frequency range (as will be discussed in Sect. 5):

$$f_T = \frac{g_m}{2 \cdot \pi \cdot C_{gg}}, \tag{1}$$

$$A_{v0} = \frac{g_m}{g_d} = \frac{g_m}{I_d} \cdot V_{EA} \neq const(f) \tag{2}$$

These key factors in turn depend on such device parameters as transconductance (g_m), drive/drain current (I_d), output conductance (g_d), Early voltage ($V_{EA} = I_d/g_d$), g_m/I_d, gate capacitance (C_{gg}), etc. One should not forget as well about parasitic capacitance and resistance elements (as will be discussed below).

If one considers the application circuit level (e.g. amplifier), then it involves gain bandwidth product, GBW, which depends in turn on the same device parameters (g_m/I_d and I_d):

$$GBW = \frac{g_m}{2 \cdot \pi \cdot C_L} = \frac{g_m}{I_d} \cdot \frac{I_d}{2 \cdot \pi \cdot C_L}, \tag{3}$$

where C_L is load capacitance.

From methodological point of view, a very useful approach for analog performance assessment of different devices is calculating g_m/I_d as a function of normalized drain current [16, 17]. Firstly, because such plot gives a complete picture of studied device, which is valid for different applications: from base-band applications, where high gain, high precision is needed, to high frequency application where high drive current is requested (Fig. 1). Secondly, as g_m/I_d is inversely proportional to the subthreshold swing, S in weak inversion regime and proportional to $\mu \cdot C_{ox}/n$ (where μ is mobility, C_{ox} is oxide capacitance and n is body factor) in strong inversion, such plot is independent of threshold voltage, V_{Th}, of substrate/back gate (or body) bias, V_{bg} and to the first order is also independent of

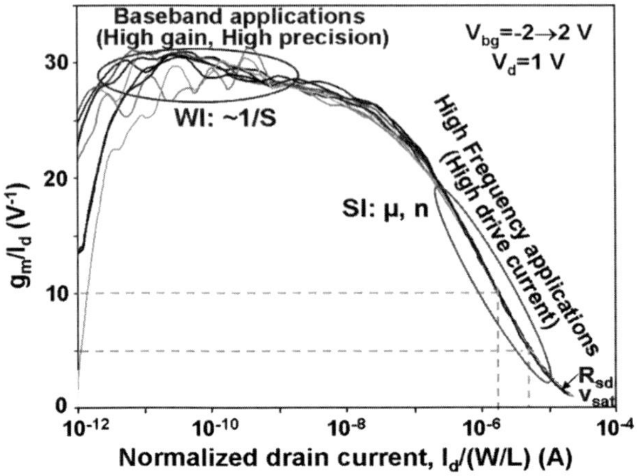

Fig. 1 g_m/I_d versus $I_d/(W/L)$ *curves* for 28 nm-long UTBB FD SOI nMOSFET measured at different V_{bg} from −2 to 2 V. $V_d = 1$ V. *Dashed lines* are drawn to represent $I_d/(W/L)$ extraction approach at a fixed $g_m/I_d = 10$ or 5 V^{-1}

L. In practice, one may fix g_m/I_d value and extract corresponding $I_d/(W/L)$ values as shown in Fig. 1. This allows assessment of device performance purely related to physical parameters as μ or body factor, thus providing a fair comparison of devices issued from different technologies, featuring different dimensions and operated at different conditions.

Figure 2 presents the complete small-signal equivalent circuit of a MOSFET. It is important to clearly distinguish between intrinsic elements, i.e. related to the device itself and extrinsic (or parasitic) elements, as access resistances, fringing and overlap capacitances. Parasitic capacitive components (for instance, capacitive coupling between source and drain through the BOX and substrate, fringing gate-to-source and gate-to-drain coupling, etc.) attain particular importance at high/RF frequencies.

With device length scaling down, importance of parasitic components increases enormously. Parasitic elements can even dominate the device performance (as will be shown in Sect. 6). Thus, ability of separate extraction of "intrinsic" and "extrinsic" elements in advanced nowadays technologies becomes crucial. Firstly, this allows for predicting "intrinsically" achievable idealistic or target values one can reach with optimization of parasitics. Secondly, for the process/configuration optimization, it is important to know wherefrom the problem comes (either intrinsic or extrinsic part). Indeed, if extrinsic part limits device performance, any innovations introduced to boost intrinsic performance (e.g. strain, orientation, etc.) will give negligible improvement to the final device performance.

Considering equivalent circuit shown in Fig. 2, MOSFET expressions for cut-off frequencies (Eq. 1) become more complex, accounting for parasitic elements and clearly reflecting SCE (through C_{gs}/C_{gd} ratio):

$$f_T \approx \frac{g_m}{2 \cdot \pi \cdot C_{gs}} \cdot \frac{1}{\left(1+\frac{C_{gd}}{C_{gs}}\right) + (R_s + R_d) \cdot \left(\frac{C_{gd}}{C_{gs}} \cdot (g_m + g_d) + g_d\right)} \tag{4a}$$

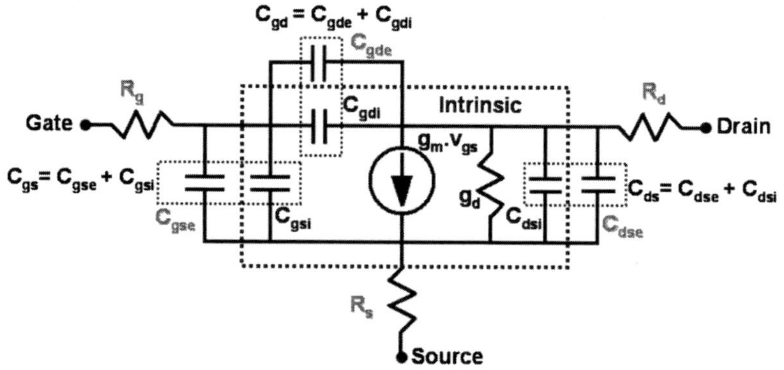

Fig. 2 MOSFET complete small-signal equivalent circuit, including both intrinsic (denoted 'i') and extrinsic (denoted 'e') elements

$$f_{\max} \approx \frac{g_m}{4 \cdot \pi \cdot C_{gs}} \cdot \frac{1}{\left(1 + \frac{C_{gd}}{C_{gs}}\right)\sqrt{g_d\left(R_g + R_s\right) + \frac{1}{2} \cdot \frac{C_{gd}}{C_{gs}}\left(R_s \cdot g_m + \frac{C_{gd}}{C_{gs}}\right)}} \tag{4b}$$

With length shrinkage, C_{gd}/C_{gs} ratio increases indicating the lost of a gate control over the channel and dominance of control from the drain (increase of capacitive coupling between gate and drain). Comparing Eqs. 4a, 4b and 1 one can see that this effect can strongly degrade 'intrinsic' cut-off frequency. Next to that, f_T and $f_{\max}$ are degraded by the parasitic access resistances, R_s and R_d. Gate resistance, R_g, however, affects only $f_{\max}$ (see Eq. 4b). According to Eq. 1, as g_m is inversely proportional to L and proportional to W, while C_{gg} is proportional to the area, f_T is expected to increase with L and be independent of W. However, this does not hold in experiments due to strong parasitic effects (as will be shown in Sect. 6).

Extraction of a complete equivalent circuit demands S-parameter measurements in a wide frequency range. Therefore, adequate structures with RF access pads must be included in the layout from a very beginning of the technology development. Details on the extraction procedure can be found for instance in [18].

4 UTBB FD SOI MOSFET Analog FoM

4.1 Benchmarking with Other MOSFETs

Extended benchmarking of analog FoM of UTBB FD SOI MOSFETs with respect to other both planar FD SOI MOSFETs and multiple-gate SOI-based FinFETs was performed in [6]. Table 1 summarizes analog FoM achievable in UTBB FD SOI with different geometries and compares them with SOI FinFET ones. One can see that UTBB FD SOI devices feature rather high values of I_d, V_{EA}, A_{v0}, g_{m_max}, which are comparable and in certain cases can be even higher than in another

Table 1 Main analog FoM for different devices

	$g_{m\ max}$, (µS/µm)	$I_d/(W/L)$ g_m/I_d = 10 V^{-1} (µA)	$I_d/(W/L)$ g_m/I_d = 5 V^{-1} (µA)	V_{EA} $V_g \sim V_{Th}$ (V)	V_{EA} $V_g \sim 1$ V (V)	A_{v0_max} (dB)
L = 30 nm						
FinFET	1,050	1.2	5.1	1.2	5.1	33.5
strained FinFET	1,410	1.3	6.4	1.5	6	34
UTBB FD SOI W = 10 µm	1,015	1.74	5.4	3.4	11	36
UTBB FD SOI W = 80 nm	1,510	2.3	7.2	5.5	10	38
strained UTBB FD SOI W = 10 µm	1,375		8			
L = 100 nm						
FinFET	700	3.1	10.8	4	12	40
strained FinFET	1,050	4.7	16.6	5	18	43
UTBB FD SOI W = 10 µm	810	4	13	8	20	46
UTBB FD SOI W = 80 nm	1,220	6	20	15	20	51
strained UTBB FD SOI W = 10 µm	985	5.6	17	5.5	16	40.5

$V_d = 1$ V. $V_{sub} = 0$ V

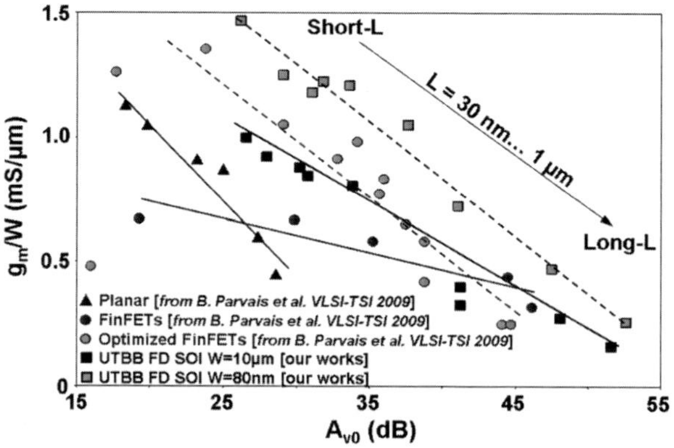

Fig. 3 g_m/W as a function of A_{vO} for different devices and processes. V_d is in saturation (1 V in our works and 1.1 V in [19]). $V_g = V_{Th} + 0.6$ V, $V_{bg} = 0$ V. $L = 30$ nm–1 µm for UTBB FD SOI MOSFETs. $L = 45$ nm–1 µm for FinFETs. Lines are intended to guide eyes

advanced architecture as FinFET. This becomes more visible in g_m/W versus A_{vo} metric plotted in Fig. 3 for various device architectures with different gate lengths. It reveals that UTBB FD SOI MOSFETs clearly outperform planar MOSFETs and non-optimized FinFETs. UTBB FD SOI MOSFETs are also very close and can even outperform optimized FinFETs particularly if narrow channel devices (offered simultaneously higher g_m/W and higher A_{vo}) are used.

It would be important to point out that UTBB FD SOI MOSFETs maintain their excellent performance in a wide temperature range, with very limited degradation of main parameters [6]. For instance, only 5 dB reduction of A_{vO} was observed over 200 °C. This makes UTBB FD SOI MOSFETs particularly attractive for high-precision analog circuits. Furthermore, A_{vo} was demonstrated [6] to be maximized in the moderate inversion regime (at $\sim V_{Th}$), which is beneficial for low-power applications.

4.2 *Effect of GP and Back Gate Bias*

A unique feature of UTBB FD SOI architecture is related to the possibility of back-gate biasing. In order to implement this back-gate biasing scheme, a highly-doped region, or ground plane, should be formed just below the BOX. Realization of GP region requires heavy implantation through the Si film and hence one can think about possible μ degradation. Next to that, GP suppresses substrate depletion, which means that there is no "BOX thickening" (provided by substrate depletion region) and hence body factor might be higher comparing to the no-GP case. Therefore, the question is "how these two facts affect Analog FoM?" Answering this question, [3, 8] demonstrate that g_{m_max} and I_d values stay almost unaffected. In the same time [8] reveals the sensitivity of the intrinsic gain to GP realization and hence a special

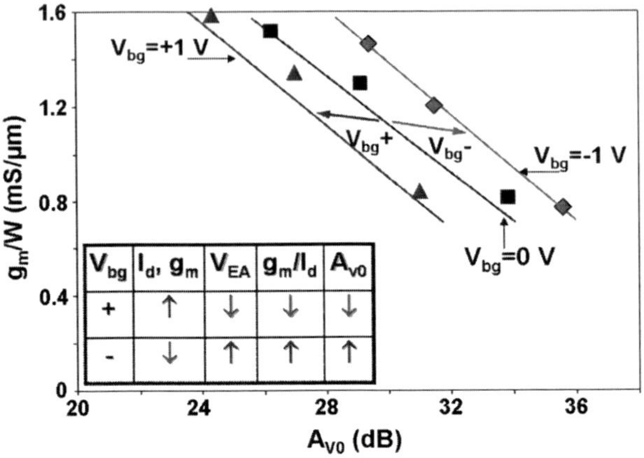

Fig. 4 g_m/W as a function of A_{v0} for different V_{bg}. $V_d = 1$ V. $V_g = V_{Th} + 0.6$ V. $L = 30$–100 nm. Lines are intended to guide eyes. Inset-table summarizes effect of back-gate bias polarity on main analog FoM

care has to be taken in order to assure high doping level just under the BOX. If the GP implantation is not well adjusted some reduction of intrinsic gain may appear (due to *DIBL* and related V_{EA} degradation) [8].

Effect of back-gate biasing on electrostatic and digital figures have been widely studied [3–5]. In this section we discuss how back-gate bias application affects analog FoM. From one side, negative back-gate bias results in V_{Th} increase and hence on-current I_{on} reduction, but in the same time it provides *S* and *DIBL* decrease (i.e. improvement). From another side, application of positive V_{bg} pushes channel centroid [5] to the bottom Si/BOX interface and hence higher g_m and I_{on} (due to higher μ values [20]) might be expected. These two trends give a clear trade-off for analog FoM. [6] demonstrates that 5–10 % enhancement of I_d and g_m may be achieved by application of positive V_{bg} and that 5–10 dB higher A_{vo} is reached in the case of negative V_{bg}. Thus, trends in I_d and A_{v0} dependence on the back-gate bias are opposite and hence the choice of "positive" or "negative" bias application for performance boosting finally depends on the target application (either high I_d or A_{v0}) (Fig. 4).

Another way to boost the device performance available in UTBB devices with GP is realization of so-called asymmetric double-gate (ADG) regime, i.e. gate to GP connection $V_g = V_{bg}$ (inset in Fig. 5), similar to DTMOS regime in PDSOI devices. [2, 8] reveals that such regime does not only allow improved SCE control, but also ~20 % performance enhancement in terms of I_d and g_m (Fig. 5). This evidently results into improved g_m versus A_{vo} metric; more details can be found in [8].

Even further performance boost can be expected in the case of so-called quasi-double gate (QDG) realization, i.e. simultaneous sweep of top and back gates with a certain coefficient *k* ($k > 1$) $V_g = k^*V_{bg}$ [12, 14]. Possibility of V_{Th} modulation together with improved SCE control accompanied by improved I_{on} and constant I_{off} achieved in QDG mode can be exploited for digital applications. Higher g_m and

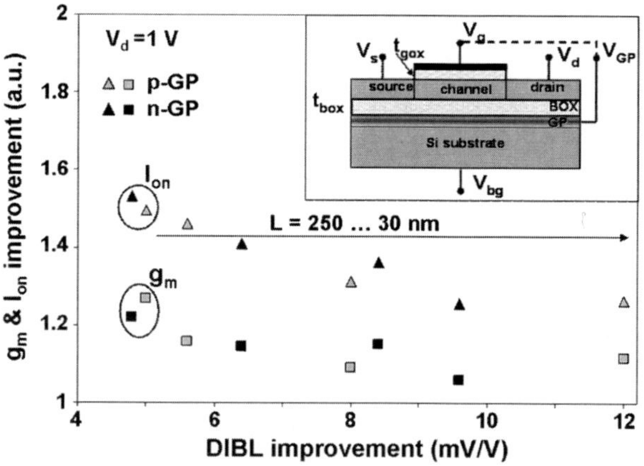

Fig. 5 g_m and I_{on} improvement in ADG mode comparing to a standard single-gate (SG) mode (calculated as g_{m_ADG}/g_{m_SG} and I_{on_ADG}/I_{on_SG}) as a function of DIBL improvement in ADG mode ($=DIBL_{SG} - DIBL_{ADG}$). Insert schematically shows ADG regime. g_m is taken at $V_g = V_{Th} + 0.6$ V and $V_d = 1$ V. I_{on} is taken at $V_g = V_d = 1$ V. $L = 30$–250 nm. BOX thickness, $T_{BOX} = 10$ nm

I_d combined with a lower DIBL (and hence higher A_{v0}) (Fig. 6) can potentially be exploited for analog applications. Thinner BOX evidently appears more promising for QDG realization [14], allowing lower k values for the same performance boost level (Fig. 6). More details on advantages of QDG mode and its exploitation for boosting the sleep transistor performances in the practical use case of a power-gated processor can be found in [13, 14].

5 Wide Frequency Band Assessment

This section points out an importance of the wide-frequency band assessment. Indeed, above-discussion was built on DC results but cited figures of merit vary over frequency. Figure 7 quasi-schematically shows simulated g_d variation as a function of frequency in FD SOI MOSFET. Increase of output conductance with frequency results in turn into decrease/degradation of intrinsic gain. This graph points out that solely DC and/or RF based extractions are not sufficient to reproduce and predict device behavior at different frequencies. Usual DC-based extractions give overestimated performance, whereas purely RF-based one, done in GHz range, underestimates device performance in MHz region. Only wide frequency band measurements allow a complete picture and thus fair analysis and comparison of different devices.

Different effects contribute to g_d variation with frequency.

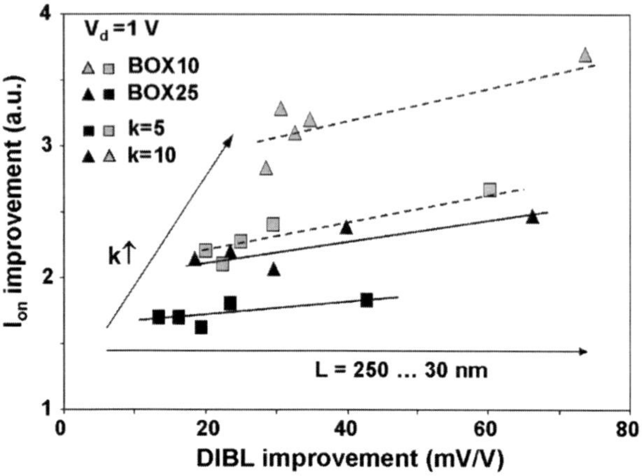

Fig. 6 I_{on} improvement in ADG mode comparing to a SG mode (calculated as I_{on_QDG}/I_{on_SG}) as a function of DIBL improvement in ADG mode ($=DIBL_{SG} - DIBL_{QDG}$). BOX thicknesses of 10 and 25 nm are considered. I_{on} is taken at $V_g \cong V_{Th}$. $V_d = 1$ V. $L = 30$–250 nm

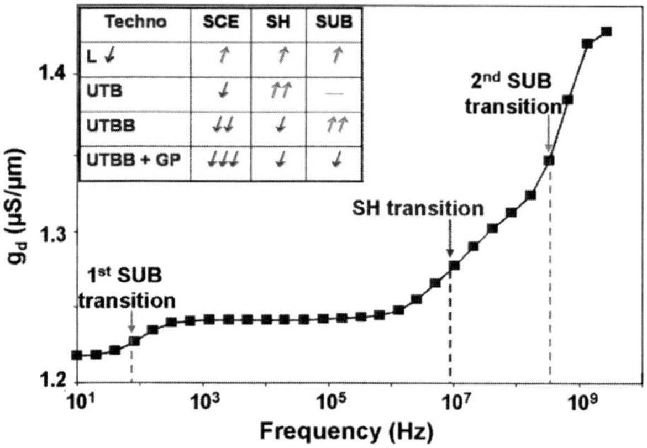

Fig. 7 Simulated g_d variation with frequency in FD SOI nMOSFET. *Dashed lines* schematically indicate transitions related to self-heating (SH) and substrate (SUB) effects. Inset-table summarizes evolution of these effects with technology advances

$$g_d(f) = g_{d\_in} + \Delta g_{d\_FB}(f) + \Delta g_{d\_SH}(f) + \Delta g_{d\_SUB}(f) \tag{5}$$

The frequency independent intrinsic term g_{d_in} is related to channel length modulation and DIBL. The second term, Δg_{d_FB} related to the floating body effect can be for the first order neglected in thin-film FD SOI MOSFETs under standard operation conditions. Two main reasons for g_d variation with frequency in advanced FD SOI devices are related to self-heating, Δg_{d_SH} and source-to-drain coupling through the substrate, Δg_{d_SUB}. Self-heating effect and its increased importance in SOI-based devices are widely known [21, 22]. Less known, substrate-related frequency dependent effect [23, 24] appears as a result of a substrate capacitance, C_{sub} variation with frequency resulting in the variation of the potential at the SUB-BOX interface, which then through g_m is translated into g_d variation [23]:

$$\Delta g_{d\_SUB} = (n-1) \cdot g_m \cdot \frac{\upsilon_{BGS}}{\upsilon_{DS}}, \tag{6a}$$

$$\upsilon_{BGS} \cong \frac{C_{BGD}}{C_{BGD} + C_{SBG} + C_{GBG} + C_{sub}} \cdot \upsilon_{DS} \tag{6b}$$

Frequency dependence of C_{sub} can be represented in a first order by two RC networks related to majority and minority carriers response [23, 24]. With frequency increase, first minority (in tens-hundreds Hz range) and then majority carriers (in a 100 MHz range) become unable to follow AC excitation so that C_{sub} decreases. Thus, two substrate-related transitions appear in g_d versus frequency curve.

Inset-Table in Fig. 7 schematically summarizes how the main effects which degrade output conductance evolve with technology advances.

Technology progress is mostly motivated by the need of SCE control and thus, first UTB, then UTBB and finally UTBB with GP were introduced.

Considering SH behavior with device downsizing and related technology advances, one can expect that

1. With *channel length reduction* phonon boundary scattering increases, thermal capacitance decreases, density of the current passing through the device increases and moreover SOI-like structures, known to suffer from enhanced self-heating start to be widely employed. All these facts contribute to the self-heating enhancement.
2. With *thinning of the Si film* (i.e. UTB devices) interface proximity enhances interface scattering resulting in thermal conductivity reduction [25] and hence thermal resistance R_{th} increase confirmed by simulations (see e.g. [26, 27]). Thus, self-heating is expected to increase with Si film thinning.
3. With *BOX thinning* one expects reduction of self-heating due to thinner thermal barrier and therefore easier heat exhaust from the channel. This assumption was also confirmed by simulations [26, 27].
4. Finally, GP introduction is not expected, in a first order, to modify SH.

Considering evolution of substrate-related effect with device downsizing and related technology advances, one can expect that

1. With *channel length shortening* source and drain become closer to each other thus naturally enhancing coupling through the substrate. Therefore, substrate-related degradation is expected to increase [28, 29].
2. There is almost no effect of *Si film thinning* on the substrate-related effect in a first order.

3. With *BOX thinning* electric field lines penetrate stronger in the substrate. Thus, coupling through the substrate is naturally expected to be enhanced. 2D simulations performed in [28, 29], indeed, confirm this hypothesis.
4. Finally, as GP acts as screen layer preventing substrate depletion and electric lines penetration, thus, coupling through the substrate and related g_d variation with frequency are expected to strongly reduced.

Figure 8 gives an example of experimental g_d variation with frequency in 30 nm long UTBB FD SOI device without GP under different bias conditions.

Firstly, one can see that both SH- and substrate-related transitions clearly appear in g_d frequency response. Secondly, it would be important to point out that g_d variation over the frequency range is very strong: low-frequency values are about three times lower than those at high frequency. Thirdly, such curves allow extracting temperature rise in the device [22]:

$$\Delta T = R_{th} \cdot I_d \cdot V_d, \text{ where } R_{th} = \frac{\Delta g_{d\_SH}}{(I_d + g_{LF} \cdot V_d) \cdot dI_d/dT_a} \qquad (7)$$

Makovejev et al. [7] reveals that despite the use of ultra-thin BOX, device temperature in UTBB MOSFETs can reach about 100 °C at high V_g and V_d. The temperature rise results in 5–7 % I_d degradation [11]. However, the main problem caused by SH in advanced UTBB devices is g_d degradation which is an important issue for analog applications. Finally, it is important to emphasize that in UTBB devices without GP, substrate-related degradation exceeds SH-related one. This gives an additional motivation for GP introduction.

Makovejev et al. [10], indeed, demonstrates that p-GP implementation allows efficient reduction of substrate-related g_d transition (Fig. 9) thanks to suppression of source-to-drain coupling through the substrate. This in turn results in significantly smaller A_{vO} reduction at high frequencies (Fig. 9) [10].

Nowadays UTBB FD SOI devices employ BOX with thickness of either 25 or 10 nm. Makovejev et al. [11] compares them in terms of SH and its effect on the Analog FoM degradation. It reveals that while temperature rise is indeed stronger in devices with 25 nm-thick BOX (see inset table in Fig. 10), there is almost no difference in g_d and I_d degradation caused by device heating in 10 and 25 nm-thick BOX devices at fixed bias conditions (Fig. 10) [11]. This suggests that thicker BOX might be used to reduce e.g. coupling through the substrate, without having much impact on the I_d and g_d degradation caused by thermal effects.

Finally, comparing A_{vO} reduction at high frequencies in FinFETs and UTBB MOSFETs with GP (Fig. 11), one can conclude that A_{vO} reduction level is very similar ($\sim$12–15 %) for these two main approaches for deeply downscaled MOSFET realization. While both SH and substrate-related effects are present in both types of devices, SH appears as a main source of A_{vO} variation ($\sim$10 %). Therefore, further developments should be focused on the optimization of device configuration/materials to ease the heat evacuation.

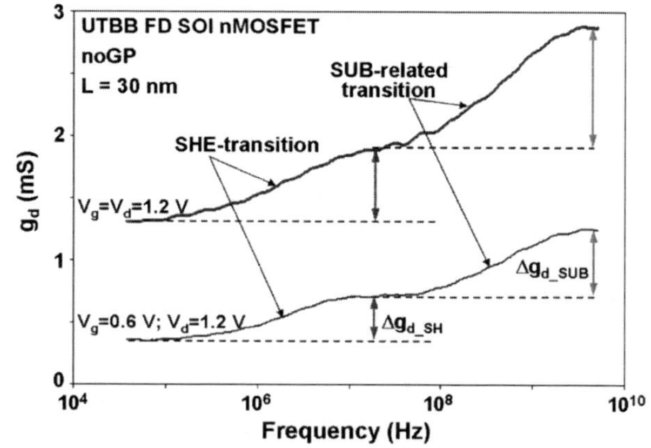

Fig. 8 The experimental g_d frequency response at $V_g = 0.6$ and 1.2 V, $V_d = 1.2$ V in the UTBB FD SOI MOSFET without GP. $L = 30$ nm. $W_{tot} = W \times N_f = 0.25 \times 80$ μm. $T_{BOX} = 10$ nm

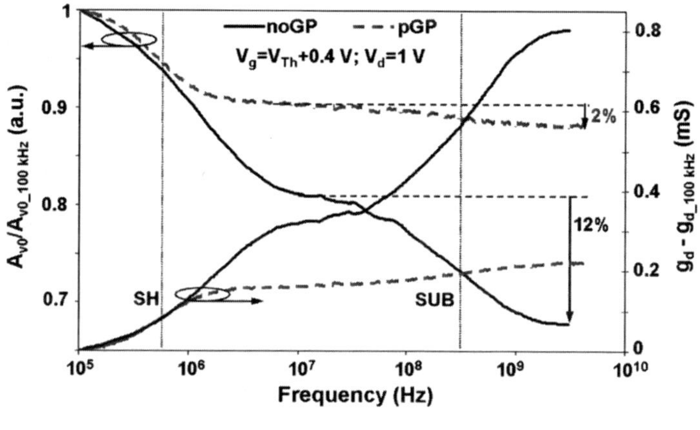

Fig. 9 g_d and A_{v0} variations with frequency with respect to their values at 100 kHz at $V_g = V_{Th} + 0.4$ V, $V_d = 1$ V in the devices with (*dashed lines*) and without GP (*solid lines*). $L = 100$ nm. $W_{tot} = W \times N_f = 1 \times 30$ μm. $T_{BOX} = 10$ nm

6 UTBB FD SOI MOSFETS RF FOM

6.1 Length and Width Dependence

This section discusses the evolution of UTBB FD SOI MOSFET RF FoM with L and W shrinking down. Thanks to the outstanding SCE control in UTBB FD SOI MOSFETs g_m/g_d and C_{gs}/C_{gd} ratios (see Eq. 4a, 4b for their effect on f_T, f_{max}) stay higher than in other technologies [9] thus allowing relatively high f_T, f_{max} (particularly for the process/devices which were not optimized for RF applications) (Fig. 12).

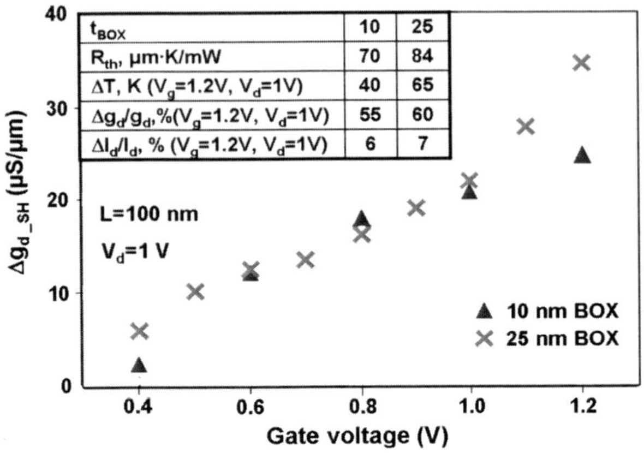

Fig. 10 Amplitude of SH-related g_d transition as a function of V_g for UTBB FD SOI MOSFETs with BOX thickness of 10 nm (*triangles*) and 25 nm (*crosses*). $L = 100$ nm. Inset table gives respective thermal resistance, temperature rise and SH-related analog FoM degradation taken at $V_g = 1.2$ V, $V_d = 1$ V

Fig. 11 Variation of the intrinsic gain with frequency with respect to its value at 100 kHz in FinFET (*solid line*) and UTBB FD SOI MOSFET (*dashed line*) with pGP

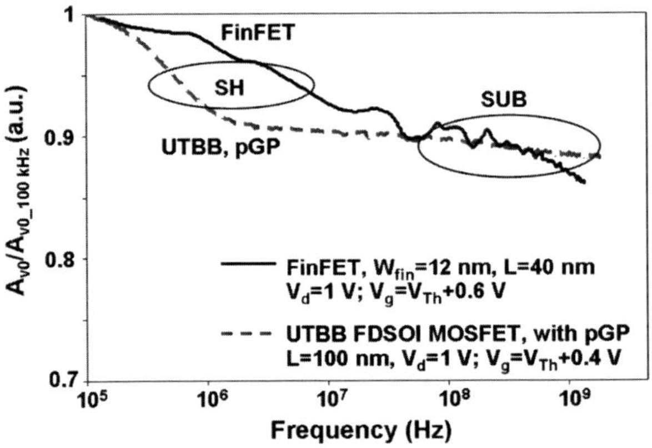

As was discussed above, 'intrinsic' cut-off frequencies are expected to increase with L scaling down and be independent of W provided that effect of parasitic elements is not strong. [9] studies L and W dependences of cut-off frequencies and different parasitic elements. It reveals that while f_T increases with length shrinking (Fig. 12), this increase is smaller then one could expect from Eq. 1. Analysis of C_{gg} and g_m dependence on L allows for concluding that f_T dependence is dominated by C_{gg} reduction (which is nevertheless lower than expected), while g_m stays almost constant. Another interesting observation to be pointed out is that, contrarily to general expectations and results for previous technology generations, in advanced devices f_T may become higher than f_{max} as a result of strong R_g effect.

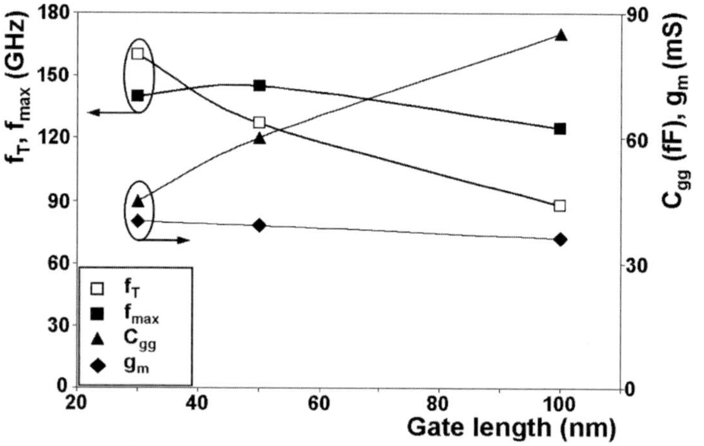

Fig. 12 f_T, $f_{\max}$, C_{gg} and g_m as a function of gate length in UTBB FD SOI nMOSFET without GP. $T_{BOX} = 10$ nm. $W_{tot} = W \times N_f = 0.5 \times 80$ μm. C_{gg} is the total gate capacitance ($C_{gg} = C_{gs} + C_{gd}$)

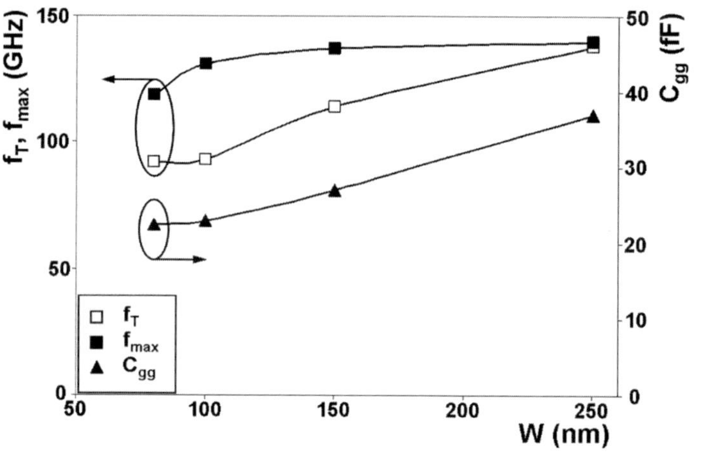

Fig. 13 f_T, $f_{\max}$ and C_{gg} as a function of gate width in UTBB FD SOI nMOSFET without GP. $L = 30$ nm. $T_{BOX} = 10$ nm. $N_f = 120$

Furthermore, [9] demonstrates that expected *W* independence of f_T and $f_{\max}$ is not maintained in UTBB FD SOI MOSFETs. Indeed, both f_T and $f_{\max}$ degrade with *W* reduction. It is important noting that this trend is opposite to the one obtained from DC measurements (see e.g. Figs. 3 and 13) thus indicating the role of parasitic elements which degrade RF performance. Effect of parasitic elements is indeed confirmed by the equivalent circuit extraction performed in [9, 15]. *W* reduction was demonstrated to result in sub-linear C_{gg} decrease (whereas g_m scales down almost linearly) suggesting strong fringing field effect at the perimeter.

Next section details effect of parasitic elements and distinguishes between 'extrinsic' and 'intrinsic' values.

6.2 Harmful Effect of Parasitic Elements on FoM

Md Arshad et al. [15] decomposes total (or as measured)C_{gg} in 'intrinsic' C_{gg_intr} (i.e. related to 'useful' device) and 'extrinsic' C_{gg_extr} (i.e. related to parasitics) parts (Fig. 14), $C_{gg} = C_{gg\_intr} + C_{gg\_extr}$. It is interesting to note that 'extrinsic' part is smaller that 'intrinsic' one in 100 nm-long devices, but it dominates in the case of 50 and 30 nm long ones. In fact, C_{gg_extr} stays almost constant independently of the device length. As a result total measured C_{gg} does not scale proportionally to L.

Next to that, [15] clarifies effect of R_{sd} on g_m extracting both 'as measured' and 'intrinsic'/ideally achievable (i.e. free from R_{sd}) g_m values. With L reduction impact of R_{sd} increases and while intrinsic g_m continues to growth, 'as measured' one stays almost invariable (Fig. 14).

Finally both these above-described effects affect cut-off frequency. [15] distinguishes 'as measured' f_T and 'intrinsic' one when first R_{sd} and then both C_{gg_extr} and R_{sd} effects are withdrawn. 'As measured' f_T is seen to be strongly degraded comparing to the intrinsically achievable one, particularly in the shortest device (Fig. 15). One can see that while 'intrinsic' f_T (with R_{sd} and C_{gg_extr} withdrawn) continues to grow strongly with L reduction, 'as measured' f_T slows down. It is worth to point out again that the process was not optimized for RF applications and specific approaches used for reduction of extrinsic C_{gg} and R_{sd} lowering were not employed. From another side 'intrinsic' f_T with R_{sd} and C_{gg_extr} withdrawn is evidently overestimated comparing to the really achievable one as parasitic elements are unavoidably present in real devices. Therefore, [15] goes further and calculates f_T which one could achieve in the case R_{sd} and C_{gg_extr} requirements imposed by ITRS are respected. It demonstrates that in this case f_T values of 310 GHz are achievable in 30 nm-long device, i.e. as high as requested by ITRS for LP applications.

Therefore, future work should be focused on the optimization (i.e. minimization) of the parasitics. There are indeed some works devoted to the optimization of parasitic elements. For example, "faceted Source/Drain" was demonstrated to be of interest for the fringing capacitance reduction [30, 31]. Thicker and/or low-k capping layer, thicker gate, etc. are also discussed as a potential solution for C_{gg_extr} reduction [32, 33]. Different silicides, both conventional ones (as NiSi, CoSi, etc.) and emerging ones (e.g. platinum, erbium or ytterbium) are studied as potential materials allowing for R_{sd} reduction [34–36]. Furthermore, different possibilities of performance improvement by geometry and overlap/underlap optimization are considered [37, 38]. However, this is a subject of separate discussion which is out of scope of this chapter. More detailed information can be found e.g. in [15].

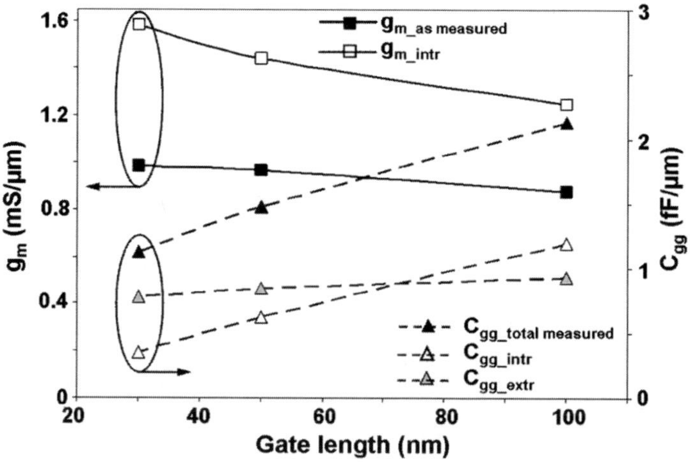

Fig. 14 Measured and intrinsic g_m ($g_{m\_as\ measured}$ and g_{m_intr}) (*squares*) as well as total measured, extrinsic and intrinsic C_{gg} ($C_{gg\_total\ measured}$, C_{gg_intr}, C_{gg_extr}, respectively) (*triangles*) as a function of gate length for UTBB FD SOI nMOSFET, without GP. $T_{BOX} = 10$ nm. $W_{tot} = W \times N_f = 0.5 \times 80$ µm

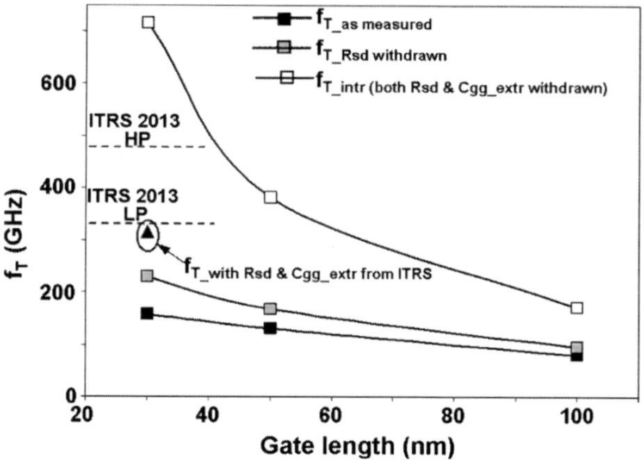

Fig. 15 Measured f_T, f_T with R_{sd} withdrawn and intrinsic f_T with both R_{sd} and C_{gg_extr} withdrawn as a function of gate length for UTBB FD SOI MOSFETs without GP. *Triangle* gives f_T estimation in the case of R_{sd} and C_{gg_extr} values from ITRS. $T_{BOX} = 10$ nm. $W_{tot} = W \times N_f = 0.5 \times 80$ µm. ITRS requirements based on 27 nm technology node FD-SOI targeted high performance (HP, $L = 22$ nm) and low power (LP, $L = 24$ nm) applications are indicated by *dashed lines*

7 Conclusions

There are two main conclusions which one should keep from this chapter:

1. First conclusion concerns *UTBB FD SOI MOSFET performance*. This device architecture features promising performance for analog/RF applications, which is comparable or even outperform other technologies, particularly in the cases when combination of high frequency with LOP/LSTP feature is needed. Furthermore, UTBB FD SOI MOSFETs possess potential for further performance improvement through the optimization of parasitics. This suggests UTBB FD SOI as a good contender for mobile/wireless applications with LOP/LSTP options.
2. Second conclusion concerns *importance of wide frequency band characterization*. We have shown that performance prediction based on DC data exclusively may be inaccurate and device benchmarking may appear even misleading. Next to that, impact of parasitics on device performance enormously increases with L downscaling and hence separation of "intrinsic" and "extrinsic" elements and related performances becomes mandatory in nowadays advanced devices. These two points call for the wide frequency band analysis for a fair FoM assessment. Therefore, adequate structures with RF access pads should be included in the layout from the very beginning of the technology development.

Acknowledgments The work has been partly funded by the FNRS (Belgium), by the FP7 NoE ''EuroSOI+'' and by Catrene "Reaching 22" projects. The authors would like to thank Olivier Faynot, Thierry Poiroux and François Andrieu from CEA-Leti, MINATEC, Grenoble, France for the provided devices and valuable discussions as well as Pascal Simon from WELCOME characterization platform of Université catholique de Louvain, Louvain-la-Neuve, Belgium for his assistance with high-frequency measurements setup.

References

1. Fenouillet-Beranger, C., et al.: Efficient multi-VT FD SOI technology with UTBOX for low power circuit design. In: Proceedings of the IEEE Symposium on VLSI Technology, pp. 65–66 (2010)
2. Andrieu, F., et al.: Low leakage and low variability ultra-thin body and buried oxide (UT2B) SOI technology for 20 nm low power CMOS and beyond. In: Proceedings of the IEEE Symposium on VLSI technology, pp. 57–58 (2010)
3. Fenouillet-Beranger, C., et al.: Impact of a 10 nm ultra-thin BOX (UTBOX) and ground plane on FD SOI devices for 32 nm node and below. Solid-State Electron. **54**(9), 849–854 (2010)
4. Monfray, S., et al.: Thin-film devices for low power applications. Solid-State Electron. **54**(2), 90–96 (2010)
5. Burignat, S., et al.: Substrate impact on threshold voltage and subthreshold slope of sub-32 nm ultra thin SOI MOSFETs with thin buried oxide and undoped channel. Solid-State Electron. **54**(2), 213–219 (2010)

6. Kilchytska, V., et al.: Ultra-thin body and thin-BOX SOI CMOS technology analog figures of merit. Solid State Electron. **70**, 50–58 (2012)
7. Makovejev, S., et al.: Impact of self-heating and substrate effects on small-signal output conductance in UTBB SOI MOSFETs. Solid-State Electron. **71**, 93–100 (2012)
8. Md Arshad, M.K., et al.: UTBB SOI MOSFETs analog figures of merit: effects of ground plane and asymmetric double-gate regime. Solid State Electron. (2013). doi:10.1016/j.sse.2013.02.051
9. Md Arshad, M.K., et al.: RF Behavior of undoped channel ultra-thin body with ultrathin BOX MOSFETs. In: Proceedings of the Silicon Monolitic Integrated Circuits in RF Systems (SiRF), pp. 105–108 (2012)
10. Makovejev, S., et al.: Comparison of small-signal output conductance frequency dependence in UTBB SOI MOSFETs with and without ground plane. In: Proceedings of the IEEE SOI Conference (2011)
11. Makovejev, S., et al.: On extraction of self-heating features in UTBB SOI MOSFETs. In: Proceeding of the 13th International Conference on Ultimate integration on silicon, pp. 109–112 (2012)
12. Kilchytska, V., Andrieu, F., Flandre, D.: On the UTBB SOI MOSFET performance improvement in quasi-double-gate regime. In: Proceedings of the Solid State Device Research Conference (ESSDERC), pp. 246–249 (2012)
13. Bol, D., Kilchytska, V., De Vos, J., Andrieu, F., Flandre, D.: Quasi-double gate mode for sleep transistors in UTBB FD SOI low-power high-speed applications. In: Proceedings of the IEEE SOI Conference (2012)
14. Kilchytska, V., Bol, D., De Vos, J., Andrieu, F., Flandre, D.: Quasi-double gate regime to boost UTBB SOI MOSFET performance in analog and sleep transistor applications. Solid-State Electron. **84**, 28–37 (2013)
15. Md Arshad, M.K., et al.: Effect of parasitic elements on UTBB FD SOI MOSFET RF figures of merit. Solid-State Electron. **97**, 38–44 (2014)
16. Flandre, D., Ferreira, L.F., Jespers, P.G.A., Colinge, J.-P.: Modelling and application of fully depleted SOI MOSFETs for low voltage, low power analogue CMOS circuits. Solid-State Electron. **39**(4), 455–460 (1996)
17. Kilchytska, V., et al.: Influence of device engineering on the analog and RF performances of SOI MOSFETs. IEEE Trans. Electron Devices **50**(3), 577–588 (2003)
18. Raskin, J.-P., Gillon, R., Chen, J., Vanhoenacker-Janvier, D., Colinge, J.-P.: Accurate SOI MOSFET characterization at microwave frequencies for device performance optimization and analog modeling. IEEE Trans. Electron Devices **45**(5), 1017–1025 (1998)
19. Parvais, B., et al.: The device architecture dilemma for CMOS technologies: opportunities and challenges of Finfet over planar mosfet. In: Proceedings of the International Symposium on VLSI Technology, Systems and Applications, pp. 80–81 (2009)
20. Rudenko, T., et al.: Experimental study of transconductance and mobility behaviors in ultra-thin SOI MOSFETs with standard and thin buried oxides. Solid-State Electron. **54**(2), 164–170 (2010)
21. Tenbroek, B.M., Lee, M.S.L., Redman-White, W., Bunyan, R.J.T., Uren, M.J.: Self-heating effects in SOI MOSFET's and their measurement by small signal conductance techniques. IEEE Trans. Electron Devices **43**(12), 2240–2248 (1996)
22. Jin, W., Liu, W., Fung, S.K.H., Chan, P.C.H., Hu, C.: Self-heating characterization for SOI MOSFET based on AC output conductance. IEEE Trans. Electron Devices **48**(4), 730–736 (2001)
23. Kilchytska, V., Levacq, D., Lederer, D., Raskin, J.-P., Flandre, D.: Floating effective backgate effect on the small-signal output conductance of SOI MOSFETs. IEEE Electron Device Lett. **24**(6), 414–416 (2003)
24. Kilchytska, V., et al.: Substrate effect on the output conductance frequency response of SOI MOSFETs. In: Hall, S., Nazarov, A.N., Lysenko, V.S. (eds.) Nanoscaled Semiconductor-on-Insulator Structures and Devices, pp. 221–238. Springer, Dordrecht (2007)

25. O'Neill, A.G., et al.: Reduced self-heating by strained silicon substrate engineering. Appl. Surf. Sci. **254**(19), 6182–6185 (2008)
26. Fiegna, C., Yang, Y., Sangiorgi, E., O'Neill, A.G.: Analysis of self-heating effects in ultrathin-body SOI MOSFETs by device simulation. IEEE Trans. Electron Devices **55**(1), 233–244 (2008)
27. Braccioli, M., Curatola, G., Yang, Y., Sangiorgi, E., Fiegna, C.: Simulation of self-heating effects in different SOI MOS architectures. Solid-State Electron. **53**(4), 445–451 (2009)
28. Kilchytska, V., et al.: Frequency variation of the small-signal output conductance of decananometer MOSFETs due to substrate cross. IEEE Electron Device Lett. **28**(5), 419–421 (2007)
29. Kilchytska, V., Flandre, D., Raskin, J.-P.: Silicon-on-nothing MOSFETs: an efficient solution for parasitic substrate coupling suppression in SOI devices. Appl. Surf. Sci. **254**(19), 6168–6173 (2008)
30. Khakifirooz, A., Cheng, K., Kulkarni, P.: Challenges and opportunities of extremely thin SOI (ETSOI) CMOS technology for future low power and general purpose system-on-chip applications. In: Proceedings of the International Symposium on VLSI Technology, System and Applications, pp. 110–111 (2010)
31. Khakifirooz, A., et al.: Scalability of extremely thin SOI (ETSOI) MOSFETs to sub-20-nm gate length. IEEE Electron Device Lett. **33**(2), 149–151 (2012)
32. Ko, C., Kuan, T., Zhang, K.: A novel CVD-SiBCN low-k spacer technology for high-speed applications. In: Proceedings of the International Symposium on VLSI Technology (2008)
33. Niebojewski, H., et al.: Extra-low parasitic gate-to-contacts capacitance architecture for sub-14 nm transistor nodes. In: Proceedings of the EuroSOI Conference (2013)
34. Ieong, M., et al.: Transistor scaling with novel materials. Mater. Today **9**, 26–31 (2006)
35. Larson, J.M., Snyder, J.P.: Overview and status of metal S/D Schottky-barrier MOSFET technology. IEEE Trans. Electron Devices **53**(5), 1048–1058 (2006)
36. Connelly, D., Clifton, P.: Ultra-thin-body fully depleted SOI metal source/drain n-MOSFETs and ITRS low-standby-power targets through 2018. In: Proceedings of the IEEE Electron Devices Meeting (IEDM), pp. 8–11 (2005)
37. Ohguro, T., Higashi, Y., Okano, K., Inaba, S., Toyoshima, Y.: The optimum device parameters for high RF and analog/MS performance in planar MOSFET and FinFET. In: Proceedings of the International Symposium on VLSI Technology, pp. 149–150 (2012)
38. Lim, T.C., Armstrong, G.A.: The impact of the intrinsic and extrinsic resistances of double gate SOI on RF performance. Solid-State Electron. **50**(5), 774–783 (2006)

Challenges and Solutions for Very Low Energy Computation

Francis Balestra

Abstract This chapter addresses the main challenges, limits and possible solutions for strongly reducing the energy per binary switching. Several paths are possible: the adiabatic logic using a slow clock, which cannot be used for high performance devices, logic stochastic resonance, feedback-controlled dynamic gate, or conventional logic with a reduction in the stored energy, therefore a decrease of device capacitance *C* (device integration) or applied bias V_{dd}, which seems to be the most promising for future ICs. The reduction of the stored energy in conventional logic can be done with a strong reduction of V_{dd} using new physics and/or devices with 60 or sub-60 mV/dec subthreshold swing *S*, in particular with the main following concepts: energy filtering (Tunnel FET, with MOS—nanowires (NW)—carbon nanotube (CNT)—or Graphene, using band-to-band tunnelling to filter energy distribution of electrons in the source), internal voltage step-up (Ferroelectric gate FET, inducing a negative capacitance to amplify the change in channel potential induced by the gate), Nano-Electro-Mechanical-Structures, or Impact Ionisation MOS devices. We will focus here on the best ones, Tunnel FETs realized with ultrathin films, multi-gates and/or alternative channel materials, which could lead to ultra short channel devices with a strong reduction of the applied bias, together with very good performance and reliability.

1 Introduction

The historic trend in micro/nano-electronics these last 40 years has been to increase both speed and density by scaling down the electronic devices, together with reduced energy dissipation per binary transition. We are facing today dramatic challenges dealing with the limits of energy consumption (static + dynamic) and heat removal, inducing fundamental tradeoffs for the future ICs.

F. Balestra (✉)
IMEP-LAHC/Sinano Institute, Grenoble INP-Minatec, 38016 Grenoble, France
e-mail: balestra@minatec.grenoble-inp.fr

A. Nazarov et al. (eds.), *Functional Nanomaterials and Devices for Electronics, Sensors and Energy Harvesting*, Engineering Materials,
DOI 10.1007/978-3-319-08804-4_3

The worldwide energy consumption has increased by 40 % these last 20 years. The ICT part represents 15 % of the electricity consumption and will increase by a factor 3 in the next 20 years. Today, in 2 days the information generated in the world corresponds to the one generated between the beginning of the ICT era till 2003.

Therefore, the researches on ultimate reduction of computation dissipation are strongly needed for the development of future very low power and high performance terascale integration and autonomous (nano) systems.

This chapter addresses the main challenges, limits and possible solutions for strongly reducing the energy per binary switching. Since the 90 nm node, V_{dd} scaling has been slowed leading to accelerated energy consumption and heating and a move from a constant field toward a constant voltage scaling. In 2005 the increase in microprocessor frequency abruptly ceased, but the integration level continued to increase and parallel processors were proposed. We are thus facing dramatic challenges and two main paths are possible for reducing the energy dissipated, which is the most critical limit for future ICs. The conventional logic with a reduction in the stored energy and the leakage currents, therefore a decrease of device capacitance C (device integration) or applied bias V as well as using novel materials and device architectures, or the adiabatic logic using a slow clock.

The ultimate limit of energy dissipation in irreversible logical operations require energy dissipation of at least $k{\cdot}T{\cdot}\ln(2)$ for each bit of information lost (corresponding to an increase of entropy [1]). Reversible operations can be performed without dissipation. In principle all computations could be performed without dissipation using only reversible operations [2], however in practice any computer will dissipate energy (e.g. due to error-correcting codes to maintain reliable operation, or due to residual resistance for non-superconducting wires and electrodes). It has recently been shown the possibility of sub-kT energy dissipation in charging a capacitor adiabatically, with charging energy of many kT delivered to the capacitor (by charging and discharging the capacitor gradually/adiabatically the power dissipation is significantly lower—at the cost of lowering the switching speed) [3]. Therefore, it is possible to use charges, and not necessarily other state variables, for reversible computation. For irreversible computation (with the least energy $k{\cdot}T{\cdot}\ln(2)$), the ultimate limit of logic switching can be obtained using the Heiseinberg uncertainty relations, leading to a minimum size of devices of 1.5 nm, a maximum density of devices of $5.10^{13}/\text{cm}^2$, a minimum switching time of 0.04 ps, inducing a power dissipation of 4.10^6 W/cm^2, which is far beyond the ultimate theoretical limit for heat removal from 2D Si surface (1000 W/cm^2) [4].

The reduction of the stored energy in conventional logic can be done with a strong reduction in V using new physics and/or devices with sub-60 mV/dec subthreshold swing S, which is the limit of MOSFETs at 300 K. The main following concepts are possible: energy filtering (Tunnel FET, with MOS—nanowires (NW)—carbon nanotube (CNT)—or Graphene-based, using band-to-band tunnelling to filter energy distribution of electrons in the source), internal voltage step-up (Ferroelectric gate FET, inducing a negative capacitance to amplify the change in channel potential induced by the gate), Nano-Electro-Mechanical-Structures, or Impact Ionisation MOS devices.

The most promising ones, TFETs, use gate-controlled p-i-n structures with carriers tunnelling through the barrier and not flowing over. In this chapter, we will address the most interesting structures that have been proposed using experiments or numerical simulations, with the aim of reducing the effective mass, the bandgap and/or the tunnelling length, for increasing I_{on} and reducing S for several decades of current.

2 Results and Discussion

Irreversible logical operations require energy dissipation of at least $k{\cdot}T{\cdot}\ln(2)$ (Laudauer limit). Today, we are very far from this limit, with a switching energy that is more than 3 decades larger than this minimum value.

The slowdown of V_{dd} scaling and the substantial increase of the subthreshold leakage lead to a dramatic enhancement of the dynamic and static power consumption. This power challenge, $P \sim I_{off}{\cdot}V_{dd}^3$ [1], is due to the subthreshold slope limit, which is 60 mV/dec at room temperature for MOSFETs. A lower limit in energy per operation exists; it is reached for (Fig. 1) [2]:

$$V_{dd\,\min} \sim S$$

with a minimum energy:

$$E_{\min} \sim C \cdot S^2$$

There are two ways for reducing the swing S:

$$S = \frac{\partial V_g}{\partial(\log I_d)} = \underbrace{\frac{\partial V_g}{\partial \psi_S}}_{m} \underbrace{\frac{\partial \psi_S}{\partial(\log I_D)}}_{n} = (1 + \frac{C_s}{C_{ins}})\frac{kT}{q}\ln 10$$

1. Decrease of the transistor body factor m, using Ultra Thin Body SOI, Multi-gate or Nanowire MOSFETs, Carbon Nanotube or Graphene channels (leading to $m \sim 1$), and Negative Capacitance FETs, or MEMS/NEMS structures (leading to $m < 1$)
2. Reduction of n, using a low temperature operation, which cannot be applied for traditional applications, or using a modification of the carrier injection mechanisms with Impact Ionisation or Band To Band Tunneling.

The best MOSFET devices with S close to its minimum value are using fully depleted [3, 4] channels (e.g. FD SOI with very thin buried oxide [5]) or fully inverted ones, with volume inversion [6], that is even better to optimize the control of the electrostatics [7] of the structure (e.g. Double-gate, Bulk or SOI Tri-gate/FinFET, Gate-All-Around MOSFET or Nanowire FET).

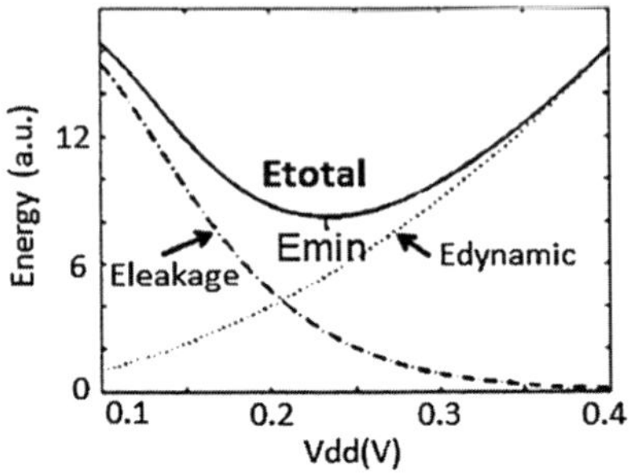

Fig. 1 Lower limit in energy per operation obtained for the optimal V_{dd}

Double-gate devices with gate underlap have also shown to lead to a very good I_{dsat}, a reduction of the *DIBL* and drain to gate tunnelling current, and a decrease of the gate capacitance, propagation delay and power [8].

Small diameter Si Nanowire FETs have demonstrated experimentally very good I_{on}/I_{off} performance [9], with the best Swing (75 mV/dec) obtained by numerical simulation down to 5 nm gate length (Fig. 2) [10].

However, other innovative devices are needed to overcome the sub-60 mV/dec barrier.

To outperform CMOS, these new devices, called the small swing/slope switches (Fig. 3), need an I_{on} in the range of hundreds of μA, an average subs-threshold swing far below 60 mV/dec for at least 4–5 decades of I_d, a ratio I_{on}/I_{off} larger than 10^5 and V_{dd} lower than 0.5 V.

The Tunnel FETs use interband tunnelling in heavily-doped p + n + junction with a control of band bending with V_g and a reversed bias p-i-n junction. In these transistors, the ambipolar effect has to be suppressed by an assymmetry in the doping level or profile, or the use of heterostructures in the channel between source and drain.

In order to optimize the Transmission probability for increasing the band-to-band-tunnelling, we need to propose structures with the aim of reducing the effective tunnel mass m^*, the source bandgap E_g and/or the tunnelling length λ, and increasing the band offset ΔΦ between the conduction/valence band of the source/channel, for enhancing I_{on} and reducing S for several decades of current.

For the optimisation of E_g, m^*, ΔΦ a change of materials is needed, and λ can be decreased by a change of device dimension, doping, gate capacitance, gate overlap on tunnel region, and bandgap using a reduced T_{ox} and t_{Si}, a high-k gate, an abrupt doping profile, a high source doping, a multi-gate structure or other materials than silicon.

It is also worth noting that L_g has little effect on I_d, a TFET with a hetero-structure is preferable (small bangap at S and large bandgap at D) and complementary TFETs are needed for logic circuits.

Figure 4a shows the experimental characteristics of SOI TFETs. A very good subthreshold swing down to 42 mV/dec is obtained for p-channel operation [11]. Using SGOI and GOI TFETs, strong improvements of the drive current, between several hundreds and thousands, are obtained compared with SOI transistors

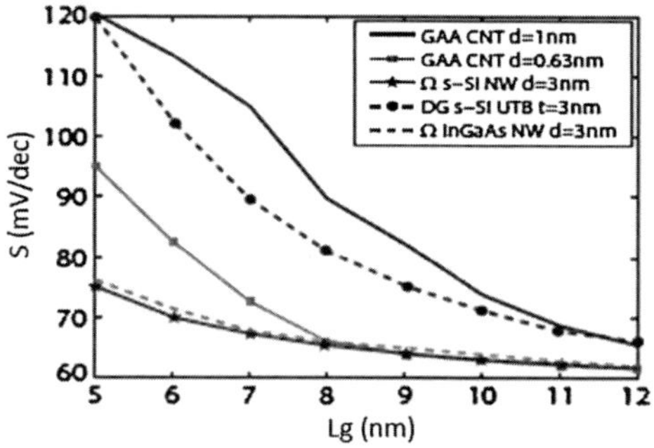

Fig. 2 Substhreshold swing obtained by quantum transport numerical simulation as a function of device architecture for gate length down to 5 nm: Double-gate strain-Si ultra-thin-body MOSFET ($t_{Si} = 3$ nm), Ω-gate strain-Si and InGaAs nanowires with 3 nm diameter, gate-all-around carbon nanotube with 1 nm and 0.63 nm diameter

Fig. 3 Comparison between the MOSFET switch, the small swing/slope switch and the ideal switch

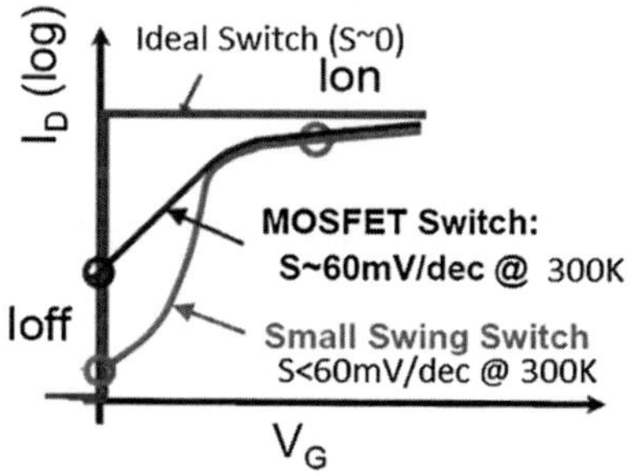

(Fig. 4b). I_{off} is very low but I_{on} is only of the order of several μA/μm for the best GOI TFETs.

As it is the case of MOSFETs, technology boosters can also be applied for TFETs, in particular: high-k, abrupt doping profile at tunnel junction, thinner body, high S doping, multi-gate, gate oxide aligned with i-region, shorter L_g/i-region. The results obtained by numerical simulations are illustrated in Fig. 5, where we can see that short channel thin Si film double-gate TFET with high-k gate dielectrics and stress at the source junction leads to the best performance with a substantial increase of I_{on} and reduction of S [12].

GeSn TFETs have also shown better performance compared with Ge TFETs (Fig. 6). A strong increase of the driving current together with a substantial reduction of the swing are obtained [13].

Feedback TFETs [14] and Z2-FET [15], with a forward biased PiN diode, have shown experimentally very small S (a few mV/dec) together with good I_{on}, however for biases larger than 0.5 V.

Interesting performance have been presented for strained Ge Double-gate TFETs with asymmetric S/D, especially with Si Drain, with a swing down to 50 mV/dec and I_{on} up to 300 μA/μm for high V_d [16]. However, these

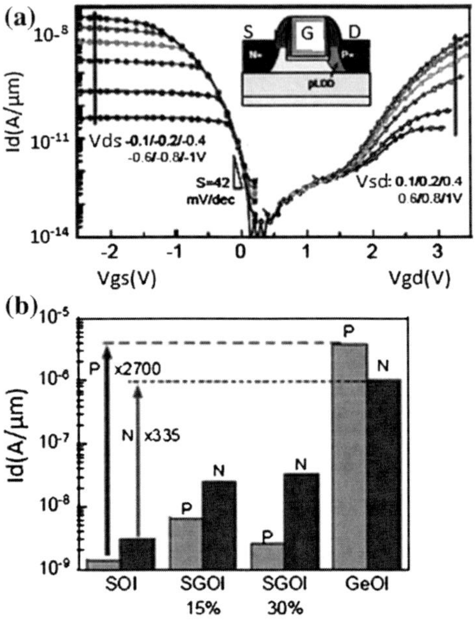

Fig. 4 **a** Transfer $I_d(V_g)$ characteristics of 100 nm gate length n- and p-channel TFETs (t_{Si} = 20 nm); **b** I_{on} of SOI, SGOI and GOI n- and p-TFETs with L_g = 400 nm (t_{SiGe} = 20 nm, t_{Ge} = 60 nm, $V_d = \pm 0.8$ V, $V_g = \pm 2$ V)

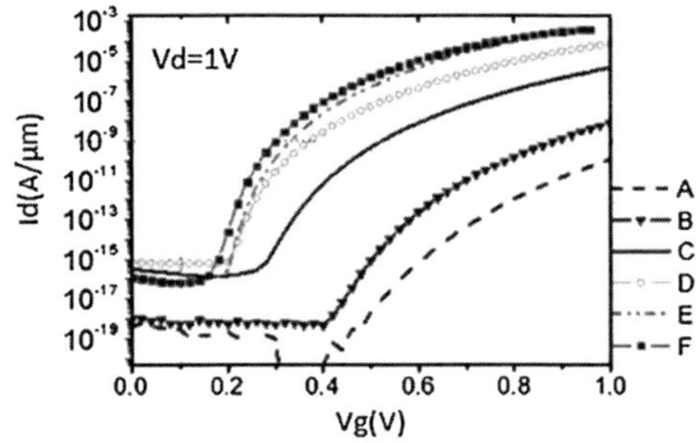

Fig. 5 **a** single gate SOI, L_g = 100, 3 nm SiO_2, **b** 4 GPa stress at source junction, **c** high-k gate dielectric, **d** double gate, **e** oxide aligned to i-region, **f** L_g = 30 nm

experimental results have been obtained with different biases applied to the front and back gates due to the different front and back gate oxide thicknesses. We have shown that an application of V_{g2}/V_{g1} proportional to t_{ox2}/t_{ox1} gives an overestimation of the performance of the devices [17].

Single-gate, double-gate and GAA InAs TFETs lead to very good simulated results (for $V_g = V_d = 0.2$ V), with S lower than 20 mV/dec for small wire diameter (down to 2 nm), due to the small effective masses and bandgap of these III–V materials.

For very short gate lengths, small wire diameters are needed, down to 5 nm or below, in order to obtain a good subthreshold swing (Fig. 7) [18].

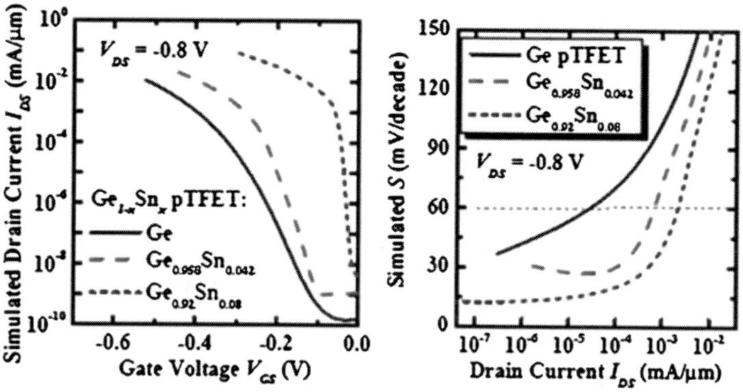

Fig. 6 Drain current versus gate bias (*left*) and substhtrehold swing versus drain current (*right*) for GeSn TFETs as compared to Ge TFETs obtained by numerical simulation

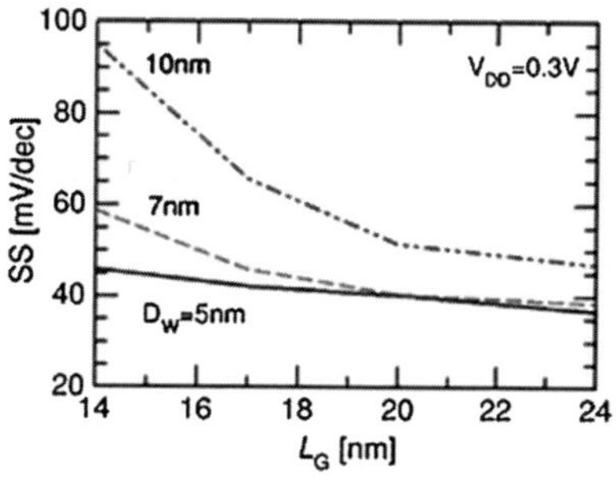

Fig. 7 Subthreshold swing versus gate length for GAA InAs nanowire TFETs for various wire diameters obtained by quantum simulation

High performances, with I_{on} up to 1 mA/μm at low V_d, have been demonstrated by quantum transport simulation on strained InAs NW TFET (Fig. 8), the best result being shown for a biaxial strain [18].

However, the on current improvements can be frustrated by the degradation of the swing in the presence of traps. Both traps and surface roughness can also be a relevant source of device variability for Tunnel-FETs [19].

The first experimental demonstration of S lower than 60 mV/dec (on less than 1 decade of I_d) using III-V materials has been shown with InGaAs Heterojunction 150 nm gate length TFET using thin EOT and high source doping, with I_{on} of a few μA/μm [20].

Carbon-based tunnel FETs were the first device showing experimentally $S < 60$ mV/dec, down to 40 mV/dec using CNT structures, but with very low I_{on} (<1 nA) [21]. Carbon materials have many advantages, in particular: small effective masses, small and direct bandgap, excellent electrostatic control (UTB), and could therefore be the best material choice. Graphene has similar properties as CNT but with planar processing compatibility. However, only theoretical studies have been performed so far. The high potential for graphene nanoribbons or bilayers have been shown by simulation [22] (I_{on} = 100 s of μA/μm, I_{off} = few

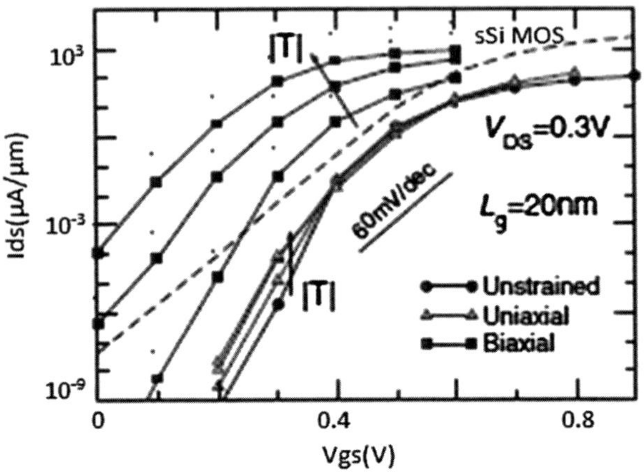

Fig. 8 20 nm gate length InAs NW TFET with 5 nm diameter and different strains obtained by quantum simulations

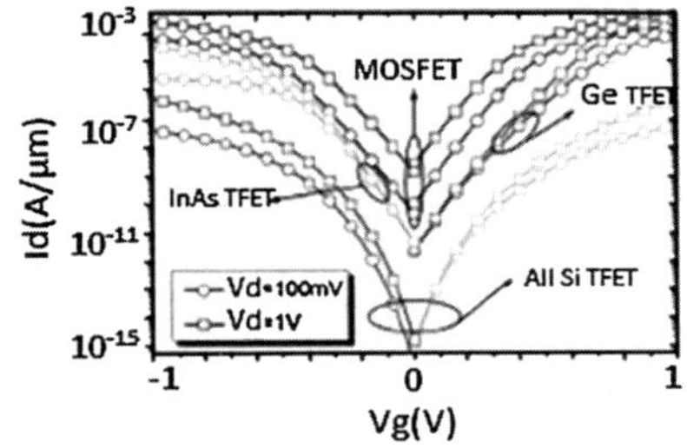

Fig. 9 InAs/Ge C-TFETs compared with all-Si TFETs and MOSFETs

pA/μm, $S = 20$ mV/dec, $E_{gap} = 200$–300 meV, $I_{on}/I_{off} > 10^3$ at $V_d = 0.1$ V) but have not been confirmed experimentally up to now.

The best trade-off obtained by simulation for Complementary TFET performance has been shown with Ge/InAs TFETs, leading to an increase if I_{on} of several hundreds compared to all-Si TFETs (Fig. 9) [23].

However, it is worth noting that to date the best TFET performance has been obtained with Si-based TFET and the experimental results are very poor compared with numerical simulation. No experimental demonstration of TFET with $I_{on} > 100$ μA/μm and $S < 60$ mV/dec has been shown so far.

Comparisons have been carried out between CMOS and TFET inverters. The Heterostructure Ge/InAs TFETs showed the most abrupt transition from 1 to 0 state with the highest differential gain and best noise margins. However, enhanced Miller effects are obtained due to the special capacitance behaviour in TFET leading to larger effective load capacitance and increased delay time. The best power-delay product has been shown so far for CMOS compared with Si TFET and Ge/InAs TFET [24].

Tunnel FETs seem up to now better in energy efficiency for applications with frequency lower than 1 GHz (Fig. 10) [25].

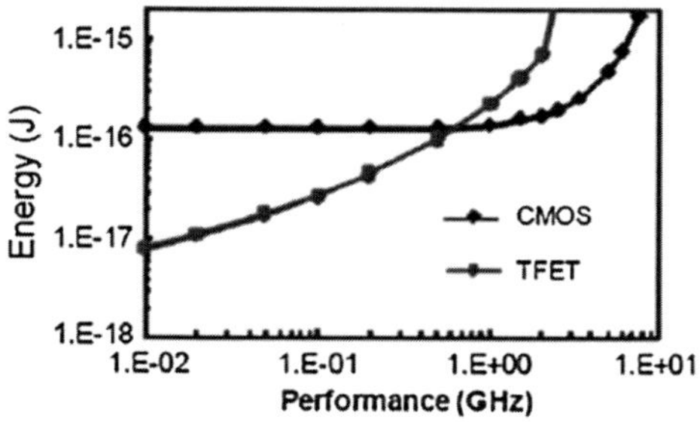

Fig. 10 Comparison of energy-frequency between CMOS and TEFT

3 Conclusion

The reduction of energy consumption is the main challenge for future electronic systems due to the very big amount of exchanged and stored data, which is exponentially increasing. A number of innovations are needed in the following fields: transistor, memory, devices and interconnects technologies, circuit design techniques, systems architectures and embedded software.

In the device domain, new physics, materials and device structures are required. This will enable to continue scaling and performance improvement.

Multi-Gate Nanowire MOSFETs with volume inversion (especially with Si, sSi) lead to the best short channel effects, S, I_{off}, V_{dd}, P, $E_{\min}$ for MOSFET architectures.

Tunnel FETs are the best Small Slope Switches up to now, with BTBT allowing for sub-60 mV/dec subthreshold swing, obtained by simulation and experimental results. TFETs simulations show promise for very good S, substantial V_{dd} reduction and high I_{on} but additional process improvements are needed to improve real device performance.

Heterostructures TFET (with Ge, III-V) could be viable solutions for $I_{on} > 100$ μA/μm, $I_{on}/I_{off} > 10^5$ and $V_{dd} < 0.5$ V with improving the technology.

All Si-TFET have shown theoretically the lowest I_{off} and poor I_{on}, but the best experimental performances have been reported to date with these devices.

Performance boosters require a good design, the best choice of materials, and integration on Si platforms.

It is also worth noting that the variability of TFET compared with CMOS is reduced for doping or gate length fluctuations, but is increased as a function of the high-k gate process, the abruptness of doping at tunnel junction, and the film thickness in Ultra Thin Body.

TFETs have a better energy efficiency at low or moderate performance level. They are ideally suited so far for low power and low standby power at moderate frequency on the order of hundreds of MHz.

Other possible TFET applications are:

- Low power SRAM (6T and 4T Cells), with a strong reduction of $I_{leakage}$ at low V_{dd}

- Excellent switch for high temperature due to the independence of the tunnel effect, I_{on} and S with T
- ULP analog ICs with enhanced temperature stability.

The biggest challenge is the improvement of I_{on} without degrading I_{off}, with $S < 60$ mV/dec for more than 4 decades of drain current. For that purpose, many technology boosters are needed together with a substantial improvement of the technology. The Heterostructure TFET offers the best performance for Complementary logic in Ultra Thin Body or Nanowire, and can be considered as an add-on Ultra Low Power device option on advanced CMOS platforms.

Acknowledgments The Sinano Institute Members and the Nanosil and Nanofunction European Networks of Excellence Partners are acknowledged.

References

1. Chang, L.: Practical strategies for power-efficient computing technologies. Proc. IEEE **98**, 215–236 (2010)
2. Hanson, S., Seok, M., Sylvester, D., Blaauw, D.: Nanometer device scaling in subthreshold logic and SRAM. IEEE Trans. Electron. Dev. **55**, 175–185 (2008)
3. Balestra, F., Brini, J., Gentil, P.: Deep depleted SOI MOSFETs with back potential control: a numerical simulation. Solid-St. Electron. **28**, 1031–1037 (1985)
4. Colinge, J.P.: Subthreshold slope of thin-film SOI MOSFET's. IEEE Electron. Dev. Lett. **7**, 244–246 (1986)
5. Andrieu, F., Weber, O., Mazurier, J., Thomas, O., Noel, J.-P., Fenouillet-Béranger, C.: Low leakage and low variability ultra-thin body and buried oxide (UT2B) SOI technology for 20 nm low power CMOS and beyond. In: Symposium of VLSI Technology (2010)
6. Balestra, F., Cristoloveanu, S., Benachir, M., Brini, J., Elewa, T.: Double-gate silicon-on-insulator transistor with volume inversion: a new device with greatly enhanced performance. IEEE Electron. Dev. Lett. **EDL-8**, 410–412 (1987)
7. Rauly, E., Potavin, O., Balestra, F., Raynaud, C.: On the subthreshold swing and short channel effects in single and double gate deep submicron SOI MOSFETs. Solid State Electron. **43**, 2033 (1999)
8. Bansal, A., Paul, B.C., Roy, K.: Impact of gate underlap on gate capacitance and gate tunneling current in 16 nm DGMOS devices. In: Proceedings of the IEEE, SOI Conference, pp. 94–95 (2004)
9. Dupre, C., Hubert, A., Becu, S., Jublot, M., Maffini-Alvaro, V., Vizioz, C., Aussenac, F., Arvet, C., Barnola, S., Hartmann, J.-M., Garnier, G., Allain, F., Colonna, J.-P., Rivoire, M., Baud, L., Pauliac, S., Loup, V., Chevolleau, T., Rivallin, P., Guillaumot, B., Ghibaudo, G., Faynot, O., Ernst, T., Deleonibus, S.: 15 nm-diameter 3D stacked nanowires with independent gates operation: ΦFET. In: Proceedings of the IEDM, p. 549 (2008)
10. Luisier, M., Lundstrom, M., Antoniadis, D.A., Bokor, J.: Ultimate device scaling: intrinsic performance comparisons of carbon-based, InGaAs, and Si FETs for 5 nm gate length. In: Proceedings of IEDM, p. 251 (2011)
11. Mayer, F., Le Royer, C., Damlencourt, J.-F., Romanjek, K., Andrieu, F., Tabone, C., Previtali, B., Deleonibus, S.: Impact of SOI, Si1-xGexOI and GeOI substrates on CMOS compatible tunnel FET performance. In: Proceedings of IEDM, pp. 1–5 (2008)
12. Boucart, K., Riess, W., Ionescu, A.M.: Asymmetrically strained all-silicon tunnel fets featuring 1 V operation. In: Proceedings of ESSDERC'2009 (2009)

13. Yang, Y., Su, S., Guo, P., Wang, W., Gong, X., Wang, L., Lu Low, K., Zhang, G., Xue, C., Cheng, B., Han, G., Yeo, Y.-C.: Towards direct band-to-band tunneling in P-channel tunneling field effect transistor (TFET): technology enablement by Germanium-tin (GeSn). In: Proceedings of IEDM'2012, p. 379 (2012)
14. Padilla, A., Yeung, C.W., Shin, C., Hu, C., King Liu, T.-J.: Feedback FET: a novel transistor exhibiting steep switching behavior at low bias voltages. In: Proceedings of IEDM, p. 171 (2008)
15. Wan, J., Le Royer, C., Zaslavsky, A., Cristoloveanu, S.: A compact capacitor-less high-speed DRAM using field effect-controlled charge regeneration. IEEE Electron. Dev. Lett. **33**, 179–181 (2012)
16. Krishnamohan, T., Kim, D., Raghunathan, S., Saraswat,K.: Double-gate strained-Ge heterostructure tunneling FET (TFET) with record high drive currents and <<60 mV/dec subthreshold slope. In: Proceedings of IEDM, p. 947 (2008)
17. Brini, J., Benachir, M., Ghibaudo, G., Balestra, F.: Subthreshold slope and threshold voltage of the volume inversion MOS transistor. IEE Proc. Part G Circ. Dev. Syst. **138**, 133 (1991)
18. Conzatti, F., Pala, M.G., Esseni, D., Bano, E., Selmi, L.: A simulation strudy of strain induced performance enhancements in InAs nanowire tunnel-FETs. In: Proceedings of IEDM, p. 95 (2011)
19. Conzatti, F., Pala, M.G., Esseni, D.: Surface-roughness-induced variability in nanowire InAs tunnel FETs. IEEE Electron. Dev. Lett. **33**, 806–808 (2012)
20. Dewey, G., Chu-Kung, B., Boardman, J., Fastenau, J.M.: Fabrication, characterization and physics of III–V heterojunction tunneling FET for steep subthreshold swing. In: Proceedings of IEDM, p. 785 (2011)
21. Appenzeller, J., Lin, Y.-M., Knoch, J., Avouris, Ph: Band-to-band tunneling in carbon nanotube field-effect transistors. Phys. Rev. Lett. **93**, 193805 (2004)
22. Fiori, G., Iannaccone, G.: Ultralow-voltage bilayer graphene tunnel FET. IEEE Electron. Dev. Lett. **30**, 1096–1989 (2009)
23. Ionescu, A., De Michielis, L., Dagtekin, N.: Ultra low power: emerging devices and their benefits for integrated circuits. In: Proceedings of IEDM, p. 378 (2011)
24. Ionescu, A., Riel, H.: Tunnel field-effect transistors as energy-efficient electronic switches. Nature **479**, 329–337 (2011
25. Kam, H., Tsu-Jae, K.-L., Alon, E., Horowitz, M.: Circuit-level requirements for MOSFET replacement devices. In: Proceedings of IEDM, p. 1 (2008)

High-Performance Tunnel FETs on Advanced FDSOI Platform

Cyrille Le Royer, Anthony Villalon, Mikaël Cassé, David Cooper, Jean-François Damlencourt, Jean-Michel Hartmann, Claude Tabone and Sorin Cristoloveanu

Abstract In this chapter, we present Tunnel FETs (TFETs) obtained with a FDSOI CMOS process flow featuring High-K Metal Gate, ultrathin body compressively strained $Si_{1-x}Ge_x$ (x from 0 to 30 %) based channels, and $Si_{0.7}Ge_{0.3}$ Raised SD. In-depth characterizations have been conducted to analyze the device structures (TEM, EELS for atom/layer identification, HAADF STEM GPA for strain) and device electrical performance ($C(V)$, $I_D(V_G)$ vs. V_{DS} and temperature, I_{ON}, S_w, tunnel extractions…). We investigate the tunneling improvements due to the different technological injection boosters: ultrathin body and gate dielectrics, strain, low band gap source, and low temperature SD anneal. The impact on $I_D(V_G)$ curves and thus on ON (and OFF) state current, subthreshold slope is presented and discussed. For the first time, TFETs with large ON current (up to 428 μA/μm) are demonstrated (with $>\times 1{,}000$ I_{ON} gain vs. SOI TFETs, and $>\times 35$ I_{ON} gain vs. best published pTFETs). Future paths towards further enhanced TFET devices are also detailed.

1 Introduction

The integrated circuit (IC) using metal-oxide-silicon field-effect-transistors (MOSFETs) has been developing rapidly for more than 4 decades. This is all driven by the scaling down of the MOSFET structure, doubling the integration density almost every 2 years according to Moore's law. The miniaturization of MOSFETs has enabled ICs with lower cost, more integrated functions and faster speed.

C. Le Royer (✉) · A. Villalon · M. Cassé · D. Cooper · J.-F. Damlencourt · J.-M. Hartmann · C. Tabone
CEA, LETI, Minatec Campus, 17 Rue Des Martyrs, 38054 Grenoble Cedex 9, France
e-mail: cyrille.leroyer@cea.fr

S. Cristoloveanu
IMEP-LAHC (UMR 5130), Grenoble INP Minatec, 38016 Grenoble Cedex 1, France

A. Nazarov et al. (eds.), *Functional Nanomaterials and Devices for Electronics, Sensors and Energy Harvesting*, Engineering Materials,
DOI 10.1007/978-3-319-08804-4_4

However, there are several physical limits on the MOSFET scaling. The short-channel effect (SCE), where the drain and source electric fields penetrate into the channel and reduce the carrier injection barrier in the OFF state, increases significantly as the gate length (L_G) is decreased. This greatly increases the leakage current and static power consumption. Though new technologies and device architectures, such as high-k/metal gate stacks, silicon-on-insulator (SOI) and multi-gate devices, have been proposed to enhance the gate controllability, the short-channel effect is still one of the main issues for the scaling of MOSFETs.

Another limit of MOSFET arises from the subthreshold slope (S_w). The S_w swing (defined in Eq. 1) is a criterion characterizing the sharpness of the switch.

$$S_w = \frac{\partial V_G}{\partial \log(I_D)} = \left(1 + \frac{C_D}{C_{ox}}\right) \cdot \ln 10 \cdot \frac{kT}{q},\ S_w \rightarrow 60\,\text{mV/dec @ T} = 300\,\text{K} \quad (1)$$

where V_G is the gate voltage, I_D is the drain current, kT/q is the thermal voltage, and C_D and C_{ox} are the depletion and the oxide capacitances, respectively. Sub kT/q values for S_w could be obtained by using new physical principles rather than thermionic injection.

In a MOSFET, the switching process involves the injection of electrons thermionic (which is temperature-dependent) over an energy barrier. This sets a fundamental limit to the steepness of the transition slope from the OFF state to the ON state.

As the gate length of transistors is reduced, improved performance requires that the supply voltage (V_{DD}) and simultaneously the threshold voltage (V_{th}) have to be lowered to keep the overdrive voltage ($V_{DD}-V_{th}$) as high as possible (Fig. 1). As a consequence, the leakage current (I_{OFF}) increases exponentially because the subthreshold slope S_w for MOSFETs is not scalable but present a minimum value of 60 mV per decade at room temperature (a +60 mV change for V_G is needed to increase the current by one order of magnitude). Typical values of S_w in advanced CMOS technology are close to 90–100 mV per decade.

The subthreshold slope (through I_{OFF}) has dramatic influence on the CMOS circuits' power consumption (which is defined by Eq. 2).

$$P = P_{dynamic} + P_{static} = k.C.V_{DD}^2.f + V_{DD}.I_{OFF} \quad (2)$$

It is highly desirable to have large I_{ON} densities without using a large supply voltage V_{DD}. It is also desirable to reduce the total load capacitance, C (including the junction capacitance of the driver devices, the gate capacitance of the driven devices, and the interconnect capacitance). Both capacitance and cost reductions provide strong motivations for reducing the size of the transistors (and therefore the size of the chip). In addition, speed has benefited from the push for smaller gate length, thinner gate dielectrics, and lower V_{th}. Power consumption has benefited greatly from the lowering of the supply voltage.

But for the sub-14 nm node technologies, the CMOS is facing fundamental limits, as illustrated in Fig. 1: Scaling down V_{DD} leads to large I_{OFF} increase (due

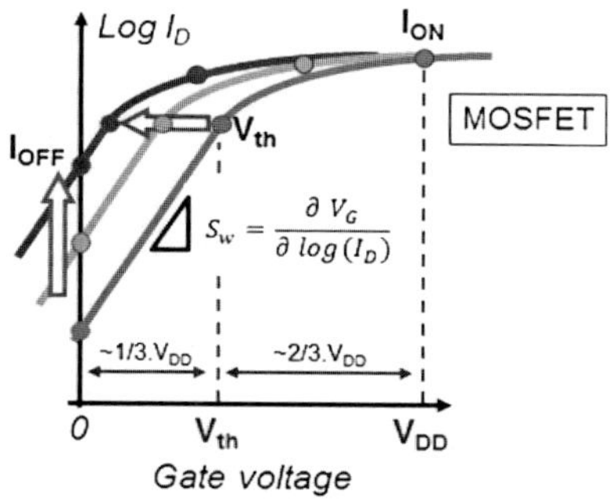

Fig. 1 Transfer characteristics (drain current I_D vs. gate voltage V_G) of a MOSFET device showing an exponential increase in OFF state current (I_{OFF}) because of an incompressible subthreshold slope, S_w. Here the scaling down of the supply voltage V_{DD}, (and thus of the threshold voltage V_{th}) leads to performance degradation: ON state current (I_{ON}) is decreased and OFF state current (I_{OFF}) is increased

to unsalable S_w), making the power consumption more and more difficult to optimize. In order to enable the scaling of V_{DD}, sharp-switching devices with S_w below 60 mV/dec are of great interest [1], as illustrated schematically in Fig. 2.

Tunnel Field Effect Transistors (TFETs) are PiN gated diodes (Fig. 3) which offer the possibility of lowering the subthreshold slope (S_w) below the kT/q limit of MOSFETs (combined with extremely low OFF currents) [2–9]. These properties make TFETs very attractive for ultra-low power applications ($V_{DD} < 0.5$ V). TFET devices present significant differences compared to conventional MOSFETs:

(i) The ON current is not limited by the electron transport but by the electron injection due to band to band tunneling [1, 10] making TFET $I_D(V_G)$ behavior independent of gate length (L_G) [5], and S_w potentially smaller than kT/q [5, 8, 9, 11];
(ii) a single TFET device can operate either in p or n channel mode, depending on the gate voltage [5, 12, 13] (Figs. 3, 4);
(iii) We can also notice that TFETs exhibit no Drain-Induced-Barrier-Lowering (*DIBL*) degradation.

However the reported performance of fabricated TFETs is poor compared to CMOS ones, especially in terms of ON currents: typical published TFET I_{ON} values are in the 0.1–1 μA/μm range [1–9, 14, 15], with perfectible subthreshold slopes. Solutions have been identified to boost the tunneling properties of TFETs [11–15]:

- electrostatics (thin body, small EOT, multigate architecture);
- low band gap (E_G) materials for homojunctions TFET or for heterostructured TFETs (low E_G in the P+ source for n mode operation, for example);
- junction optimization (position w.r.t. gate edge, abruptness).

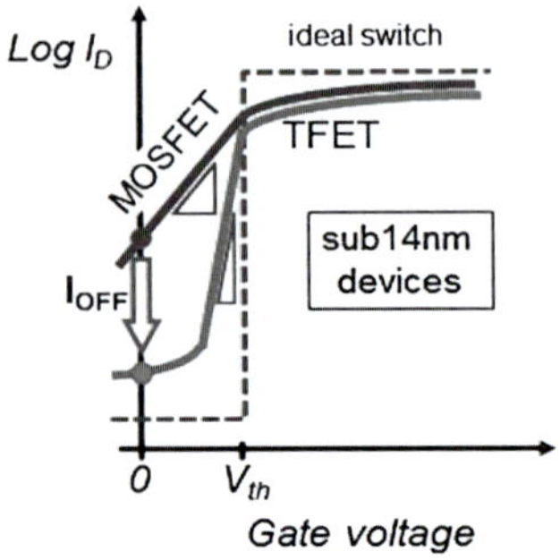

Fig. 2 Illustration of Tunnel FET benefits over MOSFET on $I_D(V_G)$ transfer characteristics: because of smaller subthreshold slope, I_{OFF} can reduced for the same threshold voltage value

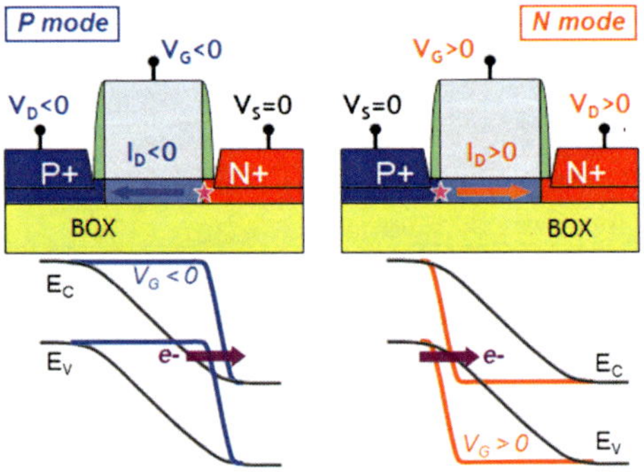

Fig. 3 Principle of TFET structures, of n and p operation modes. The corresponding band diagrams illustrate where tunnelling occurs

TFET	P mode	N mode
Source	N+ electrode	P+ electrode
Drain	P+ electrode	N+ electrode
V_S / V_D	$V_S = V_N = 0V$ $V_D = V_P < 0V$	$V_S = V_P = 0V$ $V_D = V_N > 0V$
V_{DS}	$V_{DS} = V_{PN} < 0V$	$V_{DS} = V_{NP} > 0V$
V_{GS} (ON state)	$V_{GS} = V_{GN} < 0V$	$V_{GS} = V_{GP} > 0V$
I_D (ON state)	$I_D = I_P < 0$	$I_D = I_S > 0$

Fig. 4 Summary of the n and p operation modes in terms of Source/Drain definitions, bias conditions, and current definitions

In this work, we analyze the relative contributions of different technological tunnel boosters on the electrical performances (I_{ON}, I_{OFF}, S_w) of ultrathin body compressively strained SiGe (cSiGe) based TFETs fabricated with a Fully-Depleted Silicon-On-Insulator (FDSOI) CMOS process [12, 16, 17].

2 CMOS and TFET Device Fabrication

2.1 Device Fabrication Process

We have fabricated CMOS and TFETs devices (Figs. 5, 6) on 300 mm ETSOI (Extremely Thin SOI) wafers with a FDSOI process flow.

The initial silicon top film of the SOI wafer has been first thinned down to ~3 nm. Then an epitaxy step has been performed in order to obtain a stack composed of $Si_{cap}/Si_{1-x}Ge_{x}/SOI$. The SiGe film present Ge concentration between 5 and 30 % (Fig. 6) and thickness between 3 and 8 nm. The ultrathin Si capping layer (1–2 nm) has been introduced in order to maintain a good interface between the SiGe-based channel and the gate dielectrics. We can note that the initial intrinsic $Si_{cap}/Si_{1-x}Ge_{x}$ epitaxial growth enables to obtain compressively strained cSiGe channels (see Sect. 2.2).

The device isolation is realized by patterning the $Si_{cap}/Si_{1-x}Ge_{x}/Si$ stack (MESA structures on top of the buried oxide, BOX). Then the gate stack (High-K Metal gate) is formed and patterned (Fig. 7): the PolySi/TiN/HfSiON stack leads to ultra low Equivalent Oxide Thickness, EOT (1.12–1.25 nm, see Sect. 3.1). The minimum physical gate length is around 30 nm for CMOS devices, and 80 nm for TFETs.

After first spacer definition, Raised Source-Drain are formed by epitaxial growth, with two processes compared: Si RSD and $Si_{0.7}Ge_{0.3}$ RSD. For pMOSFETs, SiGe RSD are known to increase the hole mobility and thus the device performance [17]. For TFETs, the aim was to introduce low band gap materials in the source region, in order to demonstrate heterostructured TFET with enhanced tunnelling injection [1]. The n and p-type dopants are introduced in the S-D regions with ion implantations (As and BF_2 respectively): the corresponding masks (DSN and DSP, resp.) were designed for Sources and Drains of nMOSFETs, pMOSFETs and TFETs (as illustrated in Fig. 8). Extensions (or Lightly Doped Drain, LDD) and SD (Highly doped Drain, HDD) implantations were both performed using these masks, enabling the cointegration of CMOS devices with TFET devices.

Moreover beyond the classical 1,050 °C spike SD anneal, we have investigated lower temperature spike anneal (950 °C) in order to obtain more abrupt junctions (which are beneficial for TFET injection).

The end of the process features Ni based silicidation of the RSD (and the top of gates made of PolySi), and classical Back-end (up to Metal 1).

2.2 Physical and Chemical Characterization

We have investigated the device process quality by performing different characterization techniques: Scanning TEM (STEM), EELS.

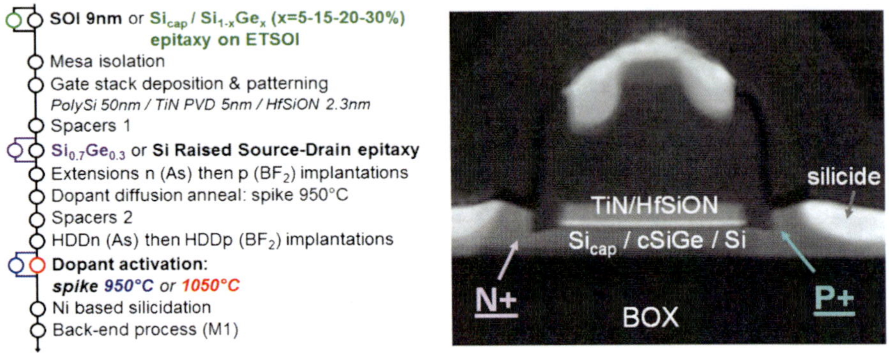

Fig. 5 Summary of the FDSOI process flow used here to fabricate CMOS and TFET devices on 300 mm ETSOI wafers (BOX = 145 nm)

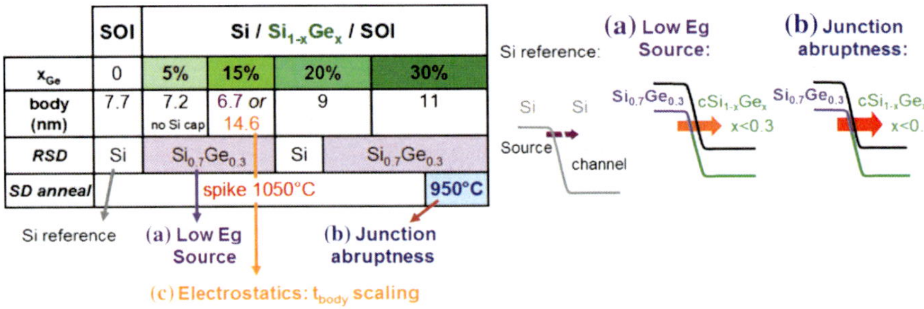

Fig. 6 Schematics of the different configurations of the fabricated TFETs (+corresponding CMOS). Tunneling optimization solutions A and B are illustrated by band diagrams

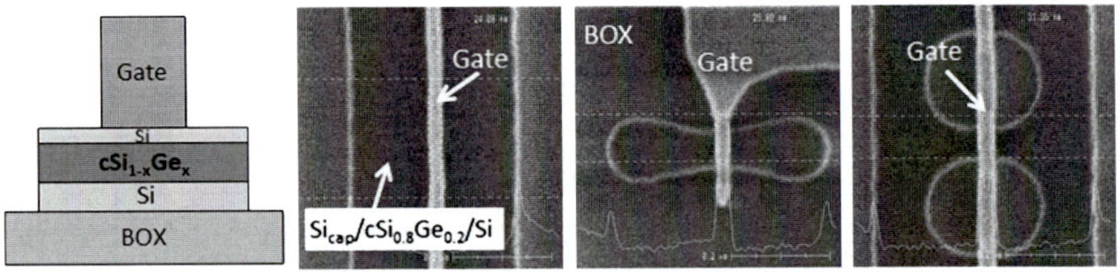

Fig. 7 *Top view* SEM pictures of fabricated devices (MOSFET here) on $Si_{cap}/cSi_{0.8}Ge_{0.2}/Si/$ BOX (after gate patterning), for wide, narrow, and multi-fingers active regions

The STEM images of Si_{cap}/cSiGe/SOI TFETs enable to check that the body is composed of capping top Si, SiGe and bottom Si layers (Fig. 9a). Moreover the Global Phase Analysis (GPA) of the HAADF STEM images (see Ref. [18] for more details on this technique) have been used to determine the 2D strain mapping in the body stack. It appears that the in-plane lattice mismatch (ε_{XX}) is zero in the

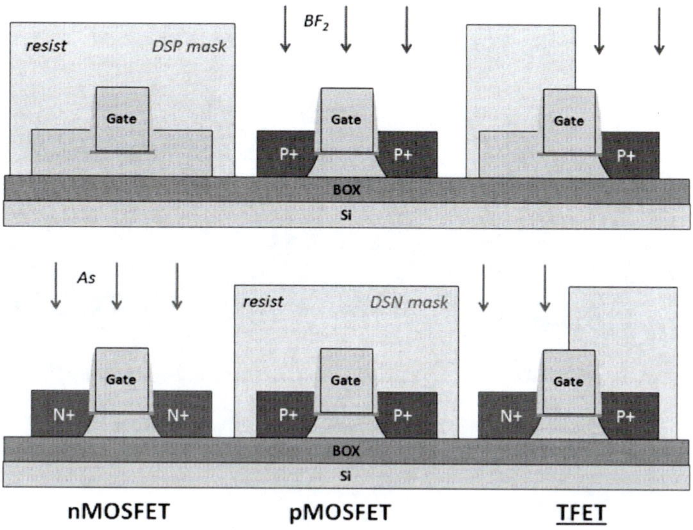

Fig. 8 Source and Drain doping: schematics of the lithography/ion implantation sequence. DSP mask is first used, then DSN (for LDD implantations). After spacers 2 formation, another DSP/DSN sequence is performed (for HDD implantations)

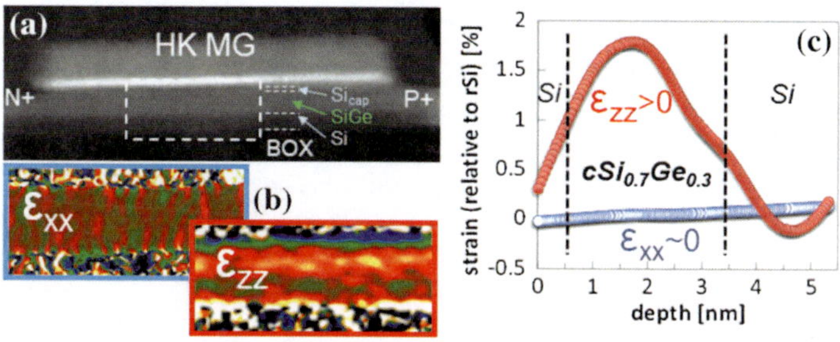

Fig. 9 **a** HAADF STEM picture of a FET with a $Si_{cap}/cSi_{0.7}Ge_{0.3}/Si$ channel stack and $Si_{0.7}Ge_{0.3}$ RSD. **b** The 2D strain mapping (ε_{XX} and ε_{ZZ}) clearly shows that the SiGe film is biaxially compressively strained. **c** Corresponding vertical profiles of in-plane ε_{XX} and out-of-plane ε_{ZZ} strain values (defined here compared with relaxed Si under the BOX)

SiGe film (with respect to unstrained Si), demonstrating thus that the SiGe film is biaxially compressively strained (Fig. 9b–c).

Electron Energy Loss Spectroscopy (EELS) characterization has been performed on a TFET device with cSiGe based channel and SiGe RSD (Fig. 10). The 2D mappings for silicon and germanium indicate the SiGe layer position in the body (with excellent interfaces) and the SiGe RSD. Moreover this technique clearly shows the silicide.

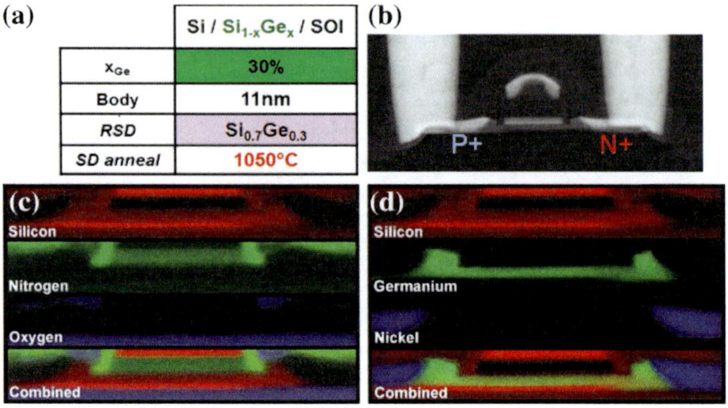

Fig. 10 **a** TFET device summary description; **b** Corresponding HAADF STEM picture (L_{mask} = 100 nm, W = 10 μm); **c–d** 2D EELS mapping for Si, N, O (**c**) and Si, Ge, Ni (**d**) of the body region of this TFET

3 Demonstration of Functional Cointegrated CMOS and TFETs

3.1 CMOS

The fabricated devices exhibit low EOT values: 1.12 and 1.25 nm for SOI and Si_{cap}/cSiGe/SOI respectively (Fig. 11). Fully depleted CMOS devices are functional even with high Ge content cSiGe channels. The corresponding p and n MOSFET $I_D(V_G)$ measurements (Fig. 12) are in line with cSiGe channels properties (mobility boost for hole and degradation for electron) [16, 17].

3.2 Tunnel FETs

Figure 13 shows $I_D(V_G)$ measurements performed on a single TFET device (from the wafer used for CMOS $I_D(V_G)$, cf. Fig. 12), emphasizing the ability of this device to operate in p or n modes (cf. Figs. 3, 4): 33 mV/dec $S_{w,lin}$ point slope is obtained for average value of 111 mV/dec over three I_D decades.

The physical phenomenon involved in the device operation has been clearly identified: the tunnelling operation of TFET is evidenced in Figs. 14 and 15. Figure 14 shows that the electrical measurements of pTFETs (with different SiGe channels) follow the tunnelling Kane's model [10] at both room temperature and high temperatures (25 and 125 °C). Figure 15 shows the same type of plots but for different V_{DS} values and for a single device operating in p and n modes.

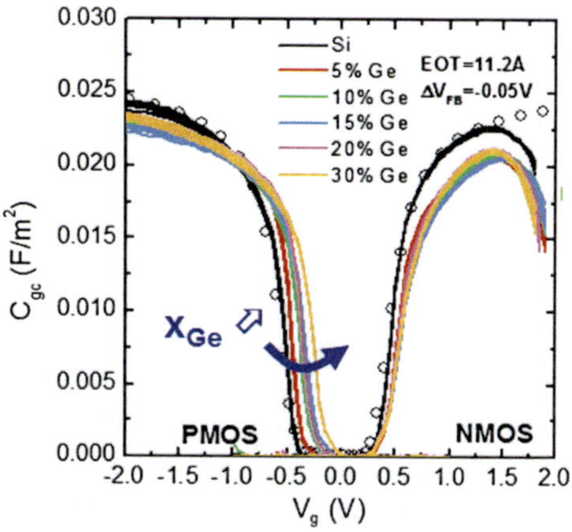

Fig. 11 $C(V)$ measurements: For reference SOI devices, the *EOT* is 1.12 nm, and for SiGe based devices the *EOT* is 1.25 nm (the 1.5 nm Si_{cap} layer is electrically equivalent to 5 Å SiO_2)

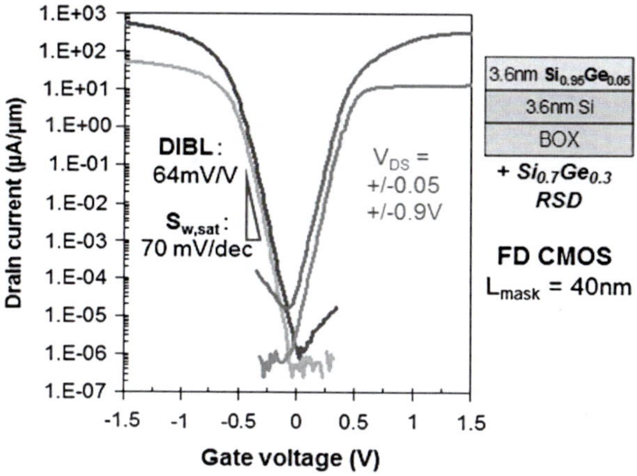

Fig. 12 Example of $I_D(V_G)$ measurements performed on CMOS devices ($W = 80$ nm) cointegrated with TFETs on $Si_{cap}/cSi_{0.95}Ge_{0.05}$/SOI wafers ($Si_{0.7}Ge_{0.3}$ RSD, 1,050 °C spike anneal)

Another TFET particularity is shown in Fig. 16: when changing temperature from 25 to 125 °C the TFET ON current increases (+15 %) through thermal activation, while the MOSFET ON current decreases (−14 %) because of mobility reduction.

4 Tunneling Boosters: SiGe, Ultrathin Body, Abrupt Junction

4.1 Low Band Gap Materials

We report the effect of tunneling boosters on experimental TFET performances, beginning with low bandgap materials [12, 14]. Figure 17a presents the effect of the channel and RSD materials on the cumulative drain current (extracted at

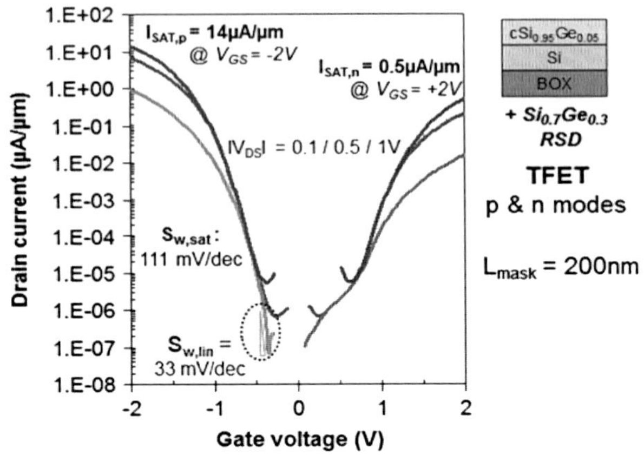

Fig. 13 Examples of measured p and n modes $I_D(V_G)$ characteristics of a single TFET device ($L_G = 200$ nm, $W = 10$ µm, 7.2 nm body $cSi_{0.95}Ge_{0.05}$/SOI channel, with $Si_{0.7}Ge_{0.3}$ RSD). I_{ON}/I_{MIN} ratio is 10^7 at $V_{DD} = 0.5$ V (for p mode)

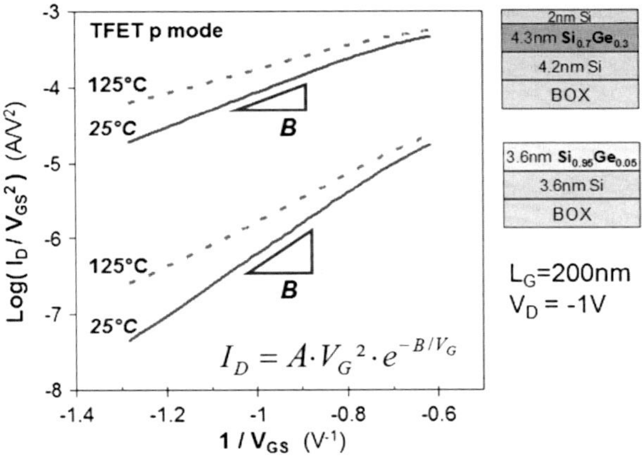

Fig. 14 Examples of TFET operation demonstration (for p mode, $V_{DS} = -1$ V): experimental data fit the Kane's tunnel model [10] at 25 and 125 °C. The $B(x_{Ge})$ slope verifies the $cSi_{1-x}Ge_x$ bandgap dependence $B \sim E_G^{1.5}$

$V_D = -1$ V) and subthreshold slope distributions of pTFETs, and shows the associated band structures. The studied devices are SOI reference TFETs with Si RSD and Si/c-$Si_{0.8}Ge_{0.2}$/SOI channel TFETs with $Si_{0.7}Ge_{0.3}$ RSD. The band diagram compares both structures. The dominant effect in these structures comes from the $Si_{0.8}Ge_{0.2}$ layer in the channel.

In a pTFET, the electrons tunnel from the valence band of the channel to the conduction band of the N^+ region, thus bandgap engineering would only provide tunneling boosters if it increases the channel E_V or lowers the N^+ side E_C. It is the case of the SiGe layer in the channel, which shifts the channel E_V up, thus increasing tunneling at the junction (+3 decades for I_{ON}: 50 µA/µm up from

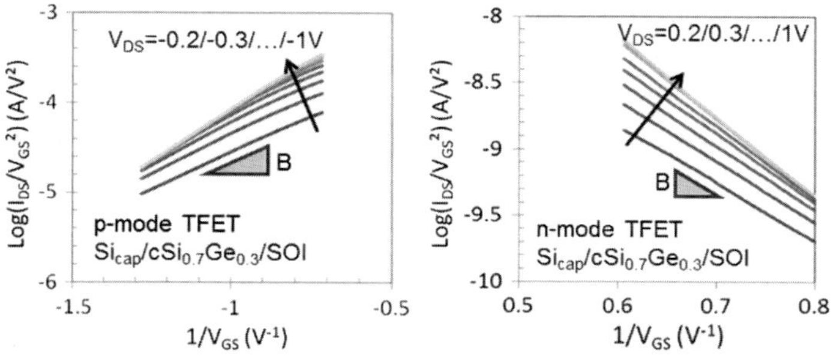

Fig. 15 Demonstration of tunneling injection in a $Si_{cap}/cSi_{0.7}Ge_{0.3}$/SOI TFET (operating in p and n modes, for different V_{DS} values): experimental data fit the Kane's tunnel model (at room temperature)

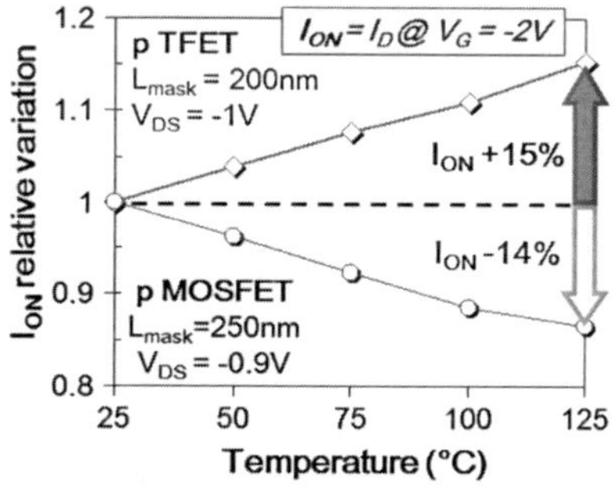

Fig. 16 Impact of temperature on the measured I_{ON} currents of SiGe based pTFETs and pFET. $I_{ON,pFET}$ is degraded because of mobility reduction, whereas $I_{ON,TFET}$ is increased (+15 %) because tunnelling is thermally enhanced

0.05 μA/μm). This current increase also leads to a decrease in the subthreshold slope (120 mV/dec down from 220 mV/dec).

Figure 17b presents the nTFET I_{ON} and subthreshold slopes, and bandgap diagram of the same structures. Here, the main booster is the presence of SiGe in the RSD (specifically, the P^+ region). Indeed, as opposed to the pTFET case, tunneling enhancement requires increasing P+ E_V or lowering channel E_C. Here, the I_{ON} gain from the low bandgap material is 1 decade (50 nA/μm up from 5 nA/μm) and we also observe a decrease in the S_w slopes (300 mV/dec down from more than 600 mV/dec).

4.2 FDSOI with Ultrathin Body

Previously reported TFETs [3, 5, 6, 8, 9] have been realized on thick body (>20 nm) SOI, leading to unoptimized tunnel injection. Figure 18 highlights the

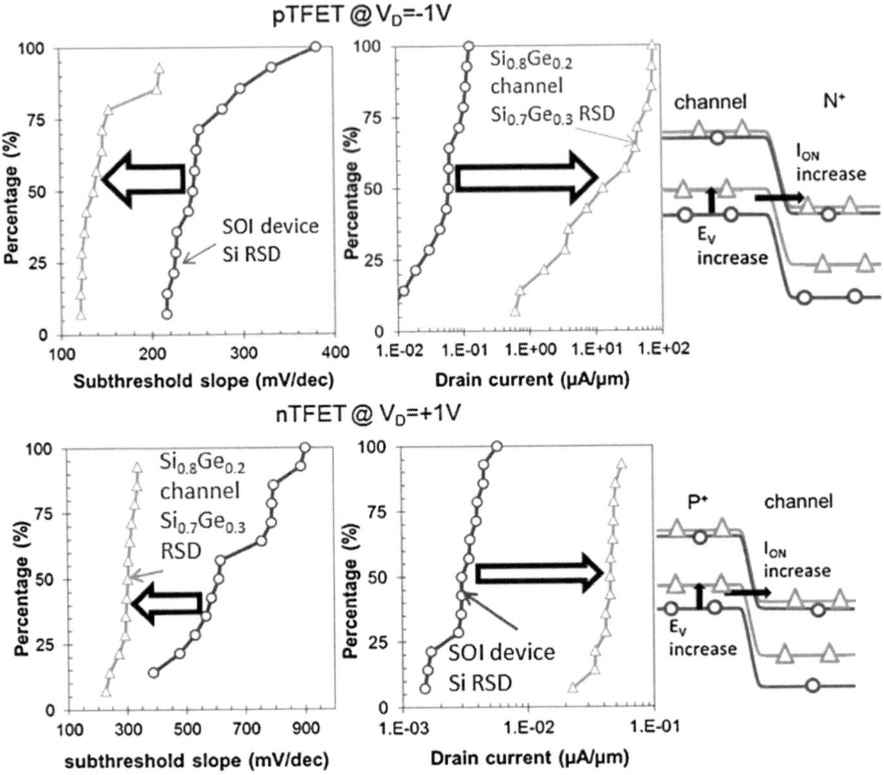

Fig. 17 Cumulative plots of the pTFET (*top*) and nTFET (*bottom*) ON-current (extracted at $V_G = \pm 2.5$ V) and of the subthreshold slopes (extracted at a specific drain current of 1 nA/µm) for the SOI reference TFET with Si RSD, and Si/c$Si_{0.8}Ge_{0.2}$/Si channel with $Si_{0.7}Ge_{0.3}$ RSD. The effect of the $Si_{0.7}Ge_{0.3}$ in the RSD (acting as low bandgap source and drain material) on the ON-current of pTFETs is clearly visible and explained in the bandgap structure. Similarly, the positive effect of the $Si_{0.8}Ge_{0.2}$ layer in the channel on the I_{ON} and S_w of nTFETs is highlighted

dramatic tunneling improvement ($I_{ON} \times 10^2$–10^4) on p and n modes of a single TFET by reducing the body from 14.6 to 6.7 nm (in the case of c$Si_{0.85}Ge_{0.15}$ based channel, with $Si_{0.7}Ge_{0.3}$ RSD). Even if the OFF currents are also increased, one can note that I_{OFF} degradation can be limited or even cancelled with junction optimization.

4.3 Low Temperature SD Anneal

Figure 19 shows for pTFET that a 950 °C SD spike anneal (vs. 1,050 °C) enables to boost I_{ON} (up to 428 µA/µm) while keeping I_{OFF} below 20 pA/µm. Moreover, the threshold voltage is reduced (by 500 mV) and S_w moves from 230 to 150 mV/dec.

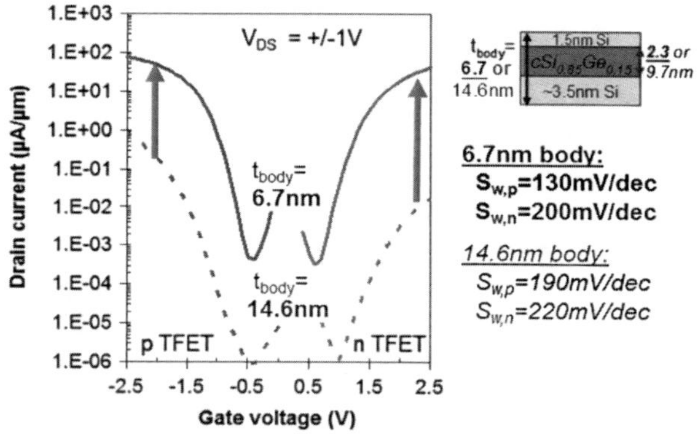

Fig. 18 Impact of the body thickness on the p and n modes $I_D(V_G)$ *curves* of $Si_{cap}/Si_{0.85}Ge_{0.15}/$SOI based TFETs with $Si_{0.7}Ge_{0.3}$ RSD ($L_G = 200$ nm, 1,050 °C spike anneal). S_W is improved from 190 to 130 mV/dec

Even if this large improvement of 950 °C TFET operating in p mode (with no change for n mode) seems in agreement with doping properties of SiGe (with B and As), it turns out that in reality the silicidation process was the physical reason. Figure 20 is an example of HAADF STEM for 1,050 and 950 °C TFET devices: whereas 1,050 °C spike anneal leads to "normal" TFET structure with silicides of SD electrodes located where they are supposed to be (far from the body), the 950 °C spike anneal leads to a different behavior for n-type doped SiGe electrode: the silicide kinetics seems to be enhanced (in nSiGe), leading to silicide penetration into the channel (from 15 to 60 nm, depending on the device structure).

Thus the p mode operation of these 950 °C annealed TFETs (structure similar to TFETs described in [8]) could be related to Schottky injection.

5 TFET Subthreshold Slope and I_{ON} (I_{OFF}) Performance Boost

5.1 S_w Figure of Merit: Drain Current versus Subthreshold Slope

In order to go beyond a basic minimum S_w value which is much smaller than the average S_w (on several I_D decades), we have plotted the subthreshold slope as function of the drain current (Fig. 21). From this $S_W(I_D)$ figure of merit we can directly compare the performance of different devices: Fig. 21 shows that the ultrathin $Si_{0.85}Ge_{0.05}$/SOI based TFET exhibits excellent point slope ($S_{w,\min} = 33$ mV/dec at $V_D = -0.1$ V) and average slope (111 mV/dec over more than 3 decades of current at $V_D = -0.2$ V), especially when compared to previous works [5]. Even if the corresponding FD pMOSFET remains better in terms of

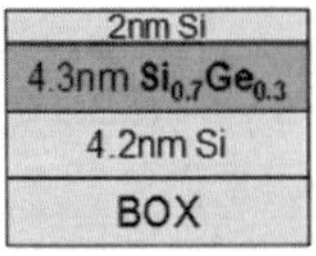

	Si / Si1-xGex / SOI	
XGe	30%	30%
Body (nm)	11	11
RSD	Si0.7Ge0.3	Si0.7Ge0.3
SD anneal	1050°C	950°C

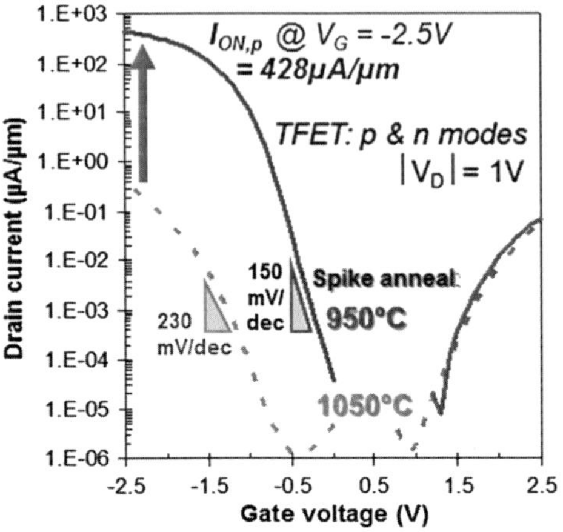

Fig. 19 Impact of the SD activation anneal temperature on the p and n modes $I_D(V_G)$ *curves* of $Si_{cap}/SiGe_{0.7}Ge_{0.3}/SOI$ based TFET (with $Si_{0.7}Ge_{0.3}$ RSD)

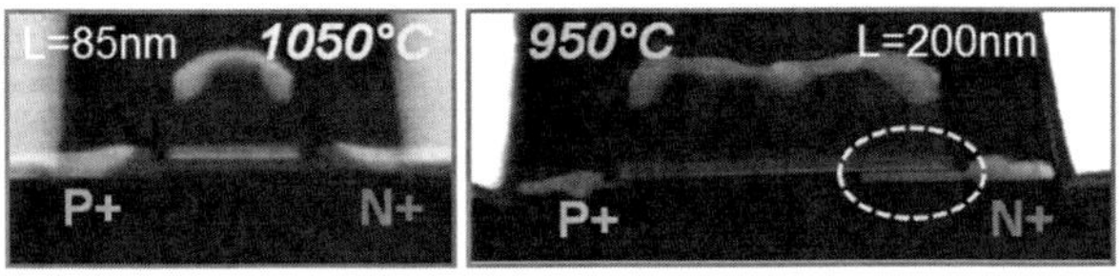

Fig. 20 HAADF STEM pictures of TFETs devices with two different activation anneal temperatures ($Si_{cap}/SiGe_{0.7}Ge_{0.3}/SOI$ body with $Si_{0.7}Ge_{0.3}$ RSD). The silicide of the N+ regions for 950 °C devices exhibit a penetration into the body under the gate

S_w and I_{ON}, we demonstrate TFETs with improved performances which are (for the first time) close to the MOSFET ones.

If we consider the $Si_{cap}/cSi_{0.7}Ge_{0.3}/SOI$ TFETs with 1,050 and 950 °C anneals (see Sect. 4.3) it appears that the low temperature case leads to best $S_w(I_D)$ characteristics, with large S_w gain for a given drain current (Fig. 22): ×3,300 versus 1,050 °C TFET (and $\times 10^6$ vs. thick SOI TFET reference with ~1,050 °C spike anneal). The minimum S_w values obtained here for the 950 °C $Si_{cap}/cSi_{0.7}Ge_{0.3}/SOI$ pTFET are slightly larger compared to the ultrathin $cSi_{0.85}Ge_{0.05}/SOI$ pTFET: 150 mV/dec versus 111 mV/dec at high V_{DS} value. This can be attributed to (i) larger body thickness for the 950 °C $Si_{cap}/cSi_{0.7}Ge_{0.3}/SOI$ pTFET (11 vs. 6.7 nm), and (ii) different tunneling mechanisms (Schottky vs. band-to-band-tunneling).

5.2 $I_{ON}(I_{MIN})$ Figure of Merit

The $I_{ON}(I_{MIN})$ plots of the SiGe RSD based TFETs in p and n mode (Fig. 23) clearly show the effect of the aforementioned tunneling boosters: low band gap

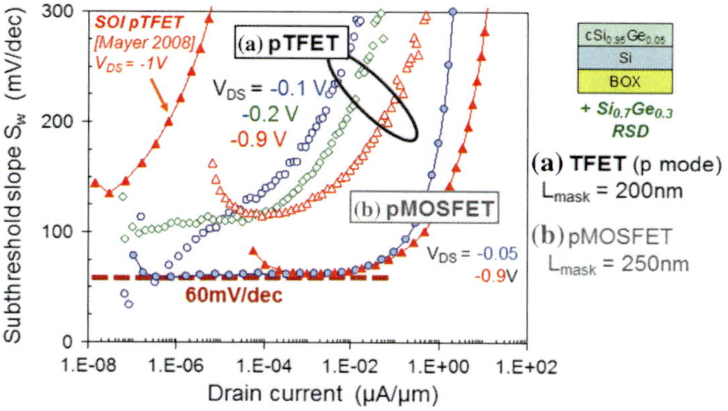

Fig. 21 Extracted subthreshold slope as function of drain current I_D for $L_G = 200$ nm pTFET and $L_G = 250$ nm pMOSFET on $cSi_{0.95}Ge_{0.05}$/SOI structure (with $Si_{0.7}Ge_{0.3}$ RSD) for different drain voltages (with comparison with previous work). Point slope of 33 mV/dec is obtained at $V_D = -0.1$ V

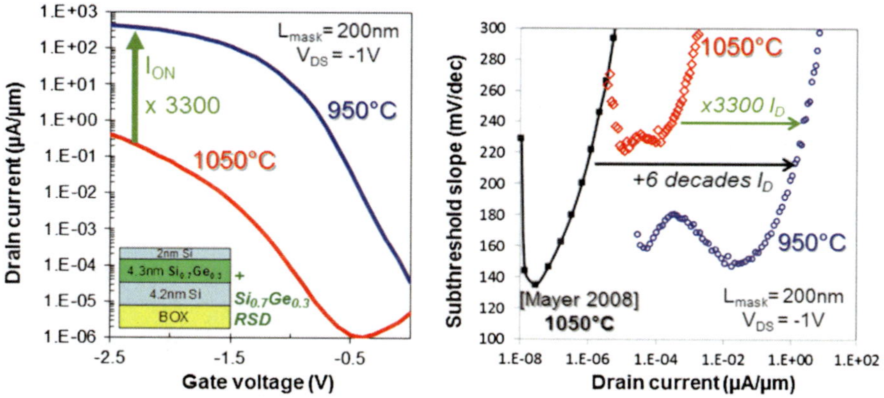

Fig. 22 *Left* $I_D(V_G)$ characteristics of Si/c-$Si_{0.7}Ge_{0.3}$/Si TFETs (10.5 nm total thickness) with 1,050 and 950 °C dopant anneal temperature. The cooler process leads to an I_{ON} 3,300 times higher than the hot process. *Right* Subthreshold slopes as a function of drain current. This graph shows the improvement in terms of slopes values and stability over a range of currents for the 950 °C anneal versus the 1,050 °C annealed structure and previous work (SOI at 1,050 °C)

materials, ultrathin body (UTB), low temperature. UTB SiGe based TFETs exhibit improved I_{ON} (+3 dec) but with a degraded I_{MIN} (+2.5 dec). The body thickness reduction induces a gain for both n and p TFETs, while anneal temperature mostly benefits the p mode TFETs: I_{ON} is enhanced by more than 3 decades while keeping I_{MIN} increase under 1 decade.

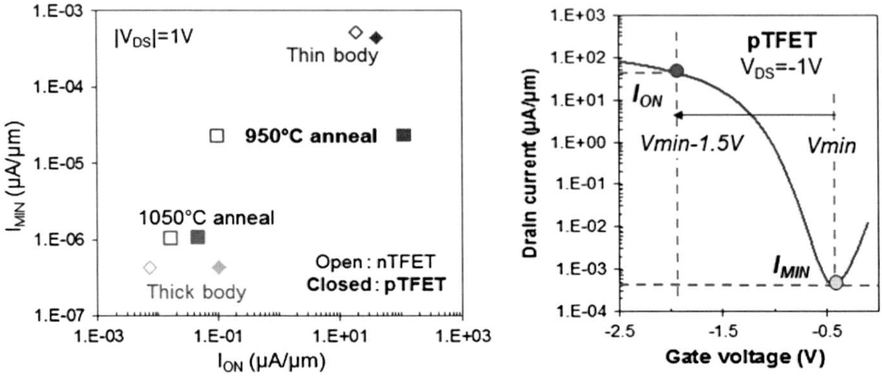

Fig. 23 *Left* cSiGe/SOI based p and n mode TFET $I_{ON}(I_{MIN})$ figure of merit at $V_{DD} = 1$ V (where $I_{MIN} = \min(I_D)$) illustrating the tunnel boost due to junction abruptness improvement by low temperature SD anneal or ultrathin body (UTB) ($L_G = 200$ nm). All TFETs present $Si_{0.7}Ge_{0.3}$ RSD. Both n (open) and p (closed) performances are represented here; *Right* Illustration of the I_{ON} and I_{MIN} definitions in the case of p mode: I_{MIN} is the minimum measured drain current, while I_{ON} corresponds to the drain current at $|V_G| = |V_{MIN}| + 1.5$ V

5.3 TFET Benchmark

Among all the TFETs reported here, Fig. 24 summarizes the best switch performance for p mode and n mode operations. From these curves, we have extracted different parameters (I_{MIN}, I_{ON}, S_w) in order to compare our results to the tunneling FET results published in literature (Figs. 25, 26).

This TFET comparison is illustrated in Fig. 25 for p mode TFET: the $I_{ON}(I_{MIN})$ plots clearly show that the studied $Si_{cap}/cSi_{0.7}Ge_{0.3}$/SOI TFETs with 950 °C anneal improve the performance state of the art (I_{ON} is 35 larger than the values reported in recent TFET papers. Figure 26 provides the detailed data for both p mode and n mode operations.

6 What Is Next?

The TFET (fabrication) studies conducted so far were based on wide device configurations ($W = 10$ µm in this work) or multi fingers structures (but without SiGe booster for example). As CMOS is moving from bulk to FD thin film architectures (FinFET, FDSOI) and from pure Si materials to SiGe channel and/or SiGe RSD (for pMOSFETs), TFET will benefit from these evolutions. This trend is illustrated in Fig. 27. The enhanced electrostatic control offered by trigate and nanowire structures will dramatically enhance the TFET performance (I_{ON}, S_w). Moreover, if we combine this architecture evolution with adapted band structures

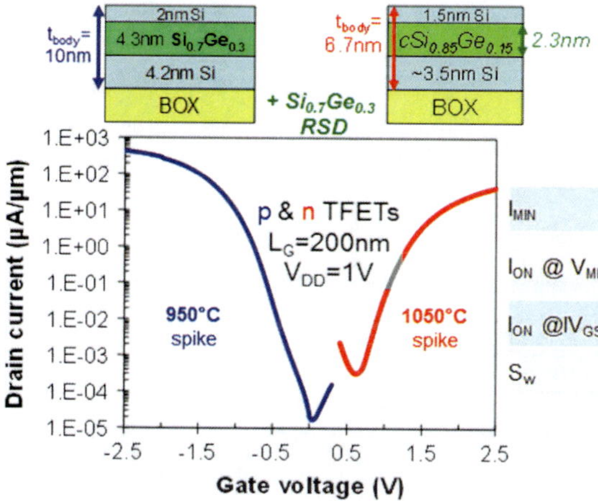

	pTFET	nTFET
I_{MIN}	0.03 nA/µm	0.3 nA/µm
I_{ON} @ V_{MIN}±1.5V	102 µA/µm	23 µA/µm
I_{ON} @|V_{GS}|=2.5V	**428** µA/µm	38 µA/µm
S_w	150 mV/dec	150 mV/dec

Fig. 24 Measured $I_D(V_G)$ of p and n mode cSiGe/SOI based TFETs with the best performance obtained in this study ($W = 10$ µm, $L_G = 200$ nm), and extracted parameters I_{MIN}, I_{ON} and S_w

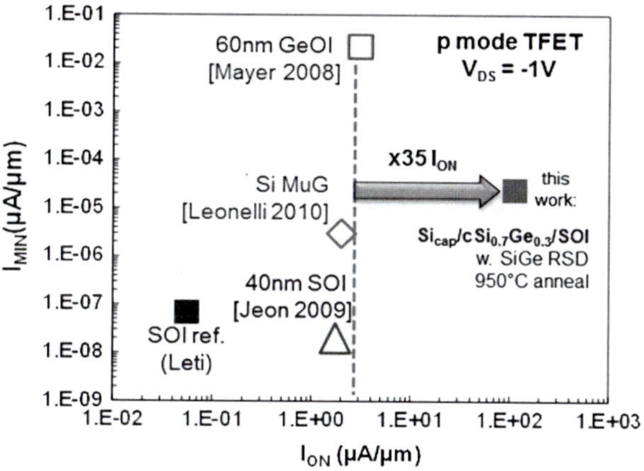

Fig. 25 p mode TFET benchmark, $I_{ON}(I_{\text{MIN}})$ figure of merit (at $V_{DS} = -1$ V) where $I_{\text{MIN}} = \min(I_D)$: the cSiGe pTFET illustrating the tunnel boost due to low temperature dopant anneal and ultrathin cSiGe body corresponds the device in Fig. 24

and junctions (Fig. 28), we can further increase the tunneling injection and thus the TFET performance. For example, *EOT* values around 1.2 nm have been considered in the previous sections, but lower *EOT* for CMOS have been already demonstrated. The situation is similar for RSD epitaxy process: we have defined junctions by ion implantations, but in situ boron doped SiGe (and in situ phosphorus doped Si) processes have been successfully demonstrated for FDSOI

Mode	p					n		
TFET	[Mayer 2008]	[Leonelli 2010]	[Jeon 2007]	[Knoll 2011]	this work		[Choi 2007]	[Tomioka 2012]
structure	SOI	Si FinTFET	SOI	sSOI planar	$Si_{0.7}Ge_{0.3}$	$Si_{0.85}Ge_{0.15}$	SOI	III-V NW
EOT (nm)	CET 2.1nm		~0.9	3nm Hf	**1.25**	**1.25**	2	
t_{body} (nm)	20	25	40	7	**11**	**6.7**	70	30nm
L_G (nm)	100	160		200	**200**	**200**	70	
V_{DS} (V)	-1	-1.2	-1	-1.1	**-1**	**1**	1	1
I_{ON} (µA/µm)	0.036	2	3	65	**112**	**18.8**	20	1
I_{MIN} (µA/µm)	$5.1 \cdot 10^{-7}$	$3 \cdot 10^{-6}$	$2 \cdot 10^{-8}$	$2.80 \cdot 10^{-4}$	$\mathbf{3.7 \cdot 10^{-5}}$	$\mathbf{5.2 \cdot 10^{-4}}$	$1 \cdot 10^{-3}$	$1 \cdot 10^{-7}$
I_{ON}/I_{MIN}	$7.1 \cdot 10^{4}$	$6.7 \cdot 10^{5}$	$1.5 \cdot 10^{8}$	$2.32 \cdot 10^{5}$	$\mathbf{3.1 \cdot 10^{6}}$	$\mathbf{3.6 \cdot 10^{4}}$	$2 \cdot 10^{4}$	$1 \cdot 10^{7}$
S_W average	120	250	60	90	**133**	**90-100**	55	57

Fig. 26 n and p modes TFET performance comparison between the devices in Fig. 24 ($cSi_{0.85}Ge_{0.15}$ TFET for n mode and $cSi_{0.3}Ge_{0.7}$ TFET for n mode) and published results (V_G excursion between I_{ON} and I_{MIN} is ±1.5 V). TFET benchmark reporting some recent published experimental Data. Very few groups have reported I_{ON} currents above 10 µA/µm

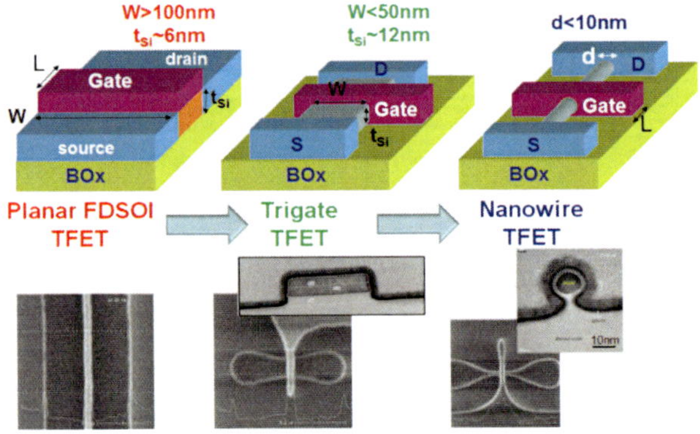

Fig. 27 TFET architecture: moving from wide device configuration (planar FD) to multigate structures (trigate, or even Gate-All-Around, GAA, nanowire) is expected to improve the tunneling injection efficiency, and thus the device performance

CMOS [17]. This kind of junctions is known to exhibit high doping levels and abrupt profiles, which will help optimizing TFET injection.

Other studies need to be investigated, like TFET based design. Very few works report on the opportunities offered by steep slope devices. Even if TFETs will not replace MOSFET in future ICs, one can imagine to combine CMOS and TFET devices for hybrid gates, in order to increase density and/or decrease the power consumption.

Band structure:	SiGe channel (Eg reduction)	SiGe RSD (Heterostructured source/channel)	strain (sSOI, SiGe RSD: Eg reduction)	
Electrostatics:	High-K/ Metal Gate (low EOT)	Planar FDSOI (thin body)	Trigate (multigate)	Nanowire (GAA)
Junction:	Ion implants+ low temp. anneal	in-situ doped S&D epitaxy (doping levels, abruptness)		

Fig. 28 Summary of the tunnelling boosters (band structure, electrostatics, junction) and corresponding key features for improving the TFET performance (in *green*)

7 Conclusion

We have reported the fabrication process and the in-depth characterizations of Tunnel FETs (cointegrated with CMOS). By using advanced FDSOI platform featuring high-K/Metal Gate, ultrathin body architecture, and innovative epitaxy processes, we are able to demonstrate the best p mode TFETs reported so far. For the first time I_{ON} up to 428 μA/μm in p mode with I_{ON}/I_{MIN} ratio large than 10^6 at $V_{DS} = -1$ V are demonstrated. Moreover we have obtained subthreshold slope S_w down to 33 mV/dec and average S_w of 111 mV/dec (over more than 3 decades of drain current).

Moreover, the different tunneling boosters have been exhaustively investigated form a technological point of view: *low band bap materials* ($Si_{1-x}Ge_x$ with x from 0 to 30 % for $Si_{cap}/Si_{1-x}Ge_x$/SOI body) and *strain* (compressively strained SiGe based channels evidenced by GPA of HAADF STEM), *thin film* architecture (FDSOI), *abrupt junctions* (low temperature anneal).

Even if the performance comparison (S_w, I_{ON}) with the cointegrated CMOS shows the gap between TFET and MOSFETs, we demonstrate here large enhancements w.r.t. previous works and detail the paths toward high performance tunneling devices (nanowire structures, sub nm *EOT*, optimized in situ boron doped SiGe…).

Acknowledgments This work was partly supported by the european "STEEPER" project (no. 257267) and the french MINOS network of excellence. The authors thank also B. Prévitali, P. Perreau, F. Allain, F. Andrieu, O. Weber, O. Faynot and T. Poiroux (from CEA Leti) for contributions in device processing, electrical characterization and mana-gerial support

References

1. Sebaugh, A.C., Shang, Q.: Low-voltage tunnel transistors for beyond CMOS logic. Proc. IEEE **98**, 2095 (2010)
2. Leonelli, D., Vandooren, A., Rooyackers, R., Verhulst A.S., De Gendt, S., Heyns M.M., Groeseneken, G.: Performance enhancement in multi gate tunneling field effect transistors by scaling the fin-width. Jpn. J. Appl. Phys. **49**, 04DC10 (2010)

3. Krishnamohan, T., Kim, D., Raghunathan, S., Saraswat, K.: Double-gate strained-Ge heterostructure Tunneling FET (TFET) with record high drive currents and <60 mV/dec subthreshold slope. In: IEEE International Electron Devices Meeting, IEDM 2008, pp. 947–950 (2008)
4. Bhuwalka, K.K., Sedlmaier, S., Ludsteck, A.K., Tolksdorf, C., Schulze, J., Eisele, I.: Vertical tunnel field-effect transistor. IEEE Trans. Electron Devices **51**, 279–282 (2004)
5. Mayer, F., Le Royer, C., Damlencourt J.-F., Romanjek, K., Andrieu, F., Tabone, C., Prévitali, B., Déléonibus S.: (2008). Impact of SOI, $Si_{1-x}Ge_xOI$ and GeOI substrates on CMOS compatible Tunnel FET performance. In: IEEE International Electron Devices Meeting, IEDM 2008, pp. 123–126 (2008)
6. Kim, S.H., Kam, H., Hu, C., Liu, T.-J.K.: Germanium-source Tunnel Field Effect Transistors with record high I_{ON}/I_{OFF}. In: 2008 Symposium on VLSI Technology, pp. 178–179 (2009)
7. Mookerjea, S., Mohata, D., Krishnan, R., Singh, J., Vallett, A., Ali, A., Mayer, T., Narayanan, V., Schlom, D., Liu, A., Datta, S.: Experimental demonstration of 100 nm channel length $In_{0.53}Ga_{0.47}As$-based vertical inter-band Tunnel Field Effect Transistors (TFETs) for ultra low-power logic and SRAM applications. In: IEEE International Electron Devices Meeting, IEDM 2008, pp. 949–952 (2009)
8. Jeon, K., Loh, W.-Y., Patel, P., Kang, C. Y., Oh, J., Bowonder, A., Park, C., Park, C.S., Smith, C., Majhi, P., Tseng, H.-H., Jammy, R., Liu, T.-J.K., Hu, C.: Si tunnel transistors with a novel silicided source and 46 mV/dec swing. In: 2010 Symposium on VLSI Technology, pp. 121–123 (2010)
9. Choi, W.Y., Park, B.-G., Lee, J.D., Liu, T.-J.K.: Tunneling field-effect transistors (TFETs) with subthreshold swing (SS) less than 60 mV/dec. IEEE Electron Device Lett. **28**, 743–745 (2007)
10. Kane, E.O.: Theory of tunneling. J. Appl. Phys. **32**, 83–91 (1961)
11. Tomioka, K., Yoshimura, M., Fukui, T.: Steep-slope tunnel field-effect transistors using III–V nanowire/Si heterojunction. In: 2010 Symposium on VLSI Technology, pp. 47–48 (2012)
12. Villalon, A., Le Royer, C., Cassé, M., Cooper, D., Prévitali, B., Tabone, C., Hartmann, J.-M., Perreau, P., Rivallin, P., Damlencourt, J.-F., Allain, F., Andrieu, F., Weber, O., Faynot, O., Poiroux, T.: Strained Tunnel FETs with record I_{ON}: first demonstration of ETSOI TFETs with SiGe channel and RSD. In: 2010 Symposium on VLSI Technology, pp. 49–50 (2010)
13. Knoll, L., Richter, S., Nichau, A., Trellenkamp, S., Schäfer, A., Bourdelle, K.K., Zhao, Q.T., Mantl, S.: Complementary-Tunnel FET inverters with ultrathin strained Si. EuroSOI conference 2013, pp. 1–2 (2011)
14. Mohata, D.K., Bijesh, R., Mujumdar, S., Eaton, C., Engel-Herbert, R., Mayer, T., Narayanan, V., Fastenau, J.M., Loubychev, D., Liu, A.K., Datta, S.: Demonstration of MOSFET-like on-current performance in arsenide/antimonide Tunnel FETs with staggered hetero-junctions for 300 mV logic applications. In: IEEE International Electron Devices Meeting, IEDM 2011, pp. 781–784 (2011)
15. Ionescu, A.M., Riel, H.: Tunnel field-effect transistors as energy-efficient electronic switches. Nature **479**, 329 (2010)
16. Hutin, L., Le Royer, C., Damlencourt, J.-F., Pouydebasque, A., Xu, C., Tabone, C., Hartmann, J.-M., Carron, V., Grampeix, H., Mazzocchi, V., Truche, R., Weber, O., Batude, P., Garros, X., Clavelier, L., Vinet, M., Faynot, O.: 20 nm gate length trigate pFETs on strained SGOI for high performance CMOS. In: 2010 Symposium on VLSI Technology, pp. 37–38 (2010)
17. Cheng, K., Khakifirooz, A., Kulkarni, P., Kanakasabapathy, S., Schmitz, S., Reznicek, A., Adam, T., Zhu, Y., Li, J., Faltermeier, J., Furukawa, T., Edge, L.F., Haran, B., Seo, S.C., Jamison, P., Holt, J., Li, X., Loesing, R., Zhu, Z., Johnson, R., Upham, A., Levin, T., Smalley, M., Herman, J., Di, M., Wang, J., Sadana, D., Kozlowski, P., Bu, H., Doris, B., O'Neill, J.: Fully depleted extremely thin SOI technology fabricated by a novel integration

scheme featuring implant-free, zero-silicon-loss, and faceted raised source/drain. In: 2010 Symposium on VLSI Technology, pp. 212–213 (2009)
18. Cooper, D., Le Royer, C., Béché, A., Rouvière, J.-L.: Strain mapping for the silicon-on-insulator generation of semiconductor devices by high-angle annular dark field scanning electron transmission microscopy. Appl. Phys. Lett. **100**, 233121 (2012)

Invariance of DC and RF Characteristics of Mechanically Flexible CMOS Technology on Plastic

Aurelien Lecavelier des Etangs-Levallois, Justine Philippe, Sylvie Lepilliet, Yoann Tagro, François Danneville, Jean-François Robillard, Christine Raynaud, Daniel Gloria, Jacek Ratajczak and Emmanuel Dubois

Abstract Combining the electrical performance of modern high frequency silicon nanoelectronics with additional properties of mechanical flexibility and stretchability continuously arouses a sustained interest for its utility in a broad range of space-weight-and-power (SWAP) constrained applications related to e.g. healthcare, structure monitoring, sport, telecommunication, security chips. However, the fabrication of transistors and circuits featuring high electrical performance independently of their deformation state (i.e. flat, folded, or stretched for instance) still constitutes an unresolved challenge. Although many different techniques based on the transfer of high mobility nanostructures or patterned thin-films onto flexible plastic foils constitute possible solutions with their respective advantages and weaknesses, little attention has been paid so far to the thinning of mature rigid technology in the ultra-thin regime followed by transfer-bonding onto a flexible handler. The basic idea developed in this chapter is to combine the advantages of a mature radio-frequency (RF) SOI-CMOS technology with mechanical flexibility provided by thinning. Moreover, performance invariance of flexible systems is another challenge requiring a careful inspection to retain function and to guarantee operation stability. A method based on silicon thinning, transfer-bonding and neutral plane engineering is therefore proposed to produce flexible devices and circuits combining high electrical and mechanical performance, in addition to

A. Lecavelier des Etangs-Levallois · J. Philippe · S. Lepilliet · Y. Tagro · F. Danneville · J.-F. Robillard · E. Dubois (✉)
Institut d'Electronique, de Microélectronique et de Nanotechnologie, IEMN-CNRS, Avenue Poincaré, 59652 Villeneuve d'Ascq, France
e-mail: emmanuel.dubois@isen.iemn.univ-lille1.fr

C. Raynaud · D. Gloria
STMicroelectronics, 850 Rue Jean Monnet, 38926 Crolles Cedex, France

C. Raynaud
CEA-LETI, Minatec, 17 Rue Des Martyrs, 38054 Grenoble Cedex 9, France

J. Ratajczak
Instytut Technologii Elektronowej, 02668 Warsaw, Poland

A. Nazarov et al. (eds.), *Functional Nanomaterials and Devices for Electronics, Sensors and Energy Harvesting*, Engineering Materials,
DOI 10.1007/978-3-319-08804-4_5

functional invariance upon deformation. In this chapter, it is demonstrated that SOI MOSFETs featuring high frequency, low noise and low power characteristics can withstand curvature radii down to the centimetre range without noticeable variation of their static and high frequency performance. The thinning and transfer-bonding of rigid technology is performed using successive chemical–mechanical lapping, wet etching and dry cleaning steps followed by room temperature bonding. Static, high frequency and noise characterization completely validate this process. Consistently with mechanical modelling, electrical measurements in bent configurations confirm the invariance of electrical performance upon flexure. Beyond the in-depth analysis of SOI-MOSFETs, CMOS circuits have been characterized to demonstrate that this technology paves the way to flexible electronic applications requiring complexity and frequency performance.

1 Introduction

Over the past decade, the field of flexible electronics has experienced a rapid development accompanying advances in both the More Moore and More-than-Moore fields (Fig. 1). This growth has been driven by an increasing market demand for electronic applications featuring new form factor, mechanical flexibility, and/or space, weight, and power constraints (SWAP). Initially developed for simple flexible devices organized in a matrix structure, as it is the case for solar cells, lightning, or displays, more complex systems combining advanced communication and signal processing capabilities are now targeted. High frequency operation is therefore now required to enable the design of smart, communicating and high-end flexible products.

Recent progresses have been achieved to combine electrical performance with mechanical bendability (Fig. 1b). For that purpose, several approaches have been recently studied to produce flexible electronic devices and circuits. Organic and printed electronics enables the fabrication of highly bendable systems [13, 22–24]. However, modest carrier mobilities and micrometer scale patterning resolution limit their electronic performance and thus restrain their use to low frequency and low complexity systems [25, 26]. The hybridization of organic flexible substrate with high-mobility thin film has also been considered as a key enabling technology for high performance bendable electronics. Along this philosophy, a first strategy has been developed to transfer bare films or nanoribbons of silicon [3, 4, 11, 12, 27], III–V compound semiconductors [5, 28–30], or carbon-based materials [1, 2, 9, 10, 15] onto an organic foil. This initial step is followed by the structuring of active devices using gentle process steps to cope with chemical and thermal constraints associated to the organic substrate. This approach is therefore severely restricted by the highest admissible thermal budget, typically <250 °C, and has led to

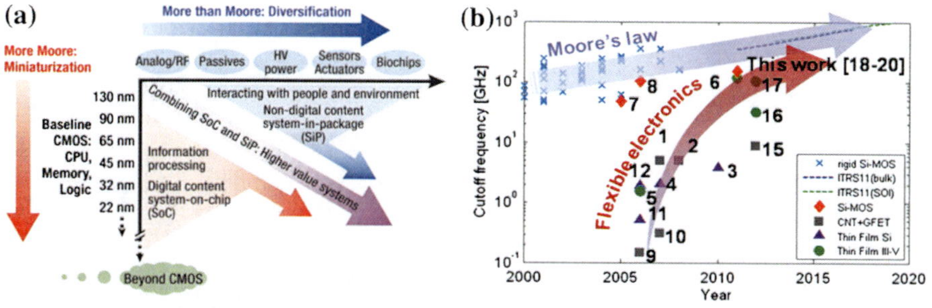

Fig. 1 **a** Dual development of the electronic field: towards miniaturization (More-Moore) and diversification (More-than-Moore). **b** Chart showing recent progress in the field of high performance flexible electronics [1–21]

demonstrations featuring only a single interconnection layer. This method remains therefore of limited utility for application classes requiring complex circuitry.

Dissociating the fabrication of transistors and circuits from the transfer-bonding step onto a plastic foil introduces considerable flexibility without suffering from the above-mentioned limitations. Starting from standard SOI-CMOS technologies, the technological approach described in this chapter relies on the extreme thinning of a processed CMOS substrate followed by the subsequent bonding of circuit dies on plastic films. This method is hereafter referred to as ultimate thinning and transfer bonding of CMOS technology (UTTB-CMOS) [18–21].

2 Flexible UTTB-CMOS Fabrication

The UTTB-CMOS fabrication sequence starts with standard CMOS circuit dies processed on standard rigid SOI wafers, thus taking advantage of the maturity of available industrial technologies. SOI technology furthermore benefits from the buried oxide layer (BOX) that provides enhanced resistance to parasitic effects, improved performance, reduced power consumption and greater integration density with respect to its bulk counterpart [31–35].

In this work, a 65 nm partially depleted SOI-CMOS technology was used, from which RF transistors are recognized for delivering a performance level suitable for high frequency, low noise and low power applications (Fig. 2a and b) [34, 35].

After initial processing, the resulting stack consists in a 800 μm thick silicon handler, a 145 nm thick buried oxide (BOX) layer, a 60 nm thick top silicon (SOI) layer, and a 5.6 μm thick back-of-the-line network featuring 6 interconnect levels. The proposed thinning-and-transfer-bonding method leads to an UTTB-CMOS stack consisting only of the BOX, SOI and interconnect layers bonded on top of a 50 μm thick polyimide flexible layer (Fig. 2c).

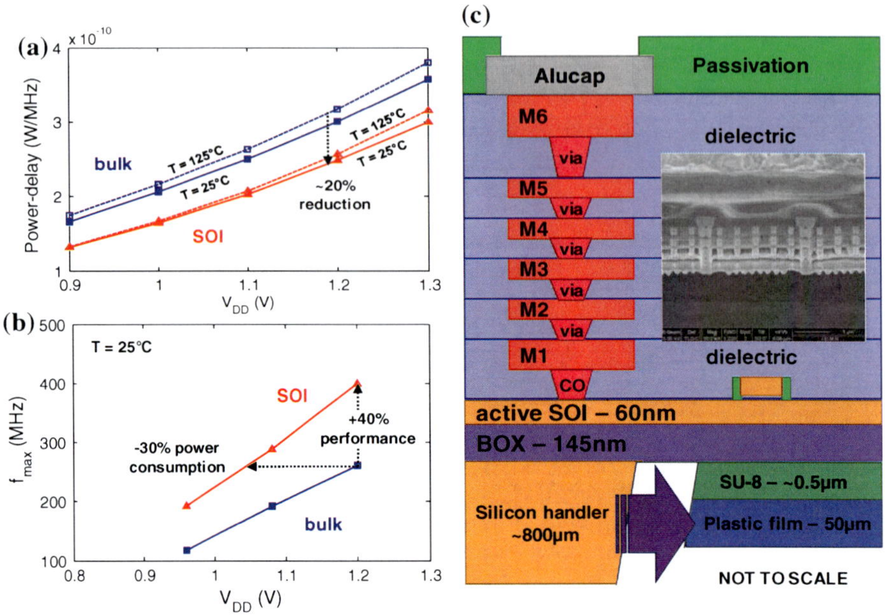

Fig. 2 Comparison between **a** power-delay products of ring oscillators, and **b** maximum access frequency of SRAM on bulk and SOI 65 nm technologies [35] **c** CMOS stack before and after thinning and transfer-bonding on a flexible handler

2.1 Thinning the Back-Side of CMOS Chips

The aim of thinning the silicon wafer is here to improve its mechanical flexibility. In a first approximation, it can be reasonably considered that thickness is the main parameter defining the minimal curvature radius that a silicon chip can withstand. As detailed in the following, the backside surface morphology should also be taken into account to obtain the maximum resistance to breakage. This is the reason why the proposed methodology relies on three steps: from rough lapping to smooth selective cleaning. Assuming the validity of the 3-point bending equations (inset of Fig. 3) in the considered range of deformation, Fig. 3 explains that for a given mechanical load F, a thinner material undergoes a larger deflection d and thus a higher level of stress σ (Fig. 3a). However, for a given deflection, a thinner wafer will be subjected to a lower level of stress (Fig. 3b). In terms of mechanical bendability, the minimal bending radius that a thin wafer can withstand is therefore smaller than for a thicker system.

The thinning methodology developed in this work is a three-step process, involving (i) chemical–mechanical lapping leaving typically 100 µm of the initial silicon handler, (ii) isotropic wet etching down to 20 µm in HNA, a mixture of

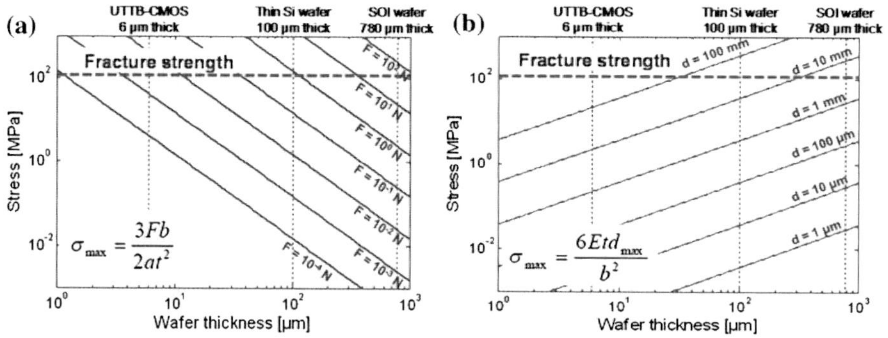

Fig. 3 Maximum stress as a function of wafer thickness t **a** for given applied forces F, and **b** for given applied deflections d. Three inches silicon wafers are considered ($a = b = 76.2$ mm) with a Young modulus $E = 150$ GPa and a fracture strength $\sigma_{lim} = 120$ MPa [36]

hydrofluoric (HF) and nitric (HNO_3) acids and (iii) dry etching of the remaining silicon by xenon difluoride (XeF_2) with high selectivity with respect to SiO_2 for smoothly landing on the BOX layer, as reported in [18, 21].

2.1.1 Chemical–Mechanical Lapping

The first step of the thinning process is performed by chemical–mechanical lapping. In this way, the silicon handler thickness is reduced from 800 to 100 µm. A rough lapping using 15 µm diameter alumina abrasive powder is first used down to 200 µm. 3 µm diameter powder is subsequently used to further thin the sample down to 100 µm, while reducing lapping scratches. A polishing step is finally performed to conclude the chemical–mechanical lapping. This method is similar to the process described in [21]. Figure 4a illustrates the rate of silicon removal during a typical chemical–mechanical lapping step.

Although chemical–mechanical lapping is an efficient and reproducible technique to quickly remove a large amount of material, a known side effect is the generation of sub-surface damages at the back-side of the thinned silicon handler. It is well established that the fracture strength of the thinned chip depends on the geometry and sharpness of the few larger scratches [37]. The best way to enhance fracture strength is therefore to apply a subsequent isotropic etch step to turn sharp V-shaped scratches into U-shaped grooves, thus reducing stress concentration [37].

Figure 4b shows the load required to break thinned silicon wafers of different final thicknesses with or without an additional stress relief step [38]. This graph first highlights the effect of the stress relief step by comparing sample that have been submitted to a sole grinding step with samples that have further been etched using a $HF:HNO_3:CH_3COOH$ solution, or a SF_6 plasma. The enhanced breaking load after a longer stress relief step consisting in etching an additional 25 µm layer from the already thinned surface is also pointed out (green symbols in Fig. 4b).

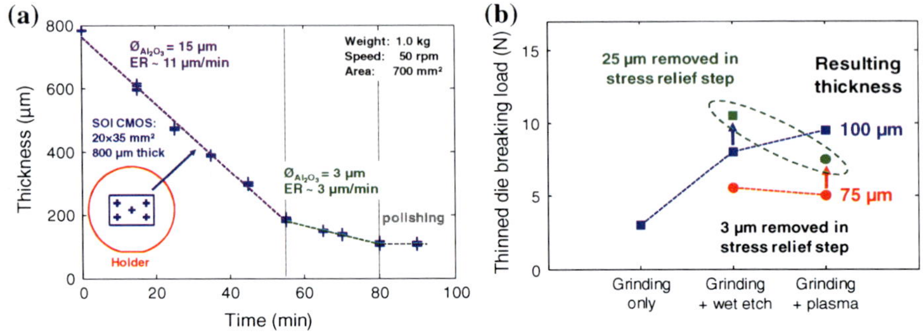

Fig. 4 **a** Typical lapping characteristic showing the thickness of a SOI CMOS chip during chemical–mechanical lapping down to 100 µm, and **b** Breaking loads measured on silicon bulk wafers thinned by grinding, with or without a subsequent stress relief step, performed by wet etching ($HF:HNO_3:CH_3COOH$, 6 µm/min), or dry plasma etching (SF_6, 3 µm/min) [38]

2.1.2 HNA Wet Etching

A mixture of $HF:HNO_3$, referred to as HNA, is extremely efficient for chemically etching silicon isotropically as shown in Fig. 5a. It is used after lapping to further thin the silicon chip from 100 µm down to 20 µm and to increase its fracture strength. Before wet etching, the edge of the CMOS chip have to be protected in order to prevent the HNA solution to etch the SOI, BOX or interconnect layers. For that purpose, an acid-resistant wax is deposited around the edges of the CMOS chips before wet etching, as shown in the thickness mapping in Fig. 5b. A drawback of this solution is the non-homogeneity of the resulting thinned chip that features a higher thickness in the centre than at the edges (Fig. 6b).

2.1.3 XeF_2 Cleaning

The last step of the thinning process consists in a highly selective etching of the remaining silicon using XeF_2 without damaging the buried oxide layer. Due to the high $Si:SiO_2$ selectivity, it is important to start this step by removing the native oxide present at the back silicon surface. This is performed by dipping the sample into diluted HF (1 v/v%) for 1 min prior to loading in the XeF_2 etching system. It can be seen in Fig. 6 that this last step compensates the thickness non-homogeneity generated by previous thinning steps. Here, thickness dispersion after XeF_2 etching is due to the interferometric measurement system unable to properly account for optical reflection on a complex stack. The complete elimination of residual silicon under the buried oxide is confirmed in Fig. 7 that shows a circuit die after bonding on a polyimide film.

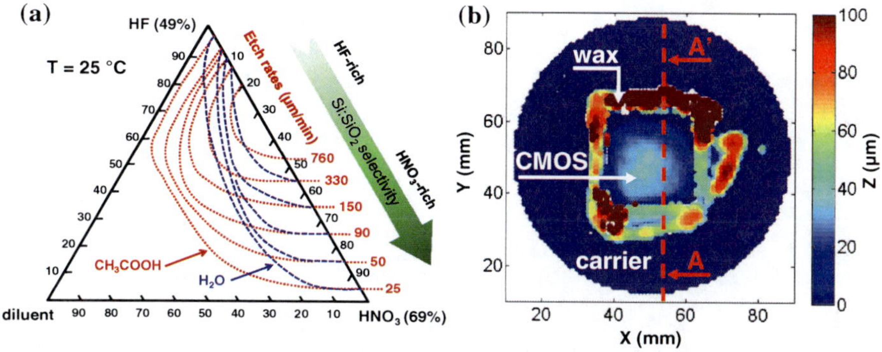

Fig. 5 **a** Silicon etch rate as a function of HF:HNO_3:diluent composition [39], and **b** *Top view* of a CMOS chip before wet etching: the edges of the sample have to be protected using a cover layer of acid-resistant wax

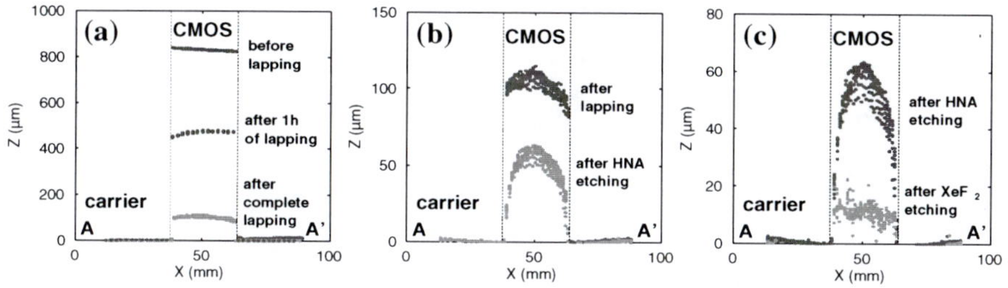

Fig. 6 Typical thickness profiles of a SOI chip bonded on a temporary carrier before and after different steps of the thinning process: **a** chemical–mechanical lapping, **b** HNA wet etching, and **c** XeF_2 dry cleaning

2.2 Bonding of UTTB-CMOS on a Flexible Foil

After thinning the back side of a CMOS chip, a low stress bonding method based on surface tension forces is used to transfer the die onto a flexible 50 μm polyimide foil. Here, the bonding method relies on the gain in surface energy resulting from wrapping of the thinned chip onto the polyimide foil that overcomes the elastic bending energy associated to the mechanical deformation of the chip [40]. This technique is simply implemented by putting the ultra-thin CMOS circuit die (Fig. 8a and b) in contact with a flexible substrate on which uncured PMMA was previously spin-coated. In contrast to previously published work [18, 19, 21], this procedure can be performed without the application neither of an external bonding pressure nor of a thermal budget, thus minimizing stress. Another distinctive advantage of this method is that the thinned CMOS chips can be bonded onto curved flexible foils, or on non-planar rigid surfaces (bending radius of 4 mm as shown in Fig. 8c).

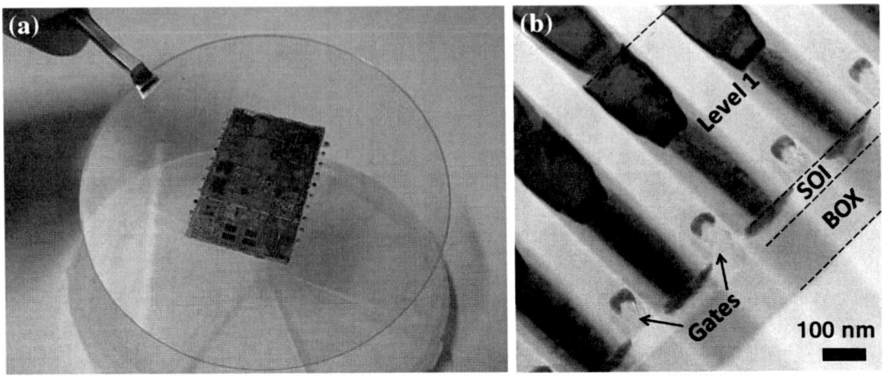

Fig. 7 **a** SOI chip bonded on the front side onto a temporary sapphire carrier to preserve the integrity of the topmost active layers during the thinning procedure. Details of the circuit layout can be observed by transparency through the BOX layer. **b** TEM image of a thinned chip transfer-bonded on a polyimide film showing that the geometry of three transistors can be clearly delineated

Fig. 8 **a** *Back side* of a thinned active SOI-CMOS sample (5.6 μm thick), the first level of metallization can be seen through the buried oxide layer, **b** same sample seen from the *front side*, bending on one corner of the sample can be noticed, and **c** thinned CMOS chip bonded onto a 3 mm curvature radius cylinder

2.3 Physical Characterization of UTTB-CMOS

Using both optical and electron microscopy characterization, no physical degradation can be noticed on UTTB-CMOS chips after thinning and transfer-bonding onto a flexible handler, as shown in Figs. 7 and 9. Complementarily, electrical characterization of UTTB-MOSFETs and CMOS circuits are discussed in the next section.

3 Electrical Performance of Flexible UTTB-CMOS

After thinning and transfer-bonding onto a 50 μm thick polyimide foil, flexible UTTB-MOSFETs have been characterized. Their static, high frequency and noise properties are discussed in this section. Electrical characteristics presented here

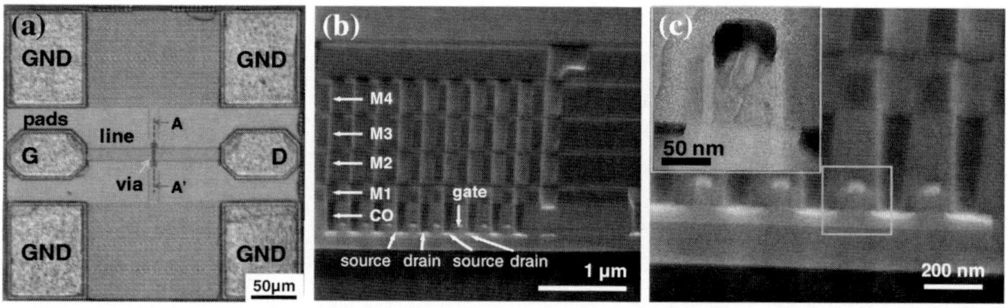

Fig. 9 **a** *Top view* of a UTTB-CMOS using optical microscopy, **b** *Cross-section view* along the AA′ line on the same sample using scanning electron microscopy, and **c** Zoom on a series of MOS transistors gate fingers obtained by transmission electron microscopy

focus on 60 nm gate long MOSFETs featuring a total gate width development of 64 μm coming from the parallel association of 0.5, 1 or 2 μm wide unitary gates.

3.1 Static Operation of Flexible UTTB-MOSFETs

Interestingly, flexible UTTB-MOSFETs feature static performance identical to their thick and rigid counterpart. Drain current and transconductance characteristics of 60 nm gate length transistors are presented in Fig. 10, in red for UTTB-CMOS and blue for rigid MOSFETs [18, 19, 21]. Although not represented here, the same excellent match between the flexible and rigid characteristics was obtained for transistors featuring varying gate lengths in the 60 nm–2 μm range.

3.2 High Frequency Operation of Flexible UTTB-MOSFETs

In addition to static characterization, high frequency measurements have been performed on UTTB-MOSFETs to evaluate the potential impact of the thinning and bonding processes on dynamic operation. S parameters measured before (blue) and after (red) thinning and transfer-bonding onto a polyimide film are compared in Fig. 11, in addition to the current gain H_{21} and characteristic frequencies f_T (unity gain cut-off frequency) and f_{max} (maximum oscillation frequency). This figure demonstrates that UTTB-MOSFETs feature high frequency operation competitive with their rigid counterparts. Only a slight decrease of the maximum oscillation frequency can be observed and explained by a concomitant increase of the gate resistance as outlined in [18].

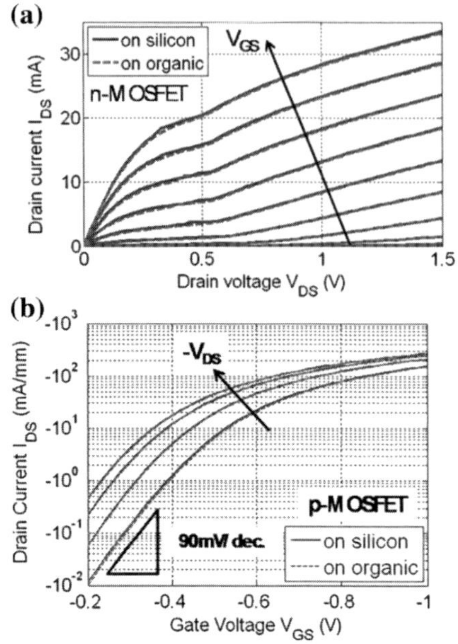

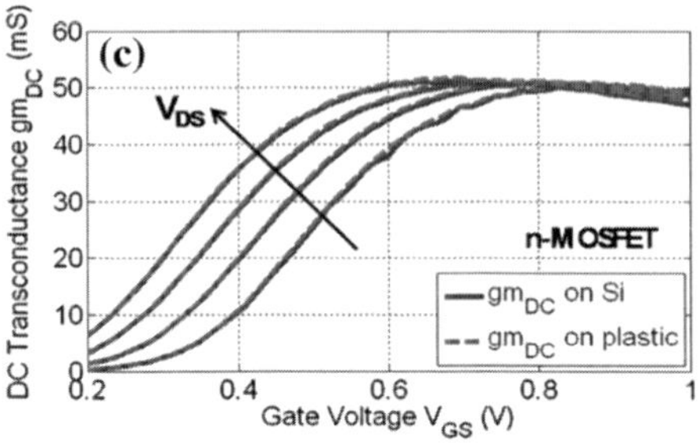

	I_{ON}	I_{OFF}	g_m^{DC}	V_{th}	DIBL	SS
	μA/μm	nA/μm	μS/μm	mV	mV	mV/dec
nR	525	13	799	490	130	90
nF	521	13	793	490	155	90
pR	-272	-2.7	499	570	90	90
pF	-270	-2.7	495	570	90	90

Fig. 10 *I–V* characteristics of **a** n-type, and **b** p-type 60 nm gate length UTTB-MOSFETs and **c** static transconductance g_m–V_{GS} of n-type flexible UTTB-MOSFETs. In addition to a table summarizing the static performance of rigid and flexible n-MOSFETs (resp. referred to as nR and nF), and rigid and flexible p-MOSFETs (resp. referred to as pR and pF)

3.3 Low Noise Properties of Flexible UTTB-MOSFETs

Over the last few years, the partially depleted 65 nm SOI-CMOS technology has been convincingly used for the demonstration of high frequency circuitry including low noise amplifiers in the 60 and 77 GHz bands [34, 35]. The noise properties of flexible UTTB-MOSFETs have therefore been investigated and analyzed for comparison with respect to those of the starting technology fabricated on a thick and rigid wafer. Figure 12 shows that no degradation of noise performance arises from the thinning and transfer-bonding processes [19].

3.4 Flexible Low Noise Amplifier

Beyond the in-depth analysis of RF transistors, a single stage low noise amplifier (LNA) has also been characterized to demonstrate that the transfer onto a plastic handler overall preserves function and performance. A specificity of the starting SOI-CMOS considered in this work is the use of a high resistivity handler to mitigate substrate losses. This characteristic is especially important to design high

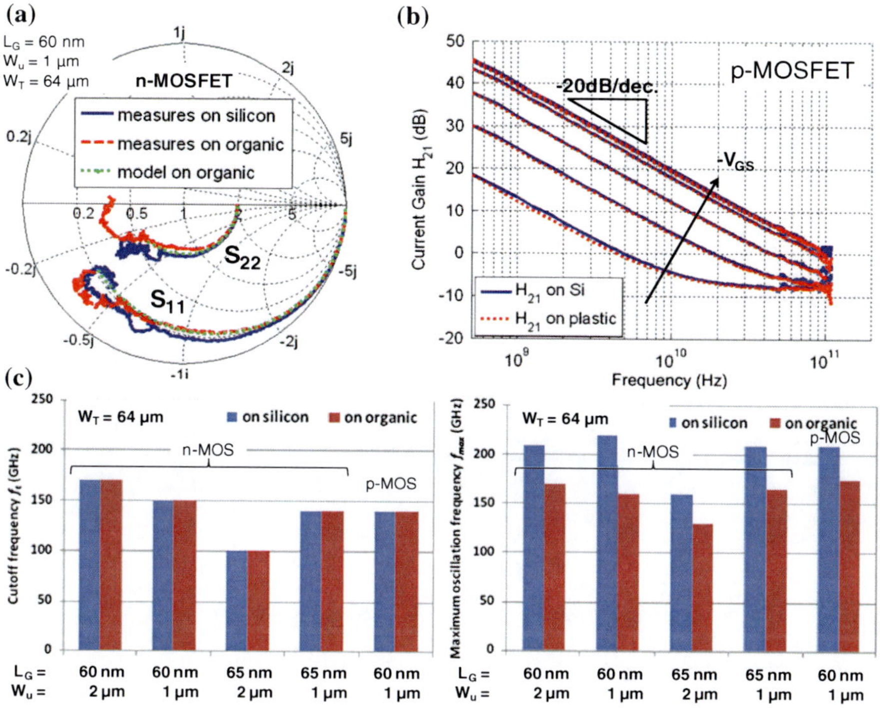

Fig. 11 **a** *S* parameters of rigid (*blue*) and flexible (*red*) n-MOSFETs, measured (*lines*), and retro-simulated (*green dots*); **b** current gain H_{21} of n-type rigid and UTTB-MOSFETs, and **c** characteristic frequencies for different transistor geometries where the unitary gate width is varied

quality passive components such as inductors and transmission lines comprised in high-frequency circuits such as LNAs. Figure 13 demonstrates that a one-stage UTTB-LNA still operates properly in the 80 GHz range even if a slight decrease of the amplification gain can be noticed. The important point to highlight here is that no specific redesign has been performed to obtain flexible circuits.

4 Mechanical Behaviour of Flexible UTTB-CMOS

4.1 UTTB-CMOS Under Bending

It has been demonstrated in previous sections that the thinning and transfer process does not degrade transistors properties in terms of static, high frequency and low noise operation as long as the flexible carrier is kept in straight flat configuration. The next step naturally consists in analyzing the impact of mechanical deformation on the electrical performance of complex flexible electronic systems and, as far as

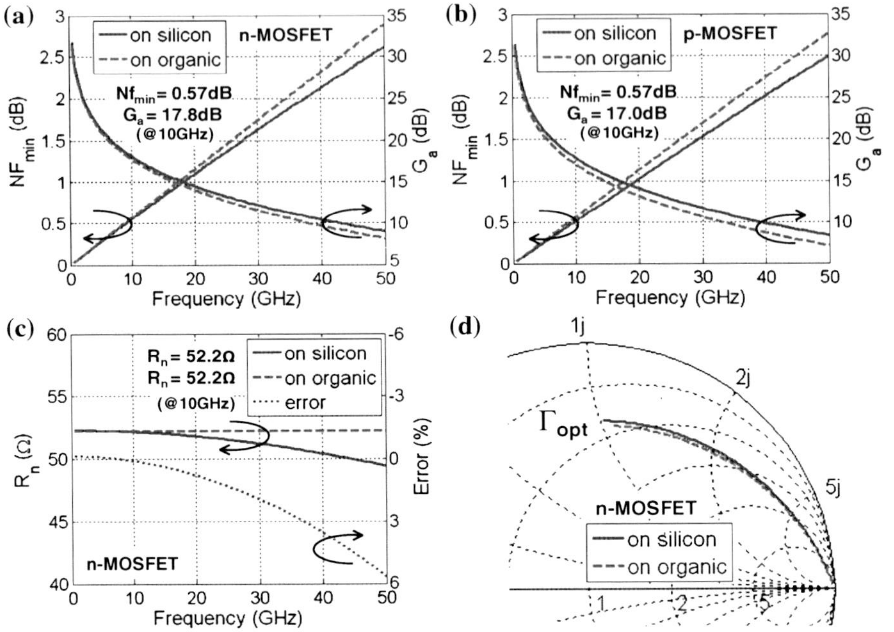

Fig. 12 Minimal noise figure NF_{min} and associated gain G_a for rigid and flexible **a** n-type, and **b** p-type MOSFETs; **c** equivalent noise resistance R_n, and **d** optimal noise reflection coefficient Γ_{opt} for a n-MOSFET on its initial rigid HR SOI wafer (*blue lines*) and after transfer onto a flexible organic film (*red dashed lines*)

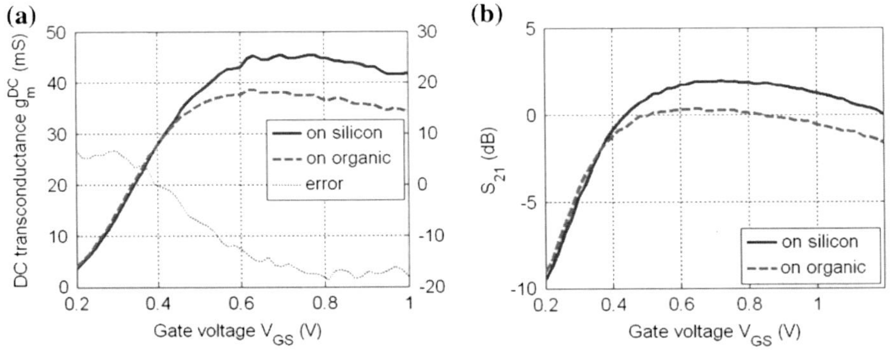

Fig. 13 **a** Static transconductance measured on a rigid (*blue lines*) and flexible (*red dashed lines*) LNA biased at a drain voltage $V_{DS} = 1.0$ V, and **b** S_{21} gain parameters measured at 80 GHz for the same LNA on both substrate flavours, i.e. rigid and flexible

possible, to minimize it. Invariance of DC and RF properties of flexible devices is highly desirable for flexible applications like wearable or foldable electronics, where strain type, direction and intensity are by essence random. However, strain

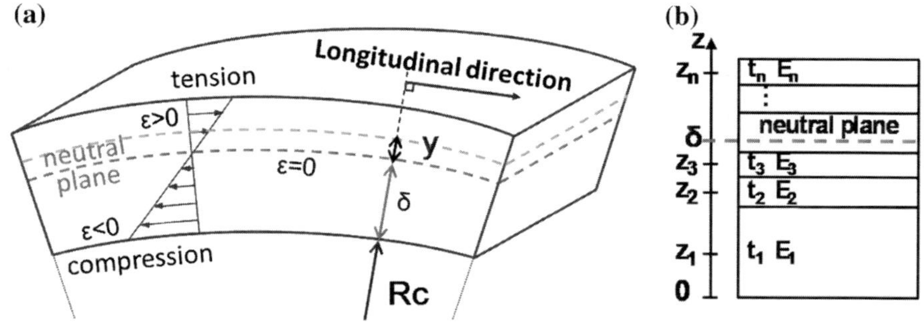

Fig. 14 **a** Schematic representation of a multilayer material subjected to cylindrical bending over a curvature radius R_C. The neutral plane location is highlighted along with the longitudinal strain distribution, and **b** Schematic of a multilayer stack with n layers featuring different thickness t_i and Young's modulus E_i. The multilayer neutral plane location δ is calculated using Eqs. 1 and 2

application to a semiconductor crystal holds the important property to modify its band structure leading to band splitting and warping that significantly modify carriers mobility [41–48]. The so-called piezoresistive effect, that describes the resistivity change with applied stress, was first studied by Smith in its seminal work on silicon and germanium [41]. As a matter of fact, this property has been widely used over the last decade to enhance carrier mobility in the 90 nm CMOS technology node and beyond [49, 50]. In this context, the legitimate question that naturally deserves a careful inspection is how mechanical deformation mechanical deformation affects UTTB-CMOS performance? The corollary question is how to cancel strain effects on electrical parameters by adopting a judicious mechanical design? Each of these pending questions is addressed in the following sections.

4.2 Neutral Plane Engineering Methodology

The idea developed to cope with the above-mentioned problem is to minimize the impact of external strain on electrical properties of flexible UTTB-CMOS devices. This can be achieved using neutral plane engineering [20]. The neutral plane is defined as the surface perpendicular to the cross-sectional plane where there is neither compression nor elongation. The strain level in layers parallel to the neutral plane is directly proportional to the distance to this plane. Positioning the multi-layer neutral plane close to the active layer therefore results in a minimization of the strain level. Following this philosophy, the variations of electron mobility due to the piezoresistive effect are, in turn, also minimized [45]. The neutral plane position depends on the thickness t_i and Young's modulus E_i of the different layers, as illustrated in Fig. 14. In a simplified one-dimensional approach, it can be written as given in the following equations [43, 51–53]:

$$\delta = \frac{\sum z_i \cdot E_i \cdot t_i}{\sum E_i \cdot t_i} \tag{1}$$

where

$$z_i = \left(\sum_{j=1}^{i} t_j \right) - \frac{t_i}{2} \tag{2}$$

Assuming that planes orthogonal to the original reference surface remain flat and orthogonal to the deformed reference surface [52], the deformed shape after bending over a cylinder can be computed. When considering a plate thickness in the micrometer range bent over a cylinder with a radius in the centimeter range, large rotations have to be taken into account. This invalidates assumptions inherent to the infinitesimal strain theory. The calculation of strain and stress distributions in the UTTB-CMOS multilayer has therefore been performed using the finite strain theory [54, 55]. Following this approach, the Green-Lagrange strain tensor $\boldsymbol{E}$ is formulated in Eq. 3, as a function of the material deformation gradient tensor $\boldsymbol{F}$ and the identity matrix $\boldsymbol{I}$. The tensor $\boldsymbol{F}$ associated to the cylindrical bending of a thin plate describes the fact that horizontal material planes are transformed into concentric tubes: $r = h(X)$, and vertical planes into radial ones: $\theta = \nu(Y)$ [54].

$$\boldsymbol{E} = \frac{1}{2}(\boldsymbol{F}^T \cdot \boldsymbol{F} - \boldsymbol{I}) \tag{3}$$

where

$$\boldsymbol{F} = \begin{bmatrix} h' & 0 & 0 \\ 0 & h\nu' & 0 \\ 0 & 0 & 1 \end{bmatrix} \tag{4}$$

The Cauchy strain tensor σ can then be computed from the deformation gradient tensor $\boldsymbol{F}$, the Green-Lagrange strain tensor $\boldsymbol{E}$ and the second Piola-Kirchhoff stress tensor $\boldsymbol{S}$ (Eqs. 5 and 6), where ν refers to the Poisson coefficient and E the to the Young modulus).

$$\boldsymbol{\sigma} = [\det(\boldsymbol{F})]^{-1} \boldsymbol{F} \cdot \boldsymbol{S} \cdot \boldsymbol{F}^T \tag{5}$$

$$\boldsymbol{S} = \frac{\nu \mathrm{E}}{(1+\nu)(1-2\nu)} \mathrm{tr}(\boldsymbol{E})\, \boldsymbol{I} + 2\frac{\mathrm{E}}{2(1+\nu)} \boldsymbol{E} \tag{6}$$

For the 65 nm SOI-CMOS technology considered in this work, the active layer is the 60 nm thick SOI film. It is only separated from the flexible plastic handler by the BOX layer (145 nm) and the adhesive layer ($\sim$500 nm). In a realistic description of the UTTB-CMOS stack, 17 uniform layers of various thicknesses

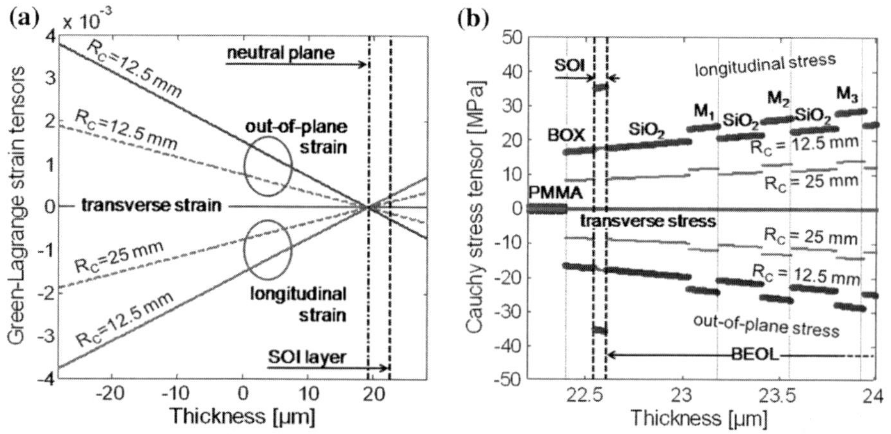

Fig. 15 **a** Green-Lagrange strain tensor, and **b** Cauchy stress tensor components simulated at various positions along the thickness of the multilayer stack for a flexible CMOS sample bent over a cylinders with 25 and 12.5 mm curvature radii. The neutral plane and active SOI layer are highlighted. **b** focuses on the front end and first layers of BEOL interconnect

and Young's moduli are taken into account for an accurate treatment of the mechanical problem. Figure 15a and b respectively show the three diagonal components of the Green-Lagrange strain and Cauchy stress tensors across the thickness of the multilayer stack after bending over a cylinder featuring a 25 and 12.5 mm curvature radius. After computing the strain and stress distributions in the UTTB-CMOS stack, the piezoresistivity theory leads to an estimation of the mobility variation in the SOI layer where carrier transport takes place. To facilitate the quantification of stress effect on current, it has been here assumed that the drain current I_{DS} is proportional to the electron mobility both in low and high field regimes [45]. It is therefore expected that the DC transconductance $g_{m,}$ the unity gain cut-off frequency f_T, and the maximum oscillation frequency f_{MAX} will follow the same variation trend.

To account for the specificity of the transport mechanism in MOSFETs, stress-dependent mobility has been calculated using piezoresistivity coefficients associated to an electron inversion layer as summarized in Table 1 [47, 56]. As the wafer orientation is (001) and transistor channels are aligned with the <110> crystal direction, piezoresistive coefficients referred to as longitudinal π_L, transverse π_T, and out-of-plane $\Pi_\perp$ can be expressed in the rotated coordinate system as given in Table 2. The variation of carrier mobility $\delta\boldsymbol{\mu}$ can subsequently be calculated using the relevant components of the stress tensor σ in the SOI layer as formulated in Eq. 7 [41, 42, 47, 56].

$$\frac{\delta\boldsymbol{\mu}_i}{\boldsymbol{\mu}_0} = \sum_{\mathrm{j}} -\boldsymbol{\pi}_{ij}\boldsymbol{\sigma}_j \tag{7}$$

Table 1 Piezoresistivity coefficients π_{ij} associated to an inversion layer in silicon

Coefficient ($\times 10^{-11}$ Pa^{-1})	n-type transport	p-type transport
π_{11}	−84	−12.5
π_{12}	34	28
π_{44}	−17	105

Table 2 Longitudinal, transverse and out-of-plane piezoresistivity coefficients

Crystal direction	Piezoresistive component	n-type transport ($\times 10^{-11}$ Pa^{-1})	p-type transport ($\times 10^{-11}$ Pa^{-1})
[110]	$\pi_L = \frac{1}{2}(\pi_{11} + \pi_{12} + \pi_{44})$	−33.5	60.25
[−110]	$\pi_T = \frac{1}{2}(\pi_{11} + \pi_{12} - \pi_{44})$	−16.5	−44.75
[001]	$\pi_\perp = \pi_{12}$	34	28

At this point, it is important to recall that electrical performance can respond differently to a stress of given type and direction, depending on the MOSFET channel type. For instance, uniaxial tensile stress applied along the channel length oriented in a $\langle 110 \rangle$ direction tends to improve electron mobility because the corresponding piezoresistive coefficient π_{Ln} is negative. Conversely, the same tensile stress will be detrimental to hole mobility because the p-type longitudinal piezoresistive π_{Lp} is positive. Both the strain level in the SOI layer and the associated variation of carrier mobility must therefore be minimized to preserve the invariance of electrical functions.

The above developments naturally suggest that the neutral plane should be placed close to the active SOI layer to cancel out mobility variations. Equation 1 shows that this can be performed by tuning the thickness of the flexible handler according to the thickness and mechanical properties of the thinned UTTB-CMOS stack. Figure 16a illustrates this point by representing the stress level generated in the SOI layer after bending a UTTB-CMOS circuit die onto various curvature radii, for different thicknesses of flexible handler. An optimal thickness close to 50 μm can be deduced from this graph. Figure 16b furthermore shows the variation of electron mobility for handler thickness values close to the optimal figure of 50 μm leading to a minimal variation of electrical properties.

4.3 Electrical Properties of UTTB-MOSFETs Under Flexure

Electrical measurements have been performed on bent UTTB-MOSFETs in order to validate the modelling approach discussed in previous section. For that sake, external mechanical strain was applied to the flexible UTTB-CMOS circuit die by

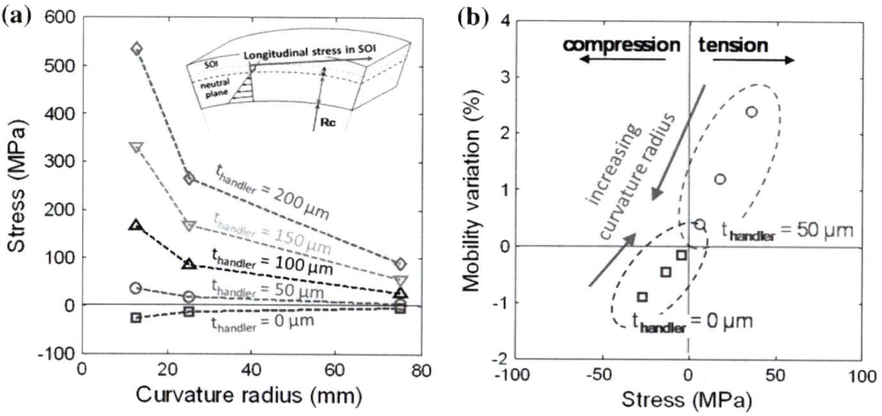

Fig. 16 **a** Longitudinal component of the stress tensor in the active SOI layer as a function of the curvature radius for plastic handlers featuring different thicknesses. **b** Variation of electron mobility in the SOI layer as a function of stress generated under flexure for three curvature radii (75, 25, and 12.5 mm), around the optimal zero-variation point

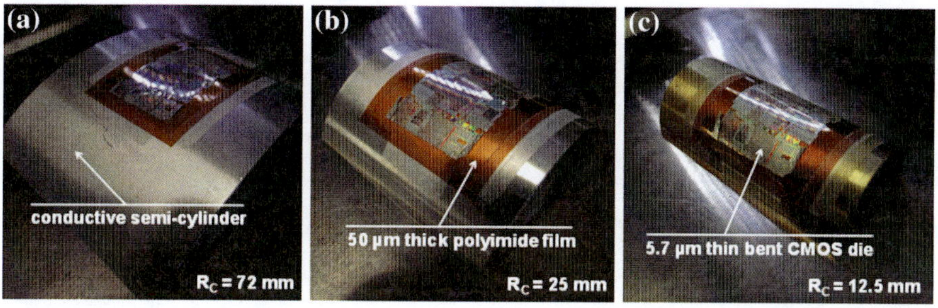

Fig. 17 Conductive semi-cylinders of known radius used to maintain the thinned and flexible UTTB-CMOS devices presented in this work in bent (i.e. strained) configuration while performing DC and RF characterization. The radius of curvature is: **a** 72 mm, **b** 25 mm and **c** 12.5 mm

bending on a semi-cylinder of known radius (Fig. 17). DC and HF characterizations of flexible n-MOSFETs were successively performed flatwise and under flexure.

UTTB-CMOS systems presented in this section are bonded onto a 50 µm thick polyimide flexible handler in order to limit the strain developed even upon aggressive bending. From the DC operation standpoint, illustration of electrical invariance is provided in Fig. 18 that shows typical current–voltage I_{DS}–V_{DS} and transconductance g_{mDC}–V_{GS} characteristics measured on a UTTB transistor both in flat and bent configurations. It can be noticed that variations of static properties are limited to less than 5 % owing to the optimal mechanical configuration of the multilayer stack even upon bending over a 12.5 mm cylinder.

Figure 19 furthermore demonstrates identical high frequency operation for UTTB n-MOSFETs characterized flatwise and upon flexure. In addition, it is

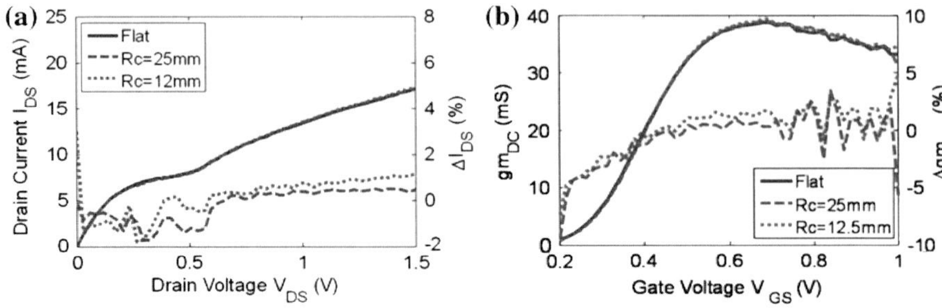

Fig. 18 **a** I_{DS}–V_{DS} characteristics, and **b** static transconductance g_m, measured flatwise (*blue lines*), under flexure on semi-cylinders with a 25 mm curvature radius (*green dashed lines*) and a 12.5 mm curvature radius (*red dotted lines*) biased at $V_{DS} = 1.2$ V and $V_{GS} = 0.8$ V corresponding to static bias conditions used to perform and extract HF figures-of-merit. Variations after bending are also plotted. The considered UTTB n-MOSFET features 32 parallel gate fingers of length $L_G = 65$ nm and unitary width $W_u = 2$ μm leading to a total gate width development $W_T = 64$ μm

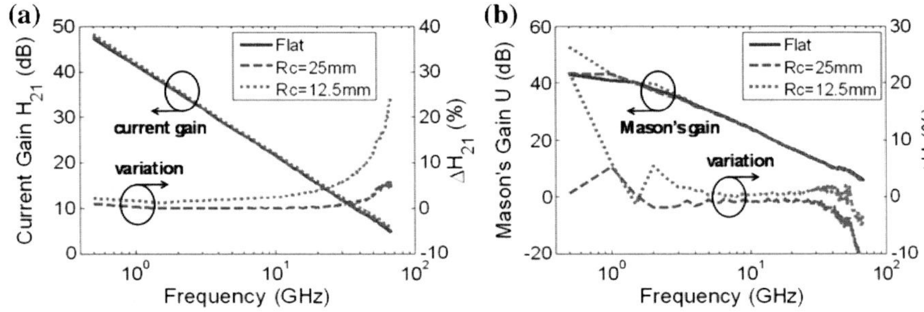

Fig. 19 **a** Unity gain H_{21}, and **b** Mason's gain U, measured flatwise, and under flexure (*same line colors as previously*); both figures demonstrate good agreement between measurements and the theoretical −20 dB/dec. slope. The considered UTTB n-MOSFET transistor features 32 parallel gate fingers of length $L_G = 65$ nm and unitary width $W_u = 2$ μm leading to a total gate width development $W_T = 64$ μm

worth noting that the theoretical slope of −20 dB/dec. is well respected in both Fig. 19a and b. This observation allows to legitimately extrapolate the characteristic frequencies f_T and f_{MAX} from gain measurements at 20 dB. Values of $f_T \sim 120$ GHz and $f_{MAX} \sim 145$ GHz are equally obtained in flat condition and after bending on cylinders with a 25 mm and a 12.5 mm curvature radii. The relative variations of H_{21} and U under bending with respect to the flat state is also given in Fig. 19, at $V_{DS} = 1.2$ V and $V_{GS} = 0.8$ V that correspond to the maximum of transconductance where best frequency performance is expected. Relative variations of H_{21} and U remain below 5 % at frequencies where f_T and f_{MAX} are extracted. This figure proves to be in excellent agreement with the theoretical estimate previously calculated.

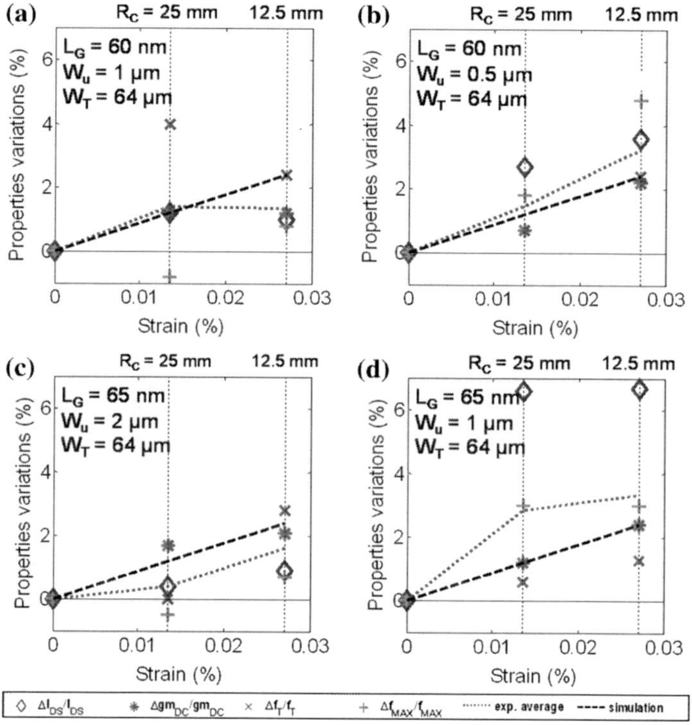

Fig. 20 Comparison of measured and simulated variations of selected electrical parameters. Relative variations of drain current I_{DS}, DC transconductance gm_{DC}, unity gain cut-off frequency f_T, and maximum oscillation frequency f_{MAX} versus strain generated in the active SOI layer of four different transistors, **a** $L_G = 60$ nm, $W_u = 1$ µm, **b** $L_G = 60$ nm, $W_u = 0.5$ µm, **c** $L_G = 65$ nm, $W_u = 2$ µm and **d** $L_G = 65$ nm, $W_u = 1$ µm ($W_T = 64$ µm for all four transistors), under flexure on 25 and 12.5 mm radius cylinders, in addition to average of experimental data points (*magenta dotted lines*) and theoretical predictions (*black dashed lines*)

The same type of measurements has been repeated on a set of four UTTB n-MOSFETs to obtain statistically representative results. These transistors feature 60 and 65 nm gate length with a total gate width development of 64 µm coming from the parallel association of 0.5, 1 or 2 µm wide unitary gates.

Figure 20 gives an extended view of relative variations of static (ΔI_{DS} and Δgm_{DC}), and high frequency (Δf_T and Δf_{MAX}) figures-of-merit. Data are presented individually for each of the four transistors where relative variations of I_{DS}, g_m, f_{MAX} and f_T, referred to *as properties variations* on the graphs, are given as a function of the strain level. It can be concluded that variations of each electrical parameter are limited to less than 5 % as expected from the strategy of neutral plane engineering. This graph globally demonstrates that each of the figures-of-merit measured in bent configuration is in excellent agreement with theoretical predictions, thereby validating the neutral plane engineering strategy.

5 Conclusions and Perspectives

The thinning and transfer-bonding methodology developed in this work leads to the realization of UTTB-CMOS devices and circuits that feature mechanical flexibility in addition to electrical performance identical to their rigid counterparts.

A novel methodology based on neutral plane engineering contributes to significantly reduce variations of electrical characteristics down to an acceptable level even under aggressive bending. This technique has been first simulated and in a second step validated by static and high frequency measurements.

It has, in particular, been demonstrated that industrial 65 nm partially depleted SOI MOSFETs thinned to the micrometer-scale and transferred onto a plastic foil leads to high performance devices with high stability under flexure. Slight variations of electrical properties (<5%) even after bending flexible UTTB-CMOS dies on a cylinder with a curvature radius in the centimetre range pave the way to promising applications for foldable HF systems taking advantage of new form factors for heterogeneous integration.

This work therefore opens new perspectives for the development of foldable devices and circuits operating above 100 GHz turning point for nomadic and SWAP constrained applications.

Acknowledgments This work was supported in part by the ST-IEMN Common Laboratory and by the region Nord-Pas-de-Calais and FEDER through the CPER-CIA project. It also contributes to the scientific and technological program of the LEAF EQUIPEX project (ANR-11-355 EQPX-0025).

References

1. Chimot, N., Derycke, V., Goffman, M.F., Bourgoin, J.P., Happy, H., Dambrine, G.: Gigahertz frequency flexible carbon nanotube transistors. Appl. Phys. Lett. **91**, 153111 (2007)
2. Vaillancourt, J., Zhang, H., Vasinajindakaw, P., Xia, H., Lu, X., Han, X., Jan-zen, D.C., Shih, W.-S., Jones, C.S., Stroder, M., Chen, M.Y., Subbaraman, H., Chen, R.T., Berger, U., Renn, M.: All ink-jet-printed carbon nanotube thin-film transistor on a polyimide substrate with an ultrahigh operating frequency of over 5 GHz. Appl. Phys. Lett. **93**, 243301 (2008)
3. Sun, L., Qin, G., Seo, J.-H., Celler, G.K., Zhou, W., Ma, Z.: 12-GHz thin-film transistors on transferrable silicon nanomembranes for high-performance flexible electronics. Small **6**, 2553–2557 (2010)
4. Yuan, H.-C., Celler, G.K., Ma, Z.: 7.8-GHz flexible thin-film transistors on a low-temperature plastic substrate. J. Appl. Phys. **102**, 034501 (2007)
5. Sun, Y., Menard, E., Rogers, J.A., Kim, H.-S., Kim, S., Chen, G., Adesida, I., Dettmer, R., Cortez, R., Tewksbury, A.: Gigahertz operation in flexible transistors on plastic substrates. Appl. Phys. Lett. **88**, 183509 (2006)
6. Cherenack, K.H., Hekmatshoar, B., Sturm, J.C., Wagner, S.: Self-aligned amorphous silicon thin-film transistors fabricated on clear plastic at 300 IEEE Trans. Electron Dev. **57**, 2381–2389 (2010)

7. Kao, H.L., Chin, A., Hung, B.F., Lee, C.F., Lai, J.M., McAlister, S.P., Samudra, G.S., Yoo, W.J., Chi, C.C.: Low noise RF MOSFETs on flexible plastic substrates. IEEE Electron Dev. Lett. **26**, 489–491 (2005)
8. Kao, H.L., Chin, A., Liao, C.C., Tseng, Y.Y., McAlister, S.P., Chi, C.C.: DC-RF performance improvement for strained 0.13 μm MOSFETs mounted on a flexible plastic substrate. In: IEEE MTT-S International Microwave Symposium Digest 2006, pp. 2043–2046 (2006)
9. Vaillancourt, J., Lu, X., Han, X., Janzen, D.C.: High-speed thin-film transistor on flexible substrate fabricated at room temperature. Electron. Lett. **42**, 1365–1367 (2006)
10. Han, X., Janzen, D.C., Vaillancourt, J., Lu, X.: Printable high-speed thin-film transistor on flexible substrate using carbon nanotube solution. Micro Nano Lett. **2**, 96–98 (2007)
11. Ahn, J.-H., Kim, H.-S., Lee, K.J., Zhu, Z., Menard, E., Nuzzo, R.G., Rogers, J.A.: High-speed mechanically flexible single-crystal silicon thin-film transistors on plastic substrates. IEEE Electron Dev. Lett. **27**, 460–462 (2006)
12. Yuan, H.-C., Ma, Z.: Microwave thin-film transistors using Si nanomembranes on flexible polymer substrate. Appl. Phys. Lett. **89**, 212105 (2006)
13. Sekitani, T., Noguchi, Y., Zschieschang, U., Klauk, H., Someya, T.: Organic transistors manufactured using inkjet technology with subfemtoliter accuracy. Proc. Natl. Acad. Sci. of U S A **105**, 4976–4980 (2008)
14. Defrance, N., Lecourt, F., Douvry, Y., Lesecq, M., Hoel, V., Lecavelier Des Etangs-Levallois, A., Cordier, Y., Ebongue, A., De Jaeger, J.C.: Fabrication, characterization, and physical analysis of AlGaN/GaN HEMTs on flexible substrates. IEEE Trans. Electron Dev. **60**, 1054–1059 (2013)
15. Sire, C., Ardiaca, F., Lepilliet, S., Seo, J.W.T., Hersam, M.C., Dambrine, G., Happy, H., Derycke, V.: Flexible gigahertz transistors derived from solution-based single-layer graphene. Nano Lett. **12**, 1184–1188 (2012)
16. Lesecq, M., Hoel, V., Lecavelier des Etangs-Levallois, A., Pichonat, E., Douvry, Y., De Jaeger, J.C.: High performance of AlGaN/GaN HEMTs reported on adhesive flexible tape. IEEE Electron Dev. Lett. **32**, 143–145 (2011)
17. Shi, J., Wichmann, N., Roelens, Y., Bollaert, S.: Microwave performance of 100 nm-gate $In_{0.53}Ga_{0.47}As/In_{0.52}Al_{0.48}As$ high electron mobility transistors on plastic flexible substrate. Appl. Phys. Lett. **99**, 203505 (2011)
18. Lecavelier des Etangs-Levallois, A., Dubois, E., Lesecq, M., Danneville, F., Poulain, L., Tagro, Y., Lepilliet, S., Gloria, D., Raynaud, C., Troadec, D.: 150-GHz RF SOI-CMOS technology in ultrathin regime on organic substrate. IEEE Electron Dev. Lett. **32**, 1510–1512 (2011)
19. Tagro, Y., Lecavelier des Etangs-Levallois, A., Poulain, L., Lepilliet, S., Gloria, D., Raynaud, C., Dubois, E., Danneville, F.: High frequency noise potentialities of reported CMOS 65 nm SOI technology on flexible substrate. In: Silicon Monolithic Integrated Circuits RF System SIRF 2012 IEEE 12th Topical Meeting 2012, pp. 89–92 (2012)
20. Lecavelier des Etangs-Levallois, A., Chen, Z., Lesecq, M., Lepilliet, S., Tagro, Y., Danneville, F., Robillard, J.-F., Hoel, V., Troadec, D., Gloria, D., Raynaud, C., Ratajczak, J., Dubois, E.: A converging route towards very high frequency, mechanically flexible and performance stable integrated electronics. J. Appl. Phys. **113**, 153701 (2013)
21. Lecavelier des Etangs-Levallois, A., Lesecq, M., Danneville, F., Tagro, Y., Lepilliet, S., Hoel, V., Troadec, D., Gloria, D., Raynaud, C., Dubois, E.: Radio-frequency and low noise characteristics of SOI technology on plastic for flexible electronics. Solid-State Electron, in press (2013)
22. Sekitani, T., Zschieschang, U., Klauk, H., Someya, T.: Flexible organic transistors and circuits with extreme bending stability. Nat. Mater. **9**, 1015–1022 (2010)
23. Sekitani, T., Iba, S., Kato, Y., Noguchi, Y., Someya, T., Sakurai, T.: Ultraflexible organic field-effect transistors embedded at a neutral strain position. Appl. Phys. Lett. **87**, 173502 (2005)

24. Uno, M., Nakayama, K., Soeda, J., Hirose, Y., Miwa, K., Uemura, T., Nakao, A., Takimiya, K., Takeya, J.: High-speed flexible organic field-effect transistors with a 3D structure. Adv. Mater. **23**, 3047–3051 (2011)
25. Schön, J., Kloc, C., Batlogg, B.: On the intrinsic limits of pentacene field-effect transistors. Org. Electron. Phys. Mater. Appl. **1**, 57–64 (2000)
26. Halonen, E., Kaija, K., Mantysalo, M., Kemppainen, A., Osterbacka, R., Bjorklund, N.: Evaluation of printed electronics manufacturing line with sensor platform application. In: European Microelectronics and Packaging Conference (EMPC) 2009 Europe, pp. 1–8 (2009)
27. Menard, E., Lee, K.J., Khang, D.-Y., Nuzzo, R.G., Rogers, J.A.: A printable form of silicon for high performance thin film transistors on plastic substrates. Appl. Phys. Lett. **84**, 5398–5400 (2004)
28. Wang, C., Chien, J.-C., Fang, H., Takei, K., Nah, J., Plis, E., Krishna, S., Niknejad, A.M., Javey, A.: Self-aligned, extremely high frequency III-V metal-oxide-semiconductor field-effect transistors on rigid and flexible substrates. Nano Lett. **12**, 4140–4145 (2012)
29. Ko, H., Takei, K., Kapadia, R., Chuang, S., Fang, H., Leu, P.W., Ganapathi, K., Plis, E., Kim, H.S., Chen, S.-Y., Madsen, M., Ford, A.C., Chueh, Y.-L., Krishna, S., Salahuddin, S., Javey, A.: Ultrathin compound semiconductor on insulator layers for high-performance nanoscale transistors. Nature **468**, 286–289 (2010)
30. Sun, Y., Kim, S., Adesida, I., Rogers, J.A.: Bendable GaAs metal-semiconductor field-effect transistors formed with printed GaAs wire arrays on plastic substrates. Appl. Phys. Lett. **87**, 083501 (2005)
31. Colinge, J.-P.: Silicon-on-Insulator Technology: Materials to VLSI. Springer, Berlin (1997)
32. Celler, G.K., Cristoloveanu, S.: Frontiers of silicon-on-insulator. J. Appl. Phys. **93**, 4955–4978 (2003)
33. Pelloie, J.L., Auberton-Hervé, A.: A new generation of IC processing: low-power, high-performance SOI CMOS. Solid State Technol. **44**, 63 (2001)
34. Martineau, B., Cathelin, A., Danneville, F., Kaiser, A., Dambrine, G., Lepilliet, S., Gianesello, F., Belot, D.: 80 GHz low noise amplifiers in 65 nm CMOS SOI. In: 33rd European Solid State Circuits Conference, pp. 348–351 (2007)
35. Raynaud, C., Haendler, S., Guegan, G., Gianesello, F., Martineau, B., Touret, P., Planes, N.: 65 nm low power (LP) SOI technology on high resistivity (HR) substrate for WLAN and Mmwave SOCs. ECS Trans. **19**, 257–264 (2009)
36. Wang, P.W.: Industrial challenges for thin wafer manufacturing. In: Fourth World Conference on Photovoltaic Energy Conversion, pp. 1179–1182 (2006)
37. McLellan, N., Fan, N., Liu, S., Lau, K., Wu, J.: Effects of wafer thinning condition on the roughness, morphology and fracture strength of silicon die. J. Electron. Packag. **126**, 110–114 (2004)
38. Jiun, H.H., Ahmad, I., Jalar, A., Omar, G.: Effect of wafer thinning methods towards fracture strength and topography of silicon die. Microelectron. Reliab. **46**, 836–845 (2006)
39. Robbins, H., Schwartz, B.: Chemical etching of silicon II. J. Electrochem. Soc. **107**, 108–111 (1960)
40. Roman, B., Bico, J.: Elasto-capillarity: deforming an elastic structure with a liquid droplet. J. Phys.: Condens. Matter **22**, 493101 (2010)
41. Smith, C.S.: Piezoresistance effect in germanium and silicon. Phys. Rev. **94**, 42–49 (1954)
42. Kanda, Y.: A graphical representation of the piezoresistance coefficients in silicon. IEEE Trans. Electron Dev. **29**, 64–70 (1982)
43. Haugerud, B.M., Bosworth, L.A., Belford, R.E.: Mechanically induced strain enhancement of metal–oxide–semiconductor field effect transistors. J. Appl. Phys. **94**, 4102–4107 (2003)
44. Maikap, S., Yu, C.-Y., Jan, S.R., Lee, M.H., Liu, C.W.: Mechanically strained strained-Si NMOSFETs. IEEE Electron Dev. Lett. **25**, 40–42 (2004)
45. Sun, Y., Thompson, S.E., Nishida, T.: Strain Effect in Semiconductors: Theory and Device Applications. Springer, New York (2009)

46. Rochette, F.: Etude et caractérisation de l'influence des contraintes mécaniques sur les propriétés du transport électronique dans les architectures MOS avancées. Ph.D. Thesis Institut National Polytechnique de Grenoble – INPG (2008)
47. Guillaume, T.: Influence des contraintes mécaniques non-intentionnelles sur les performances des transistors MOS à canaux ultra-courts. Ph.D. Thesis Institut National Polytechnique de Grenoble INPG (2005)
48. Ghani, T., Armstrong, M., Auth, C., Bost, M., Charvat, P., Glass, G., Hoffmann, T., Johnson, K., Kenyon, C., Klaus, J., McIntyre, B., Mistry, K., Murthy, A., Sandford, J., Silberstein, M., Sivakumar, S., Smith, P., Zawadzki, K., Thompson, S., Bohr, M.: A 90 nm high volume manufacturing logic technology featuring novel 45 nm gate length strained silicon CMOS transistors. In: International Electron Devices Meeting IEDM 03, Technical Digest, 11.6.1–11.6.3 (2003)
49. Thompson, S.E., Armstrong, M., Auth, C., Cea, S., Chau, R., Glass, G., Hoffman, T., Klaus, J., Ma, Z., Mcintyre, B., Murthy, A., Obradovic, B., Shifren, L., Sivakumar, S., Tyagi, S., Ghani, T., Mistry, K., Bohr, M., El-Mansy, Y.: A logic nanotechnology featuring strained-silicon. IEEE Electron Dev. Lett. **25**, 191 (2004)
50. Skotnicki, T., Fenouillet-Beranger, C., Gallon, C., Buf, F., Monfray, S., Payet, F., Pouydebasque, A., Szczap, M., Farcy, A., Arnaud, F., Clerc, S., Sellier, M., Cathignol, A., Schoellkopf, J.-P., Perea, E., Ferrant, R., Mingam, H.: Innovative materials, devices, and CMOS technologies for low-power mobile multimedia. IEEE Trans. Electron Dev. **55**, 96–130 (2008)
51. Chaboche, J.L., Lemaître, J., Benallal, A., Desmorat, R.: Mécanique Des Matériaux Solides, 3rd edn. Dunod, Paris (2009)
52. Zhao, W., He, J., Belford, R.E., Wernersson, L.-E., Seabaugh, A.: Partially depleted SOI MOSFETs under uniaxial tensile strain. IEEE Trans. Electron Dev. **51**, 317–323 (2004)
53. Kim, D.H., Ahn, J.-H., Choi, W.M., Kim, H.-S., Kim, T.-H., Song, J., Huang, Y.Y., Liu, Z., Lu, C., Rogers, J.A.: Stretchable and foldable silicon integrated circuits. Science **320**, 507–511 (2008)
54. Besson, J., Cailletaud, G., Chaboche, J.-L., Forest, S.: Non-Linear Mechanics of Materials. Springer, Dodrecht (2010)
55. Garrigues, J.: Fondements de La Mécanique Des Milieux Continus. Hermes Lavoisier (2007)
56. Canali, C., Ferla, F., Morten, B., Taroni, A.: Piezoresistivity effects in MOSFET useful for pressure transducers. J. Phys. Appl. Phys. **12**, 1973–1983 (1979)

Tri-Dimensional A2-RAM Cell: Entering the Third Dimension

Francisco Gámiz, Noel Rodriguez, Carlos Navarro, Carlos Marquez and Sorin Cristoloveanu

Abstract Chapter we present an overview of a capacitor-less DRAM cell based on a 3D multibody transistor with high scalability, low-power consumption, long retention time, non-destructive reading, and wide memory window. High performance is demonstrated on a 20 nm channel length device, including '1' to '0' current ratio larger than 10^3 (with negligible '0' current level), very low voltage bias operation and retention time longer than 20 ms at 85 °C in worst cases. Compared to previous equivalent 3D memory cells reported so far, the proposed cell shows longer retention time even though the gate length is shrunk by a factor of two. The voltages used to write and read the information are far smaller than the previously reported ones in comparable structures. We have confirmed by TCAD simulation that the improvements are attributed to an innovative operation concept: a dedicated body partitioning. This device exploits the working principle of the A2RAM memory cell recently introduced by researchers at the University of Granada and Grenoble INP. The principles of operation and key mechanisms for programming are described. The new concept of 3D (FinFET, trigate or nanowire) DRAM cell proposed features a N/P body partitioning which enables the physical separation of hole storage and sensing electron current. The hole concentration in a surrounding P-crust, controls the partial or full depletion of a N-core which short-circuits drain and source of the device. The status of the N-bridge (depleted or un-depleted) determines the two memory logic states. The cell is compatible with ultimate scaling and shows attractive performance (long retention, wide memory window, simple programming, nondestructive reading, and very low-power operation) for embedded systems.

F. Gámiz (✉) · N. Rodriguez · C. Navarro · C. Marquez
Department of Electronics, University of Granada, 18071 Granada, Spain
e-mail: fgamiz@ugr.es

C. Navarro · S. Cristoloveanu
IMEP-LAHC (UMR 5130), Grenoble INP Minatec, 38016 Grenoble Cedex 1, France

A. Nazarov et al. (eds.), *Functional Nanomaterials and Devices for Electronics, Sensors and Energy Harvesting*, Engineering Materials,
DOI 10.1007/978-3-319-08804-4_6

1 Introduction

The Dynamic Random Access Memory (DRAM) industry is making prominent efforts to address the scalability of the standard 1-Transistor + 1-Capacitor (1T + 1C) cells. The increment in complexity and the increasing dimensional difference between transistor and capacitor have motivated the search for a DRAM substitute [1–3]. One of the best-suited alternatives is the Floating-Body DRAM (FB-DRAM) [4, 5]. Floating-Body DRAMs use the majority carriers stored at one interface of the channel of a Fully Depleted Silicon On Insulator (FDSOI) transistor to modulate bipolar or minority carrier current flowing at the opposite interface [6].

The basic principle of operation of a FB-DRAM cell is summarized as follows (for more details please follow Ref. [6] and references there in). If in a partially depleted SOI MOSFET holes are injected in the floating-body of the transistor by different mechanisms ((1) MOSFET impact ionization, (2) bipolar junction transistor (BJT) impact ionization effect, (3) band-to-band (BTB) tunneling injection and (4) gate tunneling current) [6], the body potential increases. Therefore, there is a shift of the threshold voltage to lower values, leading to higher values of the drain current for the same bias conditions. This high current level (which represents the '1' state) will remain as long as the overpopulation of holes keeps in the body of the transistor. If the overpopulation of holes disappears (for example by forward biasing the drain- or source-body junction) the threshold voltage comes back to the initial state and the drain current decreases ('0' state). Figure 1 summarizes the operation of a SOI transistor as a FB-DRAM cell.

The working principle of these devices is based on the coexistence of electrons and holes in the same body, condition that requires moderately thick films ($\geq$40 nm). However, ultrathin films (<10 nm) are needed for MOSFET scaling beyond the 32 nm node; this condition also applies to embedded FB-DRAMs co-integrated with logic circuits. Unfortunately the supercoupling effect forbids the simultaneous activation of electron and hole channels, facing each other, in the same ultrathin silicon layer [7, 8]. Therefore, one of the more questioned drawbacks of FB-DRAM cells has been their scalability. As in conventional 1T-1C DRAM cell, the information relies on the charge stored and consequently on the volume or area devoted to this storage. On the other hand, the scalability of FD SOI transistors requires a decrease in the film thickness in order to suppress short-channel effects.

In the case of single gate SOI technology, the film thickness of the device should be around four times thinner than the gate length (this condition can be relaxed by a factor of two or three in the case of double- or triple-gate devices) [9]. This means that FB-DRAMs should be compatible with body thicknesses below 10 nm in order to be competitive in future technology nodes. This condition imposes a capital challenge for the FB-DRAM family. Several studies have shown severe degradation in the current margin between the memory states when the body thickness of 1T-DRAMs decreases below 30 nm [10]. This limitation, also

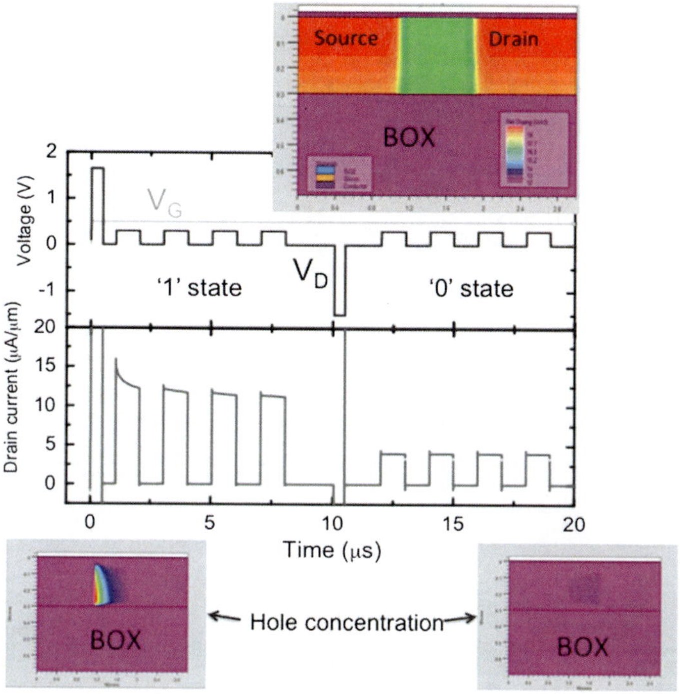

Fig. 1 Operation of a FB-1TDRAM cell. The picture shows the bias pattern (*top*) and the driven current (*bottom*). The floating body is initially charged with holes (*left bottom*) generated by impact ionization produced by a high drain voltage pulse. The cell is then read four times by using a small drain bias. A high drain current level is obtained. At $t = 10$ μs hole overpopulation is eliminated (*right bottom*) by forward biasing drain-body junction (negative bias pulse applied to the drain). The cell is then read four times by using a small drain voltage, obtaining now a much lower drain current level than in the previous '1' states. $L = 1$ μm, $T_{Si} = 300$ nm, $T_{ox} = 3$ nm, $T_{BOX} = 400$ nm, $N_A = 10^{17}$ cm^{-3}

known as super-coupling effect [7], is basically an electrostatic consequence: the thinner the body, the more difficult to achieve the potential difference in order to accommodate a high concentration of electrons at one interface and a high concentration of holes at the other interface of the same silicon slab.

In order to overcome this intrinsic limitation (for maintaining the retention and sensing margin performances despite the scaling of the film thickness), several architectures and material combinations have been proposed. The idea behind all of them is the separation of the stored carriers and the sensing carriers by creating dedicated volumes (potential wells) inside the transistor body (multi-body devices).

A first example of multibody FB-DRAM cell is the single-transistor quantum well (QW) FB-DRAM [11], Fig. 2. It uses an engineered body integrating within the Si film a thin layer of a material with a narrower band gap (i.e. SiGe). This layer serves as storage well for holes. It was theoretically demonstrated that this structure improves the current sensing margin and scalability characteristics.

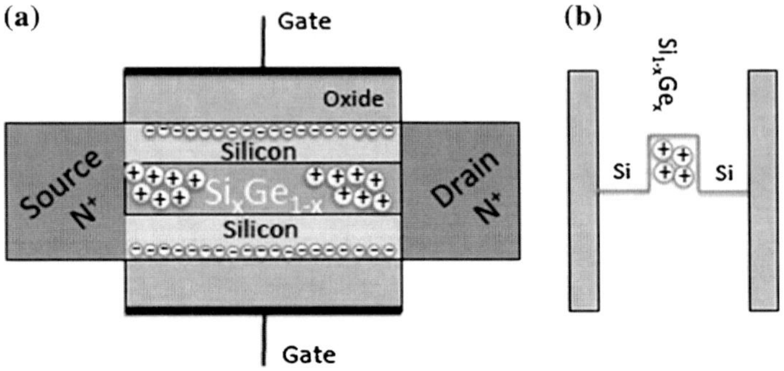

Fig. 2 **a** Schematic of the single transistor quantum well FB-DRAM [11]. **b** Detail of the potential well created by the SiGe layer, where holes are accumulated

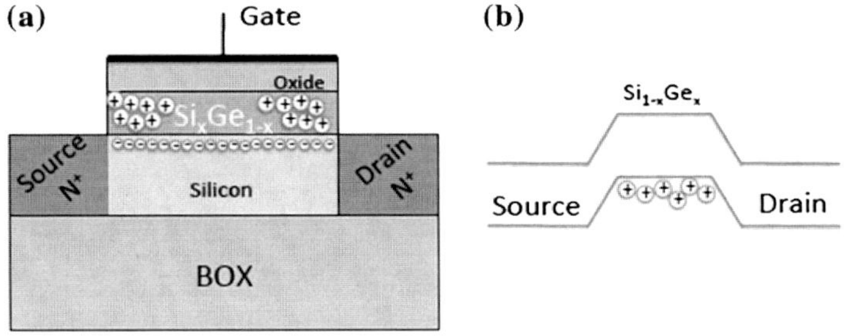

Fig. 3 **a** Schematic of convex channel FB-DRAM cell structure filled with SiGe. **b** Potential well created by the SiGe layer gate-stack

Compared with standard FB-DRAM, this QW memory has the ability to store the holes closer to the front-gate inducing an enhanced threshold voltage shift and longer retention times. Thanks to the introduction of the extra "storage room", QW devices are more scalable since the effect of the volume reduction with the channel length is lessened.

A second alternative of multibody FB-DRAM cell was proposed in 2009 by researchers at the University of Berkeley: the convex channel FB-DRAM cell [12] (Fig. 3). This cell uses the bipolar programming technique to improve the retention time of the cell [13]. Holes are stored beneath a raised gate oxide which is filled by $Si_{0.8}Ge_{0.2}$. The smaller bandgap of the SiGe in the convex channel region provides a deeper potential well and hence further improvement in retention time [13].

However, the first real multibody device, with two regions clearly isolated (a storing channel, and a sensing channel) was introduced in 2010 [14, 15]. We proposed a totally new concept (Advanced RAM, A-RAM) [16], using some of the FB-DRAM fundamentals but featuring unique architecture and electrostatic properties. The A-RAM has been designed to separate physically the majority and minority carriers even in ultrathin fully depleted (FD) SOI layers [17] by

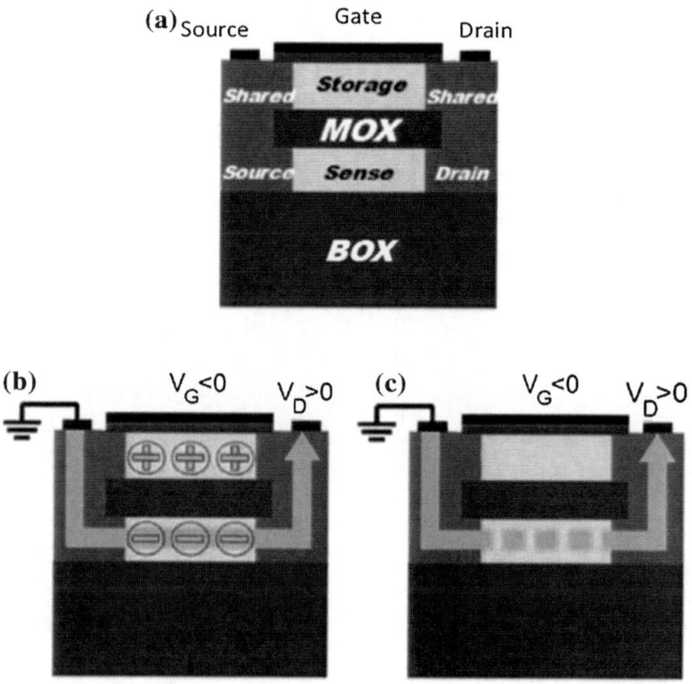

Fig. 4 **a** Schematics of ARAM cell [15]. **b** '1' state with charged semibodies. A high level current is sensed by increasing the drain voltage. **c** '0' state with discharged semibodies. Increasing the drain bias leads to a low current (subthreshold current)

eliminating the supercoupling effect. The A-RAM is an FD SOI transistor which features two ultrathin semi-bodies physically isolated by a middle oxide (MOX) but sharing the source and drain regions (see Fig. 4a).

When operated as a memory cell, the top semi-body is used for majority charge storage (holes) while the bottom semi body serves to sense the device state through an electron current. The MOX constitutes the key advantage of the A-RAM cell: electron and hole populations can be brought very close to each other due to the large electrostatic potential difference between interfaces allowed thanks to the MOX (which would not be possible in an ultrathin single body [7]). The A-RAM structure can be fabricated by the local oxidation (MOX) of the bottom semi-body, followed by the epitaxial regrowth of both the upper semi-body and the source/drain regions. The shorter the device, the better the quality of the epilayer. The Silicon-on-Nothing (SoN) process can also be envisaged: the growth of a sacrificial SiGe layer, Si epitaxy, and SiGe etch leaving a cavity to be refilled with the MOX dielectric [18]. The MOX principle is also adaptable to a vertical FinFET configuration.

Memory state '1' is programmed by charging the top semi-body with holes (see Fig. 4b) via impact ionization or band-to-band tunneling [19]. A volume inversion electron channel is activated, by electrostatic coupling, in the ultrathin bottom

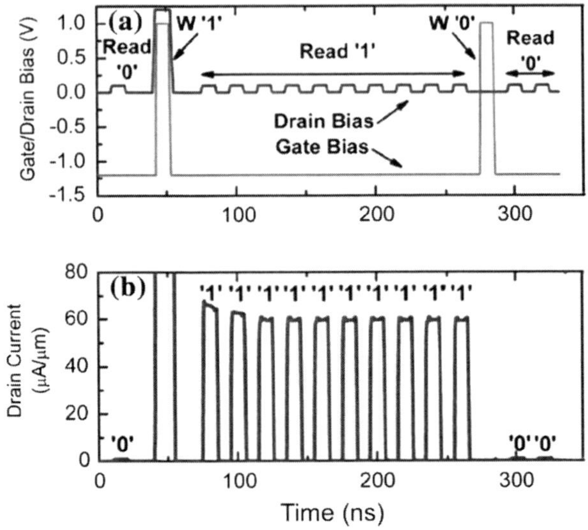

Fig. 5 TCAD simulation results for the operation of A-RAM structure (Fig. 4) as a memory cell. **a** Bias scheme for programming, holding and reading. **b** Drain current levels for '1' and '0' states. Note the low value used for the drain voltage in the reading operation, and the extremely low drain current level for the '0' state

semi-body, establishing electrical continuity between the source and drain regions. If the drain voltage is increased, a substantial current flows through the transistor. When no charge is stored in the top semi-body (storage channel), the electron concentration in the bottom semi-body (sense channel) is extremely low, defining the state '0' (see Fig. 4c). The source and drain regions are electrically isolated: even raising the drain voltage results in a negligible current. In the "retention" phase, the negative gate bias and the zero drain voltage are used to maintain the holes in the storage channel. The A-RAM operation has been validated by numerical simulations. The reading is nondestructive. The refresh is only needed to compensate for the derive of the charge in the '0' state. The '0' state is written by pulsing the gate bias to a positive value (typically, 1 V). The holes are rapidly eliminated (nanosecond) by the junctions which become forward biased due to the sudden increase in the body potential. The electron concentration in the bottom semi-body decreases and the sensed drain current returns to a subthreshold value. Note that this capacitive coupling enables the '0' state writing without changing the drain bias (the typical procedure in the FB-DRAMs is to forward bias the drain channel junction).

The Poisson and continuity equations were solved in the transient mode by accounting for the band-to-band tunneling, impact ionization, and generation-recombination mechanisms. Figure 5a shows a possible bias sequence for writing and reading the transistor states; the sensed drain current is shown in Fig. 5b. At the starting time ($t = 0$), the '0' bit has been written, and the upper semi body of the cell is discharged. The cell state, tested by slightly increasing the drain voltage (0.1 V), shows a negligible drain current. Next, the '1' state is programmed by impact ionization. A gate voltage pulse is embedded in a drain voltage pulse. This guarantees that, when the gate voltage decreases back to the negative retention

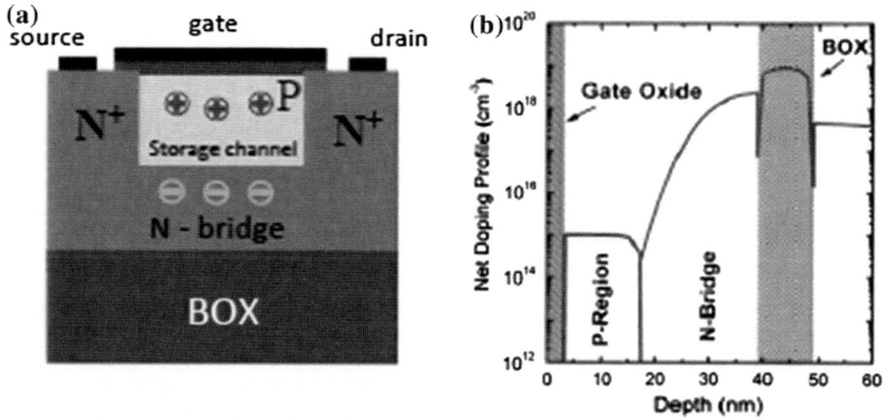

Fig. 6 **a** Schematic configuration of A2RAM cell on SOI substrate. A buried N-layer (N-bridge) connects the source and the drain regions. **b** Vertical net doping profile at the middle of the channel on SOI substrate from process simulator Silvaco ATHENA demonstrating fabrication feasibility of the A2RAM. The retrograde p–n$^+$ junction is clearly visible in a $T_{Si} = 36$ nm device according to the specifications for A2RAM operation [20]

value, the hole recombination is prevented. During the gate bias transient, the lateral field remains high enabling more holes to be generated. In Fig. 5, the '1' state has been sensed several times, showing a large enough current (60 μA/μm) and no degradation.

2 Second Generation of Advanced RAM (A2RAM)

Although physically feasible by different technological implementations (silicon-on-nothing, lateral epitaxial regrowth, etc.) the fabrication of the MOX layer adds complexity to the fabrication of the ARAM cell. As an alternative to the physical isolation provided by the middle-oxide (MOX), the electrical isolation of the two types of carriers could be also carried out by a vertical p–n junction. This idea has originated a new cell architecture, named A2RAM and illustrated in Fig. 6 [20].

To do so, a conventional MOSFET is modified by connecting the N+ source and drain regions through a buried n-type layer (named N-bridge), underneath the P-channel.

Figure 6b shows an example of a realistic doping profile at the middle of the device, perpendicular to the gate, and obtained from process simulations with ATHENA [21]. The total device thickness that can be achieved is very thin (typically in the range of 15–30 nm). Since the source and the drain have the same doping polarity as the N-bridge, the source and drain regions are, in principle, electrically short-circuited through the *bridge*. The basic idea for memory cell

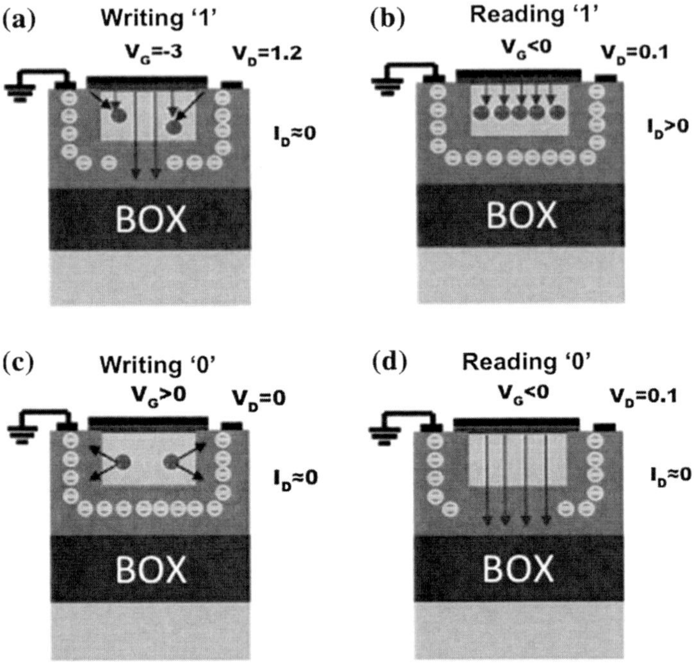

Fig. 7 Schematic operation of A2RAM memory cell operation. **a** Writing '1': the p-body is charged, screening the gate field. **b** Reading '1': majority carrier (electron) current flows through the N-bridge. **c** Writing '0': the p-body is discharged (deep depletion), and the N-bridge is fully depleted. **d** Reading '0': the current through the n-bridge is negligible

operation is to suppress/enable this short-circuiting by fully-depleting/undepleting the N-bridge through the accumulation/emptying of holes in the upper channel.

The top P-body is used as a storage node, whereas the lower body (i.e., a relatively high-doped N-bridge) serves for current sense discriminating the two memory states. The simplified sequence of states is shown in Fig. 7.

The upper body (p-type) is charged with holes generated either by band-to-band (BTB) tunneling [19] or by impact ionization (II) [22] in the MOS transistor. These holes can be retained in the top body with negative gate bias (see Fig. 7a). In this situation, the vertical electric field, originated from the negative gate bias, is screened by the positive hole charge of the P-body and has a minor effect on the majority carriers of the N-bridge. The bridge is partially depleted and a small drain bias leads to an electron current, I_1, flowing through the bridge (no current flows through the P-body) thus reading the '1' state (Fig.7b). When the top body is discharged of holes (state '0'), the gate field is no longer screened, and the N-bridge is fully depleted (Fig. 7c). The lack of holes in the N-bridge causes a very low current, I_0, if the drain bias is increased (Fig. 7d).

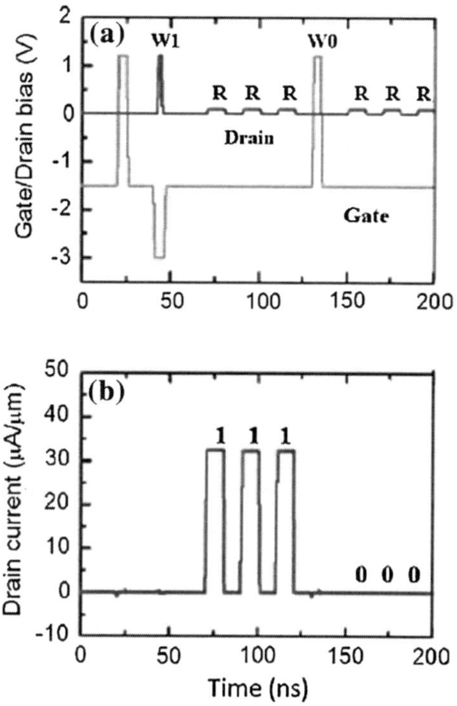

Fig. 8 Waveforms demonstrating the functionality of A2RAM cell. (*Top*) Bias pattern for gate and drain. (*Bottom*) drain current. Writing is performed by BTB tunneling. $L = 22$ nm, $\varphi_m = 4.5$ eV, $T_{ox} = 2$ nm, TP−body = 8 nm, and $T_{N-bridge} = 10$ nm

The differences with conventional FB-DRAMs are three-fold:

1. The drain current, defining the cell state, is due to electrons (majority carriers) flowing or not in the volume of the bridge.
2. The use of an insulator substrate is optional.
3. The super-coupling effect is suppressed because the vertical P–N junction ensures the coexistence of electrons and holes in the same silicon slab.

Figure 8 shows the memory operation of the cell. The gate is biased well below the threshold voltage of the transistor to generate an accumulation of holes in the channel in steady state conditions, which will determine the current level of the ‘1’ state. Initially the cell is purged by writing a ‘0’ state. To do so, a positive voltage pulse is applied to the gate. This pulse forward biases the channel-source and channel-drain PN junctions, and holes are evacuated from the channel. If the gate suddenly comes back to the negative value, the channel becomes empty of carriers.

A proof of concept for this device has been fabricated using CEA-LETI 22 nm-node technology [23] on 300 mm Unibond wafers. The design of the device has followed the rules shown in [20] to achieve functional memory operation: For N-bridge thickness of 22 nm, its doping concentration should be between 2×10^{17} and 1.5×10^{18} cm^{-3}. The retrograde doping profile was formed by 30-nm selective epitaxial growth (SEG) after arsenic implantation in the initial 6-nm-thick SOI film. The device is completed with a high-k/metal-gate stack (SiO_2/HfSiON/TiN)

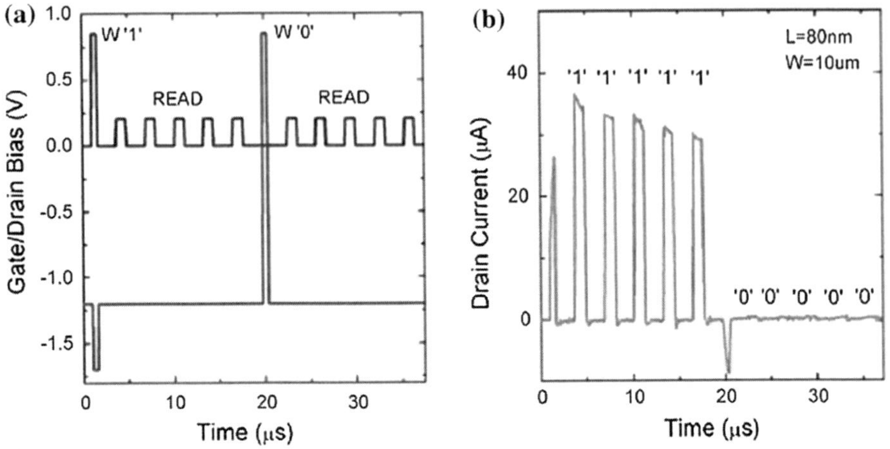

Fig. 9 **a** Gate and drain bias pattern applied to an $L = 80$ nm, $W = 10$ μm A2RAM cell fabricated on SOI. The sequence corresponds to a writing '1' pulse followed by five readings, and a write '0' pulse followed by five readings. **b** Experimental drain current readout corresponding to the sequence applied in (a) and showing the clear difference between '1' and '0' states (After Ref. 20)

(EOT = 3.1 nm). Source and drain regions are formed by epitaxy and implantation without any special optimization compared with the CMOS/SOI process for logic circuits. Finally, a standard back-end process is carried with a N_2/H_2 forming gas anneal.

The resulting net doping profile after thermal load (simulated with ATHENA) is shown in Fig. 6b. As observed, the vertical p–n$^+$ junction has been successfully achieved in a 36-nm-thick Si film resulting in a 22-nm-thick bridge below the 14-nm-thick P-body. The transfer $I_D(V_G)$ characteristics shown in Fig. 1c match the requirements for A2RAM operation. As opposed to regular MOSFETs, they exhibit large "leakage" current even for negative gate bias as a consequence of the inherent resistive connection between source and drain via the N-bridge, leading, as expected, to relatively high subthreshold swing ($SS \cong 200$ mV/dec for $L = 100$ nm).

The measurements were performed using an Agilent B1530A fast measurement/generation unit with two channels. The equipment directly drives the pulsed signals, enabling simultaneous voltage pulses and current measurement on both gate and drain terminals. All measurements were done in a SUSS-PA300PS semiautomatic probe station. FB-DRAM operation was demonstrated by applying the simple bias pattern shown in Fig. 9a. Writing the '1' state via band-to-band tunneling is achieved by increasing the electric field in the gate-to-drain overlap region ($V_G = -1.7$ V and $V_D = 0.85$ V). The storage of the generated holes in the upper P-body is possible by creating an electrostatic potential well during "hold" state ($V_G = -1.2$ V and $V_D = 0$ V). Increasing only V_D to 0.85 V while maintaining $V_G = -1.2$ V or decreasing V_G to -1.7 V while maintaining $V_D = 0$ V (not shown) does not create an electric field in the overlap region large enough for triggering the parasitic write of '1' state by band-to-band tunneling. This scheme prevents cell disturbance issues. The '0' state is achieved by pulsing the gate bias up ($V_G = 0.85$ V pulse), which forward biases the body-to-source/drain junctions and ejects the holes from the body. A deep-depletion region forms in the P-body. Reading the cell state consists in slightly increasing the drain bias and detecting whether the N-bridge is conductive or not ($V_D = 0.2$ V with $V_G = -1.2$ V).

An experimental proof of the readout currents is shown in Fig. 9b. The sequence corresponds to the following: (1) write '1' state; (2) five readings; (3) write '0' event; and (4) five more readings. It is shown that '1' state (high current) and '0' state (negligible current) are clearly discriminated. As observed, the current level of the '1' state decreases with time. This is due to the over-injection of holes during the writing '1' event; the subsequent recombination of the carriers, until thermodynamic equilibrium is achieved, leads to a decrease in the screening of the gate electric field and, therefore, a decrease in the current level of the '1' state until it becomes stable [24]. Note also that drain current during the writing events remains modest, fully compatible with low-power memory applications.

3 Tridimensional A2RAM: FinFET, Trigate and Nanowire A2RAM

Intel has recently announced the use of 3D Tri-Gate transistors in their 22 nm node, i.e., three gates wrapped around the silicon channel in a 3D structure [25]. Tri-gate transistors are expected to provide ultra-low power benefits for use in handheld devices, like smartphones and tablets, while also delivering improved performance normally expected for high-end processors. We show that the concept of multi-body partitioning 1T-DRAM cells, demonstrated so far in 2D devices, can be also transferred to 3D structures (FinFET, tri-gate and nanowire transistors), thus enabling memory cells with low voltage operation, energy efficiency, high performance, and fabrication compatibility as embedded memory in next technological nodes. A three dimensional picture of the proposed memory cell, on SOI substrate, is shown in Fig. 10. Doping sections from source to drain and perpendicular to the BOX are shown in Figs. 11a and 12a, respectively. As observed, a conventional triple-gate FET is modified by connecting the N^+ source and drain through an inner N-type wire (Fig. 10). Typical total Fin width can vary from 15 to 25 nm.

The doping profiles have been generated by process simulator ATHENA thus validating the device feasibility. The 14 nm-thick core N-bridge is 5×10^{18} cm^{-3} doped, whereas the surrounding 4-nm P-type body is maintained undoped ($N_A = 10^{14}$ cm^{-3} residual boron) (Figs. 11b and 12b). The device shares the fundamentals of conventional 1T-DRAMs, but also introduces totally new concepts. The P-type external crust that surrounds the N-type core is used as the

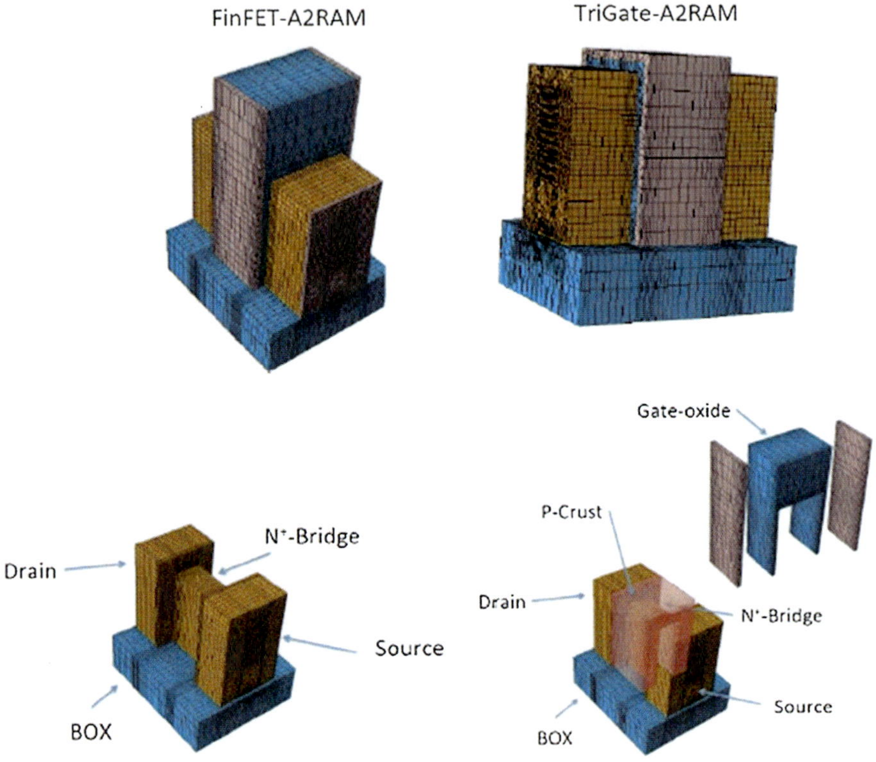

Fig. 10 Schematic representation of the 3D A2RAM memory cell

storage node while the internal core (high doped N-bridge) is used for current sense (Fig. 13a).

The memory effect relies on the following principle, adapted from the successful operation of A2RAM cell: when the P-type body is charged with holes (generated by band-to-band tunneling [19] and retained with a negative gate bias), the gate field is screened by the accumulation layer (holes, Fig. 13b) and cannot deplete the N-bridge. If V_D is increased, an electron current flows through the neutral region of the N-bridge, which behaves as a simple resistor (state '1', Fig. 13b). The current flow through the P-body is negligible (reverse biased junctions). When the outer P crust is empty of holes, the gate field is no longer screened and fully depletes the N-bridge (Fig. 13c): the drain current becomes extremely low ('0' state). As in planar A2RAM, differences with conventional 1T-DRAMs are two-fold: (i) the drain current (defining the cell state) is due to majority carriers flowing in the volume of the bridge; (ii) the coexistence of holes and electrons is ensured by the P/N junction (supercoupling suppression).

The combination of the P- and N-channels results in unconventional I_D-V_G characteristics. We have used numerical simulations to demonstrate the functionality as memory cell of the 3D A2RAM device; Poisson and continuity

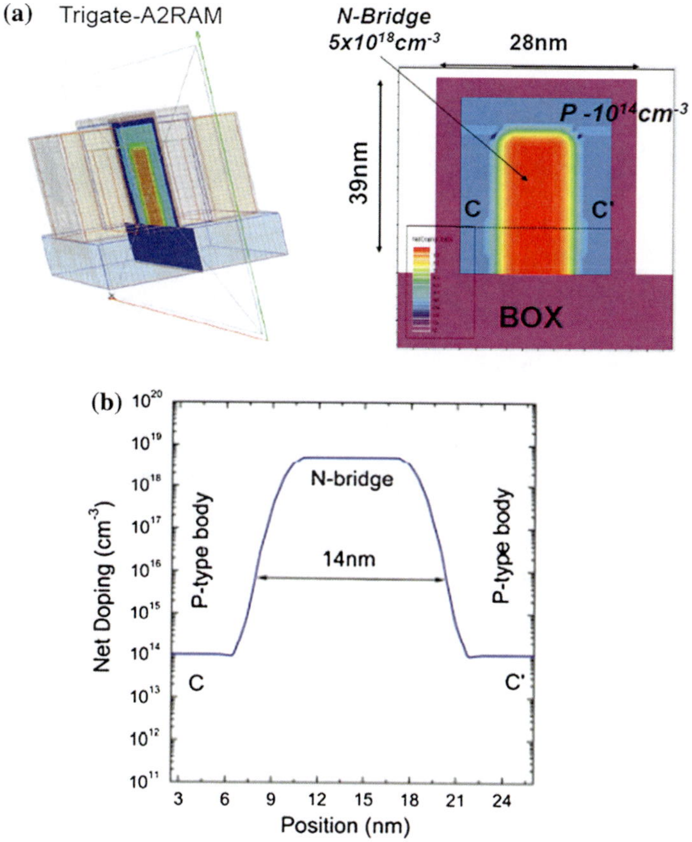

Fig. 11 **a** Doping cross-section (perpendicular to the BOX, parallel to source an drain) of the Triple-gate A2RAM. The high doped N-Bridge (5×10^{18} cm^{-3}) short-circuiting source and drain is surrounded by a low doped P-type layer (10^{14} cm^{-3}). **b** C-C' net doping cross-line in Fig. 11a. The 14 nm-thick N-bridge is surrounded by a low doped P-type layer

equations were solved self-consistently in 3-D. Firstly we studied the steady state operation of the cell: the combination of the parallel P and N channels results in unconventional I_D-V_G curves. In steady state, the N-bridge is always non-depleted regardless of the gate voltage. For negative V_G the potential difference is basically absorbed by the accumulated holes in the upper p-region: the more negative the gate voltage, the more holes accumulate, screening the electric field.

The current is weakly dependent on V_G. For positive gate bias the current flow comes from the parallel combination of the majority carriers (electrons) of the N-bridge (which behaves like a resistor), and the minority carriers (electrons) of the top MOSFET. As the gate bias is increased, the top inversion channel becomes dominant. Notice that this behaviour is different from that of a depletion-mode NMOSFET, where the conduction can be effectively cut at a certain negative gate bias.

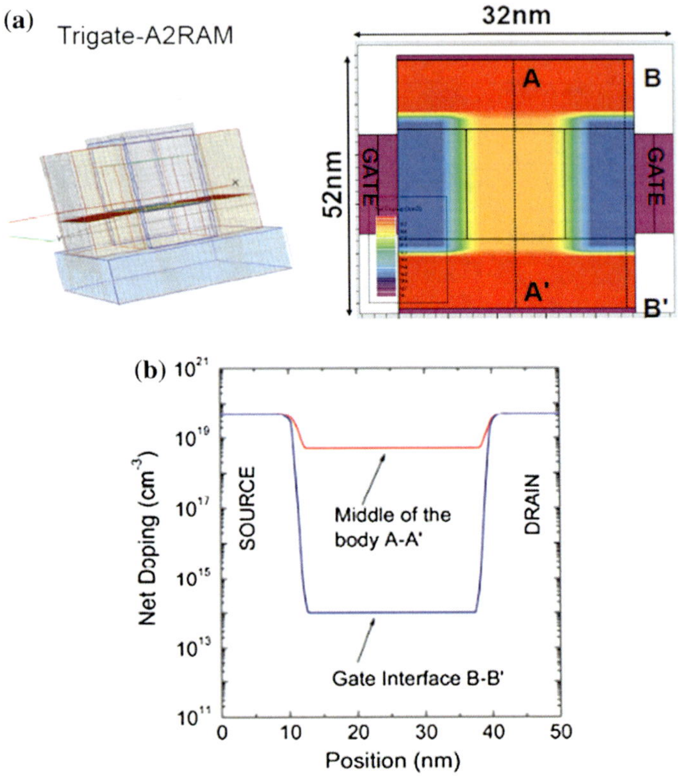

Fig. 12 2D-Doping cross-section (parallel to the BOX) of the Triple-gate A2RAM. Source to drain A-A' (N-bridge) and B-B' (P-type body) doping cross-line. The gate is 3 nm underlaped to improve retention characteristics

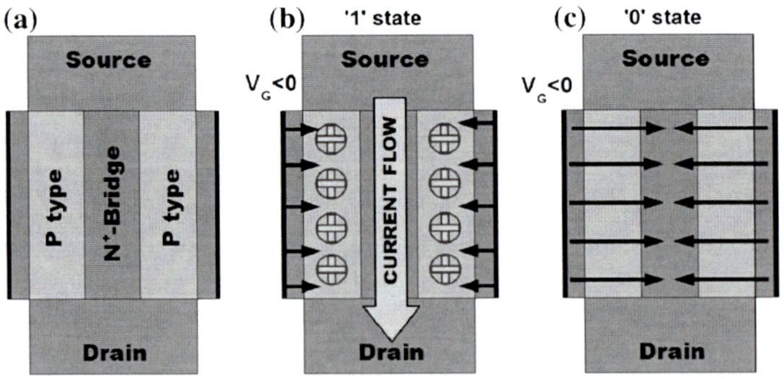

Fig. 13 Schematic memory cell operation of 3D A2RAM. **a** Conceptual multibody representation. **b** During '1' state the P-type surrounding body is populated with holes retained by a negative gate bias. The holes screen the gate field and the current can flow trough the N-Bridge. **c** In '0' state, the P-type region surrounding the body is in deep depletion and the gate field is not screened, depleting the N-bridge of carriers (electrons). The current flow is cut

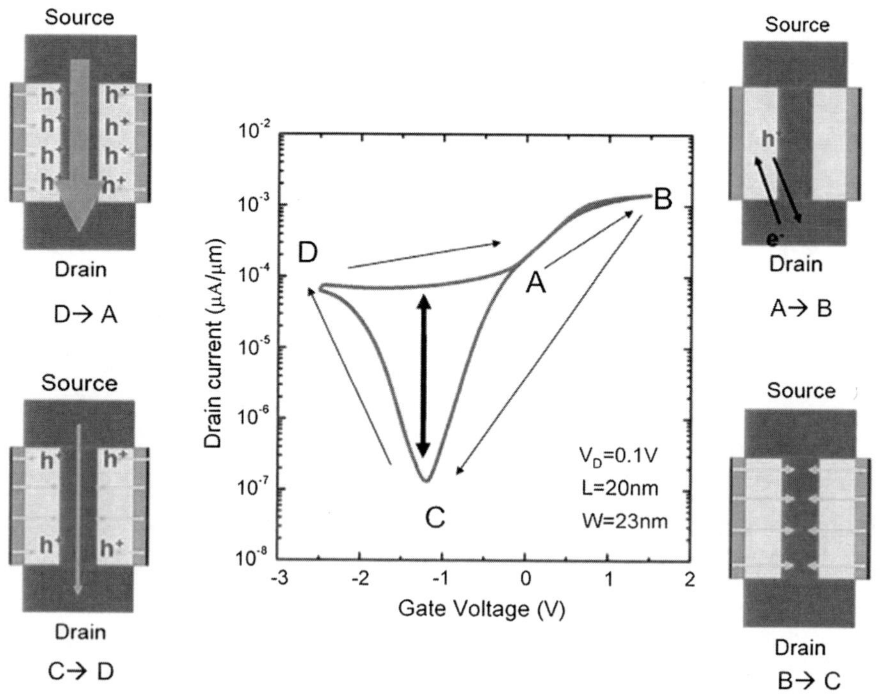

Fig. 14 Hysteresis curve demonstrating the memory effect obtained by looping the gate voltage between 1.5 V and −2.5 V

The transient behaviour is analysed in Fig. 14:

1. From A to B, a positive voltage is applied to the gate. The surrounding channel becomes inverted with electrons, and the behaviour of the device is similar to that of a MOSFET transistor.
2. From B to C, a negative voltage is suddenly applied to the gate. The channel becomes depleted of electrons, and as there are no sources of holes, the surrounding channel becomes empty of carriers. The negative electric field induced by the negative gate voltage, also depletes the N-bridge and as a consequence there is no current at all in the device.
3. If the gate voltage is decreased to even more negative values (C to D), band-to-band tunnelling starts to appear in the source-channel and drain-channel overlapped regions. This process injects holes into the channel that screen the negative electric field induced by the gate. As the negative gate electric field is now weaker, the N-bridge becomes partially un-depleted and the drain current starts to increase. The greater the hole injection, the higher the drain current.
4. If the gate voltage is reduced to zero (D to A), band-to-band tunnelling stops, no more holes are injected in the channel and the drain current becomes constant.

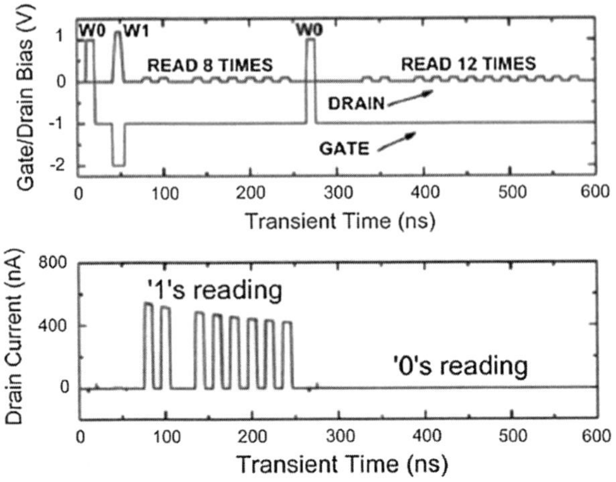

Fig. 15 Full 1T-DRAM operation of the cell. (*Top*) Drain and Gate bias pattern. The cell is initially purged (W0) then the '1' state is written by BTB tunneling and read 8 times. Next the '0' state is written by purging again the cell and read 12 times. (*Bottom*) Drain current reading

As observed in Fig. 14, under negative gate voltages, there is a current window for the same values of gate and drain voltages which, depending on the population of holes in the upper channel, allows the definition of two memory states:

1. '1' state: surrounding channel is populated with holes that screen the negative electric field. Drain current flows through the buried N-Bridge.
2. '0' state: surrounding channel is fully-depleted of carriers, i.e., there are no carriers at all in it. The negative electric field induced by the negative gate voltage depletes the N-bridge, and no current flows between drain and source. This state is non-equilibrium state. After a long time, thermal carrier generation, junction leakage, and band-to-band tunnelling will restore the hole population in the channel, and the '0' state will be corrupted.

Figure 15 shows the memory operation. The gate is biased below the threshold voltage to accommodate an accumulation of holes in the channel, which will define the '1' state current. Initially, the cell is purged by writing '0' state with a positive gate-voltage pulse. Since the pulse forward biases the channel-source and channel-drain PN junctions, the holes are expelled from the channel. When the gate returns to the negative holding bias, there are no carriers in the channel.

We propose two alternative mechanisms to write the '1' states (restore the hole population in the P-body): band-to-band (BTB) tunneling [22] by means of an over-bias pulse of the retention gate voltage, or impact ionization by applying a positive gate voltage in the gate activating the MOSFET. Nevertheless BTB tunneling is best suited for low-power embedded applications since the writing current is typically several orders of magnitude lower than with the impact ionization mechanism (during the writing time there is an additional contribution of current coming from

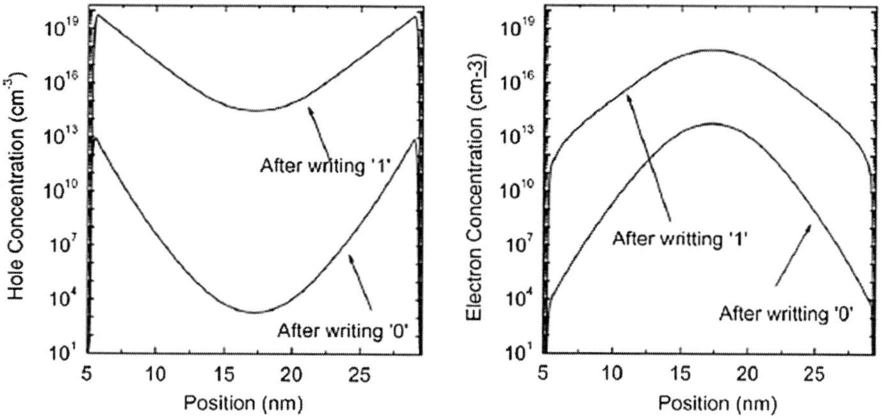

Fig. 16 (*left*) C-C' hole concentration section in Fig. 11 after writing the '1' and '0' states. The difference in hole concentration between states is more than 6 order of magnitude. (*right*) C-C' electron concentration section in Fig. 11 after writing the '1' and '0'states. The difference in electron concentration between states is more than 5 orders of magnitude

the MOSFET which is not present when using the BTB mechanism). Waveforms demonstrating the cell functionality are shown in Fig. 15 using the BTB alternative. As observed, the '0' state corresponds to zero drain current.

Cross sections of the electron and hole concentrations after writing the '1' and '0' states are shown in Fig. 16. After writing '1' state, a high concentration of accumulated holes is achieved in the crust of the fin (Fig. 16, left) which screens the gate electric field. The N-bridge remains populated with electrons (Fig. 16, right) providing a low-resistivity path between source and drain. After writing '0' state, the concentration of holes in the P-crust is six order of magnitude lower than in the previous state, and as a consequence, the concentration of electrons in the bridge decreases to values below 10^{12} cm^{-3}, more than five order of magnitude below the doping concentration.

As mentioned above, level '0' is unstable. At equilibrium the hole population is restored in the P-crust, the device leaves the deep depletion state, gate electric field is screened by the holes in the semiconductor crust, and the N-bridge starts to become undepleted, i.e., a drain current will be set if a drain voltage is applied: '0' state becomes '1' state. Three parasitic mechanisms are responsible for hole population restoration (Fig. 17a):

1. Junction leakage (drain-channel and source-channel PN junctions)
2. Gate-induce-leakage current (mainly band-to-band tunneling)
3. Thermal generation

Figure 17b shows the transient evolution of the two states under continuous reading condition. In spite of the use of a very low drain voltage to read the state of the cell, parasitic band-to-band tunneling (GIDL) at the drain edge of the channel generates holes in the channel, degrading the '0' state.

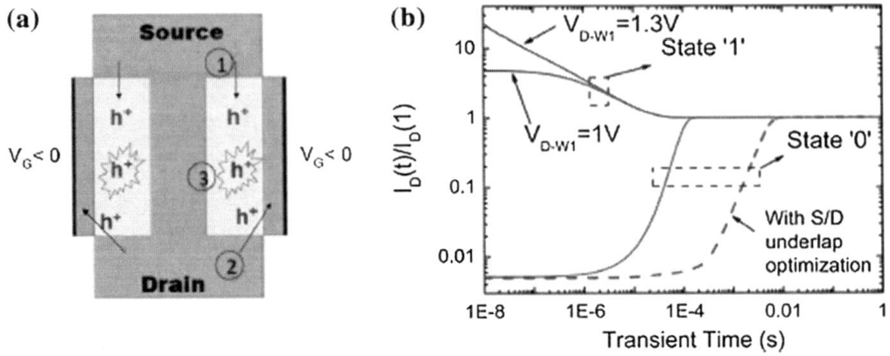

Fig. 17 **a** Mechanisms responsible for restoring /for the restoration of hole population in the P-crust. **b** Evolution of the drain current under continuous reading conditions ($V_D = 0.1$ V). '1' state presents initial current overshoot depending of the writing V_D bias due to overpopulation of holes in the P-type body. Dashed line: evolution of '0' state after source-drain underlap optimization (T=85°C)

At 300 K the retention time with worst-case BTB models [26] exceeds hundreds of miliseconds. Since junction leakage and BTB tunneling is more effective in short devices and at high temperature, we show in Fig. 17b the retention time in the 20 nm cell at 85 °C. The '1' state is stable: the initial current can be larger than the stable level if the number of injected holes by BTB is larger than the value that can be sustained by the negative V_G stationary state. The '0' state is unstable being Gate Induced Drain Leakage (GIDL) the limiting factor which tends to recover the hole equilibrium concentration in the P-body. The hole-generation rate strongly depends on the electric field at the source/drain-to-gate overlap regions. This generation can be minimized to increase retention time by careful junction engineering including a 3 nm under-lapped source/drain [27]. This solution increases retention time from initial 40 μs to more than 20 ms using worst-case models (continuous and dashed lines for '0' state in Fig. 17b).

4 Conclusion

The new proposed concept of tridimensional 1T-DRAM cell features a N/P body partitioning which enables the physical separation of hole storage and electron current. Holes concentration controls the partial or full depletion of the N-core. The cell is compatible with ultimate scaling and shows attractive performance (long retention, wide memory window, simple programming, nondestructive reading, and very low-power operation) for embedded systems.

Acknowledgments This work has been partially funded by Spanish Government through project TEC-2011-28660 and Junta de Andalucía under project TIC-2010-6209

References

1. Hu, C., King, T.J., Hu, C.: A capacitorless double-gate DRAM cell. IEEE Electron Device Lett. **23**(6), 345–347 (2002)
2. Okhonin, S., Nagoga, M., Carman, E., Beffa, R., Faraoni, E.: New generation of Z-RAM. In: IEEE International Electron Devices Meeting (IEDM), pp. 925–928. (December 2007)
3. Ban, I., Avci, U.E., Shah, U., Barns, C.E., Kencke, D.L., Chang, P.: Floating body cell with independently-controlled double gates for high density memory. In: IEEE International Electron Devices Meeting, IEDM'06, pp. 1–4. (December 2006)
4. Okhonin, S., Nagoga, M., Sallese, J.M., Fazan, P.: A SOI capacitor-less 1T-DRAM concept. In: IEEE International SOI Conference, pp. 153–154. (2001)
5. Hubert, A., Bawedin, M., Cristoloveanu, S., Ernst, T.: Dimensional effects and scalability of meta-stable Dip (MSD) memory effect for 1T-DRAM SOI MOSFETs. Solid-State Electron. **53**(12), 1280–1286 (2009)
6. Bawedin, M., Cristoloveanu, S., Hubert, A., Park, K.H., Martinez, F.: Floating-Body SOI memory: the scaling tournament. In: Semiconductor-On-Insulator Materials for Nanoelectronics Applications, pp. 393–421. Springer, Berlin Heidelberg (2011)
7. Eminente, S., Cristoloveanu, S., Clerc, R., Ohata, A., Ghibaudo, G.: Ultra-thin fully-depleted SOI MOSFETs: special charge properties and coupling effects. Solid-State Electron. **51**(2), 239–244 (2007)
8. Hurkx, G.A.M., De Graaff, H.C., Kloosterman, W.J., Knuvers, M.P.G.: A new analytical diode model including tunneling and avalanche breakdown. IEEE Trans. Electron Devices **39**(9), 2090–2098 (1992)
9. Colinge, J.-P.: In: Silicon-on-Insulator Technology: Materials to VLSI. Kluwer Academic Publishers, Boston (2004)
10. Avci, U.E., Ban, I., Kencke, D.L., Chang, P.L.D.: Floating body cell (FBC) memory for 16-nm technology with low variation on thin silicon and 10-nm BOX. In: Proceedings IEEE International SOI Conference, pp. 29–30. (October 2008)
11. Ertosun, M.G., Kapur, P., Saraswat, K.C.: A highly scalable capacitorless double gate quantum well single transistor DRAM: 1T-QW DRAM. IEEE Electron Device Lett. **29**(12), 1405–1407 (2008)
12. Cho, M.H., Shin, C., Liu, T.J.K.: Convex channel design for improved capacitorless DRAM retention time. In: IEEE International Conference on Simulation of Semiconductor Processes and Devices, SISPAD'09, pp. 1–4. (September, 2009)
13. Okhonin, S., Nagoga, M., Lee, C.-W., Colinge, J.P., Afzalian, A., Yan, R., Dehdashti Akhavan, N., Xiong, W., Sverdlov, V., Selberherr, S., Mazure, C.: Ultrascaled ZRAM cell. In: IEEE International SOI Conference, pp. 157-158 (2008)
14. Rodriguez, N., Gamiz, F., Cristoloveanu, S.: A-RAM memory cell: concept and operation. IEEE Electron Device Lett. **31**(9), 972–974 (2010)
15. Rodriguez, N., Cristoloveanu, S., Gamiz, F.: Capacitor-less A-RAM SOI memory: principles, scaling and expected performance. Solid-State Electron. **59**(1), 44–49 (2011)
16. Cristoloveanu, S.I., Rodriguez, N., Gamiz, F.: US Patent Application 13/264, 203 (2010)
17. Rodriguez, N., Cristoloveanu, S., Gamiz, F.: A-RAM: novel capacitor-less DRAM memory. In: IEEE International SOI Conference, pp. 1–2. (October 2009)
18. Jurczak, M., Skotnicki, T., Paoli, M., Tormen, B., Martins, J., Regolini, J.L., Monfray, S.: Silicon-on-nothing (SON)-an innovative process for advanced CMOS. IEEE Trans. Electron Devices **47**(11), 2179–2187 (2000)
19. Yoshida, E., Tanaka, T.: A capacitorless 1T-DRAM technology using gate-induced drain-leakage (GIDL) current for low-power and high-speed embedded memory. IEEE Trans. Electron Devices **53**(4), 692–697 (2006)
20. Rodriguez, N., Cristoloveanu, S., Gamiz, F.: Novel capacitorless 1T-DRAM cell for 22-nm node compatible with bulk and SOI substrates. IEEE Trans. Electron Devices **58**(8), 2371–2377 (2011)

21. ATHENA: User's manual. Silvac, Santa Clara, CA (2013)
22. Ranica, R., Villaret, A., Malinge, P., Mazoyer, P., Lenoble, D., Candelier, P., Skotnicki, T.: A one transistor cell on bulk substrate (1T-Bulk) for low-cost and high density eDRAM. In: IEEE Symposium on VLSI Technology, Digest of Technical Papers, pp. 128–129. (June 2004)
23. Rodriguez, N., Navarro, C., Gamiz, F., Andrieu, F., Faynot, O., Cristoloveanu, S.: Experimental demonstration of capacitorless A2RAM cells on silicon-on-insulator. IEEE Electron Device Lett. **33**(12), 1717–1719 (2012)
24. Kim, S., Choi, S.J., Moon, D.I., Choi, Y.K.: Carrier lifetime engineering for floating-body cell memory. IEEE Trans. Electron Devices **59**(2), 367–373 (2012)
25. Damaraju, S., George, V., Jahagirdar, S., Khondker, T., Milstrey, R., Sarkar, S., Subbiah, A.: A 22 nm IA multi-CPU and GPU system-on-chip. In: IEEE International Solid-State Circuits Conference (ISSCC) Digest of Technical Papers, pp. 56–57. (February 2012)
26. Hurkx, G.A.M., Klaassen, D.B.M., Knuvers, M.P.G.: A new recombination model for device simulation including tunneling. IEEE Trans. Electron Devices **39**(2), 331–338 (1992)
27. Song, K.W., Jeong, H., Lee, J.W., Hong, S.I., Tak, N.K., Kim, Y.T., Kim, C.: 55 nm capacitor-less 1T DRAM cell transistor with non-overlap structure. In: IEEE International Electron Devices Meeting (IEDM), pp. 1–4. (December 2008)

Part II
Beyond CMOS Materials, Devices and Their Diagnostic

Uniaxial Shear Strain as a Mechanism to Increase Spin Lifetime in Thin Film of a SOI-Based Silicon Spin FETs

Dmitri Osintsev, Viktor Sverdlov and Siegfried Selberherr

Abstract In this chapter we investigate spin relaxation in thin silicon films. We employ a **k**·**p** based approach to investigate surface roughness and phonon induced momentum and spin relaxation matrix elements. We show that the spin relaxation matrix elements strongly decrease with shear strain increased. In order to meet computational requirements with actual resources needed for relaxation time calculations, we demonstrate a way to find the subband wave function from the **k**·**p** model analytically. We consider the impact of the surface roughness and phonons on transport and spin characteristics in ultra-thin SOI MOSFET devices. We show that the regions in the momentum space responsible for strong spin relaxation can be efficiently removed by applying uniaxial shear strain. The spin lifetime in strained films can be improved by orders of magnitude.

1 Introduction

In order to achieve significant advantages in future microelectronic devices in comparison to present modern technology, operation principles will have to be enhanced or even modified. Spintronics is the rapidly developing technology promising to benefit from spin properties of electrons. Utilizing spin opens great opportunities to reduce device power consumption in future electronic circuits. A number of potential spintronic devices have already been proposed [1, 2]. Significant efforts are focused on developing models to study properties of future devices through simulation.

Silicon is the primary material for microelectronics. The long spin life time in silicon is a consequence of the weak intrinsic spin-orbit coupling in the conduction band and the spatial inversion symmetry of the lattice resulting in an absence of

D. Osintsev (✉) · V. Sverdlov · S. Selberherr
Institute for Microelectronics, TU Wien, Gußhausstraße 27–29/E360, A-1040 Wien, Austria
e-mail: osintsev@iue.tuwien.ac.at

A. Nazarov et al. (eds.), *Functional Nanomaterials and Devices for Electronics, Sensors and Energy Harvesting*, Engineering Materials,
DOI 10.1007/978-3-319-08804-4_7

Dresselhaus effective spin-orbit interaction [3, 4]. In addition, silicon is composed of nuclei with predominantly zero magnetic moment. A long spin transport distance of conduction electrons has already been demonstrated experimentally [5]. Spin propagation at such distances combined with a possibility of injecting spin at room temperature [6] or even elevated temperature [7] in silicon makes the fabrication of spin-based switching devices quite plausible in the upcoming future. However, the relatively large spin relaxation experimentally observed in electrically-gated lateral-channel silicon structures [3] might become an obstacle for realizing spin driven devices [4], and a deeper understanding of the fundamental spin relaxation mechanisms in silicon MOSFETs is urgently needed [8].

In this chapter we investigate the influence of the intrinsic spin-orbit interaction on the subband structure, subband wave functions, and spin relaxation matrix elements due to the surface roughness scattering in thin silicon films. We developed a semi-analytical approach allowing to analyze surface roughness and phonon induced spin and momentum relaxation in thin silicon films.

Following recent work [4], a **k**·**p** based method [9, 10] suitable to describe the electron subband structure in the presence of strain is generalized to include the spin degree of freedom. In contrast to [4], our effective 4×4 Hamiltonian considers only the relevant [001] oriented valleys with spin degree included, which produces the low-energy unprimed subband ladder. Within this model the unprimed subbands in the unstrained (001) film are degenerate, without spin-orbit effects included. An accurate inclusion of the spin-orbit interaction results in a large mixing between the spin-up and spin-down states, resulting in spin hot spots along the [100] and [010] axes characterized by strong spin relaxation. These hot spots should be contrasted with the spin hot spots appearing in the bulk system along the same directions at the edge of the Brillouin zone [4, 11]. The origin of the hot spots in thin films lies in the unprimed subband degeneracy which effectively projects the bulk spin hot spots from the edge of the Brillouin zone to the center of the 2D Brillouin zone.

Shear strain lifts the degeneracy between the unprimed subbands [10]. The energy splitting between the otherwise equivalent unprimed subbands removes the origin of the spin hot spots in a confined electron system in silicon, which substantially improves the spin lifetime in gated silicon systems.

2 Model

We numerically investigate the dependence of the matrix elements due to surface roughness induced spin relaxation in silicon films as a function of shear strain. For [001] oriented valleys in a (001) silicon film the Hamiltonian is written in the vicinity of the X point along the k_z-axis in the Brillouin zone. The basis is conveniently chosen as $[(X_1,\uparrow), (X_1,\downarrow), (X_{2'},\uparrow), (X_{2'},\downarrow)]$, where $\uparrow$ and $\downarrow$ indicate the spin projection at the quantization z-axis, X_1 and $X_{2'}$ are the basis function corresponding to the two valleys. The effective **k**·**p** Hamiltonian reads as

$$H = \begin{bmatrix} H_1 & H_3 \\ H_3^\dagger & H_2 \end{bmatrix}, \quad (1)$$

with H_1, H_2 and H_3 defined as

$$H_1 = \left[\frac{\hbar^2 k_z^2}{2m_l} - \frac{\hbar^2 k_0 k_z}{m_l} + \frac{\hbar^2 \left(k_x^2 + k_y^2\right)}{2m_t} + U(z) \right] I, \quad (2)$$

$$H_2 = \left[\frac{\hbar^2 k_z^2}{2m_l} + \frac{\hbar^2 k_0 k_z}{m_l} + \frac{\hbar^2 \left(k_x^2 + k_y^2\right)}{2m_t} + U(z) \right] I, \quad (3)$$

$$H_3 = \begin{bmatrix} D\varepsilon_{xy} - \frac{\hbar^2 k_x k_y}{M} & \left(k_y - k_x i\right)\Delta_{SO} \\ \left(-k_y - k_x i\right)\Delta_{SO} & D\varepsilon_{xy} - \frac{\hbar^2 k_x k_y}{M} \end{bmatrix}. \quad (4)$$

Here I is the identity 2 × 2 matrix, m_t and m_l are the transversal and the longitudinal silicon effective masses, $k_O = 0.15 \times 2\pi/a$ is the position of the valley minimum relative to the X point in unstrained silicon, ε_{xy} denotes the shear strain component, $M^{-1} \approx m_t^{-1} - m_0^{-1}$, and $D = 14$ eV is the shear strain deformation potential. The spin-orbit term $\tau_y \otimes (k_x\sigma_x - k_y\sigma_y)$ with

$$\Delta_{so} = 2\left| \sum \frac{\{X_1|p_j|n\}\{n|[\nabla V \times p]_j|X_{2'}\}}{E_n - E_X} \right|, \quad (5)$$

couples states with the opposite spin projections from the opposite valleys. σ_x and σ_y are the spin Pauli matrices and τ_y is the y-Pauli matrix in the valley degree of freedom space. In the Hamiltonian (Eq. 1) $U(z)$ is the confinement potential, and the value $\Delta_{SO} = 1.27$ meVnm computed by the empirical pseudopotential method (Fig. 1) is close to the one reported by Li and Dery [4].

In the presence of strain and confinement the four-fold degeneracy of the n-th unprimed subband is partly lifted by forming an $n+$ and $n-$ subladder (the valley splitting), however, the degeneracy of the eigenstates with the opposite spin projections $n\pm \Uparrow\rangle$ and $n\pm \Downarrow\rangle$ within each subladder is preserved.

The degenerate states are chosen to satisfy

$$\langle \Uparrow n \pm |f| n\pm \Downarrow \rangle = 0, \quad (6)$$

with the operator f defined as

$$f = \cos\theta\sigma_z + \sin\theta\left(\cos\varphi\sigma_x + \sin\varphi\sigma_y\right), \quad (7)$$

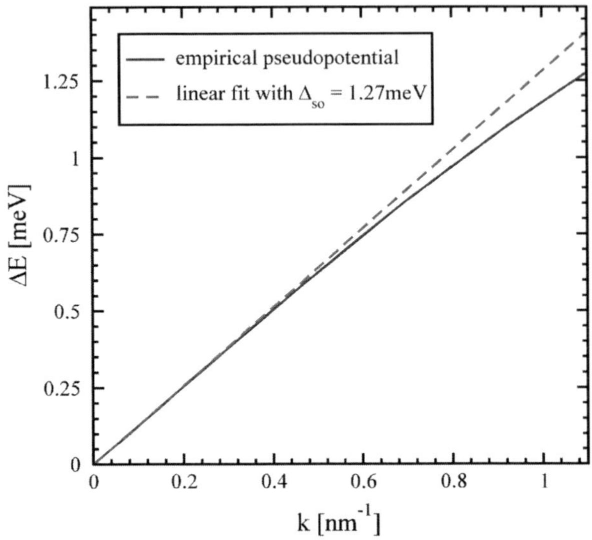

Fig. 1 Empirical pseudopotential calculations of the spin-orbit interaction strength by evaluating the gap opening at the X-point between X_1 and $X_{2'}$ for finite k_x

where θ is the polar and φ is the azimuth angle defining the orientation of the injected spin. In general, the expectation value of the operator f computed between the spin up and down states from different subladders is nonzero, when the effective magnetic field direction due to the spin-orbit interaction is different from the injected spin quantization axis

$$\bar{f} = \langle \Uparrow n \pm |f| n \mp \Downarrow \rangle \neq 0. \tag{8}$$

3 Valley Splitting

First we investigate the value of the energy splitting between the subbands with the same quantum number n but from different subsets $n+$ and $n-$ as a function of the conduction band offset at the interface, for different values of the quantum well thickness. In our calculations we assume the spin is injected along the z-direction and the components of the wave vector $\mathbf{k}$ are $k_x = 0.1\ \text{nm}^{-1}$ and $k_y = 0.1\ \text{nm}^{-1}$. Figure 2 shows the subband splitting for three values of the film width, namely 1.36, 3.3, and 6.5 nm. Figure 2 demonstrates a complicated behavior which strongly depends on the thickness value, in contrast to the valley splitting theory in SiGe/Si/SiGe quantum wells [12], which predicts that in the case of a symmetric square well without an electric field the valley splitting is simply inversely proportional to the conduction band offset ΔE_c at the interfaces. Figure 2 shows that for the quantum well of 1.36 nm width the splitting first increases but later saturates. For the quantum well of 3.3 nm width a significant reduction of the valley splitting around the conduction band offset value 1.5 eV is observed. A further increase of the conduction band offset leads to an increase of the subband splitting

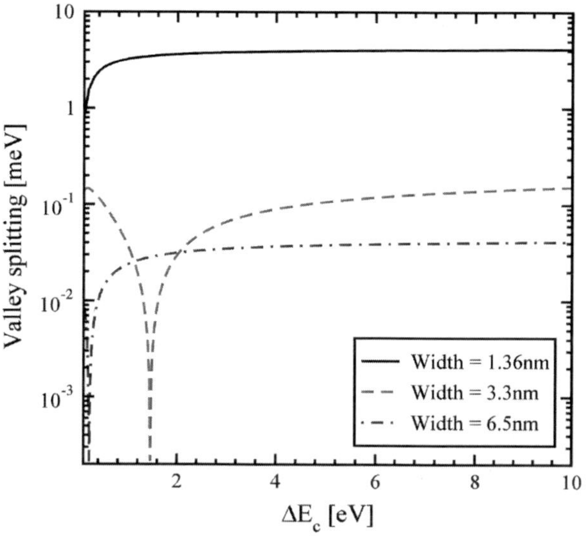

Fig. 2 Splitting between the lowest unprimed electron subbands as a function of the conduction band offset at the interface for different thicknesses for $\varepsilon_{xy} = 0$, $k_x = 0.1\ \mathrm{nm}^{-1}$ and $k_y = 0.1\ \mathrm{nm}^{-1}$

value. For the quantum well of 3.3 nm thickness the valley splitting saturates at about 0.17 meV.

For the quantum well of 6.5 nm width a significant reduction of the valley splitting is observed for a conduction band offset value 0.2 eV. The subband splitting saturates at a value 0.04 meV. Although for the values of the conduction band offset smaller than 4 eV the valley splitting depends on ΔE_c, for larger values of the conduction band offset it saturates.

The valley splitting dependence on strain as a function of the conduction band offset for the film of 3.3 nm thickness is shown in Fig. 3. Without shear strain the valley splitting is significantly reduced around the conduction band offset value of 1.5 eV. For the shear strain value of 0.25 and 0.5 % the sharp reduction of the conduction subbands splitting shifts to a smaller value of ΔE_c. However, the region of significant reduction is preserved even for the large shear strain value of 0.5 %. The value of the valley splitting at saturation for large shear strain is considerably enhanced as compared to the unstrained case.

The splitting of the lowest unprimed electron subbands as a function of the silicon film thickness for several values of the conduction band offset at the interfaces is shown in Fig. 4. The valley splitting oscillates with the film thickness increased. According to theory [9], we generalize the equation for the valley splitting in an infinite potential square well including the spin-orbit coupling as

$$\Delta E_n = \frac{2y_n^2 B}{k_0 t\sqrt{(1-y_n^2-\eta^2)(1-y_n^2)}}\left|\sin\left(\sqrt{\frac{1-y_n^2-\eta^2}{1-y_n^2}}k_0 t\right)\right|, \tag{9}$$

with y_n, η and B defined as

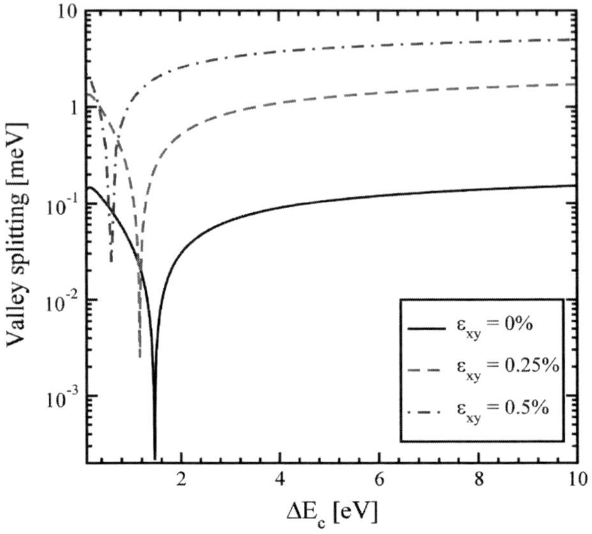

Fig. 3 Valley splitting as a function of the conduction band offset for the film thickness 3.3 nm for $k_x = 0.1\ \mathrm{nm}^{-1}$ and $k_y = 0.1\ \mathrm{nm}^{-1}$ for different shear strain values

$$y_n = \frac{\pi n}{k_0 t}, \tag{10}$$

$$\eta = \frac{m_l B}{\hbar^2 k_0^2}, \tag{11}$$

$$B = \sqrt{\Delta_{so}^2 \left(k_x^2 + k_y^2\right) + \left(D\varepsilon_{xy} - \frac{\hbar^2 k_x k_y}{M}\right)^2}. \tag{12}$$

Here t is the film thickness. As it was shown earlier the conduction band value of 4 eV provides a subband splitting value close to the saturated one. Because Eq. 9 is written for an infinite potential square well, a slight discrepancy is observed between the theoretical curve and the numerically curve calculated for the conduction band offset value 4 eV in Fig. 4. A large value of the conduction band offset demonstrates a better agreement between the theory and numerically obtained results.

Following Eq. 9, the results shown in Fig. 2 can be understood as a consequence of vanishing of the $\left|\sin\left(\sqrt{\frac{1-y_n^2-\eta^2}{1-y_n^2}}k_0 t\right)\right|$ term. Although the conduction band offset is not included explicitly in the equation for the valley splitting, it can be taken into account trough an effective film width of a finite potential well as:

$$t_{\mathrm{eff}} = t + \frac{2}{\alpha}, \tag{10}$$

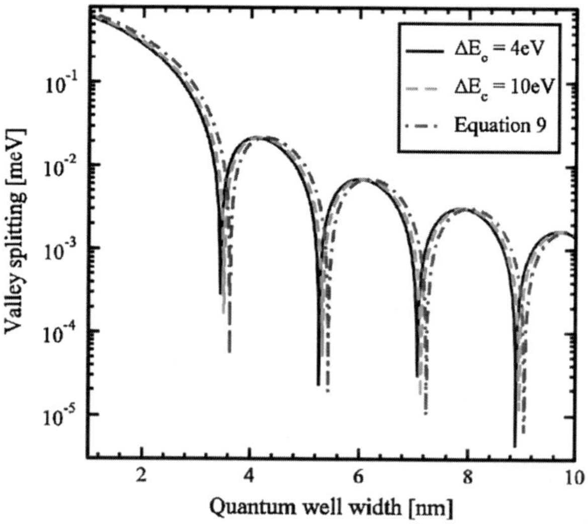

Fig. 4 Splitting of the lowest unprimed electron subbands as a function of the silicon film thickness for several values of the band offset at the interface, the shear strain value is 0.05 %, $k_x = 0.1\ \text{nm}^{-1}$, $k_y = 0.2\ \text{nm}^{-1}$

$$\alpha = \sqrt{\frac{2m(\Delta E_C - E)}{\hbar^2}}, \tag{11}$$

where E is the subband energy. Thus, increasing the potential barrier height leads to a decrease of the effective film thickness, which then results in the energy splitting dependence shown in Fig. 2.

The valley splitting reductions shown in Fig. 3 are also the result of the oscillating sine term in Eq. 9. The small increase of the shear strain leads to a decrease of the $\sqrt{\frac{1-y_n^2-\eta^2}{1-y_n^2}}$ term. This means that in order to obtain zeros of the sine term for larger shear strain values the effective quantum well thickness must be larger. A decrease in the conduction band offset leads precisely to such an increase of the effective thickness. Thus, the results shown in Fig. 3 are in very good agreement with theory.

Figure 5 shows the dependence of the energy splitting on shear strain for the in-plane wave vector $\boldsymbol{k}$ components are $k_x = 0.25\ \text{nm}^{-1}$ and $k_y = 0.25\ \text{nm}^{-1}$. The significant valley splitting reduction around the strain value 0.145 % appears to be independent of the quantum well width. According to Eq. 9, the valley splitting is also proportional to B, and the valley splitting reduction around the shear strain value 0.145 % is caused by vanishing of the $D\varepsilon_{xy} - \frac{\hbar^2 k_x k_y}{M}$ contribution. At this minimum the valley splitting is determined by the spin-orbit interaction term alone. The other valley splitting minima in Fig. 5 depend on the film thickness and are caused by vanishing values of the $\left|\sin\left(\sqrt{\frac{1-y_n^2-\eta^2}{1-y_n^2}}k_0 t\right)\right|$ term.

The valley splitting as a function of the quantum well width for different values of the effective electric field is shown in Fig. 6. Without electric field the valley splitting oscillates as shown in Fig. 4. With electric field the oscillations are not

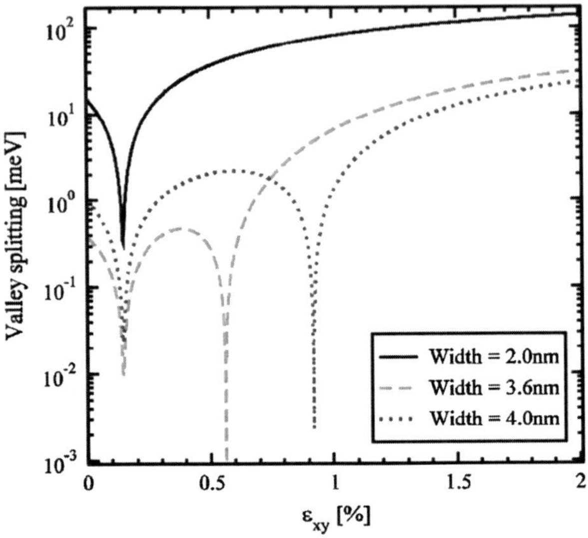

Fig. 5 Intervalley splitting as a function of shear strain for different values of the well width for $k_x = 0.25\ \mathrm{nm}^{-1}$ and $k_y = 0.25\ \mathrm{nm}^{-1}$

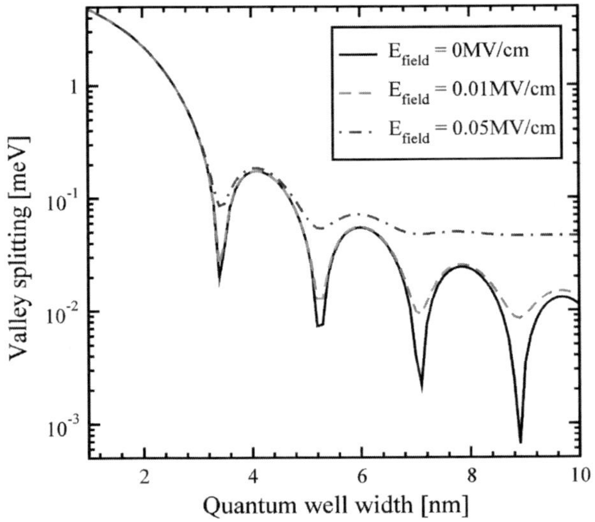

Fig. 6 Splitting of the lowest unprimed electron subbands as a function of the film thickness for different values of the effective electric field, the shear strain value is 0 %, the conduction band offset is 4 eV, $k_x = 0.1\ \mathrm{nm}^{-1}$, $k_y = 0.1\ \mathrm{nm}^{-1}$

observed in thicker films. This is due to the fact that in thick films the subband quantization is caused by the electric field. Indeed, for thin structures, when the quantization is still caused by the second barrier of the quantum well, the shape of the oscillations is similar to that in the absence of an electric field. According to [12], the condition for the independence of the valley splitting from the quantum well width is

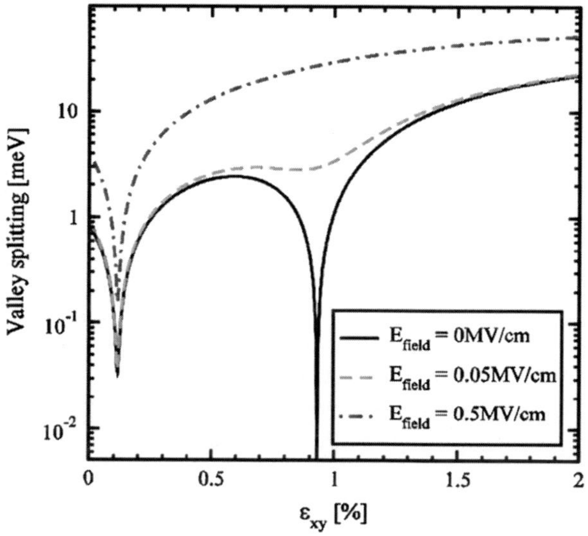

Fig. 7 Splitting of the lowest conduction subbands as a function of shear strain for different values of the electric field, the quantum well thickness is 4 nm, the conduction band offset is 4 eV, $k_x = 0.5\ \mathrm{nm}^{-1}$, $k_y = 0.1\ \mathrm{nm}^{-1}$

$$t^3 > \frac{2\pi^2\hbar^2}{m_l e E_{field}}. \tag{12}$$

For an electric field of 0.05 MV/cm the quantum well width must be larger than 6.9 nm in order to observe the valley splitting independent on the quantum well width. This value is in good agreement with the simulation results shown in Fig. 6.

Figure 7 shows the dependence of the valley splitting on strain. Without electric field the valley splitting reduces significantly around the strain values 0.116 and 0.931 % as shown in Fig. 7. With electric field applied the minimum around the strain value 0.931 % becomes smoother, however, for a strain value around 0.116 % the sharp reduction of the valley splitting is preserved. For large electric field the valley splitting reduction around the value 0.931 % vanishes completely.

For the strain value 0.116 % the sharp reduction of the valley splitting is still preserved at a minimum value only slightly affected by the electric field. As follows from Eq. 9, for $k_x = 0.5\ \mathrm{nm}^{-1}$, $k_y = 0.1\ \mathrm{nm}^{-1}$ the strain value 0.116 % causes the term $D\varepsilon_{xy} - \frac{\hbar^2 k_x k_y}{M}$ to vanish and minimizes the valley splitting, in good agreement with the first sharp valley splitting reduction in Fig. 7. Thus, the valley splitting at this strain value is solely determined by the spin-orbit interaction term. The second minimum in the valley splitting around the strain value 0.931 % in Fig. 7 is caused by vanishing of the $\left|\sin\left(\sqrt{\frac{1-y_n^2-\eta^2}{1-y_n^2}}k_0 t\right)\right|$ term. The effective electric field alters the confinement in the well and is therefore able to completely wash out the minimum in valley splitting due to the sine term. However, in agreement with Eq. 9, it can only slightly affect the first minimum due to the shear strain dependent contribution, in agreement with Fig. 7.

4 Scattering and Relaxation Matrix Elements Calculations

The surface roughness scattering matrix elements are proportional to the square of the product of the subband function derivatives at the interface [13]. The surface roughness at the two interfaces is assumed to be equal and statistically independent. It is described by a mean and a correlation length [13].

Figures 8 and 9 show the dependences on strain and electric field of the matrix elements for intrasubband and intersubband scattering due to surface roughness. The intrasubband scattering matrix elements have two decreasing features shown in Fig. 8. These features correlate with the valley splitting minima in Fig. 7. As in Fig. 7, for higher electric fields the second decreasing feature around the shear strain value of 0.9 % vanishes. For the electric field of 0.5 MV/cm the intrasubband matrix elements are sharply reduced only for the shear strain value of 0.116 %. At the same time, the intersubband matrix elements show a sharp increase around the shear strain value of 0.116 %. The electric field does not affect much the valley splitting provided by vanishing of the term $D\varepsilon_{xy} - \frac{\hbar^2 k_x k_y}{M}$, and the sharp increase in the inter-subband matrix elements is observed at higher fields as well.

Figures 10 and 11 show the dependence of the inter- and intrasubband spin relaxation matrix elements (normalized to the intravalley scattering at zero strain) on the angle between the incident and scattered wave vectors simultaneously with the valley splitting (calculated for the scattered wave vector value). As shown in Fig. 10, for small strain the sharp increases of the relaxation matrix elements are correlated with the minima in the valley splitting, which occur for the values of the angle determined by zeroes of the $D\varepsilon_{xy} - \frac{\hbar^2 k'_x k'_y}{M}$ term. This is the condition of the formation of the so called spin hot spots characterized by large spin mixing and

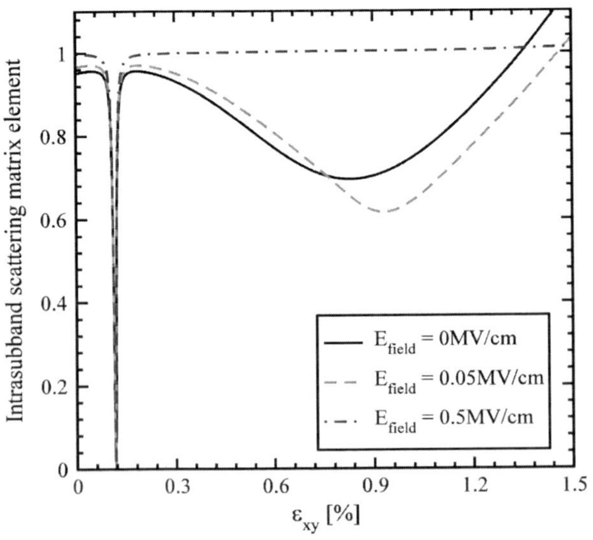

Fig. 8 Intravalley scattering matrix elements normalized by their values for zero strain as a function of shear strain for different electric field values

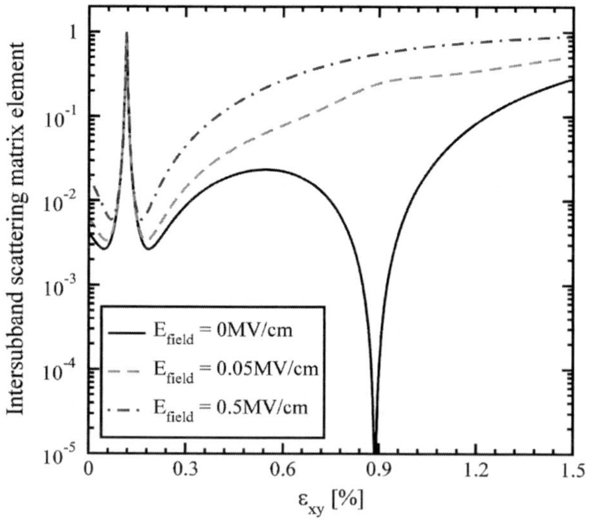

Fig. 9 Intervalley scattering matrix elements normalized to the value of the intravalley scattering at zero strain as a function of strain for different electric field values

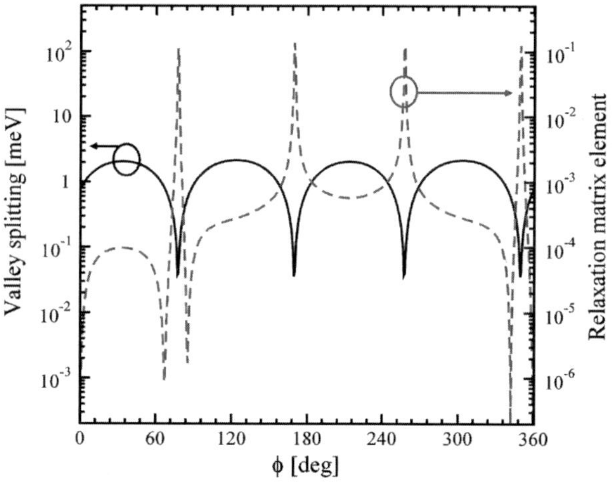

Fig. 10 Dependence of the normalized spin relaxation matrix elements and valley splitting on the angle between the incident and scattered waves for the quantum well thickness is 4 nm, the conduction band offset is 4 eV, $k_x = 0.5\ \text{nm}^{-1}$, $k_y = 0.1\ \text{nm}^{-1}$, $E_{field} = 0$ MV/cm, $\varepsilon_{xy} = 0.01$ %

relaxation. For higher shear strain values, however, the condition $D\varepsilon_{xy} - \frac{\hbar^2 k_x' k_y'}{M} = 0$ cannot be satisfied. In this case the valley splitting reduction shown in Fig. 11 is due to $\left|\sin\left(\sqrt{\frac{1-y_n^2-\eta^2}{1-y_n^2}} k_0 t\right)\right| = 0$, in contrast to Fig. 10. Correspondingly, these

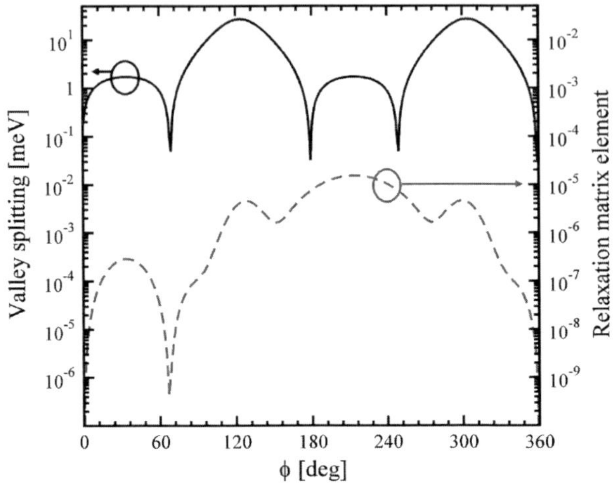

Fig. 11 Dependence of the normalized spin relaxation matrix elements and valley splitting on the angle between the incident and scattered waves for the quantum well thickness is 4 nm, the conduction band offset is 4 eV, $k_x = 0.5\ \mathrm{nm}^{-1}$, $k_y = 0.1\ \mathrm{nm}^{-1}$, $E_{field} = 0$ MV/cm, $\varepsilon_{xy} = 0.92$ %

Fig. 12 Spin relaxation matrix elements normalized to intravalley scattering at zero strain dependence on shear strain for several values of the electric field, $k_x = 0.5\ \mathrm{nm}^{-1}$, $k_y = 0.1\ \mathrm{nm}^{-1}$

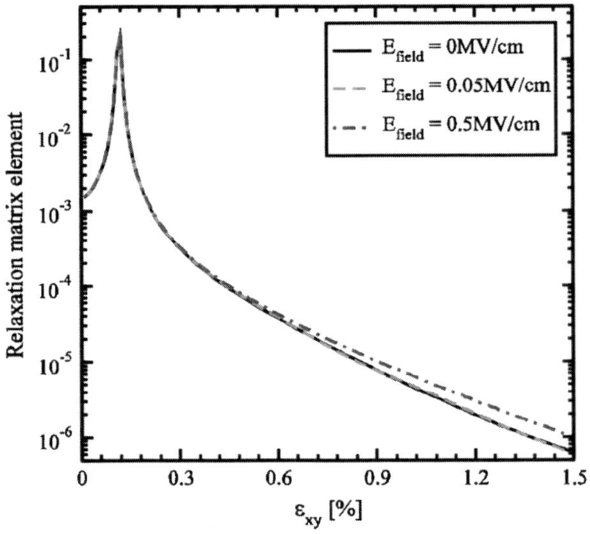

valley splitting minima do not result in sharp increases in the spin relaxation matrix elements on the angle between the incident and scattered waves.

The dependence of the intersubband spin relaxation matrix elements on shear strain for several values of the electric field is shown in Fig. 12. The spin relaxation increases up to the strain value 0.116 %, the point determined by the spin hot spot condition. Applying strain larger than 0.116 % suppresses spin relaxation significantly, for all values of the electric field. In contrast to the scattering matrix elements (Figs. 8 and 9), the relaxation matrix elements demonstrate a sharp feature only for the shear strain value of 0.116 %. Large electric fields lead to an increase of the spin relaxation matrix elements due to the additional field-induced confinement.

5 Spin Relaxation

We are considering three mechanisms which contribute to the spin and momentum relaxation: surface roughness (SR), intra- and intervalley (for spin relaxation) scattering by acoustic phonons.

The spin and momentum relaxation times are calculated by thermal averaging [4, 8, 13] as

$$\frac{1}{\tau} = \frac{\int \frac{1}{\tau(\mathbf{K}_1)} f(\varepsilon)(1 - f(\varepsilon)) \mathrm{d}\mathbf{K}_1}{\int f(\varepsilon) \mathrm{d}\mathbf{K}_1}, \tag{13}$$

$$\int \mathrm{d}\mathbf{K}_1 = \int_0^{2\pi} \int_0^{\infty} \frac{|\mathbf{K}_1|}{\left|\frac{\partial \varepsilon(\mathbf{K}_1)}{\partial \mathbf{K}_1}\right|} \mathrm{d}\varphi \mathrm{d}\varepsilon. \tag{14}$$

The surface roughness momentum (spin) relaxation rate is calculated in the following way

$$\begin{aligned} \frac{1}{\tau_{SR}(\mathbf{K}_1)} &= \frac{2(4)\pi}{\hbar(2\pi)^2} \sum_{i,j} \int_0^{2\pi} \pi \Delta^2 L^2 \frac{1}{\epsilon_{ij}^2(\mathbf{K}_2 - \mathbf{K}_1)} \frac{\hbar^4}{4m_l^2} \frac{|\mathbf{K}_2|}{\left|\frac{\partial \varepsilon(\mathbf{K}_2)}{\partial \mathbf{K}_2}\right|} \\ &\times \left[\left(\frac{d\Psi_{i\mathbf{K}_1\sigma}}{dz} \right)^* \frac{d\Psi_{j\mathbf{K}_2-\sigma}}{dz} \right]^2_{z=\pm\frac{t}{2}} \exp\left(\frac{-(\mathbf{K}_2 - \mathbf{K}_1)^2 L^2}{4} \right) \mathrm{d}\varphi, \end{aligned} \tag{15}$$

where ε is the electron energy, $K_{1,2}$ are the in-plane wave vectors before and after scattering, φ is the angle between K_1 and K_2, ε_{ij} is the dielectric permittivity, L is the autocorrelation length, Δ is the mean square value of the surface roughness fluctuations, $\Psi_{i\mathbf{K}_1}$ and $\Psi_{j\mathbf{K}_2}$ are the wave functions, and $f(\varepsilon)$ is the Fermi function, and $\sigma = +1$ is the spin projection to the [001] axis.

The momentum relaxation time is evaluated in the standard way [13, 14]. The spin relaxation rate due to the transversal acoustic phonons is calculated as

$$\frac{1}{\tau_{TA}(\mathbf{K}_1)}=\frac{4\pi k_B T}{\hbar\rho v_{TA}^2}\sum\int_0^{2\pi}\frac{d\varphi}{2\pi}\frac{|\mathbf{K}_2|}{\left|\frac{\partial\varepsilon(\mathbf{K}_2)}{\partial\mathbf{K}_2}\right|}\left[1-\frac{\frac{\partial\varepsilon(\mathbf{K}_2)}{\partial\mathbf{K}_2}f(\varepsilon(\mathbf{K}_2))}{\frac{\partial\varepsilon(\mathbf{K}_1)}{\partial\mathbf{K}_1}f(\varepsilon(\mathbf{K}_1))}\right]\frac{1}{2}$$
$$\times\int_0^t\int_0^t\exp\left(-\sqrt{q_x^2+q_y^2}|z-z'|\right)\left[\Psi^{\dagger}_{\mathbf{K}_2-\sigma}(z)M\Psi_{\mathbf{K}_1\sigma}(z)\right]^*$$
$$\times\left[\Psi^{\dagger}_{\mathbf{K}_2-\sigma}(z')M\Psi_{\mathbf{K}_1\sigma}(z')\right]\left[\sqrt{q_x^2+q_y^2}-\frac{8q_x^2q_y^2-\left(q_x^2+q_y^2\right)^2}{q_x^2+q_y^2}|z-z'|\right]dzdz',\tag{16}$$

where k_B is the Boltzmann constant, T is the temperature, $\rho = 2{,}329$ kg/m^3 is the silicon density, $\upsilon_{TA} = 5{,}300$ m/s is the transversal phonons velocity, $(q_x, q_y) = K_1 - K_2$ and M is the 4×4 matrix written in the basis for the spin relaxation rate.

$$M=\begin{bmatrix}0 & 0 & \frac{D}{2} & 0\\ 0 & 0 & 0 & \frac{D}{2}\\ \frac{D}{2} & 0 & 0 & 0\\ 0 & \frac{D}{2} & 0 & 0\end{bmatrix}.\tag{17}$$

Here $D = 14$ eV is the shear deformation potential.

The intravalley spin relaxation rate due to the longitudinal acoustic phonons is calculated as

$$\frac{1}{\tau_{LA}(\mathbf{K}_1)}=\frac{4\pi k_B T}{\hbar\rho v_{LA}^2}\sum\int_0^{2\pi}\frac{d\varphi}{2\pi}\frac{|\mathbf{K}_2|}{\left|\frac{\partial\varepsilon(\mathbf{K}_2)}{\partial\mathbf{K}_2}\right|}\left[1-\frac{\frac{\partial\varepsilon(\mathbf{K}_2)}{\partial\mathbf{K}_2}f(\varepsilon(\mathbf{K}_2))}{\frac{\partial\varepsilon(\mathbf{K}_1)}{\partial\mathbf{K}_1}f(\varepsilon(\mathbf{K}_1))}\right]\frac{1}{2}$$
$$\times\int_0^t\int_0^t\exp\left(-\sqrt{q_x^2+q_y^2}|z-z'|\right)\left[\Psi^{\dagger}_{\mathbf{K}_2-\sigma}(z)M\Psi_{\mathbf{K}_1\sigma}(z)\right]^*$$
$$\times\left[\Psi^{\dagger}_{\mathbf{K}_2-\sigma}(z')M\Psi_{\mathbf{K}_1\sigma}(z')\right]\frac{4q_x^2q_y^2}{\left(\sqrt{q_x^2+q_y^2}\right)^3}\left[\sqrt{q_x^2+q_y^2}|z-z'|+1\right]dzdz',\tag{18}$$

Here $\upsilon_{LA} = 8{,}700$ m/s is the speed of the longitudinal phonons and the matrix M is defined with Eq. 17.

The intervalley spin relaxation rate contains the Elliot and Yafet contributions [8], which are calculated in the following way

$$\frac{1}{\tau_{LA}(\mathbf{K}_1)} = \frac{4\pi k_B T}{\hbar \rho v_{LA}^2} \sum \int_0^{2\pi} \frac{d\varphi}{2\pi} \frac{|\mathbf{K}_2|}{\left|\frac{\partial\varepsilon(\mathbf{K}_2)}{\partial \mathbf{K}_2}\right|} \left[1 - \frac{\frac{\partial\varepsilon(\mathbf{K}_2)}{\partial \mathbf{K}_2} f(\varepsilon(\mathbf{K}_2))}{\frac{\partial\varepsilon(\mathbf{K}_1)}{\partial \mathbf{K}_1} f(\varepsilon(\mathbf{K}_1))}\right]$$
$$\times \frac{1}{2} \int_0^t \left[\Psi^\dagger_{\mathbf{K}_2-\sigma}(z) M' \Psi_{\mathbf{K}_1\sigma}(z)\right]^* \left[\Psi^\dagger_{\mathbf{K}_2-\sigma}(z) M' \Psi_{\mathbf{K}_1\sigma}(z)\right] dz. \quad (19)$$

Here the matrix M' is written as

$$M' = \begin{bmatrix} M_{ZZ} & M_{SO} \\ M_{SO}^\dagger & M_{ZZ} \end{bmatrix}. \quad (20)$$

$$M_{ZZ} = \begin{bmatrix} D_{ZZ} & 0 \\ 0 & D_{ZZ} \end{bmatrix}. \quad (21)$$

$$M_{SO} = \begin{bmatrix} 0 & D_{SO}(r_y - ir_x) \\ D_{SO}(-r_y - ir_x) & 0 \end{bmatrix}. \quad (22)$$

$(r_x, r_y) = K_1 + K_2$, $D_{zz} = 12$ eV, and $D_{SO} = 15$ meV/k_0 with $k_O = 0.15 \times 2\pi/a$ defined as the position of the valley minimum relative to the X-point in unstrained silicon [8].

6 Wave Function Evaluation

Because of the spin hotspots determining the strong dependence of the spin relaxation scattering matrix elements on the relative angle between the incoming and scattered waves the assumption of the independence of the subband wave functions on the in-plane momentum frequently employed to estimate momentum relaxation cannot be used to evaluate the spin lifetime. Indeed, because spin-orbit effects are linear in in-plane momentum, the calculation of the surface roughness scattering matrix elements at the center of the 2D Brillouin zone usually performed for mobility calculations would result in the complete loss of all the effects due to spin-orbit interaction. Therefore, in order to accurately compute the spin lifetime numerically, one needs to know the subband wave functions as a function of the in-plane wave vector. Numerical evaluation of the wave functions with subsequent integration makes the task prohibitively expensive. To simplify the problem, we obtain the wave functions in a semi-analytical manner. For this purpose we rotate the Hamiltonian (Eq. 1) by means of the following unitary transformation. The four basis functions $X_{1\uparrow}$, $X_{1\downarrow}$, $X_{2\uparrow}'$, $X_{2\downarrow}'$, for the two [001] valleys with spin up, spin down are transformed by Eqs. 23–30 with $\tan(\Theta) = \frac{\Delta_{SO}\sqrt{k_x^2+k_y^2}}{D\varepsilon_{xy} - \frac{\hbar^2 k_x k_y}{M}}$. The transformation effectively decouples the spins with opposite direction in different valleys.

$$\Psi_1 = \frac{1}{2}\left[\left(X_{1\uparrow} + X'_{2\uparrow}\right) + (X_{1\downarrow} + X'_{2\downarrow})\frac{k_x - ik_y}{\sqrt{k_x^2 + k_y^2}}\right], \tag{23}$$

$$\Psi_2 = \frac{1}{2}\left[\left(X_{1\uparrow} + X'_{2\uparrow}\right) - (X_{1\downarrow} + X'_{2\downarrow})\frac{k_x - ik_y}{\sqrt{k_x^2 + k_y^2}}\right], \tag{24}$$

$$\Psi_3 = \frac{1}{2}\left[\left(X_{1\uparrow} - X'_{2\uparrow}\right) + (X_{1\downarrow} - X'_{2\downarrow})\frac{k_x - ik_y}{\sqrt{k_x^2 + k_y^2}}\right], \tag{25}$$

$$\Psi_4 = \frac{1}{2}\left[\left(X_{1\uparrow} - X'_{2\uparrow}\right) - (X_{1\downarrow} - X'_{2\downarrow})\frac{k_x - ik_y}{\sqrt{k_x^2 + k_y^2}}\right], \tag{26}$$

$$\mathrm{X}_1 = \Psi_1 \cos\left(\frac{\Theta}{2}\right) - i\Psi_3 \sin\left(\frac{\Theta}{2}\right), \tag{27}$$

$$\mathrm{X}_2 = \Psi_2 \cos\left(\frac{\Theta}{2}\right) + i\Psi_4 \sin\left(\frac{\Theta}{2}\right), \tag{28}$$

$$\mathrm{X}_3 = \Psi_3 \cos\left(\frac{\Theta}{2}\right) - i\Psi_1 \sin\left(\frac{\Theta}{2}\right), \tag{29}$$

$$\mathrm{X}_4 = \Psi_4 \cos\left(\frac{\Theta}{2}\right) + i\Psi_2 \sin\left(\frac{\Theta}{2}\right). \tag{30}$$

The Hamiltonian (Eq. 1) can now be cast into a form in which spins with opposite orientation in different valleys are independent

$$H = \begin{bmatrix} H_1 & H_3 \\ H_3 & H_2 \end{bmatrix}, \tag{31}$$

H_1, H_2 and H_3 are written as

$$H_1 = \left[\frac{\hbar^2 k_z^2}{2m_l} + \frac{\hbar^2\left(k_x^2 + k_y^2\right)}{2m_t} - \delta + U(z)\right] I, \tag{32}$$

$$H_2 = \left[\frac{\hbar^2 k_z^2}{2m_l} + \frac{\hbar^2 \left(k_x^2 + k_y^2\right)}{2m_t} + \delta + U(z) \right] I, \tag{33}$$

$$H_3 = \begin{bmatrix} \frac{\hbar^2 k_0 k_z}{m_l} & 0 \\ 0 & \frac{\hbar^2 k_0 k_z}{m_l} \end{bmatrix}, \tag{34}$$

with $\delta = \sqrt{\left(D\varepsilon_{xy} - \frac{\hbar^2 k_x k_y}{M}\right)^2 + \Delta_{SO}^2 \left(k_x^2 + k_y^2\right)}$.

Following [10] we find the wave functions analytically in the same manner as for the two-band $\mathbf{k} \cdot \mathbf{p}$ Hamiltonian written in the vicinity of the X point of the Brillouin zone for silicon films under uniaxial strain.

Figure 13 demonstrates an excellent agreement between the semi-analytical and the numerically obtained results for a silicon film of 4 nm thickness for the values $k_x = 0.25\ \text{nm}^{-1}$ and $k_y = 0.25\ \text{nm}^{-1}$. For the numerical calculations a barrier of 10 eV height has been assumed.

7 Results and Discussion

Figure 14 shows the dependence of the momentum relaxation time on temperature. The contributions from the surface roughness (SR) and acoustic phonons (PH) are shown. For the film thicknesses 2.1 and 1.36 nm the contribution from the surface roughness is dominant at low temperatures. However, for a temperature around 280 K the contributions from the surface roughness and from the acoustic phonons for the film of thickness 2.1 nm are equal. Any further increase of temperature leads to higher values of the momentum relaxation time caused by acoustic phonons. Figure 14 shows that the dominant relaxation mechanism strongly depends on film thickness. The phonons limited momentum relaxation is characterized by much weaker thickness dependence and does not change significantly while the thickness decreases from 2.1 to 1.36 nm. The surface roughness limited momentum relaxation decreases by more than an order of magnitude because of the expected t^6 dependence [13, 15]. Thus, for the thickness 1.36 nm the surface roughness induced spin relaxation is the dominant mechanism for the whole range of considered temperatures.

Figure 15 shows the dependence of the different mechanisms of the momentum relaxation together with the total momentum relaxation time on shear strain. The improvement of the momentum relaxation time due to the shear strain is around 82 % for the film thickness of 2.1 nm and around 120 % for the film thickness 1.36 nm. The acoustic phonons limited momentum relaxation improves around 45 % for 2.1 nm and around 92 % for 1.36 nm. The surface roughness limited momentum relaxation time increases around 110 % for 2.1 nm and around 120 % for 1.36 nm. Because the SR mechanism is the dominant for the film thickness

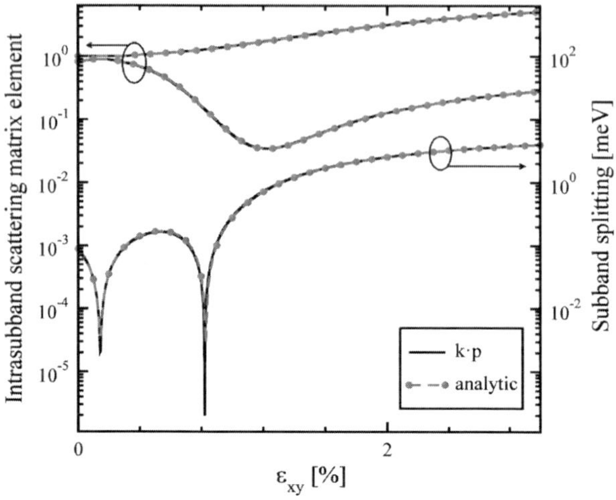

Fig. 13 Intrasubband scattering matrix elements normalized to their values at zero strain for the film thickness 1.36 nm for $k_x = 0.25\ \text{nm}^{-1}$ and $k_y = 0.25\ \text{nm}^{-1}$, and subband splitting as a function of shear strain

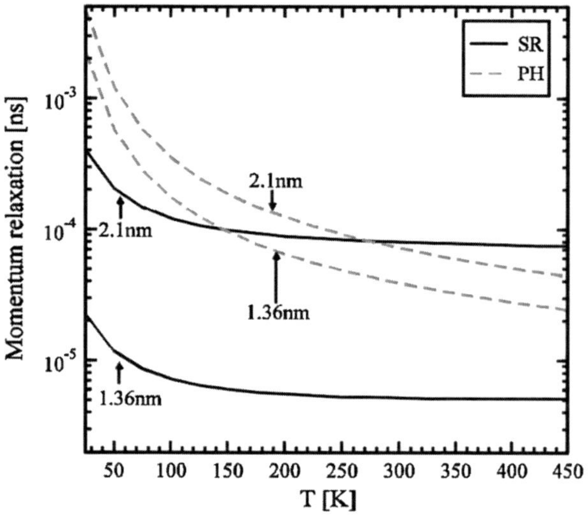

Fig. 14 Dependence of the momentum relaxation time induced by surface roughness (*SR*) and acoustic phonons (*PH*) on temperature for two different thicknesses, $\varepsilon_{xy} = 0$, and electron concentration $1.29 \times 10^{12}\ \text{cm}^{-2}$

1.36 nm, the increase of the total momentum relaxation time is higher for 1.36 nm than for 2.1 nm. We point out that the increase of the momentum relaxation time is due to the corresponding scattering matrix elements dependences' on strain. Combined with the strain induced transport effective mass decrease it should result in an even better mobility improvement supporting the use of uniaxial tensile strain as the mobility booster in fully depleted ultra-thin SOI FETs.

Figure 16 demonstrates the contribution of the inter- and intrasubband processes to the acoustic phonon and SR limited momentum relaxation. The domination of the intrasubband relaxation processes for both mechanisms of the momentum relaxation is shown, in agreement with the selection rule that the

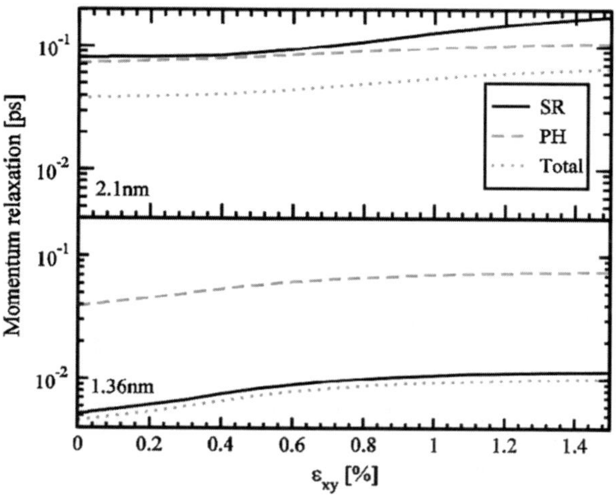

Fig. 15 Dependence of the momentum relaxation time induced by surface roughness (SR) and acoustic phonons (PH) on shear strain for 1.36 nm and 2.1 nm film thickness, for $T = 300$ K, and electron concentration 1.29×10^{12} cm^{-2}

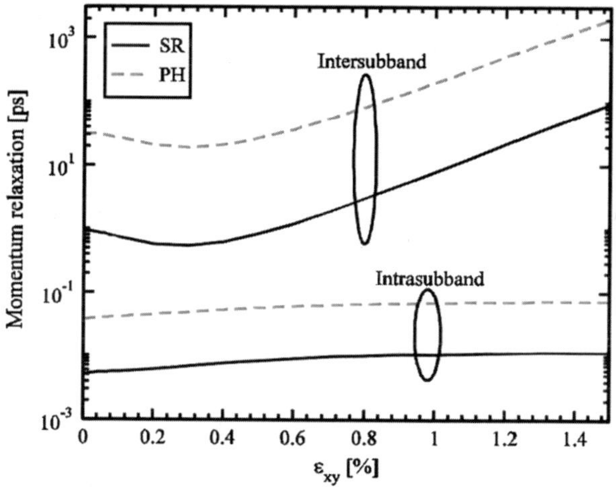

Fig. 16 Dependence of the intersubband and intrasubband components of the momentum relaxation time induced by surface roughness (*SR*) and acoustic phonons (*PH*) on shear strain for the film thickness 1.36 nm, $T = 300$ K, and electron concentration 1.29×10^{12} cm^{-2}

elastic processes result in strong intrasubband relaxation. The dominance of the SR mechanism for the film thickness 1.36 nm shown in Figs. 14 and 15 is the consequence of the high intrasubband relaxation rate.

The dependence of the spin lifetime on temperature for phonon scattering, and SR scattering for different carrier concentrations is shown in Fig. 17. The spin

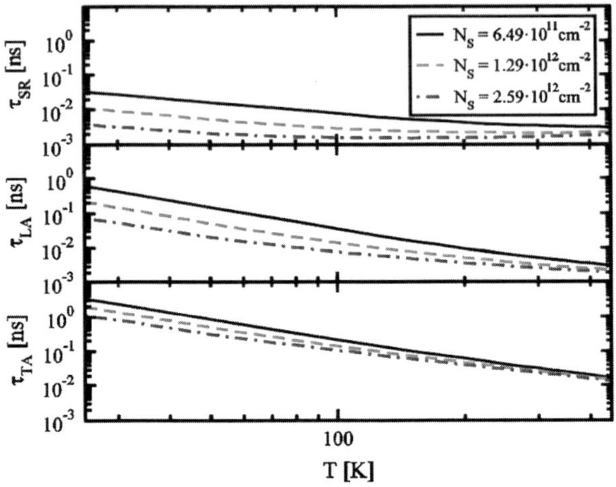

Fig. 17 Dependence of the surface roughness (*SR*), the longitudinal phonons' (*LA*), and the transversal phonons' (*TA*) contribution to the spin lifetime on temperature for different values of the electron concentration, for $\varepsilon_{xy} = 0$, and film thickness 2.1 nm

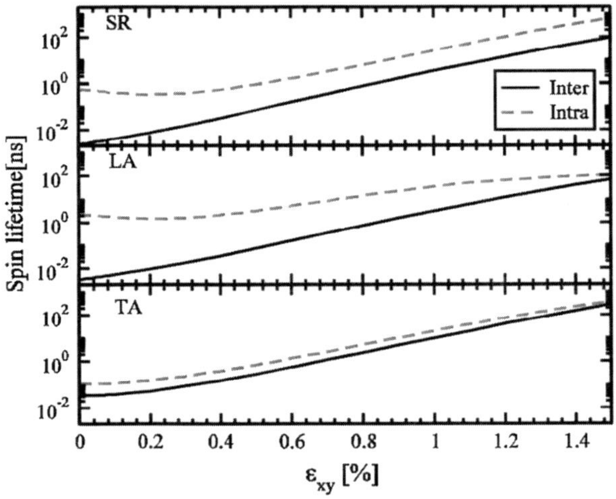

Fig. 18 Dependence of the intersubband and the intrasubband component of the spin lifetime for different spin relaxation mechanisms (surface roughness (*SR*), longitudinal phonons (*LA*), and transversal phonons (*TA*)) on shear strain for the film thickness 1.36 nm, $T = 300$ K, and electron concentration 1.29×10^{12} cm^{-2}

relaxation is more efficient for higher carrier concentrations for all three considered mechanisms. While the temperature increases, the difference between the spin lifetimes for different values of the electron concentration becomes less pronounced. Figure 17 shows that the SR mechanism dominates for all concentration

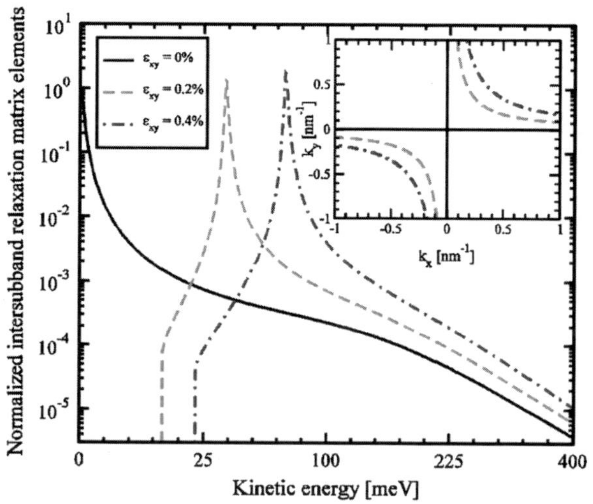

Fig. 19 Normalized intersubband relaxation matrix elements as a function of the conduction electrons kinetic energy in [110] direction. The *inset* shows the positions of the hot spots for different values of shear strain

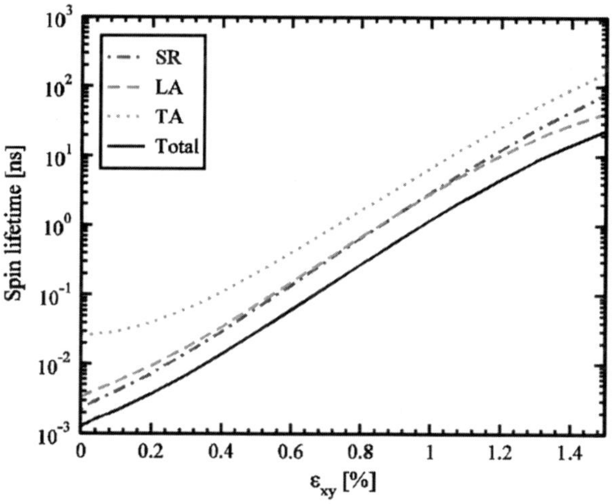

Fig. 20 Dependence of the spin lifetime on shear strain for the film thickness 2.1 nm, $T = 300$ K, and electron concentration 1.29×10^{12} cm^{-2}

values. Spin relaxation due to TA phonons is weakest among the three considered mechanisms.

Figure 18 demonstrates that the main contribution to spin relaxation comes from the intersubband processes due to the presence of the spin hot spots characterized by the sharp peaks of the intersubband spin relaxation matrix elements. Their position is shown in Fig. 19. For higher shear strain values the hot spots are pushed to higher energies away from the subband minima (inset in Fig. 19). This results in a strong increase of the spin lifetime with shear strain for SR and the phonon mechanisms as shown in Fig. 20.

8 Conclusion

We have included the spin-orbit interaction effects into the effective low-energy **k**·**p** Hamiltonian to investigate the valley splitting, scattering, and spin relaxation induced by the surface roughness and phonons in a thin film SOI MOSFET in a wide range of parameters. We have demonstrated that the valley splitting minima due to zero values of the sine term can be removed by the electric field, but the minimum due to the vanishing $D\varepsilon_{xy} - \frac{\hbar^2 k_x k_y}{M}$ term is preserved even for large electric fields. We have found that, due to the inter-subband splitting increase, the matrix elements for spin relaxation decrease rapidly with shear strain. To evaluate the wave function dependence on the in-plane momentum and spin relaxation time the **k**·**p** Hamiltonian is solved analytically. We have shown that by applying shear strain the momentum relaxation time can be improved by almost a factor of two for ultra-thin films. We have demonstrated a strong, several orders of magnitude, increase of spin lifetime in strained silicon films. Thus shear strain used to boost mobility can also be used to increase spin lifetime.

Acknowledgments This work is supported by the European Research Council through the grant #247056 MOSILSPIN. The computational results have been achieved using the Vienna Scientific Cluster (VSC).

References

1. Sugahara, S., Nitta, J.: Spin transistor electronics: an overview and outlook. Proc. IEEE **98**(12), 2124–2154 (2010)
2. Datta, S., Das, B.: Electronic analog of the electro-optic modulator. Appl. Phys. Lett. **56**, 665 (1990)
3. Li, J., Appelbaum, I.: Modeling spin transport in electrostatically-gated lateral-channel silicon devices: role of interfacial spin relaxation. Phys. Rev. B **84**, 165318 (2011)
4. Li, P., Dery, H.: Spin-orbit symmetries of conduction electrons in silicon. Phys. Rev. Lett. **107**, 107203 (2011)
5. Huang, B., Monsma, D.J., Appelbaum, I.: Coherent spin transport through a 350 Micron thick silicon wafer. Phys. Rev. Lett. **99**, 177209 (2007)
6. Dash, S.P., Sharma, S., Patel, R.S., de Jong, M.P., Jansen, R.: Electrical creation of spin polarization in silicon at room temperature. Nature **462**, 491–494 (2009)
7. Li, C.H., Van't Erve, O.M.J., Jonker, B.T.: Electrical injection and detection of spin accumulation in silicon at 500 K with magnetic metal/silicon dioxide contacts. Nat. Commun. **2**, 245 (2011)
8. Song, Y., Dery, H.: Analysis of phonon-induced spin relaxation processes in silicon. Phys. Rev. B **86**, 085201 (2012)
9. Bir, G.L., Pikus, G.E.: Symmetry and strain-induced effects in semiconductors. Wiley, New York (1974)
10. Sverdlov, V.: Strain-induced effects in advanced MOSFETs. Springer, Wien (2011)
11. Cheng, J.L., Wu, M.W., Fabian, J.: Theory of the spin relaxation of conduction electrons in silicon. Phys. Rev. Lett. **104**, 016601 (2010)

12. Friesen, M., Chutia, S., Tahan, C., Coppersmith, S.N.: Valley splitting theory of SiGe/Si/SiGe quantum wells. Phys. Rev. B **75**, 115318 (2007)
13. Fischetti, M.V., et al.: Six-band k·p calculation of hole mobility in silicon inversion layers: dependence on surface orientation, strain, and silicon thickness. J. Appl. Phys. **94**, 1079 (2003)
14. Fischetti, M.V., Laux, S.E.: Monte Carlo study of electron transport in silicon inversion layers. Phys. Rev. B **48**(4), 2244–2274 (1993)
15. Jin, S., Fischetti, M.V., Tang, T.: Modeling of electron mobility in gated silicon nanowires at room temperature: surface roughness scattering, dielectric scattering, and band nonparabolicity. J. Appl. Phys. **102**, 083715 (2007)

Quantum Noise in Nanotransistors

Alexander Orlikovsky, Vladimir Vyurkov, Igor Semenikhin and Vladimir Borzdov

Abstract The Landauer-Buttiker approach for simulation of nanometer-scale field-effect transistors is developed to incorporate a calculation of noise at arbitrary temperature and applied voltage. The method allows for both shot noise and Johnson–Nyquist (thermal) noise. The impact of random impurities in the channel and doping profile in source/drain contacts on current and noise was cleared up. The quantum noise is compared with that originating from a classical description of motion along the channel. Unlike the classical renormalized shot noise the quantum noise is smaller than the current. This discrepancy could be a crucial evidence for quantum transport in a transistor channel. From practical point of view that provides a possibility to reduce a shot noise in quantum nanotransistors.

1 Introduction

The future progress of silicon ULSI technology requires a conversion to extremely thin channels. Just those structures allow avoiding channel doping which substantially impairs a bulk transistor performance. An undoped channel wrapped by a gate (Fig. 1) results in a ballistic regime with high conductance in ON state and low leakage current in OFF state. Therefore, such transistors will exhibit higher frequency and lower power consumption. Nowadays there are two technologies

---

A. Orlikovsky · V. Vyurkov (✉) · I. Semenikhin
Institute of Physics and Technology RAS, Nakhimovsky Avenue 34, Moscow 117218, Russia
e-mail: vvyurkov@gmail.com

A. Orlikovsky · V. Vyurkov
Moscow University of Physics and Technology (State University), Doldoprudny, Moscow District 141700, Russia

V. Borzdov
Belarusian State University, 4, Nezavisimosti Avenue, 220030 Minsk, Belarus

A. Nazarov et al. (eds.), *Functional Nanomaterials and Devices for Electronics, Sensors and Energy Harvesting*, Engineering Materials,
DOI 10.1007/978-3-319-08804-4_8

close to commercialization: an extremely thin body (3–8 nm) fully depleted silicon on insulator structure (ETB FD SOI technology of IBM Corp. and STMicroelectronics) [1] and a thin channel 'raised' from a bulk substrate (3-D Tri-Gate technology of Intel Corp.) [2].

Current noise is one of the major characteristics of a transistor required for its practical application in analog as well as in digit circuits. Along with the shot noise and the Johnson–Nyquist noise (thermal noise) well-known far before, an especial quantum excess noise was revealed in nanostructures at zero temperature $T = 0$ [3, 4]. Lately, a general expression to calculate the current noise for an arbitrary temperature and voltage was derived in [5]. That expression could be used in the simulation of a field-effect transistor based on the Landauer-Buttiker approach [6, 7].

In classical physics there is a universal dependence of the Johnson–Nyquist (thermal) noise on a conductor resistance. In the same way, the shot noise is directly bound to the current. Quantum physics provides with an opportunity to make a noise weaker. Moreover, in our opinion, only the discrepancy between current and renormalized noise could be a crucial evidence of a quantum transport in a transistor channel.

The paper is organized as follows. Firstly, the developed Landauer-Buttiker formalism is described. It allows calculating a current and a noise in the same simulation. Then the classical noise is compared with the quantum one. Later the effect of random impurities and doping profile on current and noise is discussed. At last, the most intriguing question of *"how to attain the highest ON-current and the lowest noise in a nanotransistor?"* is discussed.

2 Developed Landauer-Buttiker Formalism: Current and Noise in the Same Simulation

As a carrier wave length becomes commensurable with a channel size the all-quantum simulation of such small devices becomes challenging. The effective T-matrix method for solution of a scattering problem for the stationary Schrödinger equation in a transistor channel of nanometer length with an arbitrary potential relief was developed in [5, 8, 9]. The method allows fulfilling a simulation of a field effect transistor based on Landauer-Buttiker approach for calculation of current. It has a distinct physical sense clearly based on the conception of a transistor channel as a quantum wire or quantum wave-guide. The necessary transmission coefficients T are determined via a self-consistent solution of Schrödinger and Poisson equations. Finally, the current is calculated via the relation

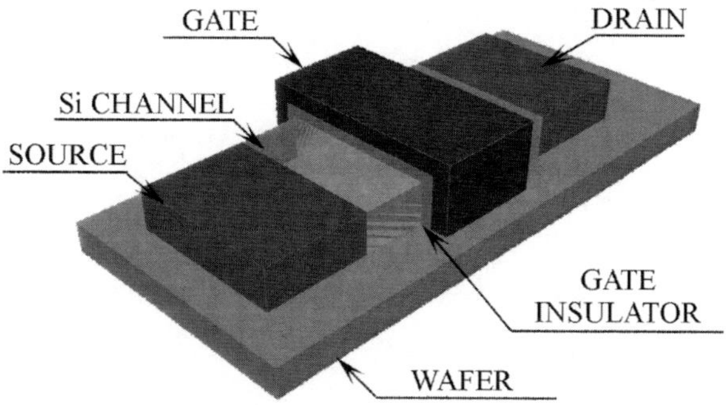

Fig. 1 SOI MOSFET structure to be simulated. The spacers are removed for clarity

$$\begin{aligned} I(V) &= \frac{2e}{h}\sum_{\nu}\sum_{i}\sum_{j}\int dET_{ij\nu}(E)[f_S(E)-f_D(E)] \\ &= \frac{2e}{h}\sum_{\nu}\sum_{i}\int dET_{i\nu}(E)[f_S(E)-f_D(E)], \end{aligned} \tag{1}$$

where E is a total energy consisting of transversal quantization energy and longitudinal motion energy, there is a summation over all wave-guide modes i and j involved into simulation (possible transformations between modes are incorporated), υ stands for band valleys, f_S and f_D are Fermi-Dirac distribution functions in a source and drain contact shifted by drain bias eV: $f_S(E) = f_D(E\text{-}eV)$, the pre-summation factor originates in the quantum wire conductance quantum for spin-unpolarized current $G_0 = 2e^2/h$, where h is Planck constant.

For better understanding the supervening results we simplify the Eq. 1 by eliminating all summations, although the simulation presented further was performed correctly:

$$I(V) = \frac{2e}{h}\int dET(E)[f_S(E)-f_D(E)] \tag{2}$$

The general idea of derivation of quantum noise in [5] was guided by the up-to-date theory of quantum measurements. Indeed, the source and drain contacts measure the state of an electron in the channel. The wave function could collapse in one contact or in another, contributing or not contributing to the current. Here we only sketch out a plan of quantum noise deduction. With that aim, the wave function of a particular electron coming from the source contact into the channel is presented as a superposition:

$$|\psi> \; = r|S> \; + t|D> , \tag{3}$$

where r and t are reflection and transmission amplitudes dependent on electron energy E, $R = |r|^2$ и $T = |t|^2$ are the reflection and transmission coefficients, respectively, providing $|r|^2 + |t|^2 = 1$. The wave function (Eq. 3) could be regarded as a solution of an appropriate Hamiltonian in second quantization representation with operators of creation and annihilation of an electron in contacts. The transmission coefficients correspond to the probability to find the electron in source (S) and drain (D) contact. In the first case the current is $I = 0$, in the second case the current is $I = 1$ (in arbitrary units). Hence, the probability function of current $f(I)$ is as follows $f(1) = T$, $f(0) = R = 1 - T$. Then, one arrives at the mean current equal to $<I> = 1 \cdot f(1) = T$, and the mean square current $<I^2> = 1^2 \cdot f(1) = T$. After that, the standard deviation of current reads:

$$\sigma = <(<I> - I)^2> = <I^2> - <I>^2 = T(1 - T) \tag{4}$$

This is very close to the famous Lesovik's formula [3] (see also [4]) for quantum shot noise (excess noise):

$$dI_n^2 = \frac{2e^2}{h} \Delta v |eV| T(E)(1 - T(E)), \tag{5}$$

where Δv is a frequency bandwidth. This formula was derived for zero temperature and infinitesimal bias V. It should be emphasized that the bracket (1 − T) has no connection with the Pauli exclusion principle. For low transmission through the channel $T \ll 1$, the renormalized noise $I_n^2/e\Delta v$ coincides with the current in accordance with the well-known classical result. Otherwise, the renormalized noise is smaller than current. Interestingly, for the transparent channel ($T = 1$) the noise vanishes. To some extent, it already alludes to the possibility to lower a noise at non-zero temperature.

To generalize the Eq. 4 for an arbitrary temperature and voltage one should take into account averaging over the Fermi-Dirac distribution function and the Pauli exclusion principle. The latter means that an electron coming from the source can occupy only an empty state in the drain, and vice versa. Then one arrives at the general expression for noise power including both shot noise and Johnson–Nyquist (thermal) noise [5]:

$$I_n^2 = \frac{2e^2}{h} \Delta v \int dE [T_S(1 - T_S) + T_D(1 - T_D)] \tag{6}$$

Here the probability T_S to find an electron coming from the source contact to the drain contact and the analogous probability T_D for the drain electrons are introduced:

$$T_S = T(E) f_S(1 - f_D), \tag{7}$$

$$T_D = T(E) f_D(1 - f_S), \tag{8}$$

where the Pauli exclusion principle establishes the statistical correlations between the source and drain contacts absent in classical description. A validity of integration in the Eq. 6 was justified in [5]. For zero bias the resultant expression gives rise to the Johnson–Nyquist noise in quantum limit. This noise is also roughly proportional to the temperature. For a non-zero drain bias there is a blend of the Johnson–Nyquist noise and the shot noise. Previously, the shot noise was used in measurement of fractional charge at very low temperature [10–12]. However, this discussion is beyond the scope of the present paper.

It should be noted that initially the Dirac exclusion principle was also involved into derivation of current according to the Landauer-Buttiker approach. However, in the final expression it was eliminated:

$$I = \frac{2e}{h}\int dE[T_S(E) - T_D(E)] = \frac{2e}{h}\int dET(E)[f_S(E) - f_D(E)] \tag{9}$$

A similar situation occurs in quantum collision integrals.

3 Classical and Quantum Noise

One of the goals of present communication is to reveal how the classical longitudinal motion in the channel or quantum one affects the noise power I_n^2 and current *I*. Classical current is calculated in a similar way as the quantum one:

$$I(V) = \frac{2e}{h}\int dET_i(E)[f_s(E) - f_d(E)], \tag{10}$$

where the transmission coefficient *T(E)* yields only two values: 0 or 1. Consequently, the renormalized (reduced) classical noise reads

$$\tilde{I}_n^2 = \frac{2e}{h}\int dET(E)[f_s(E) + f_d(E)], \tag{11}$$

where the Pauli exclusion principle establishing correlations between source and drain contacts is ignored, meanwhile, it is evidently included in the Fermi-Dirac statistics.

For classical one-dimensional motion the efficient program [9] was exploited instead of the common Monte Carlo simulation. The leading idea of that method originates in a fact that a particle wave vector *k* is immediately bound to the potential $\varphi(x)$ along the channel:

$$E = \varepsilon_i + \frac{\hbar^2 k(x)^2}{2m_x} - e\varphi(x) \tag{12}$$

That allows doing without solving the motion equations. Moreover, the calculation of electron density in the channel can be carried out with the help of the

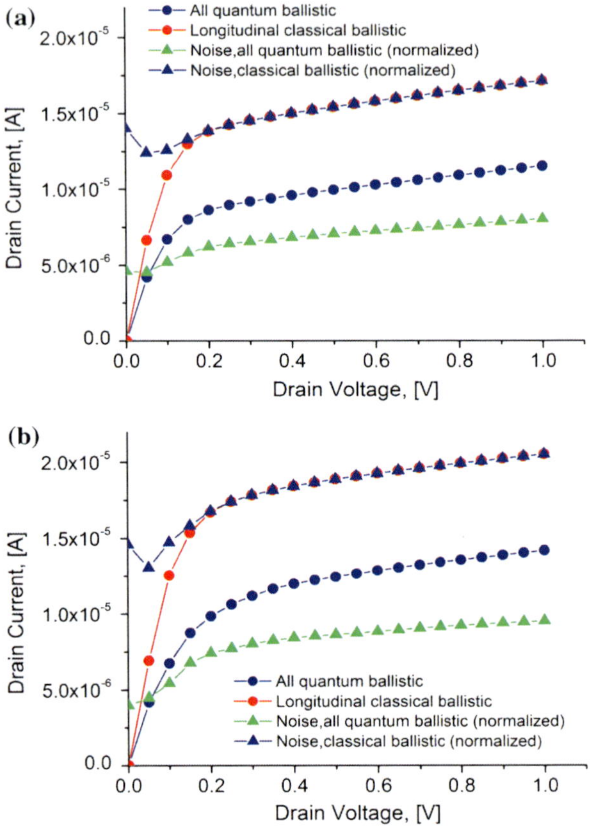

Fig. 2 Classical ballistic current and renormalized classical noise (*upper curves*); quantum ballistic current and renormalized quantum noise (*lower curves*) for perfect channel (**a**) and one random impurity in the channel (**b**)

quasi-classical (WKB) density of states. The electron density is required for the self-consistent solution of Poisson equation. After that the Landauer-Buttiker formula (Eq. 10) can be used for calculation of current.

The extremely thin fully depleted silicon-on insulator field-effect transistor (ET FD SOI FET) in simulation is as follows: a gate length is 10 nm, spacers are 3 nm, channel thickness is 3 nm, channel width is 10 nm, equivalent gate oxide thickness is 1.5 nm, and source/drain contact doping is 10^{20} cm^{-3}. In fact, according to the foregoing dimensions the structure under simulation could be regarded at the same moment as ET SOI FET and fin-FET.

Figure 2 depicts the current and the renormalized noise $I_n^2/e\Delta\nu$. Upper curves correspond to classical regime whereas the lower ones correspond to quantum regime. There is no perfect current saturation owing to the DIBL effect (drain-induced barrier lowering). The gate voltage counted off from the threshold value is equal to zero.

The thermal (Johnson–Nyquist) noise is clearly visible at zero drain voltage. For higher drain voltage the current and renormalized shot noise coincide for classical calculations. In the quantum limit they diverge. Comparison of curves results in the estimation of the mean quantum transmission coefficient for conducting electrons which turns out to be about 0.7 for the sample structure. This is just a cause of lower quantum current. Hence, to reach higher values of ON-current one should augment the transmission coefficient which is discussed later.

Worth mentioning a small dip in curves at low drain bias is not an *artifact*. This is an intrinsic property of a ballistic transport. At low bias the channel behaves as a conductor. Low energy electrons coming from the drain contact are scattered by the potential and return back to the contact. They contribute neither to the current nor to the noise. At higher bias the transistor-like behavior of the structure much matters, especially, the self-consistency.

4 Effect of Random Impurities and Doping Profile on Current and Noise

The effect of channel imperfections on current was elucidated in [5]. Here we simply remind that for the sake of completeness. Figure 2a concerns a perfect channel with no charged impurities inside. In Fig. 2b a single random impurity was inserted in the channel. To avoid misunderstanding it should be noted that a positively charged impurity merely shifts the threshold voltage, therefore, both currents in Fig. 2b are a bit greater compared with that in Fig. 2a. A charged center may originate from defects at silicon/dielectric interface or inside dielectric. It could be a native impurity of bulk silicon. The most likely it penetrates to the channel during annealing. It should be outlined that all kinds of imperfections have much stronger impact on the current when they are located near the source region.

To begin with the influence of the doping profile let us look at the Fig. 3. There, the calculated transmission coefficients are depicted for sharp and smooth doping at zero gate voltage. Here the 'sharp doping' implies that the heavily doping of source/drain contacts abruptly stops at the spacer ends. The 'smooth doping' means that the doping level falls down under the spacers with the slope equal to 1 decade per 1 nm. Obviously, the potential profile is not the same way sharp as the doping.

The transmission coefficients (Fig. 3a) at zero drain voltage markedly oscillate with respect to energy because of interference between the channel ends. The main reason lies in the fairly steep built-in potential profiles near source and drain contacts where heavily/low doping junctions exist. At relatively high drain voltage the interference disappears because the carriers acquire a high energy by the drain end and undergo no quantum reflection. Thereafter, the curve is monotonic and determined by quantum reflection only from the source end. The Fig. 3 demonstrates a tendency to classical behavior as the potential profile becomes smoother. In classical ballistic case the transmission coefficient abruptly jumps from zero up to unity.

The Fig. 4 show that with smoothing doping the increase in current surpasses the increase in noise.

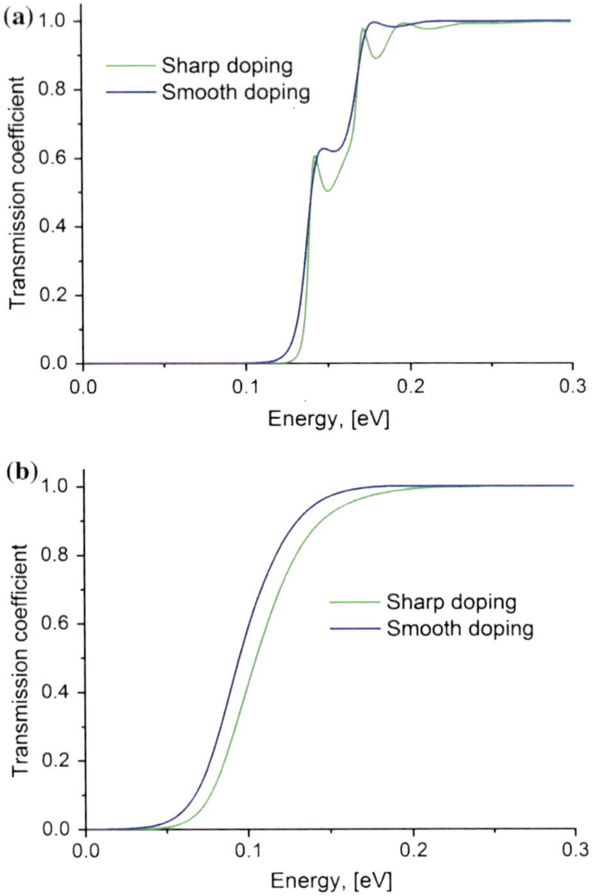

Fig. 3 Calculated transmission coefficient vs. electron energy E for sharp and smooth doping. Electrons coming into the channel belong to the lowest transversal mode of [100] oriented conduction band valley. The drain voltage is equal to zero (**a**) and 0.5 V (**b**)

5 How to Attain the Highest ON-Current and the Lowest Noise in a Nanotransistor?

Here we proceed to the most captivating topic. According to the Eq. 6, an obvious way to make a noise lower is to make a frequency of a transistor performance lower. However, a quantum behavior yields the other possibility to reduce a noise in nanotransistors.

To achieve the maximum ratio of a saturated current to renormalized noise one should press to (1) the transmission coefficient approaching unity ($T(E) \rightarrow 1$) and (2) the degenerated gas inside the channel. The first goal could be achieved with the help of a fairly smooth potential profile in the channel. The second one could be achieved by the sufficiently high gate voltage. In the situation, electrons in the range of several kT around the Fermi level mainly contribute to the noise (here T is

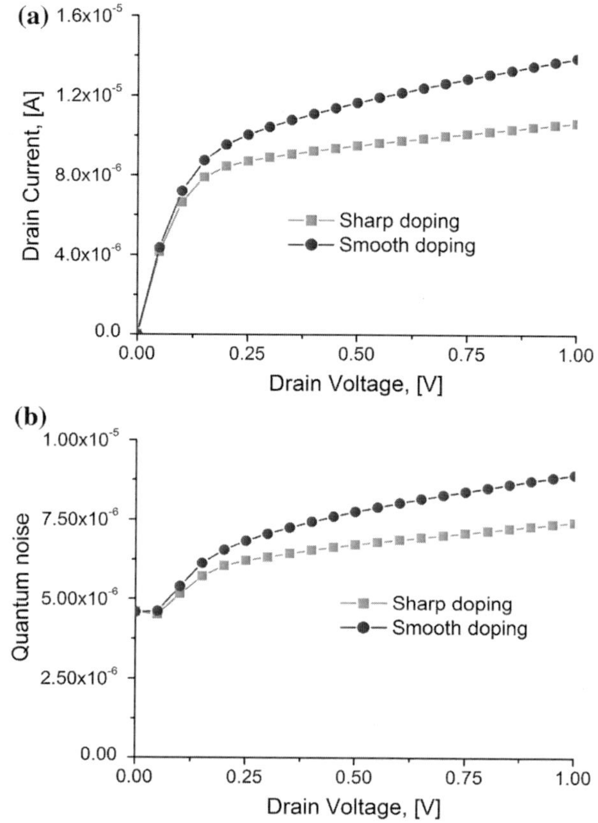

Fig. 4 Drain current (**a**) and noise (**b**) versus drain voltage for sharp and smooth doping, gate voltage is equal to zero

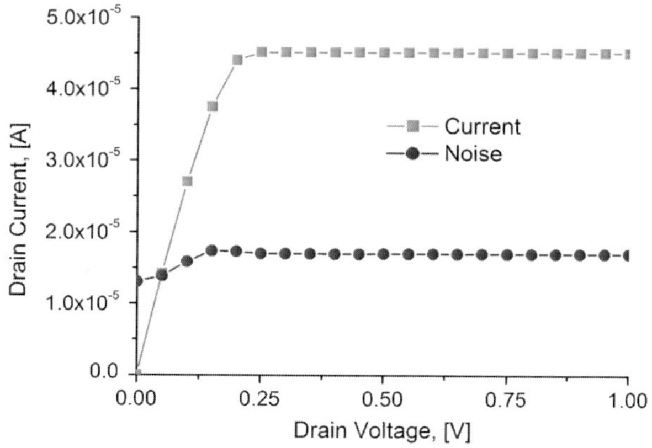

Fig. 5 Drain current as a function of drain voltage for sharp and smooth doping, gate voltage is equal to 0.5 V

a temperature and k is the Boltzmann constant). Roughly speaking, only those electrons are 'noisy'. In Fig. 5 the gate voltage was equal to $V_G = 0.5$ V and the abovementioned ratio rises up to 3. For sufficiently large Fermi energy ε_F the maximum ratio of a saturated current to renormalized noise that one can ever obtain is approximately equals to ε_F/kT.

One possible way to obtain a degenerated gas inside the channel at moderate gate voltage is to use semiconductors with a small effective mass where the density of states is also small. For example, they could be compound semiconductors. In the sense, graphene-based channels with zero effective mass of carriers (although not zero density of states) look like the most prospective for the low-noise electronics.

6 Conclusion

The Landauer-Buttiker approach for simulation of nanometer-scale field-effect transistors was developed to incorporate a calculation of noise at arbitrary temperature and applied voltage. The method allows for both shot noise and Johnson–Nyquist (thermal) noise calculations. The derivation is based on the theory of quantum measurement.

The effect of random impurities in the channel and doping profile in source/drain contacts on the current and noise was elucidated.

The quantum noise is compared with that originating from a classical description of motion along a transistor channel. Unlike the classical renormalized shot noise the quantum noise is smaller than the current. This discrepancy could be a crucial evidence for quantum character of transport in a transistor channel.

From the practical point of view, the quantum physics provides a possibility to reduce a noise to current ratio in quantum nanotransistors.

Acknowledgments The research was supported by the grants of Russian Academy of Sciences and the Russian Foundation for Basic Research (grants #11-07-00464-a and #12-07-90011-Bel_a).

References

1. Celler, G.K., Cristoloveanu, S.: Frontiers of silicon-on-insulator. J. Appl. Phys. **93**, 4955 (2003)
2. Ferain, I., Colinge, C.A., Colinge, J.-P.: Multigate transistors as the future of classical metal–oxide–semiconductor field-effect transistors. Nature **479**, 310–316 (2011)
3. Lesovik, G.B.: Quantum excess noise in two-dimensional ballistic microcontacts. JETP Lett. **49**, 594–596 (1989)
4. Buttiker, M.: Scattering theory of thermal and excess noise in open conductors. Phys. Rev. Lett. **65**, 2901–2904 (1990)
5. Vyurkov, V., Semenikhin, I., Filippov, S., Orlikovsky, A.: Quantum simulation of an ultrathin body field-effect transistor with channel imperfections. Solid-State Electron. **70**, 106–113 (2012)

6. Landauer, R.: Transport as a consequence of the incident carrier flux. In: Bergmann, G., Buynseraede, Y. (eds.) Localization, Interaction and Transport Phenomena, p. 38. Springer, Heidelberg (1985)
7. Buttiker, M.: Symmetry of electrical conduction. IBM J. Res. Dev. **32**, 317 (1988)
8. Semenikhin, I., Vyurkov, V., Orlikovsky, A.: Advanced T-matrix method for quantum simulation of nanoscale field-effect transistors. IEEE Xplore. In: Proceedings of 14th International Workshop on Computational Electronics (IWCE), Pisa, Italy, 27–29th October 2010
9. Orlikovsky, A., Vyurkov, V., Lukichev, V., Semenikhin, I., Khomyakov, A.: All quantum simulation of ultrathin SOI MOSFET. In: Nanoscaled Semiconductor-on-Insulator Structures and Devices, pp. 323–340. Springer, Berlin (2007)
10. de Picciotto, R., Reznikov, M., Heiblum, M., Umansky, V., Bunin, G., Mahalu, D.: Direct observation of a fractional charge. Nature **389**, 162–164 (1997)
11. Kirczenow, G.: Composite fermion edge states, fractional charge and current noise. arXiv:cond-mat/9803089 (1998)
12. Heiblum, M.: Fractional charge determination via quantum shot noise measurements. In: Aharony, A., Entin-Wohlman, O. (eds.) Perspectives of Mesoscopic Physics, p. 115. World Scientific, Singapore (2010)

Non-volatile Memory of New Generation and Ultrafast IR Modulators Based on Graphene on Ferroelectric Substrate

Maksym V. Strikha

Abstract A review of recent achievements in graphene-on-ferroelectric systems is presented. These systems have several unique features. Among them are: the possibility to obtain the high carrier concentration ($\sim 10^{12}$ cm^{-2}) for the moderate gate voltages (of ~ 1 V) and the existence of hysteresis (or anti-hysteresis) in the dependence of the graphene channel resistance on the gate voltage. The use of ferroelectric substrates for graphene had enabled the construction of the robust elements of non-volatile memory of new generation. These elements operate for more than 10^5 switches and preserve information for more than 1000 s. Graphene-on-ferroelectric systems can be characterized theoretically by the ultrafast rate of switching (~ 10–100 fs). It was also demonstrated theoretically, that the effective, fast and small modulators of the middle- and near-IR radiation for different optoelectronic applications can be constructed on the base of graphene on the $Pb(Zr_xTi_{1-x})O_3$ ferroelectric substrate.

1 Introduction

Although graphene was obtained for the first time in 2004, the interdisciplinary "graphene physics" on the border of solid state physics, high energy physics, physical chemistry and engineering had been already formed as a new discipline with a dynamic progress [1–5]. On the first stage, graphene itself had attracted all the interest of researchers (substrate and gate had played an additional role only, enabling graphene's "doping" with electrons or holes). An understanding, however, had appeared soon, that graphene can have practical application together with other components only. Therefore the study of graphene in its interaction with

M.V. Strikha (✉)
V. Lashkaryov Institute of Semiconductor Physics, NAS of Ukraine,
Nauky Ave., 41, Kyiv-28 03650, Ukraine
e-mail: maksym_strikha@hotmail.com

A. Nazarov et al. (eds.), *Functional Nanomaterials and Devices for Electronics, Sensors and Energy Harvesting*, Engineering Materials,
DOI 10.1007/978-3-319-08804-4_9

substrate, contacts, phonon and photon thermostat, and other factors that determine the transport peculiarities became important.

The examination of graphene on the substrates with high permittivity k had developed intensively (see e.g. [6, 7]). Indeed, the substitution of the traditional quartz by the substrates with high k (AlN, Al_2O_3, HfO_2, ZrO_2) allows for getting higher carrier concentrations for the same gate voltages. Moreover, there exists an assumption that the effective screening of the Coulomb field of the charged impurities in the substrate and on its surface would decrease the scattering of carriers in graphene by these impurities and therefore would increase the carrier mobility.

However, these expectations did not realize. As it was demonstrated in [7], the decrease of Coulomb scattering by impurities in real systems is accompanied by essential growth of scattering by surface phonon modes. Moreover, the highest permittivity one can get for such traditional linear dielectrics is 24 (ZrO_2).

Therefore, the linear dielectric substrate was suggested to be substituted by a ferroelectric one with orders of magnitude higher permittivity. The use of ferroelectric substrates like organic one or $Pb(Zr_xTi_{1-x})O_3$ (PZT), which behave at low voltage as dielectrics with extremely high permittivity (up to 3850), can be beneficial for the construction of modulators of near- and mid-IR range, that can be perspective for low-voltage devices with optical on-chip interconnections [8]. Moreover, the existence of hysteresis in ferroelectric polarization dependence on the applied field make them promising for the construction of memory units [9]. Despite the first study focused on graphene-on-ferroelectric device [9] had appeared in 2009, the number of such works had increased within recent years, and the two review articles on the problem [10, 11] are available now.

2 Graphene on Ferroelectric Substrate: Transport Characteristics and Non-volatile Memory Structures Construction

As it was mentioned previously, the main drawback of traditional quartz (SiO_2) substrate is its relatively low permittivity ($k = 3.9$), which limits the possibility to get high carrier concentration in graphene. The concentration of carriers in gated graphene n depends linearly on the gate voltage V_g, permittivity of substrate k, and is inversely proportional to the substrate thickness d:

$$n\left(cm^{-2}\right) = 7.2 \cdot 10^{10}\left(\frac{300}{d(nm)}\right)\left(\frac{\kappa}{3.9}\right)V_g(V) \tag{1}$$

Figures in Eq. 1 are normalized according to characteristics of SiO_2 substrate with thickness 300 nm used in the first works on graphene (see [1–4]). Therefore, the highest values of concentration (and conductivity) are determined by substrate breakdown field, which is about 0.5 V/nm for SiO_2 substrate.

The use of ferroelectric substrates theoretically allows for reaching carriers concentration in graphene of $\sim 6 \cdot 10^{14}$ cm^{-2}, which is 100 times higher than for the traditional SiO_2 substrate. This corresponds to Fermi energy of $\sim$ 1 eV; the band spectrum is no longer linear for such energy. Moreover, the use of ferroelectric substrates with spontaneous polarization allows for creating the non-volatile memory units.

Such kind of device was firstly constructed by a Singapore group [9]. States «0» of «1» of binary system were maintained by a high and low resistivity of graphene conducting channel in field effect transistor (FET). Switching between these two states occurred due to a change of thin ferroelectric film polarization under the gate voltage variation. These hybrid graphene-ferroelectric devices had supported the non-volatile resistance variation of about 200 %.

The scheme of device realized in [9] is presented in Fig. 1. The graphene layer with mobility of ~ 1500 cm^{2}V^{-1}s^{-1} was placed onto traditional quartz substrate on Si gate. Approximately 0.7 μm-thick layer of liquid ferroelectric PVDF-TrFE was placed over the graphene sheet and the top Au gate was placed on it. The dependence of the graphene channel resistance (between S and D electrodes, see Fig. 1) on the Au top gate voltage V_{TG} had the form presented in Fig. 2 (for all the 15 samples under examination). This dependence features completely symmetrical hysteretic loop for the top gate voltage sweeps from 0 to +85 V, then from +85 to –85 V, then from –85 V back to 0. The relative change of resistivity $\Delta R/R = (R_{max}-R_{min})/R_{min}$ was as large as 3.5 for some samples.

Such form of $R(V_{TG})$ dependence is determined by the peculiarities of ferroelectric polarization dependence on applied field. The concentration of carriers in graphene is determined both by the field, created by the top gate voltage, and by the field of dipoles on graphene-ferroelectric interface. Therefore graphene can remain *n*-type (*p*-type) even at negative (positive) gate voltage, when this voltage is not high enough to over-polarize the dipoles of ferroelectric film. The authors of [9] had defined the state with maximum resistance as «1», and with minimal resistance as «0», and proposed possible routes of switching between these two states by asymmetric gate voltage sweeps.

Such hybrid memory on graphene-ferroelectric base can potentially feature extremely high switching rate. The graphene-on-ferroelectric channel can have the

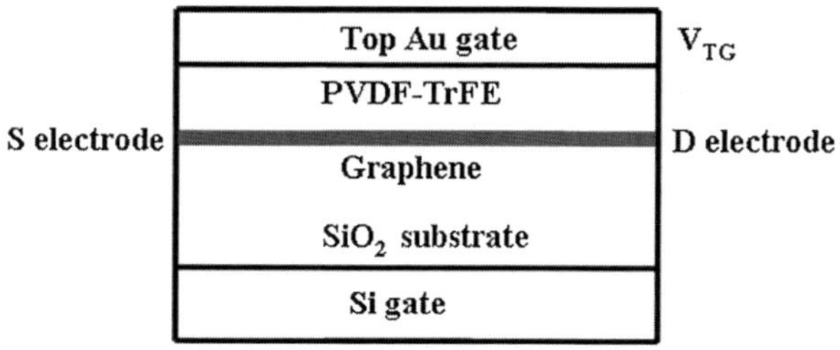

Fig. 1 Principal scheme of non-volatile memory device on graphene-on-ferroelectric base realized in [9]

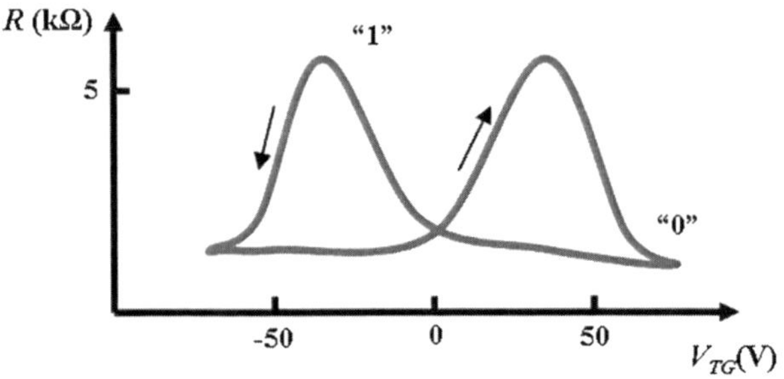

Fig. 2 Hysteretic dependence of graphene channel resistance on the top gate voltage (according to experimental data of [9])

upper theoretical limit of carrier mobility of 200,000 $cm^2\ V^{-1}\ s^{-1}$. Thus, for switching voltages of $\sim$ 1 V and for the channel length of μm the switching time can be less than 10^{-13} s. Therefore, such device can be a promising candidate for the principal element of super-fast non-volatile memory of new generation.

It the next work [12] the same group had obtained essentially higher values of mobility ($\sim$4200 $cm^2\ V^{-1}\ s^{-1}$), relative change of resistivity ($\sim$500 %) and reproduction of non-volatile memory within 10^5 cycles of switching for the device geometry presented in Fig. 1. Similar system with the back and top gates and 200 nm film of organic P(VDF-TrFE) ferroelectric was studied very recently in [13]. The group from Switzerland had demonstrated a high retention performance for both memory states with fully saturated time-dependence of the graphene channel resistance. This behavior is in contrast with ferroelectric-polymer-gated silicon field-effect-transistors, where the gap between the two memory states continuously decreases with time.

Before reaching saturation, the current decays exponentially as predicted by the retention model based on the charge injection into the interface-adjacent layer. The drain current saturation attests a high quality of the graphene/ferroelectric interface with low density of charge traps.

In [14] the authors from USA had reported the construction of FET on *N*-layer graphene ($N = 2$–15) on the thin film of $Pb(Zr_{0.2}Ti_{0.8})O_3$. Figure 3 presents schematically the resistivity of 7-layer graphene as a function of gate voltage at 300 K. The ferroelectric field permittivity (determined from Eq. 1 and Hall measurements of concentration) was $k \sim 100$. The $R(V_g)$ dependence had definitely different form for low and high voltages. For the sweep range $|V_g| < 2$ V (left maximum curve on Fig. 3), carriers concentration and resistivity were governed by Eq. 1, and the sweeps from positive voltages to negative ones were reversible.

For the sweep-range $V_g > 2$ V the dependence of resistance on gate voltage had a hysteretic form, somewhat similar to one observed in [9]. For the backward voltage sweep from maximum value to zero the resistance had changed according to the right maximum curve, presented in Fig. 3. The point of the resistance saturation of the forward sweep had corresponded to the electro-neutrality point for the backward sweep and vice versa.

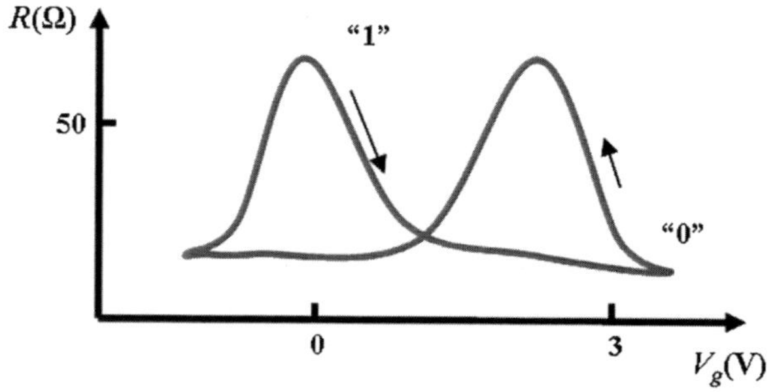

Fig. 3 Resistivity of 7-layer graphene channel on the film $Pb(Zr_{0.2}Ti_{0.8})O_3$ as the dependence on gate voltage (according to experimental data of [14])

The microscopic mechanism of the observed effect had remained, however, rather unclear, because the fields in ferroelectric film, corresponding the applied gate voltages, were much smaller than coercive field, that could change the direction of ferroelectric polarization. Moreover, the direction of hysteresis itself was opposite to one that could be predicted from the ferroelectric polarization dependence on external field. This behavior (reproduced later in all the works on graphene on PZT substrate) was called in [14] the "anti-hysteretic" one.

However this anti-hysteresis was reproducible and characterized by a large relaxation time. The curves with left maximum in Fig. 3 proved to be more stable in a low gate voltages range, and the curves with the right maximum are more stable in the upper voltages range. With a slow voltage sweep from 0 to 2 V, resistivity changes according to the "left" curve. However, as soon as the voltage was fixed in this point, the resistivity had slowly increased up to the value, which corresponds to the curve with the "right" maximum. This relaxation occurred according to the exponential curve, where time constant τ was 6 h for 300 K and 80 days for 77 K. Considering this process as thermal relaxation between the two meta-stable states, the authors of [14] had estimated the activation energy ΔE from the Arrhenius formula:

$$\frac{1}{\tau} \sim \exp\left[-\frac{\Delta E}{kT}\right] \tag{2}$$

The value of this energy was within 50–110 meV range.

The authors of [14] had also proposed to use the observed anti-hysteretic effect for construction of memory units. The obtained correlation between minimal and maximal resistances (states « 0 » and « 1», see Fig. 3) $\Delta R/R = (R_{\max} - R_{\min})/R_{\min}$ is within a range of 2–3.5 and depends on the number of graphene layers. The obtained high values of mobility make the high rate of switching possible.

In [15], a team from Singapore and Korea had also rejected the configuration with two gates and studied the FET with single- and bi-layer graphene, fabricated by chemical vapor deposition (CVD) method on Cu, placed on 360 nm-thick $Pb(Zr_{0.3}Ti_{0.7})O_3$ substrate. The high permittivity ($k \approx 400$) enabled to get carriers concentration of $\sim 10^{13}$ cm^{-2} for the gate voltages of ~ 1 V. A correlation $n = \alpha V_g$ with $\alpha = 6.1 \times 10^{12}$ cm^{-2} V^{-1} was valid within linear regime. Beyond the linear range, at $V_g > 1.1$ V the ferroelectric polarization had caused essential hysteresis in resistance on gate voltage dependence. The increase of ferroelectric polarization leads to increase of distance between the two maxima.

In [16], American and Korean teams had reported on the fabrication of the robust non-volatile memory based on a single-layer graphene on ferroelectric PZT substrate ($k \sim$ 400–500). Electrical measurements were performed in vacuum (1.1×10^{-6} Tor) at a constant source-drain voltage V_{ds}. The dependence of current in source-drain circuit I_d on gate voltage V_g for exfoliated graphene had a characteristic form with two minima for the sweep voltage $V_{\text{g(sweep)}} > 1$ V. The Fermi level crosses the Dirac point at these voltages, and the electrostatic potential of dipoles in ferroelectric is balanced by the potential of interface absorbed impurities.

The theoretical upper limit of "memory window" ΔV_M (the width of the hysteresis loop) is determined by V_C voltage, which corresponds to the ferroelectric coercive force: $\Delta V_M = 2V_C$. In [16], with the increase of the sweep voltage memory window had reached saturation at approximately 7 V. This value is essentially smaller for ordinary semiconductor FETs. The two states «0» and «1» could be preserved in the system under examination for approximately 1000 s.

3 Mechanism of Anti-hysteresis in Graphene-on-PZT Resistance

The qualitative mechanism of anti-hysteresis $R(V_g)$ behavior (observed for all graphene-on-PZT systems) was proposed in [16] and included interface states. The explanation correlates with one, proposed in [17] for so called "inverse" hysteresis in graphene on non-ferromagnetic substrate. A general quantitative model (for substrate of different nature) for competing hysteretic mechanisms in graphene channel resistivity considering a direct one (caused by absorption with dipole moment on surface, e.g. water) and an inverse one (caused by capture of free carriers onto localized states on graphene-substrate interface) was proposed recently in [18].

Later in this chapter we will focus on the inverse hysteresis (or anti-hysteresis in terms of [14]) common for graphene on ceramic ferroelectric substrate. Quantitative model of anti-hysteresis behavior of graphene on PZT substrate resistance was developed in [19, 20]. The gated single-layer graphene was examined there with Fermi energy depending on concentration as:

$$E_F = \hbar \cdot v_F \cdot (\pi \cdot n)^{1/2}, \tag{3}$$

where $v_F = 10^8$ cm/s. It was supposed that an interface state exists at the energy E_T. On the forward V_g sweep, when $E_F < E_T$, the carriers' concentration is governed by a simple correlation:

$$n = k \cdot V_g / 4 \cdot \pi \cdot e \cdot d, \tag{4}$$

where d is substrate thickness. If the sweeps of gate voltage are small enough (the condition $E_F < E_T$ is valid) no hysteresis in the $n(V_g)$ dependence appears. However, when $E_F = E_T$, electrons from gated graphene are captured by the interface states with a high 2D density n_T. As it was demonstrated in [20], for the gate voltages range of

$$\frac{4 \cdot \pi \cdot e \cdot d}{k} \cdot \frac{E_T^2}{\pi \cdot \hbar^2 \cdot v_F^2} \le V_g < \frac{4 \cdot \pi \cdot e \cdot d}{k} \cdot \frac{E_T^2}{\pi \cdot \hbar^2 \cdot v_F^2} + \frac{4 \cdot \pi \cdot e \cdot d \cdot n_T}{k} \tag{5}$$

the occupation of the localized states by the free electrons from graphene occurs. Negative localized charges screen the field in the substrate. Therefore, within the range defined by Eq. 5 the free electrons concentration in graphene remains constant:

$$n = \frac{E_T}{\sqrt{\pi} \cdot \hbar \cdot v_F} \tag{6}$$

For the further forward sweep of V_g the concentration of carriers in the gated graphene is governed by obvious correlation:

$$n = k \cdot V_g / 4 \cdot \pi \cdot e \cdot d - n_T \tag{7}$$

The next assumption is that the lifetime of electrons on interface states is much larger than the switching time of the system. Therefore above correlation (Eq. 7) is valid for the backward sweep as well, and the general dependence of n on gate voltage has the hysteresis form, presented in Fig. 4.

Concentration n reaches now the Dirac point at some gate voltage V_{DP}, determined by the interface states concentration n_T:

$$V_{DP} = 4 \cdot \pi \cdot e \cdot d \cdot n_T / k \tag{8}$$

Note that left to the Dirac points Fig. 4 presents the holes concentrations. At large negative V_g the trapped electrons recombine with the holes in graphene sheet and n is again governed by Eq. 4.

The total resistivity of graphene sheet according to a Mattiessen's rule is:

$$\rho(V_g) = 1/\sigma(V_g) + 1/\sigma_{\text{int}}(T) + 1/\sigma_{\min} \tag{9}$$

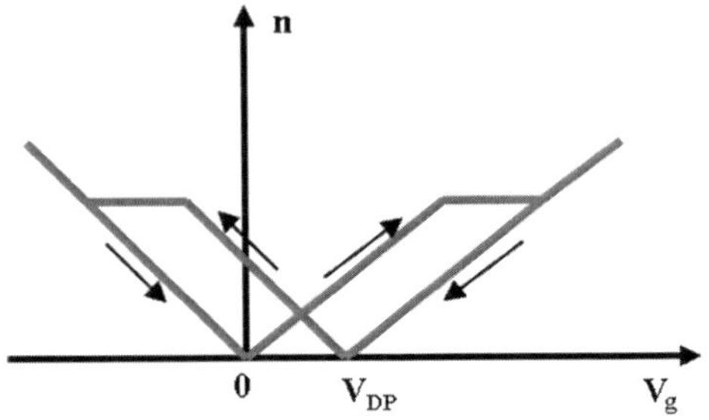

Fig. 4 Anti-hysteresis in graphene-on-PZT dependence of carriers concentration on gate voltage: theory [20]. *Arrows* indicate the sweep directions

Here the denominator of the first term in the right part represents the gated graphene conductivity, which changes linearly with V_g and n; the denominator of the second term is the intrinsic graphene conductivity (intrinsic carriers concentration in graphene at room temperature is $\sim 10^{11}$ cm^{-2}); the denominator of the third term is the minimal quantum graphene's conductivity at the Dirac point [3]. Accounting for $n(V_g)$ dependence (presented in Fig. 4), Eq. 9 leads to ρ (V_g) dependence similar to the experimental one (presented in Fig. 2).

The distance between the Dirac points in Fig. 4 is determined by interface states concentration through Eq. 8 and does not depend in this rough approximation on E_T. The substitution of experimental values from [14] into Eq. 8 yields $n_T = 2.7 \times 10^{12}$ cm^{-2}. This value seems to be natural for the interface between graphene and ferroelectric, because much higher concentrations of surface states in ferroelectrics were observed [21]. The relaxation of the anti-hysteresis observed experimentally [14] can be explained by the finite lifetime of electrons on these states.

The simple model developed in [19] explains the anti-hysteresis behavior of graphene-on-PZT resistivity by special interface states with a high 2D concentration (they can be related to HO^- anions, captured by Pb^{2-} sublattice of PZT, thus creating the localized states with binding energy of $\sim$200 meV [22], which corresponds within the order of magnitude to the thermal activation energy, observed in [14]). [19, 20] neglect the hysteresis of PZT itself. This can, however, be done for the small range of V_g, which corresponds the range of n, which is much smaller than the nominal 2D charge density, corresponding to the polarization of PZT $\sim 3{\cdot}10^{14}$ cm^{-2} [14].

In [20] a non-zero energy width of localized states band was taken into account, which had allowed for explaining the memory window increase and further saturation with the increase of the gate voltage sweep observed in [15, 16]. Indeed, the case when the switching voltage V_{sweep} is within the interval defined by Eq. 5, i.e. when the occupation of the localized states is not accomplished yet, is of special interest. In this case, the separation between electro-neutrality points on Fig. 4 increases with increase of V_{sweep}. The distance between them (i.e. the memory window) is determined now by:

$$V_{DP}\left(V_{sweep}\right) = 4 \cdot \pi \cdot e \cdot d \cdot n_T\left(V_{sweep}\right)/k, \tag{10}$$

where $n_T(V_{sweep})$ is the concentration of interface states occupied by electrons at the moment of gate voltage switching from forward to backward sweep. One can see (Eq. 10) that $V_{DP}(V_{sweep})$ increases with V_{sweep} when V_{sweep} is in the interval defined by Eq. 5 and saturates at higher voltages. This situation was observed experimentally in [15, 16] and obtained $V_{DP}(V_{sweep})$ dependences allow for estimating the interface centres energy and concentration [20]. The estimations can be useful in fabrication of non-volatile memory of new generation, based on the bi-stable system, where logical "0" corresponds to one value of graphene's resistance, and "1" corresponds to the other value.

4 Modulation of Radiation of Near- and Middle-IR Range by Graphene on Ferroelectric Substrate

The essential feature of the graphene optical properties is it's substantial interaction with radiation in the wide spectral range, from far-IR up to UV, due to effective interband transitions (see [23–25] and refs therein). The amplification of response in the Fresnel system: "graphene—300 nm-thick quartz substrate—Si gate" had permitted to make graphene optically visible in the first work [4]. The example of the application of the graphene's unique optical properties is a fabrication of graphene-based saturable absorber for ultrafast lasers in telecommunication range (see [26]).

Recently, modulation [27] and polarization [28] of IR radiation were observed experimentally in graphene structure, integrated with waveguide. Namely, a graphene-based optical modulator for near-IR range (1.35–1.6 μm) was realized in [27]. Principal scheme for such a modulator is presented in Fig. 5.

It was demonstrated, that such a modulator can be perspective for the devices with on-chip optical interconnections, and it's modulation efficiency is already comparable to, if not better than, traditional semiconductor materials such as Si, GeSi, InGaAs, which is orders of magnitude larger in active volume. In that work, a single layer graphene was placed on 7 nm-thick Al_2O_3 substrate over the Si gate, which had also served as a waveguide for near-IR radiation; the device footprint was as small as 25 μm^2 and the operation speed as high as 1.2 GHz.

The general theory of the carrier-induced modulation of radiation by a gated graphene has been developed in [29]. It was demonstrated that the carriers contribution modifies essentially the graphene response due to the Pauli blocking effect, when absorption is suppressed at $\hbar\omega/2 < E_F$, where E_F is the Fermi energy. At low temperatures or at high doping levels, the threshold frequency for the jump of absorption (when absorption becomes essential) is determined by following condition (see Fig. 6):

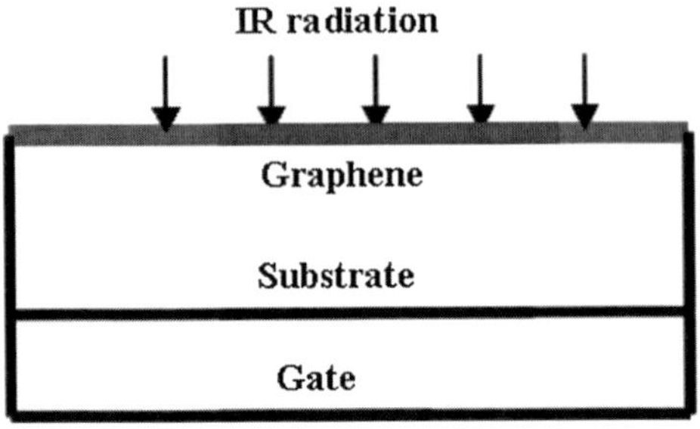

Fig. 5 Graphene-based IR radiation modulator (principal scheme)

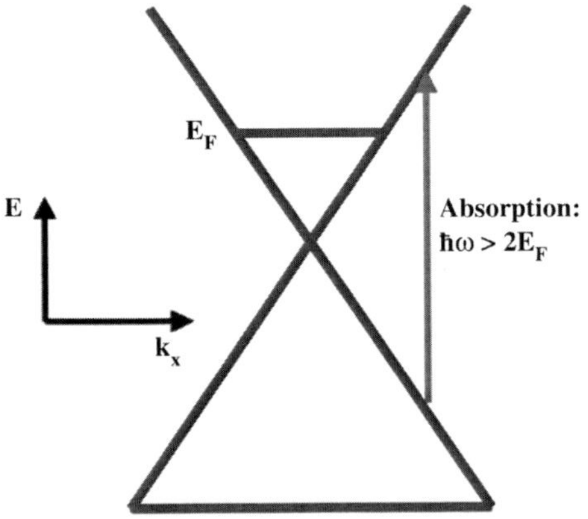

Fig. 6 Optical transition in graphene is permitted for the photon energy at $\hbar\omega/2 > E_F$

$$\hbar\omega_{th} = 2E_F \sim \sqrt{n}, \tag{11}$$

where concentration n is determined by Eq. 1.

As it was mentioned above, the use of high-k substrates (AlN, Al_2O_3, HfO_2, ZrO_2) enables for obtaining higher concentrations for the same gate voltages. This is important, because for the threshold wavelength λ_{th} corresponding threshold frequency, Eqs. 1,3,4,11 yield:

$$\lambda_{th} \sim \sqrt{\frac{d}{V_g}} \equiv 1/\sqrt{E_s}, \tag{12}$$

where E_s is homogeneous gate-induced electric field intensity in the substrate.

Equation 12 yields that in order to get modulation for the shorter wavelength of radiation (i.e. in visible range) one needs stronger fields (and higher gate voltages), which finally can cause the breakdown of the substrate.

The transmissivity and reflectivity coefficients of the "graphene layer—substrate—Si gate" system can be written as (see [29]):

$$T_\lambda = \sqrt{k_{Si}(\lambda)}\frac{|E_t|^2}{E_{in}^2};R_\lambda = \frac{|E_r|^2}{E_{in}^2} \tag{13}$$

where k_{Si} is Si gate permittivity, dependent on wavelength. The correlation between the amplitudes E of the incident (*in*), reflected back into vacuum (*r*) and transmitted into Si gate (*t*) waves can be obtained by solving the system of wave equations in vacuum, substrate, and gate, with the proper boundary conditions, taking into consideration the absorption due to interband carriers transitions in graphene layer.

The calculations performed in [29] demonstrate that the modulator for telecommunication in near-IR range ($\sim$1.5 μm) can be based on single-/multi- layer graphene, placed over high-*k* substrate (in fact, this was the case, realized in [27]). The effective modulation of near-IR radiation by the gated graphene can be realized in the case of high-*k* substrates for the applied fields of $\sim$5 MV/cm. For the case of low-*k* SiO_2 substrate the field should be essentially stronger $\sim$20 MV/cm, which is comparable to the breakdown value. However, graphene-on-quartz can modulate efficiently the middle-IR radiation. The highest values of E_s, reached in [27], were in fact on the order of 5 MV/cm. This required, however, the extremely high accuracy of substrate preparation (7 nm-thick Al_2O_3 substrate was deposited over the Si gate, which had also served as a waveguide, by atom layer deposition).

The ferroelectric substrates with extremely high permittivity can be useful for the further development of the gated graphene-based modulators. It is important, that at low voltages ($V < V_c \sim$ 1–2 V) the epitaxial ferroelectric $Pb(Zr_xTi_{1-x})O_3$ (PZT) thin films behave as a high-*k* dielectric with $k = 73$ (x = 0.2 [14]), $k = 400$ (x = 0.3 [15]). This allows for using them in low voltage mid-IR gated graphene-based modulators.

The critical value for the field in PZT substrate, under which this substrate still behaves as a high-*k* dielectric, can be obtained from [14] ($d = 300$ nm, $V_{cr} \sim 2$ V) and [15] ($d = 360$ nm, $V_{cr} \sim 1$ V). This yields $E_{cr} = 67$ kV/cm (x = 0.2), $E_{cr} = 28$ kV/cm (x = 0.3). Figure 7 presents the threshold wavelength (Eq. 12) dependence on the field in the substrate for the substrates with different permittivities: SiO_2 ($k = 3.9$, curve 1), Al_2O_3 (12.53, 2), ZrO_2 (24.0, 3), PZT 20/80 (73, 4), PZT 30/70 (400, 5).

As one can see from Fig. 7, for telecommunication range ($\lambda = 1.55$ μm, horizontal line) the fields which correspond to λ_{th} are several times higher than E_{cr} for PZT. However, for mid-IR range (namely for $\lambda = 10.6$ μm, corresponding the CO_2 laser wavelength) modulation can be realized for the fields essentially lower than E_{cr}, when PZT behaves as high-*k* dielectric with extremely high permittivity.

Reflectivity and transmissivity coefficients, calculated in [8] according to Eq. 13 for single layer and 5-layers graphene on PZT substrates with x = 0.3 for the films of different thickness demonstrate a strong jump in absorption at approximately 2.5–3 kV/cm for single layer graphene, and at 13–15 kV/cm for 5-layers graphene on PZT with x = 0.3 (where permittivity $k \approx 400$), which leads to essential jump in reflection and transmission. The deepness of modulation can

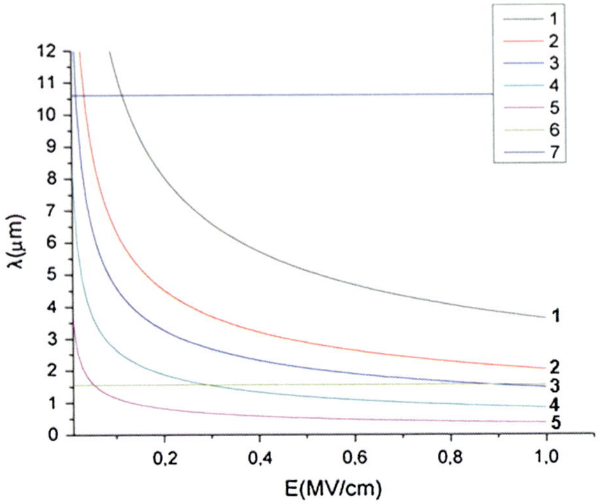

Fig. 7 Dependence of threshold wavelength (Eq. 12) on field in substrate for different substrates: SiO_2 (curve 1), Al_2O_3 (2), ZrO_2 (3), PZT (x = 0.2, 4), PZT (x = 0.3, 5). *Lines* 6 and 7 correspond to 1.55 and 10.6 μm

be of ~20 % (~2 % for each layer) for the fields, which are much lower than the critical ones, at which the ferroelectric hysteresis phenomena start.

A comparison of results from [29] and [8] demonstrates that electric fields in the substrate, under which an effective modulation of middle-IR radiation occurs is two orders of magnitude smaller for graphene on ferroelectric substrate than for graphene on quartz. Moreover, the "modulation edge" is much more acute in this case. This is of great importance for the devices that operate at low switching voltages.

The results obtained in [8] demonstrate the possibility of fabrication of mid-IR range modulator based on low voltage gated graphene on PZT ferroelectric substrate. The advantage of such a modulator in comparison with one realized in [27] can be a relative simplicity of the epitaxial PZT film substrate preparation. This modulator can potentially operate at 500 GHz, because carries recombination and generation times (i.e. speed limiting processes) in graphene are in picoseconds range. Moreover, as the modulation of middle IR radiation is efficient in this case not only for transmissivity, but also for reflectivity, the geometry of modulator can be much simpler than one presented in Fig. 5 (where the gate plays the role of waveguide in [27]).

The modulation of near-IR can be based on the same mechanism, but for the substrates of PZT with higher k. It is known, that the PZT features an extremely large k at the morphotropic phase boundary near $x = 0.52$. The dielectric constant of PZT can be as high as 3850 depending upon orientation and doping. The authors of [16] had observed $k = 2000$ by substitutional doping of Pb by La and by fine tuning of the ratio between Zr and Ti. Such a value can be sufficient for near-IR modulation.

It has been predicted theoretically in [30], that anti-hysteresis behavior of concentration in graphene-on-PZT, presented in Fig. 4, can also appear in optics. Note that the threshold wavelength λ_{th} (Eq. 12) can be written through Eqs. 3, 4 as:

$$\lambda_{th} = \frac{\sqrt{\pi}c}{v_F\sqrt{n}} \tag{14}$$

where c is light velocity in vacuum.

In the case, when $n = n_T$, the modulation edge in V_g scale corresponds to the V_{DP} point. The reflectivity R in this point rapidly increases equaling approximately to 0,023·N, where N is number of graphene layers. With the further increase of V_g, when E_F becomes equal to E_T, the electrons are trapped on the interface levels; the concentration of electrons in graphene is governed by Eq. 7; E_F decreases and can become smaller than the value, corresponding to the V_{DP} according to Eq. 8. This means that the graphene layer can no longer modulate the radiation with λ wavelength, because the interband direct transitions are now forbidden due to Pauli blocking effect. Therefore the reflectivity R decreases to its initial level. This hysteresis in R can be observed experimentally and used for creation of a fast bistable system for the new non-volatile memory devices with on-chip optical interconnection.

5 Conclusion

Graphene-on-ferroelectric system has several unique features. Among them are: the possibility to obtain high concentrations of $\sim 10^{12}$ cm^{-2} for the moderate gate voltages (of the order of 1 V) and the existence of hysteresis (or anti-hysteresis) in the dependence of the graphene channel resistance on the gate voltage. The use of ferroelectric substrates for graphene had enabled the realization of the robust non-volatile memory elements of new generation (see rev. [10, 11] and refs therein). These elements work for more than 10^5 switches and preserve information for more than 1000 s. Such systems can be characterized theoretically by the ultrafast rate of switching ($\sim$10–100 fs).

A quantitative model was proposed in [19, 20] to explain the anti-hysteresis behavior of graphene on $Pb(Zr_xTi_{1-x})O_3$ ferroelectric substrate resistance on gate voltage sweep. The model takes into consideration a screening of electric field in the substrate by electrons captured by the states on graphene-ferroelectric interface and it explains experimental data obtained previously. The estimations can be useful for fabrication of non-volatile memory of new generation, based on the bistable system, where logical "0" corresponds to one value of graphene's resistance and "1" corresponds to other value.

A general quantitative model for competing hysteretic mechanisms in resistivity of graphene channel on a substrate of different nature including a direct one (caused by adsorption with dipole moment on surface, e.g. water), and an inverse one (caused by capture of free carriers on localized states on graphene-substrate interface) was proposed recently in [18].

It was also demonstrated theoretically, that the efficient, fast and small modulators of the middle- and near-IR radiation for different optoelectronic applications can be realized based on graphene on the $Pb(Zr_xTi_{1-x})O_3$ ferroelectric substrate [8]. The depth of modulation for 5-layers graphene for the range of gate voltages, for which thin PZT epitaxial film behaves as high-*k* dielectric can be on the order of 10 %.

A model for hysteresis behavior in reflectivity *R* of the "graphene—$Pb(Zr_xTi_{1-x})O_3$ (PZT) ferroelectric substrate—gate" system with gate voltage variation including electrons trapping on graphene-PZT interface states was analyzed in [30]. It was demonstrated, that this hysteresis in *R* can be observed experimentally for the telecommunication range radiation $\lambda = 1.55$ μm at low gate voltages and can be used for a creation of a fast bi-stable system for the new non-volatile memory devices with on-chip optical interconnection.

Acknowledgments This work was supported by State Fundamental Research Fund of Ukraine (Grant 53.2/006).

References

1. Geim, A.: Graphene: status and prospects. Science **324**, 1530–1934 (2009)
2. Novoselov, K.: Graphene: materials of the Flatlands. Uspekhi Fizicheskih Nauk. **181**, 1298–1311 (2011). (In Russian)
3. Das Sarma, S., Adam, Shaffique, Hwang, E.H., Rossi, E.: Electronic transport in two dimensional graphene. Rev. Mod. Phys. **83**, 407–470 (2011)
4. Novoselov, K., Geim, A., Morozov, S., Jiang, D., Zhang, Y., Dubonos, S., Grigorieva, I., Firsov, A.: Electric field effect in atomically thin carbon films. Science **306**, 666–669 (2004)
5. Peres, N.M.R.: Colloquium: the transport properties of graphene: an introduction. Rev. Mod. Phys. **82**, 2673–2700 (2010)
6. Kim, S., Nah, J., Jo, I., Shahrjerdi, D., Colombo, L., Yao, Z., Tutuc, E., Banerjee, S.: Realization of a high mobility dual-gated graphene field effect transistor with Al_2O_3 dielectric. Appl. Phys. Lett. **94**, 062107 (2009)
7. Konar, A., Fang, T.: Depdeep Jena. Effect of high-κ gate dielectrics on charge transport in graphene-based field effect transistors. Phys. Rev. B. **82**, 115452 (2010)
8. Strikha, M.: Modulation of a mid-IR radiation by a gated graphene on ferroelectric substrate. Ukr. J. Phys. Optics. **12**, 162–165 (2011)
9. Zheng, Y., Ni, G.-X., Toh, C.-T., Zeng, M., Chen, S., Yao, K., Özyilmaz, B.: Gate-controlled nonvolatile graphene-ferroelectric memory. Appl. Phys. Lett. **94**, 163505 (2009)
10. Strikha, M.: Non-volatile memory and IR radiation modulators based upon graphene-on-ferroelectric substrate. A. Rev. Ukr. J. Phys. Opt. **13**, S5–S27 (2012)
11. Hong, X., Zou, K., DaSilva, A., Ahn, C., Zhu, J.: Integrating functional oxides with graphene. Solid State Commun. **152**, 1365–1374 (2012)
12. Zheng, Y., Ni, G.-X., Toh, C.-T., Tan, C.-Y., Yao, K., Özyilmaz, B.: Graphene field effect transistors with ferroelectric gating. Phys. Rev. Lett. **105**, 166602 (2010)
13. Raghavan, S., Stolichnov, I., Setter, N., Heron, J.-S., Tosun, M., Kis, A.: Long-term retention in organic ferroelectric-graphene memories. Appl. Phys. Lett. **100**, 023507 (2012)

14. Hong, X., Hoffman, J., Posadas, A., Zou, K., Ahn, C., Zhu, J.: Unusual resistance hysteresis in *n*-layer graphene field effect transistors fabricated on ferroelectric $Pb(Zr_{0.2}Ti_{0.8})O_3$. Appl. Phys. Lett. **97**, 033114 (2010)
15. Zheng, Y., Ni, G.-X., Bae, S., Cong, C.-X., Kahya, O., Toh, C.-T., Kim, H., Im, D., Yu, T., Ahn, J.H., Hong, B.H., Ozyilmaz, B.: Wafer-scale graphene/ferroelectric hybrid devices for low voltage electronics. Europ. Phys. Lett. **93**, 17002 (2011)
16. Song, E.B., Lian, B., Kim, S.M., Lee, S., Chung, T.-K., Wang, M., Zeng, C., Xu, G., Wong, K., Zhou, Y., Rasool, H., Seo, D., Chung, H.-J., Heo, J., Seo, S., Wang, K.: Robust bi-stable memory operation in single-layer graphene ferroelectric memory. Appl. Phys. Lett. **99**, 042109 (2011)
17. Wang, H., Wu, Y., Cong, C., Shang, J., Yu, T.: Hysteresis of electronic transport in graphene transistors. ACS Nano **4**, 7221–7228 (2010)
18. Kurchak, A., Morozovska, A., Strikha, M.: Rival mechanisms of hysteresis in the resistivity of graphene channel. Ukr. J. Phys. **58**, 473–479 (2013)
19. Strikha, M.: Mechanism of anti-hysteresis behavior in graphene-on- $Pb(Zr_xTi_{1-x})O_3$ substrate resistance. JETP Lett. **95**, 198–200 (2012)
20. Kurchak, A., Strikha, M.: Antihysteresis of the electrical resistivity of graphene on a ferroelectric $Pb(Zr_xTi_{1-x})O_3$ substrate. JETP **116**, 112–117 (2013)
21. Ohtomo, A., Muller, D.A., Grazul, J.L., Hwang, H.Y.: Artificial charge-modulationin atomic-scale perovskite titanate superlattices. Nature **419**, 378–380 (2002)
22. Fong, D.D., Kolpak, A.M., Eastman, J.A.: Stabilization of monodomain polarization in ultrathin PbTiO3 Films. Phys. Rev. Lett. **96**, 127601 (2006)
23. Nair, R.R., Blake, P., Grigorenko, A., Novoselov, K., Brooth, T.J., Stauber, T., Peres, N.M.R., Geim, A.: Fine structure constant defines visual transparency of graphene. Science **320**, 5881 (2008)
24. Strikha, M.V., Vasko, F.T.: Electro-optics of graphene: field-modulated reflection and birefringence. Phys. Rev. B. **81**, 115413 (2010)
25. Orlita, M., Potemski, M.: Dirac electronic states in graphene systems: optical spectroscopy studies. Top. Rev. Semicond. Sci. Technol. **25**, 063001 (2010)
26. Vasko, F.T.: Saturation of interband absorption in graphene. Phys. Rev. B. **82**, 245422 (2010)
27. Liu, M., Yin, X., Ulin-Avila, E., Geng, B., Zentgraf, T., Ju, L., Wang, F., Zhang, X.: A graphene-based broadband optical modulator. Nature **474**, 64–67 (2011)
28. Bao, Q., Zhang, H., Wang, B., Ni, Z., Haley, C., Lim, X., Wang, Y., Tang, D.Y., Loh, K.P.: Broadband graphene polarizer. Nat. Photonics **5**, 411–415 (2011)
29. Strikha, M.V., Vasko, F.T.: Carrier-induced modulation of light by a gated graphene. J. Appl. Phys. **110**, 083106 (2011)
30. Strikha, M.V.: Bi-stable optical system, based on hysteresis in graphene-on-$Pb(Zr_xTi_{1-x})O_3$ reflectivity. Ukr. J. Phys. Optics. **13**, 45–50 (2012)

Scanning Probe Microscopy in Practical Diagnostic: 3D Topography Imaging and Nanometrology

Petro M. Lytvyn

Abstract This overview presents a common approach of practical atomic force microscopy (AFM) diagnostic of surfaces at the sub-micrometer and nano-meter levels. A common metrological model of AFM and sources of uncertainty of measurements are analyzed. Procedures for scanner and tip calibration are presented. Application of precise topometry concerning geometrical sizes of surface features and its metrological traceability are illustrated using original data of systematic AFM diagnostics applied to semiconductor nano-structures with quantum dots grown by molecular beam epitaxy. A number of weighty results important to understand physics of processes during structural ordering in low-dimensional semiconductor systems has been described. Physical, methodological and experimental parts have been presented without extended details that could be found in the complete list of references.

1 Introduction

In actual scientific and technological investigations, methods of scanning probe microscopy (SPM) comprise one of the advanced positions. For several decades of its existence, SPM was developed as a separate area of scientific and engineering explorations. Realized for the first time by IBM collaborators in 1981, the tunnel microscope [1–3] became the father of a new generation of microscopes based on the idea of local diagnostics of surface properties by using the probe body (probe, sensor) with the size of operation area (tip) close to several unities or tens of nanometer.

P.M. Lytvyn (✉)
V. Lashkarev Institute of Semiconductor Physics, National Academy of Sciences of Ukraine, 41, Nauky Pr, Kyiv 03028, Ukraine
e-mail: plyt@isp.kiev.ua; peter.lytvyn.isp@gmail.com

A. Nazarov et al. (eds.), *Functional Nanomaterials and Devices for Electronics, Sensors and Energy Harvesting*, Engineering Materials,
DOI 10.1007/978-3-319-08804-4_10

From the historical viewpoint, the principal ideas of scanning probe microscopy can be dated back to 1928, when E. H. Synge proposed a theoretical approach to overcome the diffraction limit in conventional optical microscopy. Scanning with a small nanoscale probe (sub-wavelength aperture) over a sample in proximity to its surface was suggested. The idea to use tunneling current to control stylus-surface distance was described by Russell Young in 1966 [4]. The first instrument, where the probe scanned over surface and measured height of single atomic steps using tunneling current was introduced by his scientific group in 1971 [5]. Ten years later, group of Binnig (Nobel Prize in 1986) demonstrated first surface image with atomic resolution obtained by STM [3]. Then, Binnig et al. realized the scanning probe microscope operating due to local force interaction in 1986 [6, 7].

In a general case, current scanning probe microscopes are high-tech diagnostic facilities combining in one device the whole complex of means for performing surface diagnostics. As a rule, these methods are based on physical effects of electric and force interaction between the probe and surface. The above mentioned scanning tunnel microscopy (STM), atomic force microscopy (AFM) and its derivatives: electrostatic force microscopy (EFM) [8, 9], scanning force Kelvin-probe microscopy (SFKPM) [10, 11], magnetic force microscopy (MFM) [12] are related to these methods. Also widely used are SPM methods for local diagnostics of surface electrical properties where the AFM method provides control of force interaction "probe—surface" as well as mapping the relief, while independent channels of measurements register current flow (conductive atomic force microscopy [13, 14]), local resistance (scanning spreading resistance microscopy [15]) or the capacitance "probe—surface" (scanning capacitance microscopy [16]).

In addition to mapping the surface properties, the above methods allow for obtaining respective spectroscopic data in a chosen point. For example, atomic force spectroscopy (that enables to measure the dependence of the interaction force on the distance probe—surface) possesses a sufficient range of sensitivity to obtain important information on specificity of intermolecular interaction [17, 18], on the one hand, and perform nanomechanical investigations of local surface properties [19], on the other hand. In their turn, conducting AFM and EFM allow obtaining the current-voltage and capacitance-voltage characteristics of surfaces with a high spatial resolution (area of the contact is close to 10–100 nm^2) [20–22]. Besides, most of the series SPM models are rather efficiently used in intentional modification of a surface in nano-probe lithography, manipulation and preparation of nano-objects [23, 24].

A cogent advantage of SPM is its capability to obtain reliable data about surface micro- or nano-relief both in vacuum and in ambient atmosphere as well as in liquid medium. The objects of investigations do not require any special preparation as it takes place in some other methods. It is natural that preparation of the samples is an important stage of SPM diagnostics, but in this case it is directed not to modification of the object, which provides usability of this diagnostic method, but to provide specific conditions for measuring the necessary properties. For example, it is essential to remove random contaminations and adsorbent layer from nano-structured surface before mapping their surface or application of more complex protocols for preparation of biological objects, namely: separation, extraction, immobilization on the substrate and functionalization of the AFM probe tip. Due to the extremely wide spectrum of diagnostic methods, SPM is efficiently used in various scientific and technical areas, starting from fundamental investigations in physics of surfaces, applied materials science, diagnostic of functional elements in nano-electronic devices, and up to nano-medicine and biosensor technologies.

Particularity of SPM application sets respective requirements on the way of their hardware realization. Fundamental investigations at the atomic level require ultrahigh-vacuum systems providing purity and stability of the studied surface as well as high sensitivity and speed of measurements. For solving most of tasks in applied diagnostics of surfaces at micro- and nano-levels, SPM methods operating in air and liquid media and possessing the field of view up to $100 \times 100\ \mu m^2$ are the most suitable. In biomedical investigations, the main priority in the SPM construction is the possibility to perform measurements in various liquid media as close as possible to the native ones and convenience to operate with biomaterials.

However, despite relative simplicity in performing SPM investigations, obtaining reliable data and their scientific interpretation require understanding of physical processes of probe-surface interaction, knowledge on how one should separate head and minor factors that influence formation of SPM images in dependence of measurement conditions. Not less important component is metrological aspects of SPM diagnostics. Here, main questions are those concerning calibration of scanners, shape of the probe tip, determination of mechanical parameters describing cantilevers in measurements based on the AFM method, usage of respective test structures in electric-force and capacitance measurements.

Despite the fact that SPM is one of the most simple and convenient tools for obtaining quantitative topographical data in the micro- and nano-scale range and providing better accuracy than other microscopic techniques, determination of other exact quantitative surface parameters (chemical, mechanical, magnetic, electrical) is mated with considerable methodical difficulties. In most of these cases, SPM is used for qualitative and semi-quantitative estimation, and currently is not recommended for metrological purposes. However, even rude SPM measurements and mapping the above parameters are of a great scientific and applied interest. Besides, hardware and methodical bases in SPM are continuously perfected, which constantly lowers uncertainties in measurements.

A lot of monographs and informative reviews are devoted to generalization and systematization of theoretical approaches in considering the features of probe-surface interaction in diversity of SPM methods as well as methodical and applied SPM aspects of diagnostics. However, there is a lack of examples of application of systems approach in solution of diagnostical tasks, which could combine various methodical and analytical approaches with account of metrological peculiarities inherent to SPM diagnostics. We hope that description of our own experience concerning complex application of SPM diagnostics in a number of applied tasks, peculiarities of performing measurements and interpretation of the data obtained will be useful both for specialists and beginners in this field.

2 Atomic Force Microscopy and Spectroscopy

2.1 Atomic Force Microscopy: Physical Principles and Technical Realization

Let us remind the basic principles of functioning and hardware realization of the scanning atomic force microscopy, as this method is most often used in diagnostic of functional materials and device structures. The AFM method is based on one of the most universal interactions in nature—attraction and repulsion between bodies. The basic set-up of the modern SPM for scientific and applied investigations can be represented by the following components (Fig. 1): tip, scanner for displacement of the tip, system for registration of parameters corresponding to tip-surface interaction and the feedback loop, console for control and visualization of measurement results, system for vibration and noise isolation.

In the atomic force microscope, a special monitoring system performs precise raster displacement of solid probe (micromachine in the form of tip with the radius of 5–30 nm that is fixed to elastic cantilever) above the studied surface. The force of probe-surface interaction is kept constant due to changing the probe height above the surface. Values of voltages on piezoelements of the three-coordinate displacement system (scanner), by using preliminary calibration, are converted by AFM software into 3D map of the surface.

Bearing in mind the way providing force interaction probe-surface, the modes of AFM operation can be separated by three groups, namely: continuous contact, periodical contact ("intermittent contact mode") and noncontact ones. Latter two modes use modulation methods, where the probe vibrates with the frequency of its mechanical resonance (or near it) and changes in amplitude, frequency or phase inherent to these vibrations are monitored in the feedback loop. These methods are also named the dynamical ones. When mapping a relief in the contact mode, it is important to keep the constant value of the probe cantilever deflection, which corresponds to the state of equilibrium of all the forces acting on the probe from the surface and the force caused by elastic deformation of the cantilever. Measurements of absolute deflection values are not necessary when mapping the surface. By analogy, in the other modes one should monitor only changes in the vibration amplitude or shift of phase/frequency in topometric measurements.

However measurements of the absolute value for the deflection of the probe cantilever are important in force spectroscopic investigations. If sensitivity of the measuring system to probe deformation, cantilever spring constant and the vertical scanner translation are known, one can obtain quantitative dependence of the force value for probe-surface interaction on the distance. A typical measurements scheme and the force-distance curve are shown in Fig. 2. In the position I, the probe is far from surface, and interaction between them is missed. In the position II, the probe is approached to the surface so close (movement along a curve is indicated by arrow), that jumps to contact with the surface due to action of van der Waals attraction forces. Thus, if the measurement takes place in air, the layer of

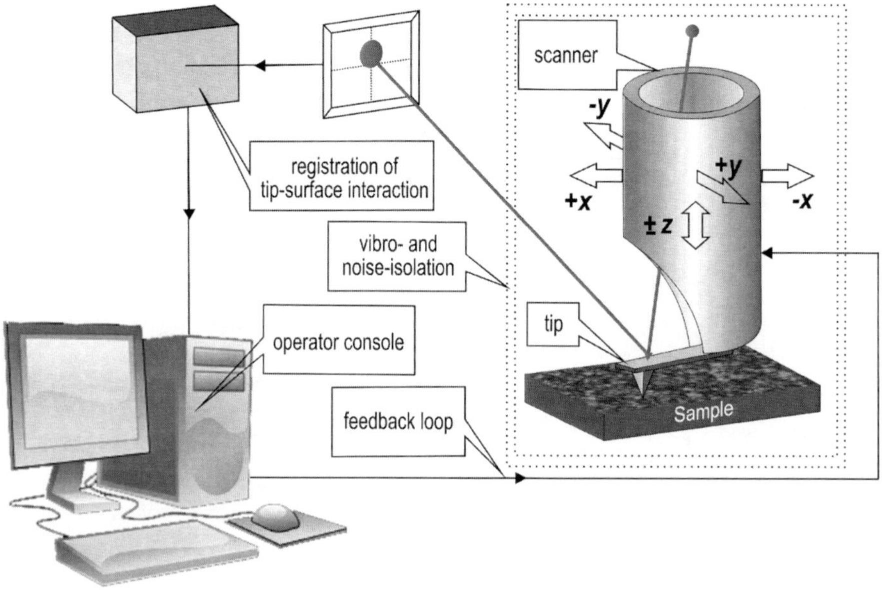

Fig. 1 General functional setup of a scanning probe microscope

liquid condensed on the surface can play a significant role. In the position III, the probe reaches the given maximum of the repulsion force and is withdrawn from the surface. Hysteresis occurs due to adhesive attraction forces. In the position IV, the elastic deformation force of deflected cantilever exceeds the adhesive force and probe released by surface. In the position V, the AFM probe cantilever returns to the equilibrium state.

In terms of distance, we have the following situation. When the AFM probe approaches to surface, it begins to perceive long-range electrostatic and magnetic interactions starting from the distance close to 1 μm. At the distances of 10–100 nm, the main force interactions will be long-range van der Waals interactions. Even closer, in ambient conditions, water bridges can appear between tip apex and surface due to capillary condensation. Charge transfer via tunneling appears, and van der Waals forces become dominant at the distances of 1–10 nm. At contact, the Coulomb repulsion takes place. Deformation of AFM tip or sample can occur at high values of applied forces.

There are various physical models describing tip-surface interaction depending on acting forces [25, 26]. The simplest model for an interatomic force that covers both the short range repulsive and long-range attractive interactions is based on the Lennard-Jones (LJ) potential. It can be described by the following formula:

$$V = 4\varepsilon\left[\left(\frac{\sigma}{r}\right)^{12} - \left(\frac{\sigma}{r}\right)^{6}\right]$$

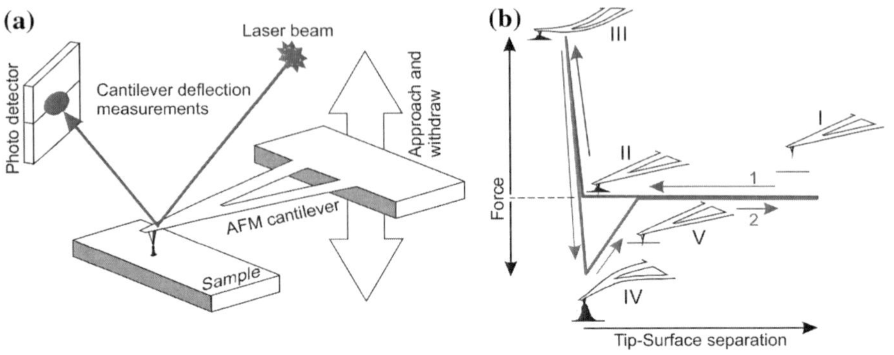

Fig. 2 Scheme of measurements (**a**) and schematic view of force curves when measuring in air (**b**): approaching curve (*1*), withdrawing curve (*2*)

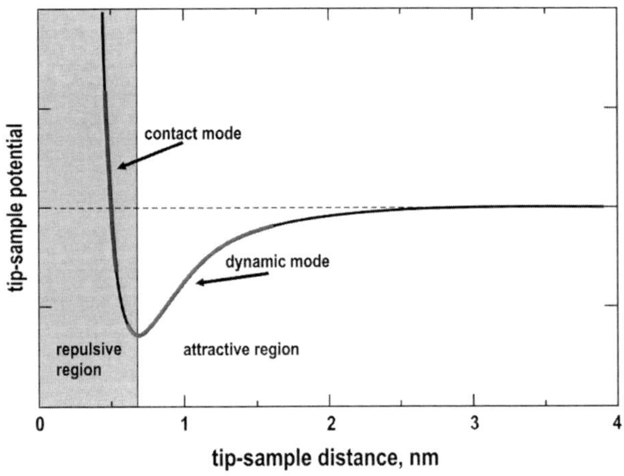

Fig. 3 Typical tip-sample interaction potential dependence on distance with marked regions of contact and non-contact mode performance

where ε corresponds to the depth of the potential well (reflected interaction strength) and σ is the interatomic distance where the potential is zero, r is an interatomic distance for a system of two LJ particles. Schematically, the dependence of the LJ potential on the tip-sample distance is shown in Fig. 3. Marked in the figure are areas where contact and dynamic AFM modes are realized.

A good starting point to understand force curves can be found in B. Cappella review [27]. The raw AFM measured force curve does not reproduce a clear tip-surface interaction forces. The sum of interaction forces F(D), elastic force of cantilever and its deflection are hidden in the force-displacement data. Besides, in the vicinity of the extreme observed for the F(D) function, there exists an ambiguity leading to difference between curves of approaching the probe to and removing it from the surface. Figure 4 shows the dependence of the interatomic

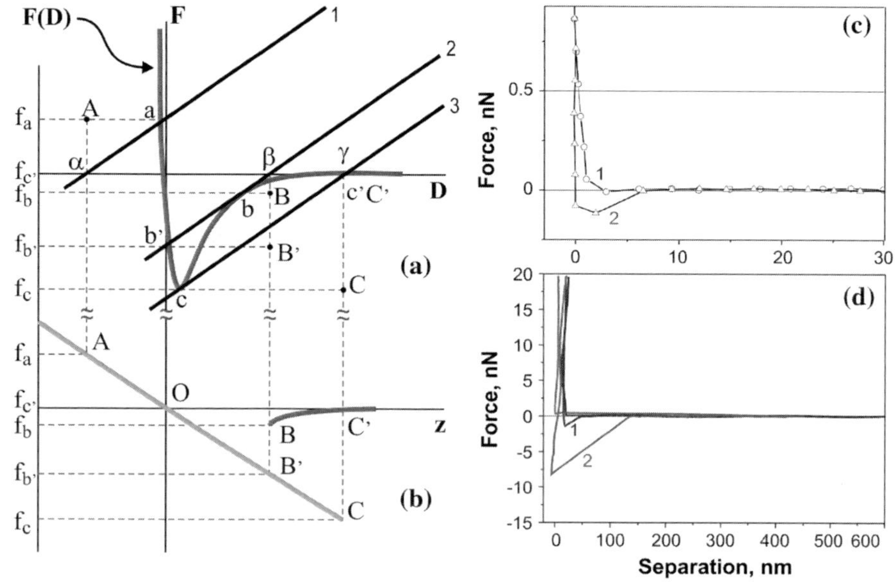

Fig. 4 Construction of experimental force-distance curve. Dependence of the Lenard-Jones force on distance (**a**, curve) and the force of elastic deformation of tip cantilever (**a**, lines 1, 2, 3). The respective curve measured with AFM (**b**). Experimental force—tip surface separation curves (**d**) recorded in air (*1*) and under water buffer (*2*). Jump-to-contact of corresponding curves shown in (**c**)

Lennard-Jones force on the distance [27]. In these coordinates, the force of tip cantilever elastic deformation is represented with a straight line with the slope equal to the cantilevels spring constant in accord with Hooke's law. The force of tip-surface interaction is balanced by the force of cantilever elastic deformation in every point of the force curve that corresponds, for example, the intersection points in the F(D) dependence and straight lines 1, 2 and 3. It is seen that moving the straight line from the right to the left (curve of approaching), in the region between c' and b' (Fig. 4a) we obtain three intersections and hence three equilibrium positions. Two of these positions (between c' and b and between c and b') are stable, while the third position (between b and c) is unstable because of two possible points of balance. At the stage of approaching, the tip follows the trajectory from c' to b and then "jumps" from b to b' (i.e., from the force value f_b to $f_{b'}$,). During retraction, the tip follows the trajectory from b' to c and then jumps from c to c' (i.e., from f_c to $f_{c'}$).

These jumps correspond to the discontinuities BB' and CC' in the Fig. 4b. Thus, the region between b and c is not measured. The difference in path between approach and withdrawal curves is usually called "force—displacement curve hysteresis". The two discontinuities in force values are called "jump-to-contact" in the approach curve (BB') and "jump-off-contact" in the withdrawal curve (CC').

Caused by the above reasons, the force curve measured with the microscope has the look shown in Fig. 4c, d. The ambiguity in the vicinity of CB points can be avoided by increasing the stiffness of the probe cantilever. However, on the other hand, it will cause a loss in sensitivity. Therefore, the choice of elastic parameters inherent to the cantilever depends on peculiarities of the solved task. It is noteworthy that for most of routine spectroscopic AFM measurements minimization of "jump-to-contact" and "jump-off-contact" effects is not critical. Curves bring sufficient information to recover parameters of real probe-surface interaction after separation of the probe elastic component. Besides, as seen from Fig. 3, dynamic AFM methods operate in the very "problem" range. In these modulation methods, they use more complex models to analyze the obtained data [28, 29].

2.2 *Atomic Force Microscopy 3D Metrology for Assessment Surface Topography*

From the viewpoint of using SPM for maintenance of up-to-date nanotechnologies, metrological traceability of measurements is very important [30–33]. Although, nanotechnology now should be understood as science and technology of the structures, in which sizes of separate elements lie within the range 0.1–100 nm, nanometrology essentially covers this diapason. Measurements should be performed with the accuracy lying inside this or less dimensional diapason. To solve these tasks, scanning probe microscopes are ideal candidates that are capable to provide 3D-measurements of geometrical sizes and diverse physical and chemical properties of objects in the dimensional scale from parts of angstroms up to hundreds of micrometers.

On the one hand, flexibility and multipurpose character of SPM methods provides their wide application in various fields of science and technique (materials science, electronics, optics, energetic, food industry, biology, pharmacology, medicine, etc.), but on the other hand, it awfully complicates development of joint standards for SPM measurements. Probe microscopy of various purposes essentially differs by its hardware realization, list and level of fulfilling the measurement methods, analytical software, and so on.

However, to verify and calibrate SPMs, one should use common unified approaches that could provide worldwide comparability of measurement results and metrological traceability. The most accepted instructions are usually documentary standards developed in committees of the International Standardization Organization ISO. The ISO committee in charge of drafting standards for SPM is the Technical Committee ISO/TC 201 Surface Chemical Analysis, mainly its subcommittee SC 9 Scanning Probe Microscopies established in 2004. Up to date, about a dozen of normative documents are under development, with the first ones already published or nearing completion [34].

Standards regulate both using the respective terminology and set procedures for testing the microscope units as well as parameters of probes. Developed in addition are also the standards that regulate usage of separate methods for SPM measurements (see, for instance [35]). Beside these standards, important are those that regulate determination of respective quantitative characteristics by the data of SPM measurements. First positions in the list of these standards can be occupied by the standards for measuring the geometrical parameters of surfaces [36, 37].

Except standardization of the very SPM, respective technical committees develop standards that regulate performing the nano-technological measurements, because transfer of technologies from the stage of researches through fabrication up to commercial market requires neatly defined estimation criteria. Metrology is called on not only control production but provide solution of matters concerning the ecological safety, legal and ethical aspects. The results of analytical researches made in Europe and USA indicate that standardization is one of key issues that bound commercial realization of micro- and nanotechnologies [38]. The respective international organizations work at these tasks. In particular, created in 2005, in the International Standard Organization (ISO), are the technical committee (TC) 229 "Nanotechnologies", and in the International Electrotechnical Commission (IEC)—TC 113 "Nanotechnology standardization for electrical and electronic products and systems".

At the national level, adaptation (implementation) of these standards is executed by respective authorized bodies. Leaders in development of SPM standards are Asian countries, in particular, Japan and Taiwan. In Europe, the largest activity is demonstrated by Germany, and among the countries of Former Soviet Union—Russia and Kazakhstan.

The above mentioned standards allow formulation of requirements to SPM parameters providing solution of specific tasks in diagnostics of functional materials, or, on the other hand, to determine the range of tasks that can be solved using the specific SPM method. It should be noted that technical parameters announced by a producer can be considered in this case only as an approximate qualitative indicator and must be tested [39]. Like to any measuring device, estimation of SPM should begin from construction of its metrological model. First of all, one should determine sources of uncertainties in measurements and characterize them in accordance with adopted standards. In what follows, we shall show an example of metrological estimate for the probe microscope NanoScope IIIa Dimension 3000.

There are many error and uncertainty sources in SPM, however, basic errors and uncertainties that can be observed in any SPM could be classified as follows: external influences, intrinsic influences and the operator-related ones. External factors are determined by surrounding where this facility operates. It implies stabilization of climate conditions for exploitation (temperature, humidity), protection from noise and vibrations, quality of power supply and grounding, etc. Internal factors are determined by construction features of the device and quality of their accomplishment. The operator-related ones include quality of positioning and device calibration, optimum in the choice of probes, accuracy of adjustment of

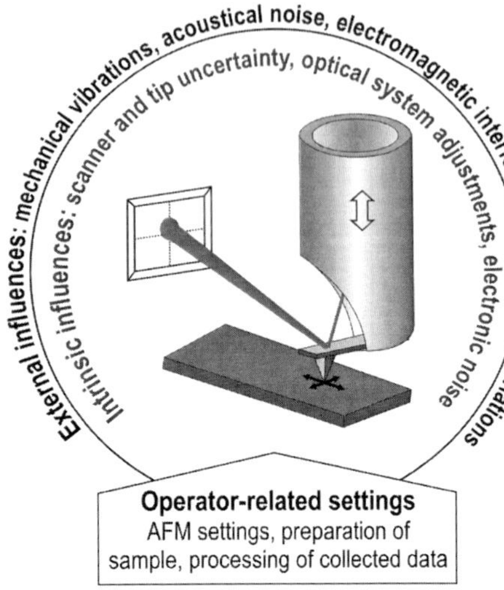

Fig. 5 Scheme of the AFM metrological model

measurement parameters, choice of algorithms for mathematical processing the data, etc. All these uncertainties can be summarized as the structural scheme shown in Fig. 5.

It should be noted that a considerable part of external and internal sources of uncertainties is constant and can be rather efficiently minimized. For instance, usage of the systems for air cleaning and climate-control, especially in the case of SPM operating in ambient air, active or passive systems for vibro- and noise-protection, individual electrometric grounding are one of the main ways to minimize external influences.

Among the intrinsic factors, the common for SPM source of uncertainties is a scanner. Such its characteristics as sensitivity, drift and creep should be the objects of special attention. In general, metrological traceability of AFM in tasks of mapping the surface lies in piezo-scanner calibration (i.e., determination of the dependence for the displacement value on the applied voltage).

The method consists of two main parts: calibration of the scanner movement within the XY plane as well as along the Z axis. It is realized using special calibration test-structures that are made, as a rule, applying technologies of modern semiconductor electronics. Figure 6 shows an example of results obtained when calibrating the scanner of NanoScope IIIa Dimention 3000. Linearity of the scanner and accuracy of measurements were estimated using the test-structure Au/Si (received from the manufacturer of this facility) with the depth of elements 180 ± 3 nm and period 10 μm. In the image, absence of distortions in the shape of square recessions in various sections of image is indicative of the linearity in the scanner's movement, while correspondence in sizes confirms accuracy in calibration.

Analysis of these values should be carried out by using both separate profiles of cross-sections (Fig. 6b) and spectra of the spatial frequencies that can be simply

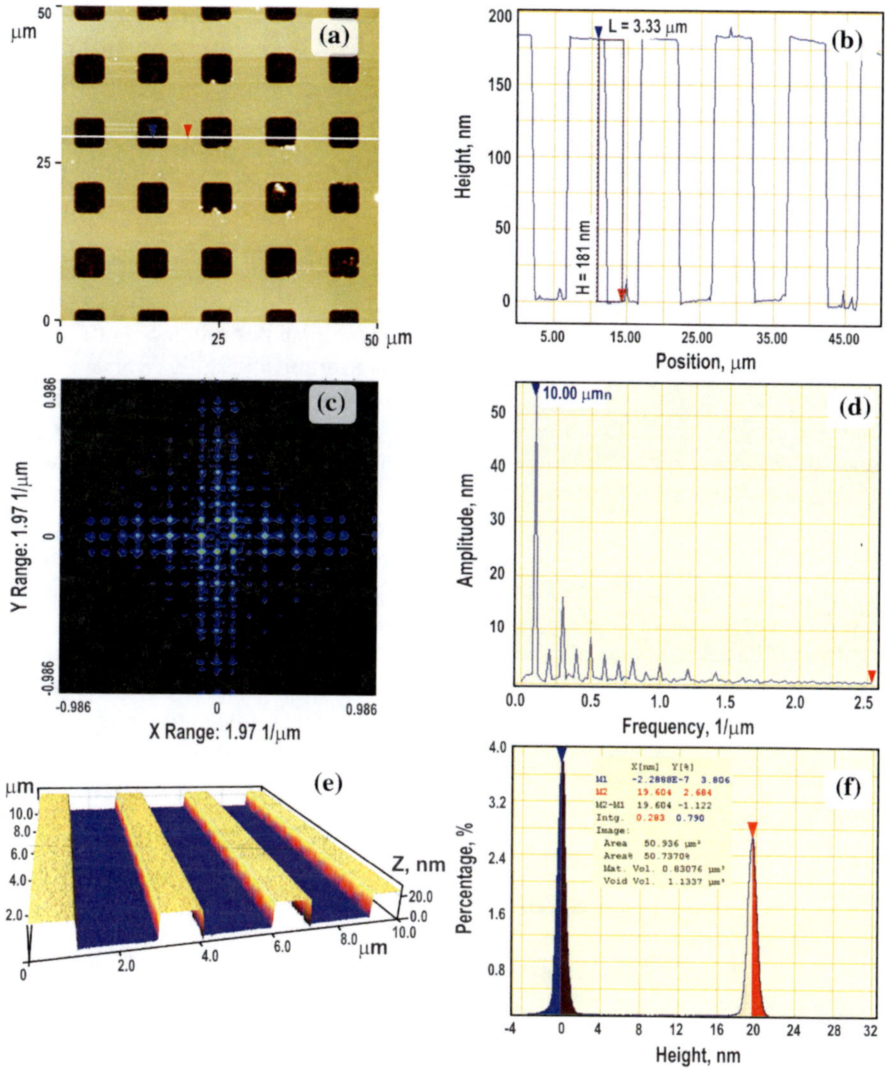

Fig. 6 Results of AFM measurements for test structures: **a**—Au/Si. AFM map of the surface, profile of the surface along a chosen line, histogram of heights, and Fourier-transform for the height profile in this line; **b**—TGZ1 grating. 3D image for the surface map and results of the profile analysis

obtained applying the Fourier analysis (Fig. 6c, d). From the statistical viewpoint, this analysis is more reliable, as it comprises all the points of the image and not the single separated line. By analogy, vertical calibration of the scanner is based on calibration gratings TGZ made by the NT-MDT (Russia) from silicon. Shown in Fig. 6e, f are the results obtained when verifying the measurement accuracy of vertical dimensions for surface elements by using the test grating with the height of 19 ± 1 nm. Again, the analysis of results should be performed with account of local profile measurements and the histogram of heights over the whole AFM image (Fig. 6f).

2.2.1 The Uncertainty Related with the Piezo-Drive (Scanner)

As it follows from tests made within various dimensional ranges, boundaries of deviations for the measured values in sizes of test-structures in horizontal and vertical planes correspond to those claimed in technical performances of this SPM, which means that the accuracy of AFM measurements is no worse than the accuracy in manufacturing the test-structures. In particular, the results adduced in Fig. 6 clearly illustrate that dimensionality and orthogonality both in the XY plane and along the vertical direction in nanometer diapason of sizes are kept. The histogram of heights for the Au/Si test-object shows the most statistically probable thickness of the gold film of 181 nm (distance between two adjacent maxima). As follows from the Fourier-transform for the relief, the most typical value for the period is close to 10.00 μm. Respectively, the height of steps in the test grating TGZ1 averaged by 20 profiles is close to 19 nm.

Being based on the performed calibration measurements, the values for boundaries of deviations in dimensions of test-structures can be adopted as respective boundaries of deviations in the nanometer range of measured sizes $\Delta_{scan} = \pm 1$ nm. Thereof, the standard uncertainty of the scanner is equal:

$$u_{scan} = \frac{\Delta_{scan}}{\sqrt{3}} = \frac{1}{\sqrt{3}} = 0.577\,\text{nm}$$

2.2.2 The Uncertainty Related with an AFM Tip

A real (non-digital) spatial resolution of AFM images is determined by geometrical shape and sizes of the probe tip, as it is the probe that interacts with a sample, and the measuring system reconstructs the surface profile by using the coordinates of its tip apex. As mentioned above, in relation with that, the curvature radius of the tip apex can be commensurable with sizes of surface elements (that furthermore do not possess any axis-symmetric shape), the AFM image is a "convolution" of the tip shape and real surface relief. Depicted in Fig. 7 are 3D-AFM images of quantum wires InGaAs/GaAs recorded using the typical probe [20] with the radius of the apex close to 10 nm (Fig. 7a) and using the ultra-sharp tip [21], the radius of which is less than 1 nm (Fig. 7b). The range of vertical dimensions is 5 nm. It is seen that, due to convolution, there missed is information on a fine structure of wires. Besides, their transverse sizes are overstated. Deviations can reach 60 % of real heights. Thus, based on statistical analysis of AFM images obtained using typical and

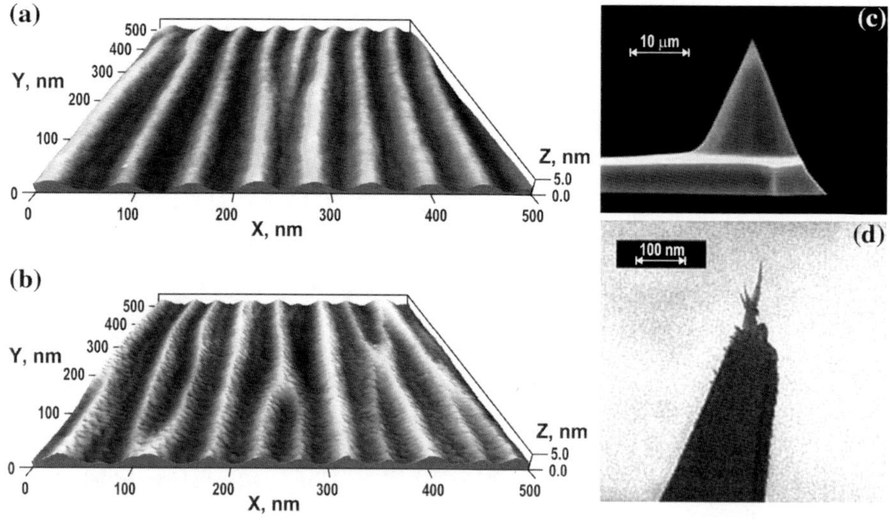

Fig. 7 3D-AFM image of In-GaAs/GaAs quantum wires obtained with a typical tip (**a**) and image of the same surface obtained with an ultra-sharp tip (**b**). Scanning electron microscope images of a typical (**c**) and an ultra-sharp tip (**d**)

ultra-sharp tips, the calculated value of the uncertainty for the typical probe with the apex radius of 10 nm in the sub-nanometer range (below 1 nm) is equal to

$$u_{tip} = \frac{\Delta_{tip}}{\sqrt{3}} = \frac{0.6}{\sqrt{3}} = 0.346\,\text{nm}$$

2.2.3 Uncertainties Related with an Optical System, Digital Electronic Parts and External Factors

When constructing the metrological model of AFM, it was noted that results of measurements can be distorted by external influences, namely: vibrations, acoustical noises, changes in the temperature of ambient medium, etc. Besides, "digital" noise and variations in the sensitivity of laser optical system can introduce some errors. The uncertainties introduced by the above factors can be estimated using the test of AFM noise amplitude on condition that effects of piezo-drive and probe are excluded. This test was realized by scanning the area with dimensions of 1 × 1 nm (which means the practically static mode for the scanner with the maximum scanned area of 100 × 100 μm) on a freshly cleaved mica surface. The test was performed within the time range equivalent to the typical time of measurements (10 min). In these tests, it was ascertained that the noise amplitude equals to 0.05 nm.

So, it was obtained that the total uncertainty related with digital electronics, optical system of AFM, systems of vibro- and noise-protection, as well as systems of conditioning, shielding and electrometric grounding does not exceed:

$$u_{opt} = \frac{\Delta_{vibr}}{\sqrt{3}} = \frac{5 \times 10^{-2}}{\sqrt{3}} = 0.029\,\text{nm}$$

Thus, the maximum total uncertainty of measurements calculated with account of the most essential above mentioned components (which have no correlation bonds) is equal in this case to:

$$u_{total} = \sqrt{u_{scan}^2 + u_{tip}^2 + u_{opt}^2} = 0.675\,\text{nm}$$

In other conditions, the value of the measurement uncertainty will have another meaning. In this case, the ratio of sizes typical for the tip of the probe and for elements of the studied surface can be essential. For instance, in topometric investigations of nanostructural surface elements, the resolution value of the very SPM should be distinguished from that of the image. The real image resolution is determined by the relationship of scan step size and that of probe tip radius. For the apparatus accuracy of horizontal positioning in the SPM scanner higher than 0.5 nm, the image resolution (scan step) on the area 500 × 500 nm, which is recorded into the data array 512 × 512 points, will be close to 1 nm.

Starting from simple geometric considerations, the probe with the tip radius of 10 nm is able to distinguish two surface points with a dimple of 0.1 nm between them, if beginning from the minimum distance of 3 nm between them

$$d_{\min} = (4\Delta z(2R - \Delta z))^{0.5},$$

where R is the radius of the probe tip, Δz—depth of the dimple. The respective resolution will reach 1.5 nm. It is clear that lowering the only scan step (for example, when recording the image of 250 × 250 nm), one cannot reach a higher image resolution, which is caused by too large tip radius. At the same time, the super-sharp probe with the tip radius of 1 nm is able to provide resolution of 0.5 nm, at the scanned area of 300 × 300 nm adequate to it (Fig. 8).

As seen from Fig. 8a, in the case of densely located SnTe nano-islands of small sizes (close to 5–20 nm), there takes place a considerable "tip effect" that includes both expansion of nano-islands and impossibility for the tip to penetrate into narrow dimples between them [40]. In its turn, using the ultra-sharp tip enables to eliminate both causes for distortion of data (Fig. 8b). Comparison of the measurement data obtained with the standard and ultra-sharp tips has shown that the error in determination of such important for analyses of growth processes parameter as the form-factor (ratio of the nano-islands height to the diameter of its basis) can reach 55 %, when scanning with the standard tip.

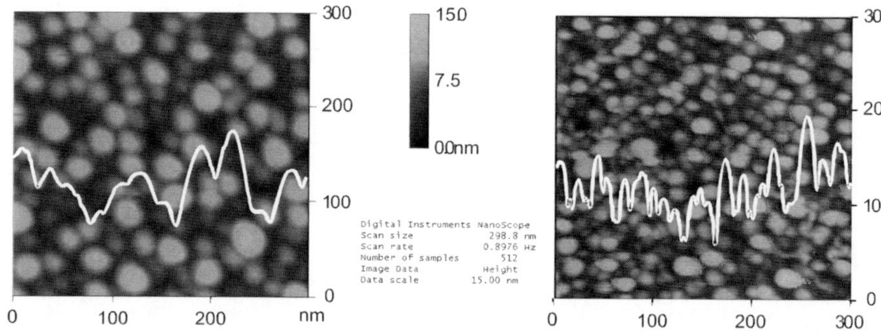

Fig. 8 AFM image of SnTe nano-islands on BaF_2 obtained with standard (*left*) and ultra-sharp (*right*) tips. It is the case of dense location of nano-islands on the surface. White curves are used to illustrate transverse sections of images

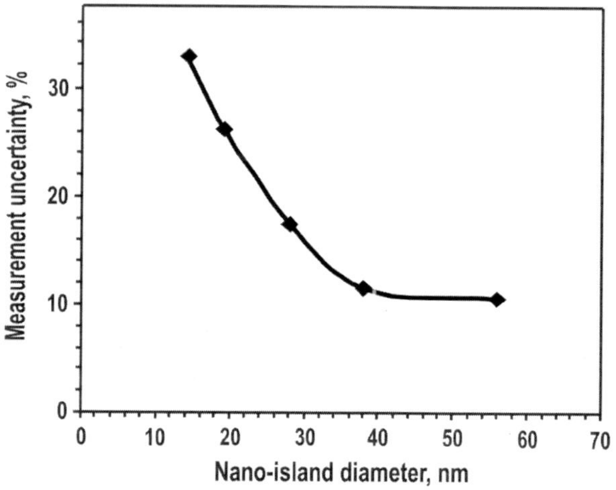

Fig. 9 Errors in measurements of diameters inherent to surface elements in dependence on their lateral sizes. Data were obtained for the silicon tip (apex radius 10 nm) using the images in Fig. 8

However, this situation is changed if investigating nano-islands with the sizes 15–60 nm, if nano-islands are located at a sufficient distances one from another (Fig. 9). The difference between images obtained using ordinary and ultra-sharp tips is related only with expansion of nano-islands (under movement, the tip penetrates up to the substrate). Analyzing the results of grain size measurements in Fig. 8, one can plot the dependence of errors in nano-islands diameters on their sizes for the case of ordinary silicon probes with the nominal tip radius 10 nm (Fig. 9).

The dependence we obtained coincides well with the approximated formula for probe effect correction [41]:

$$D = d\sqrt{(1 - 8Rh/d^2)},$$

where d, h are the diameter and height of nano-islands derived from the AFM image.

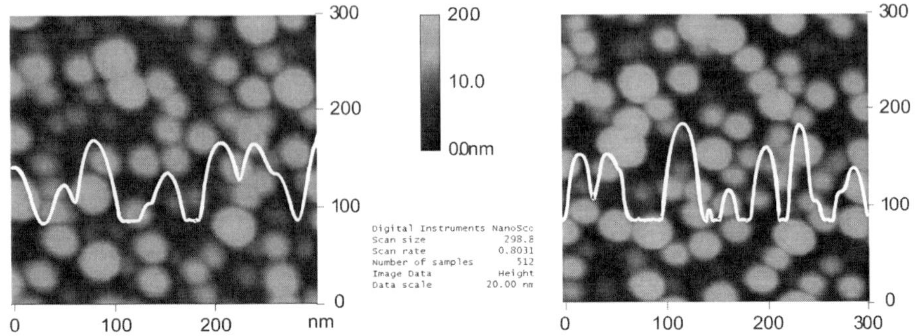

Fig. 10 AFM image of SnTe nano-islands on BaF_2 obtained with standard (*left*) and ultra-sharp (*right*) tips. The case of free location of nano-islands on the surface

This equality is valid only in the case when the vertical resolution of AFM images does not depend on the probe radius, i.e., when the distance between adjacent surface elements is larger than its radius. Thus, in this case (Fig. 10) there is no necessity to use ultra-sharp tips, it is sufficient to use the standard tip for a correct analysis of surface elements.

2.3 Elimination of AFM Probe Geometry on the Surface Image by Using the Method of Computer Reconstruction

The minimization of the probe effect on SPM topometry results can be solved by two main ways. First, one can use probes with a small tip radius and a little apex angle. In this case, various technologies for sharpening ordinary silicon probes as well as technologies for additional growing up the wire-like crystals from diverse materials or carbon nanotubes at the top of probe tip are often used [42–44].

This approach provides an increase of image resolution up to the molecular level. However, in most of cases, ultra-sharp probes are efficient on surfaces with the range of heights up to 20 nm (it does not concern special probes for measurements at the vertical walls of parts with a relief of micrometer height). Besides, their high commercial cost and very limited resource prevent wide usage.

The second way is the computer software reconstruction of experimentally obtained AFM images, when tip contribution is excluded from the image, if the shape of a tip is known [45, 46]. In this approach, the key task is to ascertain the shape of the tip operation part. In this case, except a simple approach with a second order surface [45], the real shape of the tip is usually determined using special test structures [47, 48] or the so-called "blind" reconstruction based on a preliminary measured image [49–51]. The method based on the approach of the tip shape with the second order surface gives the highest error. Two other methods have their own advantages and deficiencies. For example, usage of test structures is optimal

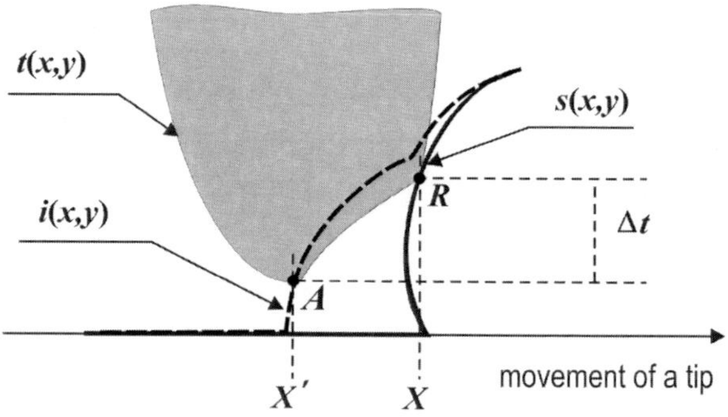

Fig. 11 To reconstruction of a real surface by using the known shape of the tip and recorded AFM image. s(x, y) is the real surface, t(x, y)—tip surface, i(x, y)—suface in non-reconstructed AFM data. A—tip apex (the point monitored by AFM at image recording), R—point of tip-surface contact (the point determined tip-surface interaction)

when their geometrical parameters are not only less than the probe tip ones but are commensurable with components of the relief that should be studied. Besides, this way for testing the tip shape requires additional test measurements before and after topometric surface investigations and is inefficient if the tip shape is changed in the course of measurements. In its turn, the tip shape reconstructed using the blind method has a maximum possible size that still provides obtaining the image under reconstruction [49]. The scheme of method is shown in Fig. 11. Main idea is to find minimal tip-surface distance within the region under the tip (X'X) and correct the height value in this point by Δt.

Thus, the known ways to minimize the probe effect are not self-sufficient, and correct results of topometric investigations of nanostructured surfaces is possible only being based on clear understanding their advantages, deficiencies and limits of usability.

Among the mentioned above methods for obtaining the tip shape, blind reconstruction is the most efficient one. This method does not require any additional measurements, there is no necessity in export/import of data that set the tip surface shape, reconstructed part is the most essential imaging tip part, etc. However, with all its advantages, this method has definite limitations that can essentially influence the results of reconstruction. For instance, if the surface has a regular shape (diffraction gratings, etched monocrystalline surfaces, quantum dots of close sizes and shape, and so on), then the set of possible limits for tip sizes is limited, which can result in imaging the regular relief elements in the shape of sharp peaks; when analyzing smooth surfaces (dispersion of heights Z for which is close to 2–3 nm), the set of possible limits for tip sizes comprises only a very small range [40]. It leads to an unlikely value of the tip radius; apparatus effects of various kinds, which are not related with the probe but take their place in the image (local spikes, noises, scanner drift, too large or too small force of probe-surface interaction, etc.) and can contribute to the tip shape reconstructed by the software.

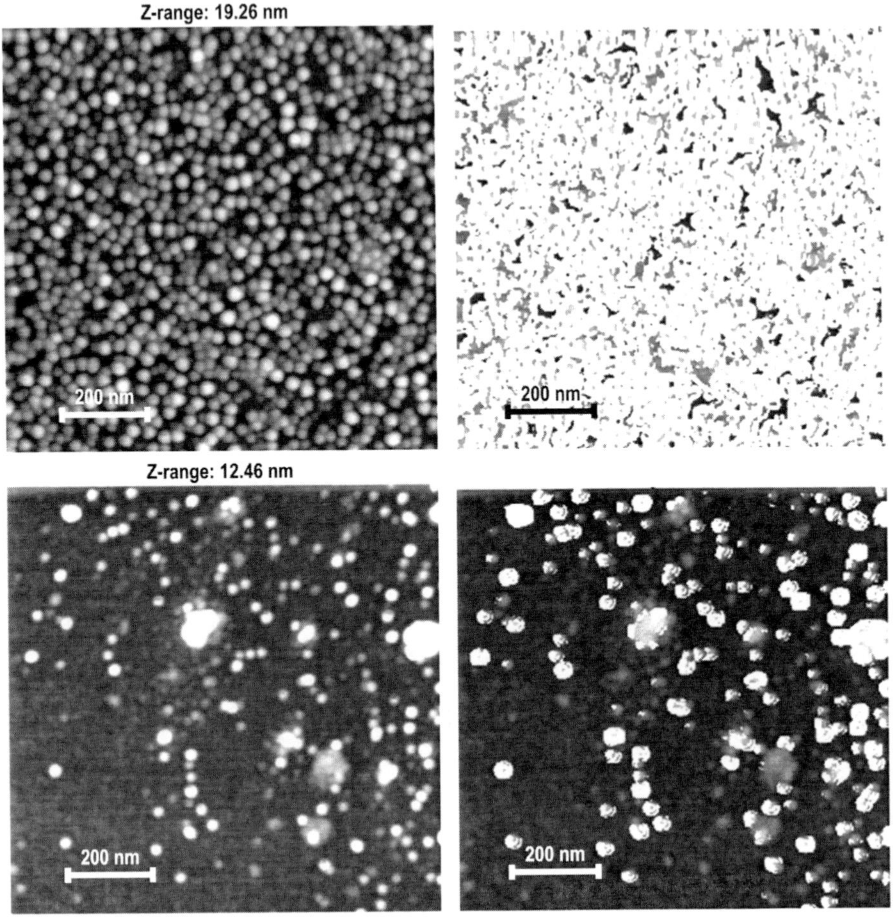

Fig. 12 Non-reconstructed AFM images (*left*) and their uncertainty maps (*right*) for surfaces with different location and shape of nano-islands

It should be also noted that the reconstructed image, independently of the way used to obtain the tip shape, can exactly coincide with the real sample surface under condition that in the scanning process the tip touches only one point of the surface in each time moment. As a consequence, the tip can touch every point of the surface in this process. In other case, there remain the so-called "blind areas" on the surface, and their correct reconstruction is impossible. Their shape and sizes depend on the tip geometrical shape. In particular, for the ordinary silicon tip with the quadrangular pyramid shape, these blind areas are those located with the angle less than the angle at the pyramid apex 11° with respect to the vertical, or pores that are narrower than the tip. The algorithm for reconstruction of AFM images allows to easily depict these surface blind areas as uncertainty maps (Fig. 12), which can serve as a criterion of usability for the tip with given geometry. If the area of blind zones (marked with white color) reaches more than 60 % of the total image area, then the reconstruction program is incorrect, and more sharp probes should be used.

The example of usability of the software image reconstruction for a surface with the area of blind zones close to the acceptable limit in the sample of SnTe with nano-islands is shown in Fig. 13. In this case, the zone at the basis of nano-islands is blind, therefore, reconstruction does not lead to changes in the sizes of the nano-island basis. In relation with it, when analyzing nano-island sizes as lateral characteristics, it is more correct to use the diameter of the nano-island section at the half of its height (see insets in Figs. 13a and b), because at this level reconstruction approaches to the sizes and shape of the real ones, and, if necessary, the basis size can be calculated using extrapolation of the nano-islands shape.

For comparison, shown in Fig. 13c is the image of the same sample obtained in the scanning electron microscope Zeiss Ultra 55 with the resolution close to 1 nm. It is seen that contrary to the non-reconstructed image (where nano-islands have the shape of hemispheres), nano-islands in the reconstructed one are depicted more correctly—as triangular pyramids (for a good layout, see the fragment of 3D representation), which fully coincides with the electron-microscopic data. Some comparison of scanning electron microscope and AFM imaging could be found in [52].

3 AFM Investigations of Semiconductor Quantum Dots Shape and Surface Ordering

Topometrical AFM investigations provide an important information concerning peculiarities of growth processes at nanostructures fabrication. For example, we investigated self-assembled nano-islands in heteroepitaxial GeSi systems grown by molecular beam apitaxy [53]. Strain-induced self-assembled nano-islands in heteroepitaxial GeSi systems have attracted much attention because they offer the possibility of realization of new optoelectronic devices based on the well-developed Si technology. The large lattice mismatch between Ge and Si (4 %) results in the growth of ultrathin Ge layers on Si substrates being driven by the Stranski–Krastanov mechanism. However, this mechanism gets complicated under Ge–Si composition transformation at certain growth conditions.

We have used AFM and micro-Raman scattering to study how the density, volume, shape and composition of Ge islands change depending on the thickness of the deposited Ge layer and the Si substrate temperature. The assertion that bimodal island size distribution is related with two (pyramid-like and dome-like) possible shapes of their equilibrium configuration is confirmed. The island composition is shown to be of mixed Ge_xSi_{1-x} -like type due to surface diffusion of Si atoms from the substrate. This process is strongly enhanced when temperature

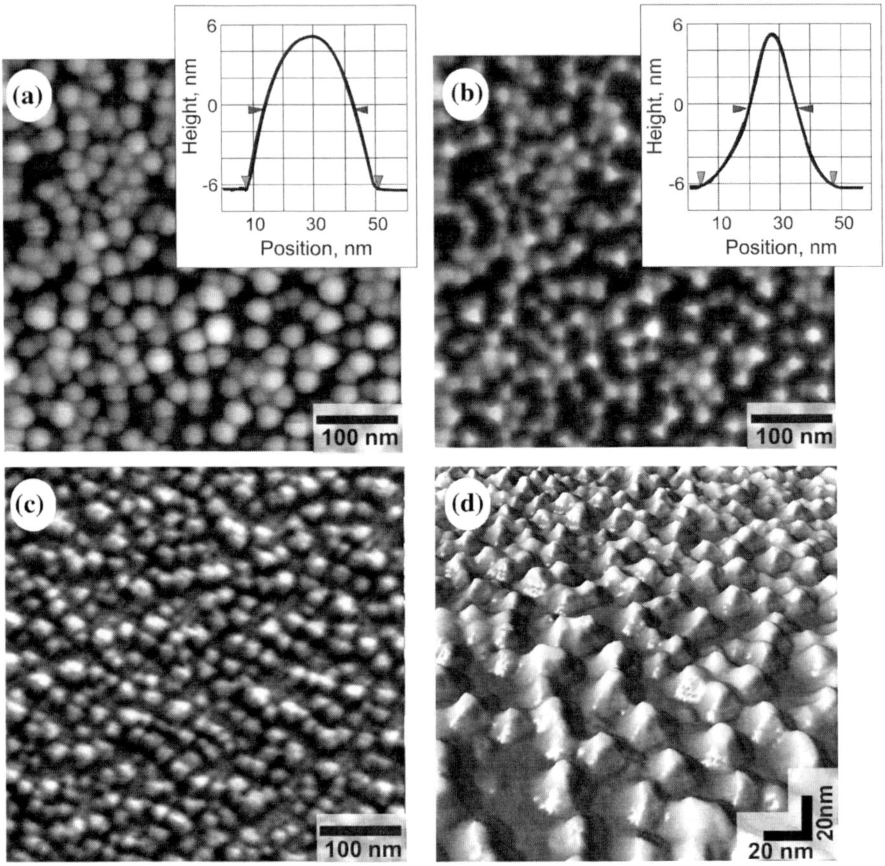

Fig. 13 AFM images of the surface fragment and nano-islands section for the system SnTe/BaF_2 before (**a**) and after (**b**, **c**) computer reconstruction; (**d**)—image of the surface obtained in the scanning electron microscope with high resolution ($\sim$1 nm)

increases. As a result, the stability range of the pyramid-shape island volumes at the measured deposition rate becomes substantially wider. Thus, it is only at low temperature that one can obtain a high concentration of Ge islands on Si(100) surfaces with a narrow size distribution.

The nature of the bimodal size distribution can be explained in two ways. In the first model, one considers that a specific minimum energy configuration corresponds to each particular shape of strained islands, and an activated transition can occur between the two configurations [54]. The key idea of the second model is that the chemical potential of an island undergoes an abrupt change as the equilibrium shape changes from pyramid-like to dome-like [55]. This occurs at a well-defined volume, the one at which the energy of the dome becomes lower than the energy of

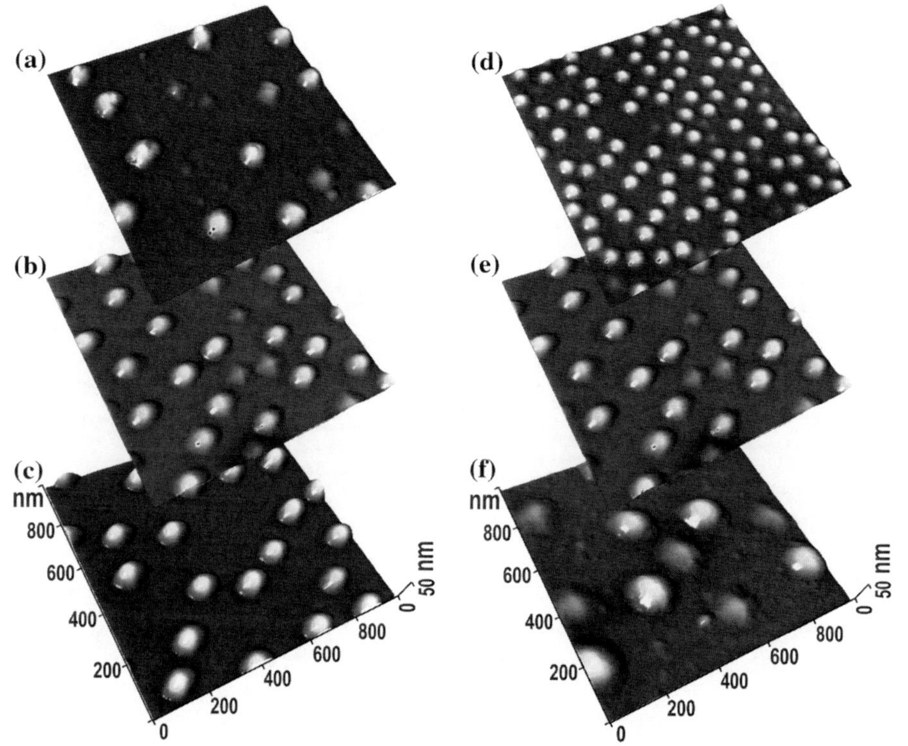

Fig. 14 3DAFM images of self-organized nano-islands grown at 700 °C with the nominal thickness of Ge wetting layers 5.5 ML (**a**), 9 ML (**b**) and 11 ML (**c**). The set of AFM images of nano-islands grown at different temperatures (600, 700 and 750 °C, correspondingly d, f and g) from 9 ML Ge wetting layer

the corresponding pyramid. In any case, the experimental data concerning the distribution of both the shape and size of the islands are of great importance.

AFM images taken for Ge layers of three different thicknesses are presented in Fig. 14a–c. The growth temperature was 700 °C and the growth rate was 0.015 nm/s. Another set of AFM data taken from a series of Ge layers with the nominal thickness of 9 ML, but grown at different substrate temperatures is shown in Fig. 14d–f. The presence of pyramid-like and dome-like islands is clearly shown in Fig. 15 for the 9 ML sample grown at 700 °C.

Plotted in Fig. 16 are the volume distributions of islands presented in Fig. 13. A comparison of Figs. 16a–f demonstrates an important qualitative difference between the island shape transformations occurring in these two series of samples. When the Ge layer thickness increases from 5.5 up to 9 ML (Fig. 16a–c), the critical volume at which the pyramid–dome transformation happens does not change. This volume is about 4×10^4 nm^3, and is shown by an arrow in the figure.

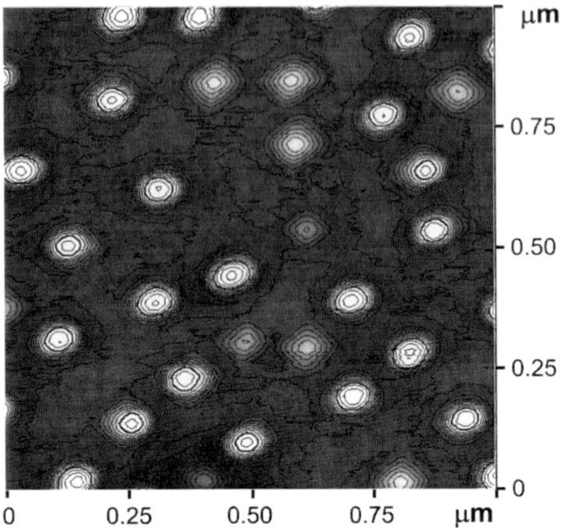

Fig. 15 2D AFM image of the 9 ML sample grown at 700 °C. Relief is shown by isolines with 2 nm height steps

One may also consider this critical volume to remain constant even for $d_{Ge} = 11$ ML, when all pyramids disappear. In this case, all the islands become dome-shaped and have a big average volume (about 12×10^4 nm^3).

When the deposition temperature is increased, the situation becomes quite different Fig. 16d–e). The critical volume of the pyramid–dome transformation shifts strongly toward bigger volumes, and at 750 °C the pyramid-shaped islands are predominant. We believe that this change in the island shape distribution is due to surface diffusion of Si atoms from the substrate to the bottom of the islands [56]. As a result, the islands take on a mixed Ge_xSi_{1-x} composition. The presence of shallow grooves near the bases of the dome-shaped islands that can be seen in the AFM images may serve as support of the mechanism for alloy formation in the islands.

Using Raman Scattering, we have studied the relationship between the island Ge–Si composition and their strains. The results averaged over each sample including islands of different shapes are presented in Table 1. The island compositions for samples with different thicknesses grown at the same temperature are very close. At the same time, as the temperature is increased, the silicon content in the islands grows considerably due to easier diffusion of Si atoms. Since the high stress near the bases of the islands favors intense Si diffusion, this mechanism is very important when the incipient pyramids develop.

To control the strain value in the Si-Ge heterosystem, the $Si_{1-x}Ge_x$ buffer sublayer is used [57–60]. In this case, the areal density of the islands increases with increasing Ge content in the $Si_{1-x}Ge_x$ buffer layer. As the areal density of the nano-islands increases, the spacing between the islands becomes comparable to their size, and lateral interaction between the elastic strain fields of neighboring islands can occur. This interaction promotes in-plane ordering of the islands. On the other hand, the presence of spatially nonuniform elastic strain fields leads to the enhanced importance of interdiffusion processes causing an anomalously

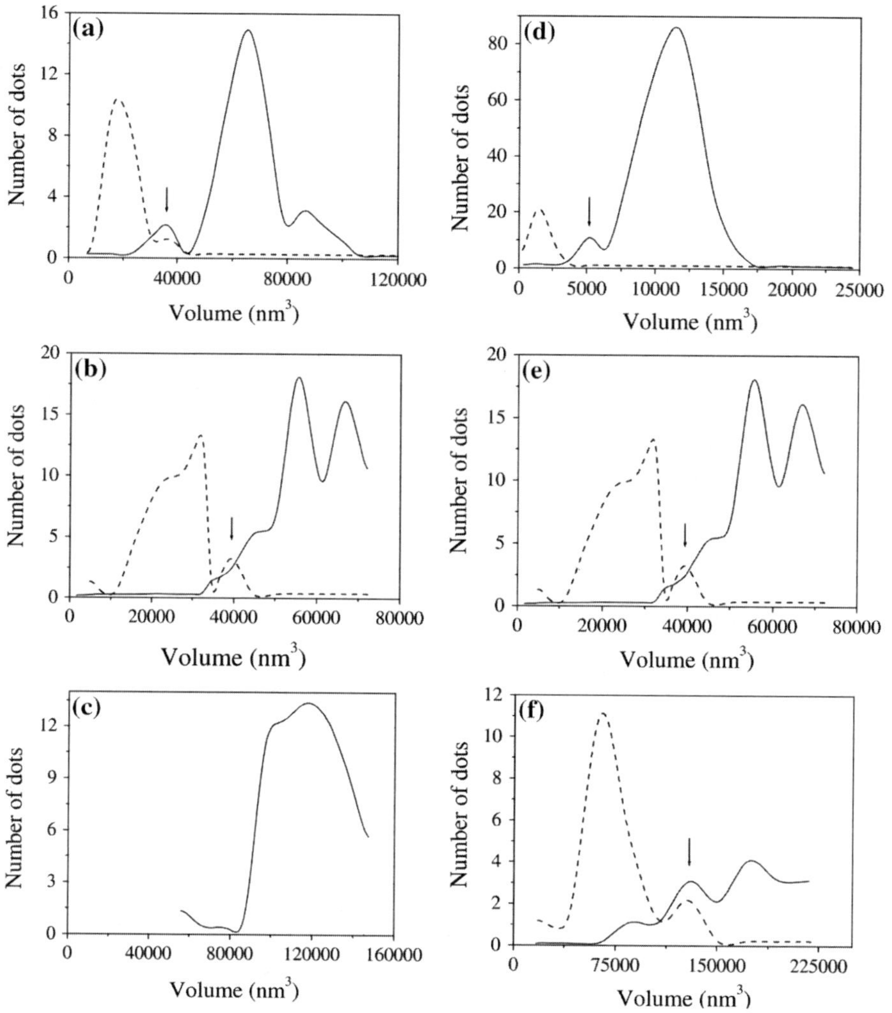

Fig. 16 The dependence of pyramid-shaped (*dashed curve*) and dome-shaped (*solid curve*) island number on volume for samples grown at 700 °C using various thickness of Ge wetting layer (**a** 5.5 ML; **b** 9 ML; **c** 11 ML) and grown from the 9 ML Ge wetting layer at 600, 700 and 750 °C (d–f, correspondingly)

intense atomic flux from the buffer sublayer to the islands, in which partial relaxation of elastic strains takes place.

AFM data presented in Fig. 17 illustrate the lateral self-ordering in single layers of SiGe nano-islands grown on strained $Si_{1-x}Ge_x$ buffer layers of different thicknesses. Surface images and corresponding statistical analysis for the control spots corresponding to 9–11 MLs of Ge deposited during the process of island formation are shown.

Table 1 Composition (x) and strain (ε) of the Ge_xSi_{1-x} islands investigated, as it follows from RS measurements

ML	T (°C)	x, Ge content (%)	ε, strain (%)
9	600	0.73 ± 0.04	−1.3 ± 0.3
9	700	0.63 ± 0.01	−0.8 ± 0.1
9	750	0.43 ± 0.02	−1.0 ± 0.2
5.5	700	0.57 ± 0.01	−1.0 ± 0.1
11	700	0.56 ± 0.02	−1.2 ± 0.2

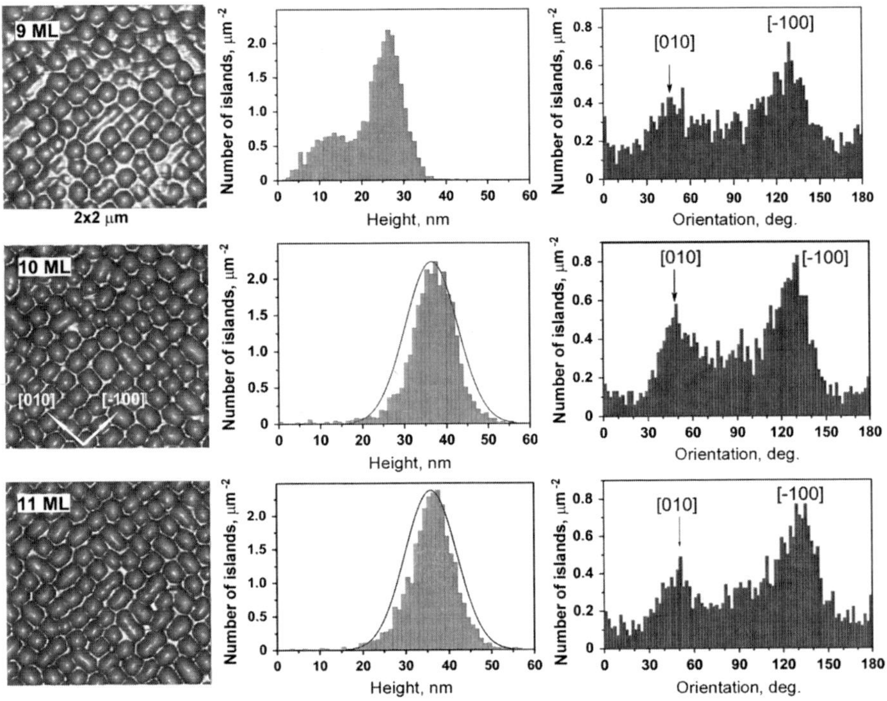

Fig. 17 AFM images of the surface of the structure and histograms for the island heights and the major axis orientation of the island base ellipse for 9–11 MLs of deposited Ge

For 9 MLs of Ge, we find a bimodal distribution of the sizes and shapes of the islands, which are of the hut-cluster and pyramid types. An increase in the nominal thickness of deposited Ge causes transition to a unimodal distribution of dome-shaped islands for 10 MLs of Ge and a further narrowing of this distribution for 11 MLs of Ge. To some extent, the anisotropy in the shape of the island base is indicative of the intensity of elastic interaction between the islands and of the diffusive mass transfer. Thus, for 11 MLs of deposited Ge, the peaks in the distribution of island orientations (i.e., orientations of the major axes of the ellipses approximating the shape of the island bases) are more pronounced (Fig. 17). This is an evidence for a higher degree of anisotropy of the diffusion processes taking place in the process of island formation. It is also important to note that the two maxima in the island orientation distributions are separated by ~82; i.e., the lateral orientation of the islands deviates somewhat from the [−100] and [010] crystallographic directions.

The number and arrangement of peaks in two-dimensional (2D) autocorrelation functions (Fig. 18) built over the 10 × 10 μm scans give clear evidence of the formation of a characteristic two-dimensional grid in the arrangement of nanoislands. The islands are oriented along directions close to [010] and [−100]. The occurrence of three peaks in the profiles taken along the two directions shown in Fig. 18b is indicative of the short-range order in the mutual arrangement of the islands up to the third nearest neighbor. It is most pronounced for 10 MLs of deposited Ge. In this case, a fourth order 2D autocorrelation peak is observed, which gives the evidence of a better defined periodicity in the island arrangement. The spacing between the peaks in the autocorrelation function profiles corresponds to the average distance between the islands in a given direction.

The role of diffusion during the process of island formation could be estimated using analysis of island volumes with increasing thicknesses of deposited Ge. The amount of material in the islands exceeds the nominally deposited amount of Ge by factors of 3.3 and 5 corresponding to 9 and 11 MLs, respectively. It means that up to 60 % of the strained SiGe buffer sublayer is transferred to the islands. Note that in the case of conventional high temperature (≥500 °C) epitaxy on top of the Si buffer, this difference, caused by the diffusion of Si from the buffer layer, can be as large as tens of percent only. Such a huge flux of diffusing atoms at a temperature considerably lower than the material's melting point can only be explained with account of the stimulating role of a nonuniform elastic strain field (the Gorsky effect [61]), the gradient of which in the studied structures with nanoislands may be very high. This strong diffusion of material into the islands during the process of their formation leads to considerable changes in the nominal composition of the layers and strains in the system. It is also evident that kinetics of the diffusion process considerably affects the resulting structural morphology.

Strain-driven self-assembly has matured into a promising method for fabrication of quantum dot nanostructures for semiconductors of various types. Epitaxy of III-V system has attracted more attention due to the unique physical properties of QDs as a zero-dimensional quantum confined system and the variety of applications in electronic and optoelectronic devices (tunable, high-efficient QD lasers [62, 63], single or multicolored QD photodetectors [64–66], etc.).

However, more complex multi-component systems as well as strongly anisotropic structures are not completely described by the simplified basic model of Stranski-Krastanov growth mode. Elastic properties of the bulk crystal lattice and surface mass transport have a significant impact on the QD growth [67, 68].

So, in the case of multi-layer structures of InGaAs QDs, additional effects must be considered namely: the strain distribution through the spacer layers, In migration on the surface during the capping process, and possibly surface roughening

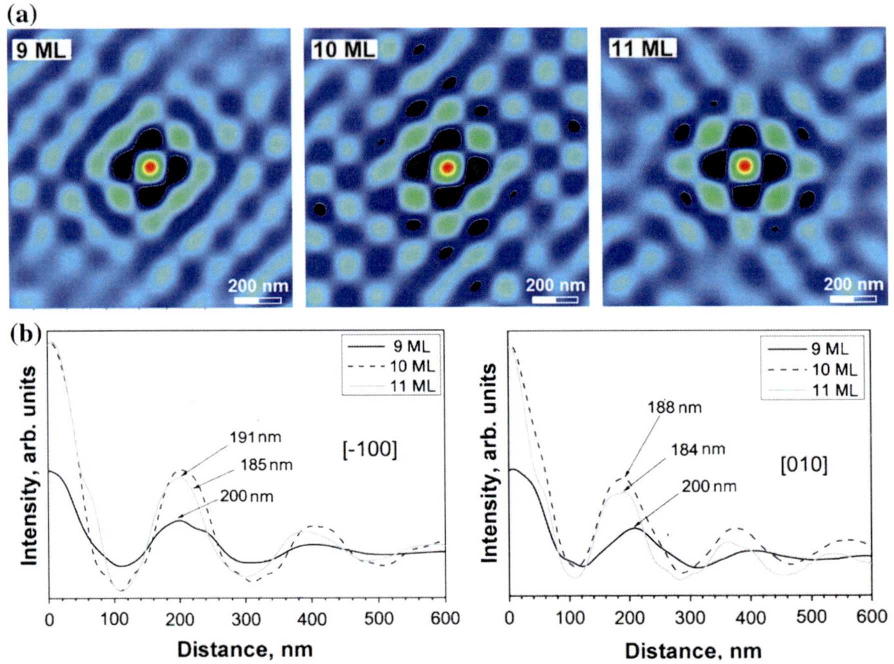

Fig. 18 **a** Autocorrelation maps obtained from AFM scans of the structures under study for 9–11 MLs of deposited Ge **b**. Profiles of the autocorrelation maps along the directions [010] and [−100]

throughout the deposition of the spacer layer. The AFM surface diagnostic of these structures plays a very impotent role for optimization of technological processes and fundamental investigations in physics of low-dimensional structures.

The good starting point for surface anisotropy estimation by AFM could be micro-size defects, known as oval defects typical for MBE grown III-V structures [69]. We investigated surface morphology of micro-size defects on the surface of various high-index GaAs substrates [70]. The investigated surfaces were the top layer of 1- and 17-period $In_{0.45}GaAs_{0.55}$/GaAs structures with quantum dots. These structures were characterized by formation of oval defects on (100) surfaces, and micro-size defects possessing the shape of multifaceted pits and hillocks on (n11)A/B (n = 7, 5, 4, 3) surfaces. We have illustrated that their distribution and density do not depend on the substrate orientation, while the shape and orientation of the micro-size defects depend on the crystallographic orientation of the substrate. This dependence was determined to be the result of anisotropy of surface diffusion and surface elastic properties. Anisotropy of elastic properties of high-index surfaces was found to be the dominating factor in determining the shape of micro-size defects.

Oval defects (ODs) are the main type of morphological defects found on GaAs films grown using the MBE method on GaAs substrates with (100) orientation.

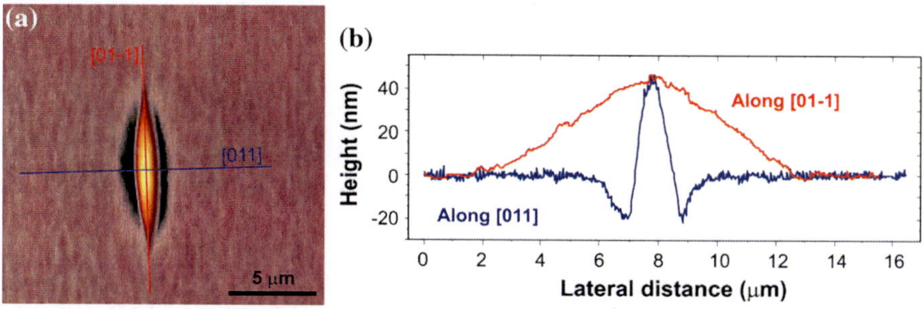

Fig. 19 AFM image emphasizing one oval defect (**a**); corresponding height profiles of the oval defect (**b**)

When using a substrate with (100) orientation, the long axis of the OD lies along the <011> direction [69] (Fig. 19). Typical dimensions of these defects are of the order of several micrometers depending on the epitaxial layer thickness and their density can reach 10^4 cm^{-2}, in our case.

On the (*n*11)A/B surfaces, micro-size defects form as dimples having a multifaceted shape, and in all cases, possess the (100) facet (Fig. 20). As we change the growth surface from (311) to (711), the observed micro-size defects are basically similar to those observed on the (311) surface. However, the tail at the defect base is reduced in length, while the angle at the apex, formed by the defect facets, is increased. Crystallographic orientations of micro-size defects were determined using X-ray diffraction in symmetric and asymmetric reflections. It confirms the constant orientation of defects tails.

As seen from Fig. 21, the crystallographic orientation of micro-size defects coincides with the direction of the largest elastic constants. In multilayer structures grown on GaAs(1 0 0) substrates, the micro-size defects are oriented in the direction [01−1], which coincides with the direction of the largest In/Ga adatom diffusion [71, 72]. When changing the surface orientation from (100) to (711), the anisotropy of the surface elasticity modulus is considerably reduced. In our opinion, it is the effect, which is responsible for the change in the micro-size defect shape. That is, for the (711)B surface, the relative magnitude of anisotropy of the elasticity modulus is smaller than that for (100) and (311), which results in round shape defects (Fig. 21c). On the other hand, a decrease in the defect tail length with increasing n-index on GaAs (*n*11) substrates can be explained stemming from purely geometrical grounds. The defect front facet coincides with the (100) crystallographic plane that has been neither smoothed nor overgrown in the course of the structure growth. It is most probable that the opposite defect facet is also limited by a crystal plane, but it is strongly subjected to diffusion smoothing and partial overgrowing. However, the angle between the front and back facets of multifaceted defects on (*n*11) surfaces remains approximately the same. As a consequence, when the *n*-index reaches high values, the (*n*11) surface will cut a larger and larger part of the defect tail, and, as a result, the defect shape will be more rounded, which is what observed experimentally.

We observed a clear tendency in multilayer structures to form laterally ordered arrays of QDs (Fig. 21). Depending on the substrate surface orientation, one can obtain QDs ordered along the [01–1] direction (QD chains) (Fig. 21a) [73] or laterally ordered networks of nearly equal-distant QDs [74] (Figs. 21b, c). Here, the defect orientation and symmetry have a one-to-one correlation with the position of QDs on the surface. The cell dimension of the QD lateral network on the less anisotropic surface (711)B is less than that on the (311)B surface, and the shape of the lateral cell is close to the rhombus one. Lateral self-ordering of the QDs in multilayer structures is most probably caused by formation of periodically changing field of elastic strains as well as an accompanying redistribution of the impurity-defect composition of the growth surface. The influence of an elastic strain far-acting field on QD lateral ordering is indirectly confirmed by Fig. 21b where the defect strain field causes fluctuations in QD lateral ordering. In the defect core region (i.e., in the ranges of the largest strain and compositional gradients), several QD rows exactly follows along the defect facet orientation directions, that is the lateral QD arrangement ''feels'' the shape of the strain field distribution around the defect.

The strain fields developed in the growth direction of multilayer $In_{0.45}Ga_{0.55}$ As/GaAs structures determine the size and lateral arrangement of (In,Ga)As QDs, and their value considerably exceeds the strain fields and concentration fluctuations formed around micro-size defects. The latter is developed during QD growth on the micro-size defect surface and confirmed by the appearance of small fluctuations of their lateral arrangement (QD chains and networks cover defects without significant transformations). The studied micro-size defects can be used to determine surface crystallographic orientation, as indicators of anisotropy inherent to surface physical (elastic) properties. Meanwhile, sets of microsurfaces with various orientations can be used as a playing field to explore QD growth processes.

Further, we have investigated in details a lateral ordering scenario of QDs as a function of substrate orientation and the number of vertical periods. Because of the statistical nature of Stranski-Krastanov growth mode, self-assembled dots are sometimes not very uniform in size, shape, and interdot spacing. This fact poses significant limitations for device applications. While self-assembled QD multilayers have shown that the vertical alignment throughout subsequent layers can be engineered to be nearly perfect, the lateral ordering tendency was found to be much less pronounced. A different type of arrangement, i.e., an anticorrelation of dots on subsequent layers, has been observed for II–VI, IV–VI, and III–V systems [75, 76].

It was further demonstrated that the elastic anisotropy of the materials plays a crucial role for the lateral and vertical self-organizations in QD superlattices [77]. It includes both anisotropy effects of the strain fields and adatom diffusion, as well as the elastic interaction of neighboring QDs. Generally, the balance between strong repulsive elastic interaction of adjacent initial dots and, on the other hand, the minimization of the total strain energy acts as the main driving force for the

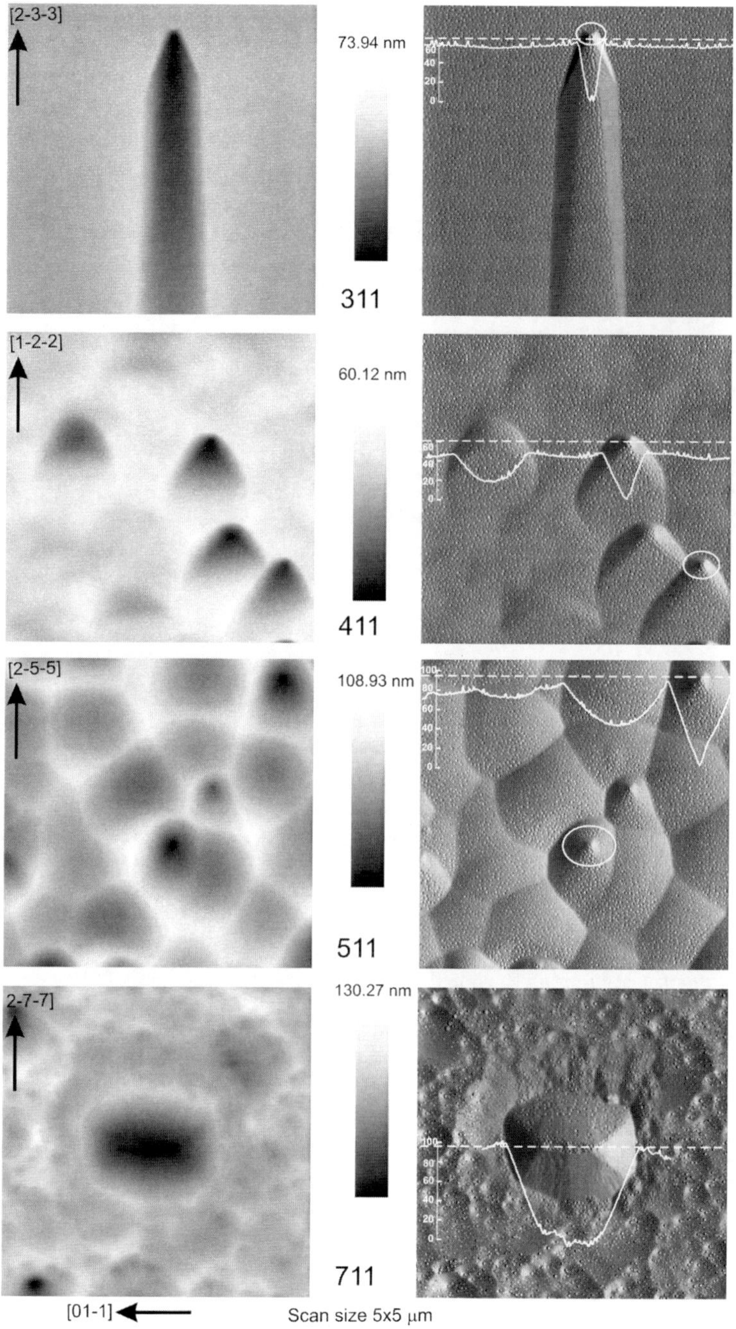

Fig. 20 Surface height maps (*on the left*) and derivatives (dz/dx) of AFM images (*on the right*) of typical microsize defects on the surface of $In_{0.45}Ga_{0.55}As$ (7 ML) layer on GaAs (*n*11)B substrate. There pointed are the sections (profiles) of surfaces along the dotted lines. Empty ovals point the defect cores. Quantum dots can be seen on the surfaces and on the defects

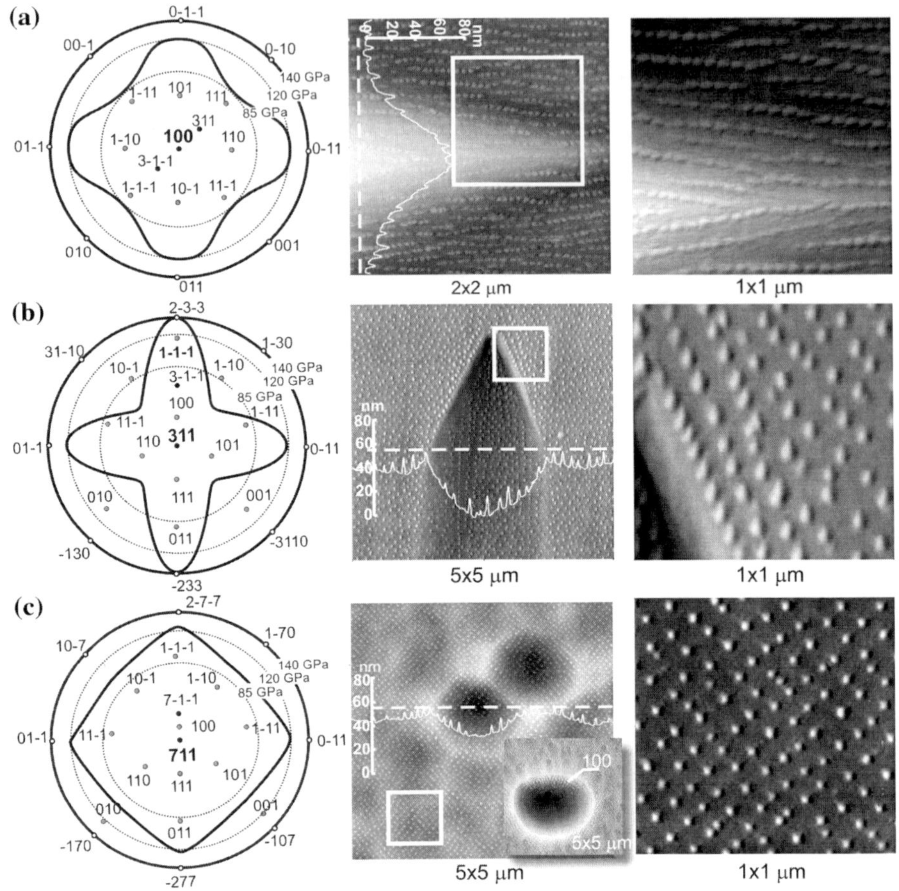

Fig. 21 Schematic representation of stereographic projections with indication of the elasticity modulus value distribution along crystallographic directions (*left*) and micro-size defect AFM images with cross-sections along dotted lines for the surface of 17-period $In_{0.45}Ga_{0.55}As/GaAs$ structure when the substrate orientation is: **a** (100), **b** (311)B and **c** (711)B. Images of an enhanced resolution that illustrate formation of QD patterns are shown at the *right*

lateral and vertical self-assembling. Investigating the strong impact of high index surfaces on formation of ordered QD arrays is expected to provide more detailed understanding of the underlying growth kinetics. Thus, it might help to improve physical properties of low-dimensional structures.

Our AFM studies indicate no well-ordered QD arrangement at the surfaces of the 1.5 period structures [78]. At the same time, the differences in density, shape, and size of QDs grown on differently oriented substrates were well pronounced. A variation of the growth surface from (100) to (911)B, (711)B, (511)B, and (311)B (increase of the angle of surface (*n*11) deviation from the surface (100)

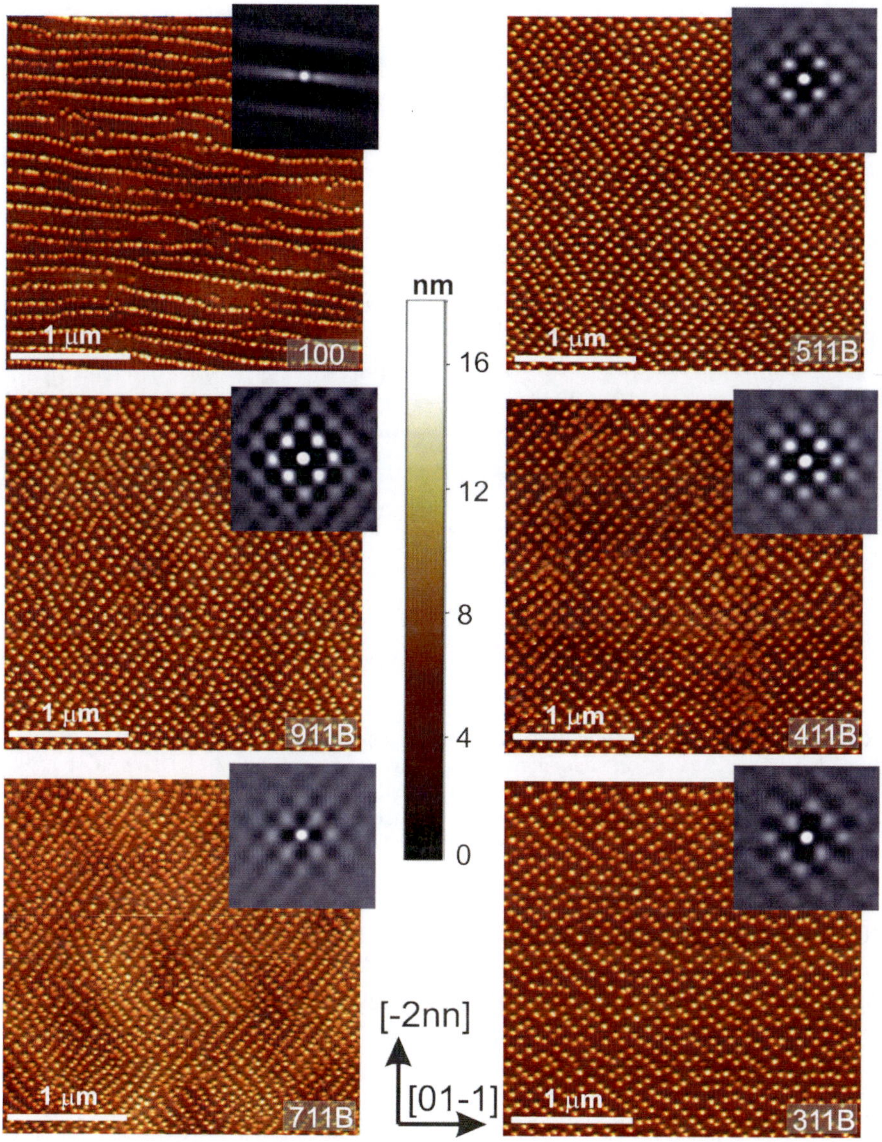

Fig. 22 AFM images of $In_{0.4}Ga_{0.6}As/GaAs$ QDs 16.5 periods grown on GaAs substrates of the following orientations: (100) (**a**), (911)B (**b**), (711)B (**c**), (511)B (**d**), (411)B (**e**), and (311)B (**f**). The 2D autocorrelation functions are shown in insets

toward (111)) yields the general trend to increase the QD size, while the QD density decreases.

While increasing the number of periods up to 16.5, the degree of the QD arrangement considerably improves for all probed substrate orientations. Figure 22

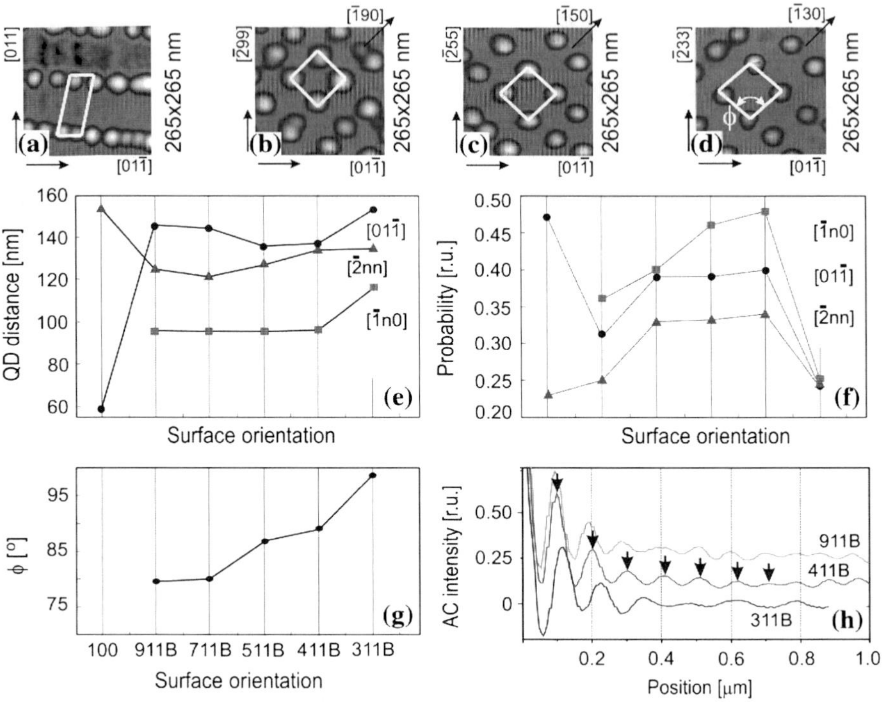

Fig. 23 **a–d** depict surface unit cells on GaAs(100) and (n11)B, where $n = 9, 5, 3$; **e–g** plots as a function of the surface orientation the dot-dot distance, probability of QD nearest neighbor occupation along the directions of preferred ordering, and characteristic angle ϕ; **h** plots the autocorrelation functions of the QDs curves are offset for a better view

depicts corresponding AFM images of QDs on high index surfaces, as well as their autocorrelation analysis insets. The presence of peaks up to the fourth order in 2D autocorrelation functions proves highly correlated QD ensembles. Further, on the autocorrelation analysis indicate a lateral QD arrangement along two preferred directions at the surface. In order to quantify this effect, we have defined a surface unit cell (Fig. 23) by taking into account a coordination shell with the three nearest neighbors of QD.

However, the most obvious feature of the QD multilayers seems a systematic impact of the substrate orientation onto the QD lateral ordering. Figure 23e presents mean QD-QD distances at the top surface along [01–1], [$-2nn$], and [$-1n0$], whereas these most preferred directions and the QD-QD distances have been extracted from the autocorrelation analysis. Interestingly, we found that the characteristic directions of the QD ordering at the surface correspond to the form of fourfold symmetrical anisotropic distribution of Young's modulus (Fig. 24), calculated by us for each of the surfaces according to [79].

Fig. 24 Angular distribution of the elasticity modulus for different GaAs(n11) substrate orientations (n = 7, 5, 4, 3) and GaAs(100)

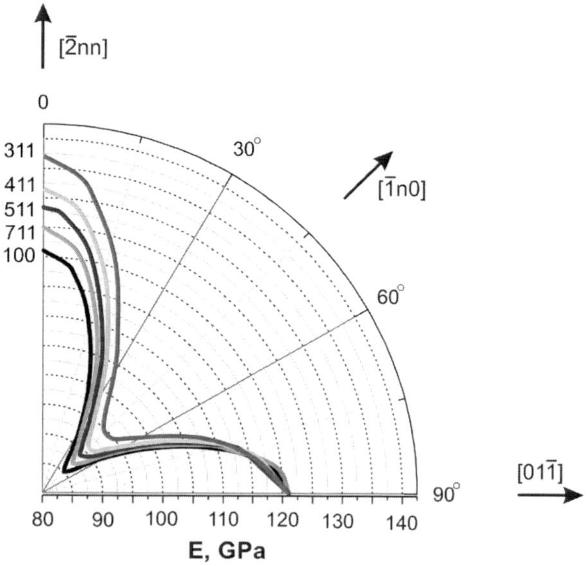

The probability of this kind of QD positioning determined from autocorrelation data is shown in Fig. 23f. It can be clearly seen that the 2D QD lattice of a good quality should be formed on (411) and (511) high index surfaces. The elastically softer $[-1n0]$ direction demonstrates the best QD ordering for all investigated substrates and accordingly stiffer directions have a lower probability of QD ordering. It seems that surface elastic anisotropy dominates the QD lateral ordering, whereas QD's elastic interaction in dense arrays and QD shape and size play a second role.

From the comparison of Fig. 23g, which shows the opening angle ϕ of the defined unit cell, and Fig. 24, one can see that the QD-QD distance along the elastically stiffer $[-2nn]$ direction (the largest Young modulus values) increases when going from (911) to (311). Along the [0–11] direction, the QD-QD distance demonstrates a trend to decreasing, while remaining practically the same along the elastically softer direction $[-1n0]$ (the smallest Young modulus values). However, again, the above regularities brake down in the case of (311)B substrate due to the largest anisotropy among the investigated orientations.

All of the above mentioned self-assembled QDs become ordered due to the interacting strain fields of successive QD layers and surface diffusion. It is apparent that new possibilities in 3D self-directed QD ordering could be achieved if strain and anisotropic diffusion can be controlled separately. One of the known effective ways to change surface diffusion is the deposition of a few monolayers of another material on a growth surface [80] or a change in the source gas composition [71]. Improvement of InAs QDs optical properties was reported, where As_2 flux was used instead of usual As_4 [81, 82], but differences in physical processes of lateral and vertical QDs ordering were not under investigation. We use either As_4 or As_2 as the arsenic source gas for growth of InGaAs/GaAs QD superlattices to study the role of both surface diffusion and elastic strain in formation and development of 3D ordering of the QDs in a GaAs matrix [83]. In particular, our findings show an influence of As flux type on multilayered growth of (In, Ga)As QDs on GaAs (100). This provides an excellent opportunity to vary and control the symmetry of the diffusion and strain pattern in each layer with the aim to optimize the spatial ordering of nanostructures with identical sizes and shapes in multilayers of QDs.

It is known that, for a given growth conditions, the maximum sticking coefficient is only 0.5 for As_4, but it can reach 1.0 for As_2. However, for sufficient overpressure of As with the same growth conditions, the sticking coefficient for Ga and In is very readily 1.0 [84]. Therefore, the amount of material grown is completely determined by the group III flux. The use of the As_2 background for effective manipulation of QD shape, positioning and deformation can be understood by considering surface diffusion. Due to the nature of the (2 $\times$ 4) GaAs(100) surface reconstruction, the adatoms diffusion length along [0–11] direction is much larger in comparison with that for [011] direction.

This anisotropic surface diffusion leads to elongation of the QDs in each layer along the direction of higher mobility. This elongation in turn creates anisotropic strain fields in each capping layer (strain in the [0–11] direction is smaller than in [011] [85]), which then enhances the elongation of the QDs subsequently forming chains or wires. The difference between the As_4 and As_2 appears to be in terms of limiting the ultimate diffusion lengths. A microprobe-RHEED/SEM study has shown that the lateral flow of Ga atoms is reduced under As_2 flux in comparison with As_4 [86]. Since the As_2 does not need to be cracked in order to incorporate into the crystal [87], having that as the arsenic source provides a lower energy barrier for incorporation and thus a shorter diffusion length for the adatoms [88]. As can be seen from the AFM images of Figs. 25 and 26, it helps to keep the QDs as separate entities forming chains of QDs instead of wires.

Figure 25 shows AFM topographic images of single layers and multilayers for $x = 0.4$, which were grown under As_4 and As_2 fluxes. QDs in the single layer structures have weak lateral ordering for both As_4 and As_2 fluxes (Figs. 25a and b).

However, it is apparent that the sample grown using As_2 (Fig. 25b) is already more uniform and somewhat more ordered than the one using As_4 (Fig. 25a), just after the first layer. After further growth, completing the full 15 period structure, well defined periodic dot-chains are observed for both As_4 and As_2 growth (Figs. 25c and d). At the same time, it could be seen that the use of either As_4 or As_2 during the growth causes significant differences in the density and size of the QDs. These data imply a generally higher surface mobility for QD growth using As_2 as compared to As_4.

The dependence of QD ordering on composition and nominal thickness of deposited wetting layer both for As_4 and As_2 fluxes are illustrated in Fig. 26. The same thickness of GaAs spacer layer of 60 MLs was kept in the $In_xGa_{1-x}As$/GaAs

Fig. 25 AFM images of the $In_{0.4}Ga_{0.6}As$ QD's on single layer structures ((**a**), (**b**)) and 15.5 period structures ((**c**), (**d**)) grown using As_4 ((**a**), (**c**)) and As_2 ((b), (d)) background fluxes under identical conditions. Insertions in (**a**)–(**d**) show the cross-sections (linear profiles) along the [011] direction of most typical QDs determined from size distribution functions (vertical and horizontal scale bars correspond to 6 and 40 nm). Height scale bar corresponds to 15 nm

structures (x = 0.3 for 15.5 MLs and x = 0.5 for 5.7 MLs). For each of these compositions, the dot layers exceed the critical thickness for relaxation by 25 %. As it is seen from Fig. 26, composition increasing from x = 0.3 to 0.5 leads to InGaAs nano-feature transformations from wire-like to closely packed dot-chains in the case of As_4 flux (Figs. 26c and a), and from large elongated QDs to small well separated QDs in the case of As_2 flux (Figs. 26d and b).

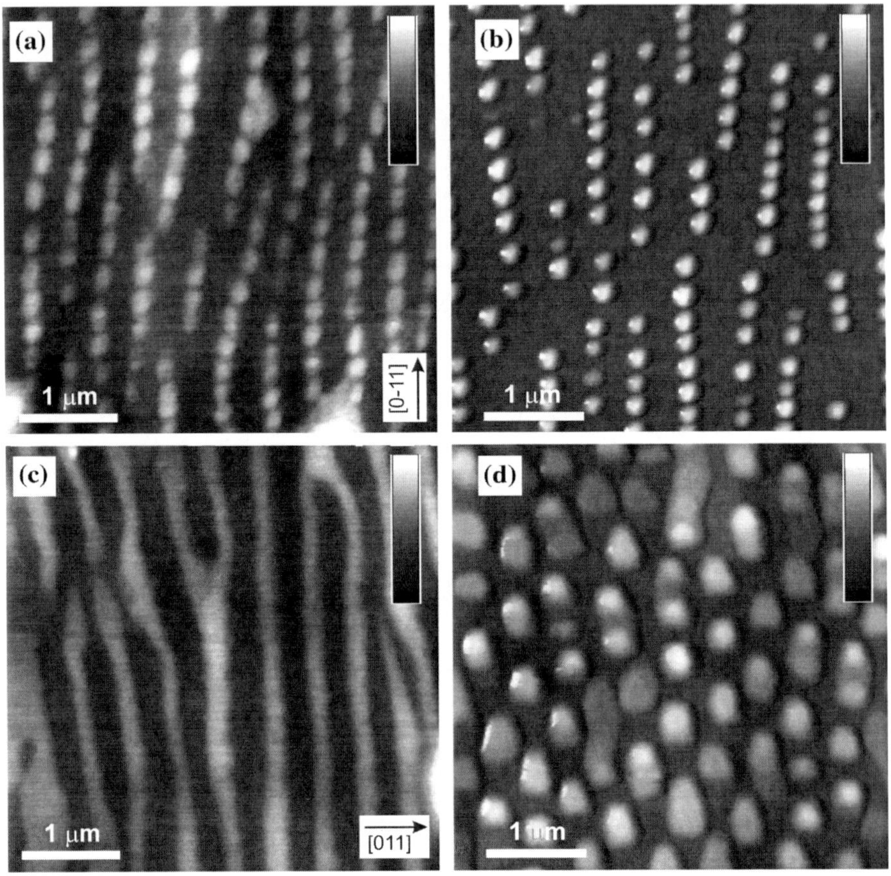

Fig. 26 AFM images of $In_{0.5}Ga_{0.5}As$ ((**a**), (**b**)) and $In_{0.3}Ga_{0.7}As$ ((**c**), (**d**)) nano-features in multilayered structures grown using As_4 (*left column*) and As_2 (*right column*) gas fluxes. Height scale bar corresponds to 20 nm

4 Conclusion

In conclusion, we tried to describe a common approach to organization of practical AFM diagnostics of surfaces at the sub-micrometer and nano-meter levels. Considerable attention was paid to one of the main AFM applications—topometry of geometrical sizes of surface elements and its metrological traceability. As an example of a system approach to diagnostics of nano-structured surfaces, we adduced our original investigations of growth processes in semiconductor nanostructures with quantum dots. Careful observance of protocols for calibration of AFM as well as tests of uncertainty of measurements added by a shape of the probe tip enabled us to obtain a number of weighty results important for understanding physics of processes of structural ordering in low-dimensional semiconductor systems.

Topometry of surfaces is a most frequently used in practice AFM method with well developed theoretical and metrological foundation. Their description in detail can be found in the list of references. At the same time, these methods are continuously developed, respective facility is upgraded, which enables to considerably enhance their information capability and to widen their application scope.

References

1. Binnig, G., Rohrer, H.: Scanning tunneling microscope. US Patent 4,343,993, 10 Aug 1982
2. Binnig, G., Rohrer, H., Gerber, C., Weibel, E.: Surface studies by scanning tunneling microscopy. Phys. Rev. Lett. **49**(1), 57 (1982)
3. Binnig, G., Rohrer, H., Gerber, C., Weibel, E.: 7 × 7 reconstruction on Si(111) resolved in real space. Phys. Rev. Lett. **50**(2), 120 (1983)
4. Young, R.D.: Field emission ultramicrometer. Rev. Sci. Instrum. **37**(3), 275 (1966)
5. Young, R., Ward, J., Scire, F.: The topografiner: an instrument for measuring surface microtopography. Rev. Sci. Instrum. **43**(7), 999 (1972)
6. Binnig, G., Quate, C.F., Gerber, C.: Atomic force microscope. Phys. Rev. Lett. **56**(9), 930 (1986)
7. Binnig, G.K.: Atomic force microscope and method for imaging surfaces with atomic resolution. US Patent 33,387, 16 Oct 1990
8. Belaidi, S., Girard, P., Leveque, G.: Electrostatic forces acting on the tip in atomic force microscopy: modelization and comparison with analytic expressions. J. Appl. Phys. **81**(3), 1023 (1997)
9. Martin, Y., Abraham, D.W., Wickramasinghe, H.K.: High-resolution capacitance measurement and potentiometry by force microscopy. Appl. Phys. Lett. **52**(13), 1103 (1988)
10. Nonnenmacher, M., o'Boyle, M., Wickramasinghe, H.: Kelvin probe force microscopy. Appl. Phys. Lett. **58**, 2921 (1991)
11. Jacobs, H., Leuchtmann, P., Homan, O., Stemmer, A.: Resolution and contrast in Kelvin probe force microscopy. J. Appl. Phys. **84**(3), 1168 (1998)
12. Martin, Y., Wickramasinghe, H.K.: Magnetic imaging by ''force microscopy'' with 1000 Å resolution. Appl. Phys. Lett. **50**(20), 1455 (1987)
13. Gröning, O., Küttel, O., Gröning, P., Schlapbach, L.: Field emission from DLC films. Appl. Surf. Sci. **111**, 135 (1997)
14. Zhang, L., Sakai, T., Sakuma, N., Ono, T., Nakayama, K.: Nanostructural conductivity and surface-potential study of low-field-emission carbon films with conductive scanning probe microscopy. Appl. Phys. Lett. **75**(22), 3527 (1999)
15. De Wolf, P., Clarysse, T., Vandervorst, W., Hellemans, L., Niedermann, P., Hanni, W.: Cross-sectional nano-spreading resistance profiling. J. Vac. Sci. Technol., B **16**(1), 355 (1998)
16. Williams, C.C., Slinkman, J., Hough, W.P., Wickramasinghe, H.K.: Lateral dopant profiling with 200 nm resolution by scanning capacitance microscopy. Appl. Phys. Lett. **55**(16), 1662 (1989)
17. Baŕo, A.M., Reifenberger, R.G. (ed.): Atomic force microscopy in liquid. Biological applications. Wiley, Germany, 2012
18. Noy, A. (ed.): Handbook of molecular force spectroscopy. Springer, New York, 2008
19. Bhushan, B. (ed.) Nanotribology and nanomechanics: an introduction. Springer, Heidelberg, 2008
20. Foster, A., Hofer, W.A.: Scanning probe microscopy–atomic scale engineering by forces and currents. Springer, Heidelberg, 2006

21. Kalinin, S.V., Gruverman, A. (ed.): Scanning probe microscopy (2 vol. set): electrical and electromechanical phenomena at the nanoscale. Springer, New York, 2007
22. Kalinin, S.V., Gruverman, A. (ed.): Scanning probe microscopy of functional materials: nanoscale imaging and spectroscopy. Springer, New York, 2011
23. Samor, P. (ed.) Scanning probe microscopies beyond imaging: manipulation of molecules and nanostructures. Wiley, Germany, 2006
24. Landis, S. (ed.) Nano-lithography (Wiley-ISTE, New York, 2011)
25. Israelachvili, N.: Intermolecular and surface forces, 3rd ed. (Academic Press, New York, 2011)
26. Butt, H.-J., Cappella, B., Kappl, M.: Force measurements with the atomic force microscope: technique, interpretation and applications. Surf. Sci. Rep. **59**(1), 1 (2005)
27. Cappella, B., Dietler, G.: Force-distance curves by atomic force microscopy. Surf. Sci. Rep. **34**(1), 1 (1999)
28. Rabe, U., Janser, K., Arnold, W.: Vibrations of free and surface-coupled atomic force microscope cantilevers: theory and experiment. Rev. Sci. Instrum. **67**(9), 3281 (1996)
29. García, R., Pérezb, R.: Dynamic atomic force microscopy methods. Surf. Sci. Rep. **47**(6–8), 197 (2002)
30. Whitehouse, D.J: Handbook of surface and nanometrology. CRC Press, Boca Raton, 2010
31. Gao, W.: Precision nanometrology: sensors and measuring systems for nanomanufacturing. (Springer, London, 2010
32. Leach, R.: Fundamental principles of engineering nanometrology. William Andrew, Amsterdam, 2009
33. Klapetek, P. (ed.): Quantitative data processing in scanning probe microscopy: SPM applications for nanometrology (micro and nano technologies). William Andrew, Norwich, 2013
34. ISO/TC 201/SC 9 - Scanning probe microscopy. International organization for standardization, 2013, http://isostore.sii.org.il/sii/home/catalogue_tc/catalogue_tc_browse.htm?commid=354756. Accessed 22 Nov 2013
35. ISO/DIS 13083. Standards on the definition and calibration of spatial resolution of scanning spreading resistance microscopy and scanning capacitance microscopy. International organization for standardization, 2013. http://isostore.sii.org.il/sii/home/catalogue_tc/catalogue_detail.htm?csnumber=52691. Accessed 22 Nov 2013
36. International Organization for Standardization: Geometrical product specifications - Surface texture: Areal - Part 2: Terms, definitions and surface texture parameters, (ISO 25178-2:2012) (2012)
37. ASME Intl., Surface Texture (Surface Roughness, Waviness, and Lay) ASME B46.1 (2009)
38. Dowling, A., Clift, R., Grobert, N., Hutton, D., Oliver, R., O'neill, O., Pethica, J., Pidgeon, N., Porritt, J., Ryan, J.: Nanoscience and nanotechnologies: opportunities and uncertainties. London: The royal society and The royal academy of engineering report, 61 (2004)
39. Wilkening, G., Koenders, L.: Nanoscale calibration standards and methods: dimensional and related measurements in the micro and nanometer range. Wiley, Weinheim, 2005
40. Lytvyn, O.S., Lytvyn, P.M., Prokopenko, I.V., Sheremeta, T.I.: Peculiarities of nanostructures topometry studied by atomic force microscopy methods. Nanosyst. Nanomater. Nanotechnol. **6**(1), 33 (2008)
41. Ferreira, S.O., Neves, B.R.A., Magalhães-Paniago, R., Malachias, A., Rappl, P.H.d.O, Ueta, A., Abramof, E., Andrade, M.S.: AFM characterization of PbTe quantum dots grown by molecular beam epitaxy under Volmer–Weber mode. J. Cryst. Growth **231**(1), 121 (2001)
42. Minhua, Z., Vaneet, S., Haoyan, W., Robert, R.B., Jeffrey, A.S., Fotios, P., Bryan, D.H.: Ultrasharp and high aspect ratio carbon nanotube atomic force microscopy probes for enhanced surface potential imaging. Nanotechnology **19**(23), 235704 (2008)
43. Tay, A.B.H., Thong, J.T.L.: High-resolution nanowire atomic force microscope probe grownby a field-emission induced process. Appl. Phys. Lett. **84**(25), 5207 (2004)
44. Givargizov, E., Stepanova, A., Obolenskaya, L., Mashkova, E., Molchanov, V., Givargizov ME., Rangelow, I.: Whisker probes. Ultramicroscopy **82**(1), 57 (2000)

45. Keller, D.: Reconstruction of STM and AFM images distorted by finite-size tips. Surf. Sci. **253**(1), 353 (1991)
46. Bonnet, N., Dongmo, S., Vautrot, P., Troyon, M.: A mathematical morphology approach to image formation and image restoration in scanning tunnelling and atomic force microscopies. Microsc. Microanal. Microstruct. **5**(4–6), 477 (1994)
47. Markiewicz, P., Goh, M.C.: Atomic force microscope tip deconvolution using calibration arrays. Rev. Sci. Instrum. **66**(5), 3186 (1995)
48. Hübner, U., Morgenroth, W., Meyer, H.G., Sulzbach, T., Brendel, B., Mirandé, W.: Downwards to metrology in nanoscale: determination of the AFM tip shape with well-known sharp-edged calibration structures. Appl. Phys. A **76**(6), 913 (2003)
49. Villarrubia, J.S.: Algorithm for scanned probe microscope image simulation, surface reconstruction, and tip estimation. J. Res. Natl. Inst. Stan. Technol. **102**, 425 (1997)
50. Dongmo, L.S., Villarrubia, J.S., Jones, S.N., Renegar, T.B., Postek, M.T., Song, J.F.: Experimental test of blind tip reconstruction for scanning probe microscopy. Ultramicroscopy **85**(3), 141 (2000)
51. Tranchida, D., Piccarolo, S., Deblieck, R.: Some experimental issues of AFM tip blind estimation: the effect of noise and resolution. Meas. Sci. Technol. **17**(10), 2630 (2006)
52. Castle, J., Zhdan, P.: Characterization of surface topography by SEM and SFM: problems and solutions. J. Phys. D Appl. Phys. **30**(5), 722 (1997)
53. Krasil'nik, Z.F., Lytvyn, P., Lobanov, D.N., Mestres, N., Novikov, A.V., Pascual, J., Valakh, M.Y., Yukhymchuk, V.A.: Microscopic and optical investigation of Ge nanoislands on silicon substrates. Nanotechnology **13**(1), 81 (2002)
54. Shchukin, V.A., Ledentsov, N.N., Kop'ev, P.S., Bimberg, D.: Spontaneous ordering of arrays of coherent strained Islands. Phys. Rev. Lett. **75**(16), 2968 (1995)
55. Ross, F.M., Tersoff, J., Tromp, R.M.: Coarsening of self-assembled Ge quantum dots on Si(001). Phys. Rev. Lett. **80**(5), 984 (1998)
56. Krasil'nik, Z.F., Dolgov, I.V., Drozdov, Y.N., Filatov, D.O., Gusev, S.A., Lobanov, D.N., Moldavskaya, L.D., Novikov, A.V., Postnikov, V.V., Vostokov, N.V.: The elastic strain and composition of self-assembled GeSi islands on Si (001). Thin Solid Films **367**(1), 171 (2000)
57. Valakh, M.Y., Lytvyn, P.M., Nikolenko, A.S., Strelchuk, V.V., Krasilnik, Z.F., Lobanov, D.N., Novikov, A.V.: Gigantic uphill diffusion during self-assembled growth of Ge quantum dots on strained SiGe sublayers. Appl. Phys. Lett. **96**(14), 141909 (2010)
58. Strelchuk, V.V., Nikolenko, A.S., Lytvyn, P.M., Kladko, V.P., Gudymenko, A.I., Valakh, M.Y., Krasilnik, Z.F., Lobanov, D.N., Novikov, A.V.: Effects of the lateral ordering of self-assembled SiGe nanoislands grown on strained $Si_{1-x}Ge_x$ buffer layers. Semiconductors **46**(5), 647 (2012)
59. Lobanov, D.N., Novikov, A.V., Vostokov, N.V., Drozdov, Y.N., Yablonskiy, A.N., Krasilnik, Z.F., Stoffel, M., Denker, U., Schmidt, O.G.: Growth and photoluminescence of self-assembled islands obtained during the deposition of Ge on a strained SiGe layer. Opt. Mater. **27**(5), 818 (2005)
60. Vostokov, NV., Drozdov, Y.N., Krasil'nik, Z.F., Kuznetsov, O.A., Novikov, A.V., Perevoshchikov, V.A., Shaleev, M.V.: $Si_{1-x}Ge_x$/Si (001) relaxed buffer layers grown by chemical vapor deposition at atmospheric pressure. Phys. Solid State **47**(1), 42 (2005)
61. Geguzin, Y.E.: Ascending diffusion and the diffusion aftereffect. Sov. Phys. Uspekhi **29**(5), 467 (1986)
62. Fafard, S., Hinzer, K., Raymond, S., Dion, M., McCaffrey, J., Feng, Y., Charbonneau, S.: Red-emitting semiconductor quantum dot lasers. Science **274**(5291), 1350 (1996)
63. Jin, C.Y., Liu, H.Y., Groom, K.M., Jiang, Q., Hopkinson, M., Badcock, T.J., Royce, R.J., Mowbray, D.J.: Effects of photon and thermal coupling mechanisms on the characteristics of self-assembled InAs/GaAs quantum dot lasers. Phys. Rev. B **76**(8), 085315 (2007)
64. Liu, H.C., Gao, M., McCaffrey, J., Wasilewski, Z.R., Fafard, S.: Quantum dot infrared photodetectors. Appl. Phys. Lett. **78**(1), 79 (2001)

65. Maimon, S., Finkman, E., Bahir, G., Schacham, S.E., Garcia, J.M., Petroff, P.M.: Intersublevel transitions in InAs/GaAs quantum dots infrared photodetectors. Appl. Phys. Lett. **73**(14), 2003 (1998)
66. Ye, Z., Campbell, J.C., Chen, Z., Kim, E.-T., Madhukar, A.: Voltage-controllable multiwavelength InAs quantum-dot infrared photodetectors for mid- and far-infrared detection. J. Appl. Phys. **92**(7), 4141 (2002)
67. Schmidbauer, M., Wang, Z.M., Mazur, Y.I., Lytvyn, P.M., Salamo, G.J., Grigoriev, D., Schafer, P., Kohler, R., Hanke, M.: Initial stages of chain formation in a single layer of (In, Ga) As quantum dots grown on GaAs (100). Appl. Phys. Lett. **91**(9), 093110 (2007)
68. Hanke, M., Dubslaff, M., Schmidbauer, M., Wang, Z.M., Mazur, Y.I., Lytvyn, P.M., Lee, J.H., Salamo, G.J.: On the complex behavior of strain relaxation in (In, Ga) As/GaAs (001) quantum dot molecules. Appl. Phys. Lett. **95**(2), 023103 (2009)
69. Mehta, S.K., Muralidharan, R., Sharda, G.D., Jain, R.K.: Some investigations on oval defects in MBE-grown GaAs. Semicond. Sci. Technol. **7**(5), 635 (1992)
70. Lytvyn, P.M., Prokopenko, I.V., Strelchuk, V.V., Mazur, Y.I., Wang, Z.M., Salamo, G.J.: Microsize defects in InGaAs/GaAs (N11) A/B multilayers quantum dot stacks. J. Cryst. Growth **284**(1), 47 (2005)
71. Itoh, M.: Atomic-scale homoepitaxial growth simulations of reconstructed III–V surfaces. Prog. Surf. Sci. **66**(3), 53 (2001)
72. Apostolopoulos, G., Herfort, J., Däweritz, L., Ploog, K.H., Luysberg, M.: Reentrant mound formation in GaAs(001) homoepitaxy observed by ex situ atomic force microscopy. Phys. Rev. Lett. **84**(15), 3358 (2000)
73. Wang, Z.M., Mazur, Y.I., Salamo, G.J., Lytvin, P.M., Strelchuk, V.V., Valakh, M.Y.: Persistence of (In, Ga) As quantum-dot chains under index deviation from GaAs (100). Appl. Phys. Lett. **84**(23), 4681 (2004)
74. Wang, Z.M., Seydmohamadi, S., Lee, J.H., Salamo, G.J.: Surface ordering of (In, Ga)As quantum dots controlled by GaAs substrate indexes. Appl. Phys. Lett. **85**(21), 5031 (2004)
75. Wang, X.D., Liu, N., Shih, C.K., Govindaraju, S., Holmes, A.L.: Spatial correlation-anticorrelation in strain-driven self-assembled InGaAs quantum dots. Appl. Phys. Lett. **85**(8), 1356 (2004)
76. Strassburg, M., Kutzer, V., Pohl, U.W., Hoffmann, A., Broser, I., Ledentsov, N.N., Bimberg, D., Rosenauer, A., Fischer, U., Gerthsen, D., Krestnikov, I.L., Maximov, M.V., Kop'ev, P.S., Alferov, Z.I.: Gain studies of (Cd, Zn)Se quantum islands in a ZnSe matrix. Appl. Phys. Lett. **72**(8), 942 (1998)
77. Shchukin, V.A., Bimberg, D., Malyshkin, V.G., Ledentsov, N.N.: Vertical correlations and anticorrelations in multisheet arrays of two-dimensional islands. Phys. Rev. B **57**(19), 12262 (1998)
78. Lytvyn, P.M., Strelchuk, V.V., Kolomys, O.F., Prokopenko, I.V., Valakh, M.Y., Mazur, Y.I., Wang, Z.M., Salamo, G.J., Hanke, M.: Two-dimensional ordering of (In, Ga)As quantum dots in vertical multilayers grown on GaAs(100) and (n11). Appl. Phys. Lett. **91**(17), 173118 (2007)
79. Sirotin, Y.I., Shaakolkoya, M.P.: Fundamentals of crystal physics. MIR Publishers, Moscow, 1982
80. Cirlin, G.E., Tonkikh, A.A., Ptitsyn, V.E., Dubrovskii, V.G., Masalov, S.A., Evtikhiev, V.P., Denisov, D.V., Ustinov, V.M., Werner, P.: Influence of antimony on the morphology and properties of an array of Ge/Si (100) quantum dots. Phys. Solid State **47**(1), 58 (2005)
81. Sugaya, T., Amano, T., Komori, K.: Improved optical properties of InAs quantum dots grown with an As_2 source using molecular beam epitaxy. J. Appl. Phys. **100**(6), 063107 (2006)
82. Sugaya, T., Numakami, O., Oshima, R., Furue, S., Komaki, H., Amano, T., Matsubara, K., Okano, Y., Niki, S.: Ultra-high stacks of InGaAs/GaAs quantum dots for high efficiency solar cells. Energy Environ. Sci. **5**(3), 6233 (2012)
83. Lytvyn, P.M., Mazur, Y.I., Marega Jr, E., Dorogan, V.G., Kladko, V.P., Slobodian, M.V., Strelchuk, V.V., Hussein, M.L., Ware, M.E., Salamo, G.J.: Engineering of 3D self-directed

quantum dot ordering in multilayer InGaAs/GaAs nanostructures by means of flux gas composition. Nanotechnology **19**(50), 505605 (2008)
84. Kaldis, E. (ed.): Current topics in materials science (Elsevier Science Ltd, New York, 1981)
85. Wang, Z.M., Holmes, K., Mazur, Y.I., Salamo, G.J.: Fabrication of (In, Ga)As quantum-dot chains on GaAs(100). Appl. Phys. Lett. **84**(11), 1931 (2004)
86. Ogura, T., Kishimoto, D., Nishinaga, T.: Effect of As molecular species on inter-surface diffusion in GaAs MBE for ridge structure fabrication. J. Cryst. Growth **226**(2), 179 (2001)
87. Morgan, C.G., Kratzer, P., Scheffler, M.: Arsenic dimer dynamics during MBE growth: theoretical evidence for a novel chemisorption state of As_2 molecules on GaAs surfaces. Phys. Rev. Lett. **82**(24), 4886 (1999)
88. Sugaya, T., Komori, K., Yamauchi, S., Amano, T.: 1.3 μm InAs quantum dots grown with an As_2 source using molecular-beam epitaxy. J. Vac. Sci. Technol., B **23**(3), 1243 (2005)

Part III
New Functional Nanomaterials and Nanoscaled Devices for Energy Harvesting, Light Emission, Optoelectronics and THz Range

Towards Self-Powered Systems: Using Nanostructures to Harvest Ambient Energy

Gustavo Ardila, Anne Kaminski-Cachopo, Marco Pala, Alessandro Cresti, Laurent Montès, Vincent Consonni, Ronan Hinchet, Jérôme Michallon, Mehdi Daanoune, Mauro Zanuccoli, Claudio Fiegna and Mireille Mouis

Abstract In this chapter, we present the advantages of semiconducting nanostructures (nanowires) for energy harvesting applications. Three sources of energy are considered: mechanical inputs, light and thermal energy. Different simulation approaches are used to discuss the prospects of these energy transduction solutions at nanoscale. Some guidelines are brought out for the improvement of energy conversion efficiency by nanowires, when integrated into functional devices.

1 Introduction

The combination of new materials integration, 3D processing and low-power circuits enable the development of autonomous systems. These systems are typically used on Wireless Sensors Networks (WSN) applications, with the objective to monitor human health, environment, or structures such as airplanes or buildings [1]. Three main issues need to be addressed to improve the performance of the autonomous systems: (i) the reduction of the overall energy consumption to ideally

G. Ardila (✉) · L. Montès · R. Hinchet · J. Michallon · M. Daanoune · M. Mouis
IMEP-LAHC/Minatec (CNRS-Grenoble INP, UJF), 3 Parvis Louis Neel, Grenoble, France
e-mail: ardilarg@minatec.inpg.fr

A. Kaminski-Cachopo · M. Pala · A. Cresti · V. Consonni
LMGP/Minatec (CNRS-Grenoble INP, UJF), 3 Parvis Louis Neel, Grenoble, France
e-mail: anne.kaminski@phelma.grenoble-inp.fr

M. Pala
e-mail: pala@minatec.inpg.fr

M. Zanuccoli · C. Fiegna
Department of Electrical, Electronic, and Information Engineering "Guglielmo Marconi", Advanced Research Center on Electronic Systems, University of Bologna & Italian Universities Nano-Electronics Team, 47512 Cesena (FC), Italy

A. Nazarov et al. (eds.), *Functional Nanomaterials and Devices for Electronics, Sensors and Energy Harvesting*, Engineering Materials,
DOI 10.1007/978-3-319-08804-4_11

less than about 100 μW [2], (ii) the reduction of the general size and/or surface for easier integration and (iii) the augmentation of the power autonomy or battery lifetime. To solve this last issue, ambient energy harvesting is a promising solution. In addition, this solution allows the maintenance cost associated to battery replacement and/or charging to be cut down or even suppressed.

An autonomous system is composed of several subsystems [2] for computing, communication, sensing, energy management and energy harvesting, respectively. Basically, the energy from the harvester must be processed before been stored in a capacitor or a rechargeable battery, and later used in the system. This chapter will only deal with the strictly speaking energy harvesting subsystem.

To date, several approaches have been proposed to harvest energy from different energy sources, such as thermal, solar, RF, or mechanical sources [3], using thin films and MEMS technologies. With the advent of ultra-low power circuits, the energy needed for autonomous systems can be harvested by even smaller structures and, eventually, nanostructures. Most importantly, some properties of nanostructures can be controlled and improved compared to bulk [4, 5].

In this paper we present the advantages and prospects of using semiconducting nanostructures (nanowires) for energy harvesting applications from three different sources: mechanical inputs, solar and thermal energy sources. The chapter is divided in three main sections focalized on each energy conversion. Each section provides first a brief review of the concepts at the macro and micro scales, before discussing the main advantages at the nano scale from a theoretical point of view, supported by simulation and modeling results. The chapter ends by the conclusions and perspectives.

2 Mechanical Energy Harvesting Using Piezoelectric Nanostructures

Many approaches have been proposed to harvest ambient mechanical energy: using the variation of electromagnetic or electrostatic fields and using piezoelectric materials [3]. All these approaches have been largely studied at the macro and micro scale leading to some commercial devices. This section is focalized on piezoelectric materials, where electrical charges are generated when a mechanical load is applied (direct effect), or which get strained when an electric field is applied (reverse effect). This property is quantified by the piezoelectric coefficients (d), measured in C/N or pm/V, respectively. In what follows, we will firstly sum up briefly the approach used at the macro scale to harvest energy from mechanical inputs, before introducing the different approaches that can be used with nanoscale piezoelectric materials.

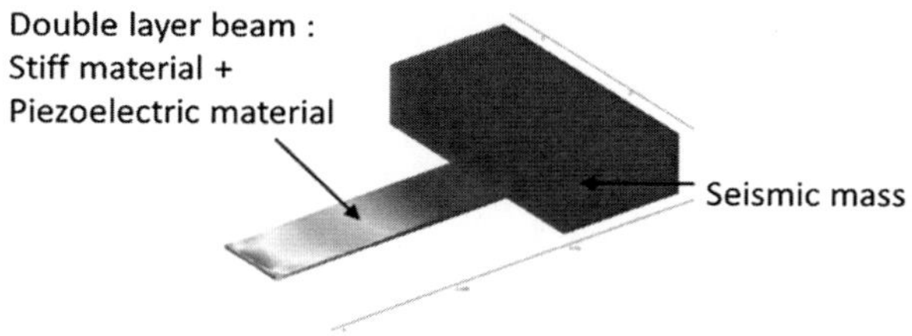

Fig. 1 Schematic structure of a resonant piezoelectric cantilever illustrating the location of the maximum stress (fixed end). The dimension of the cantilever and the weight of the seismic mass are adjusted to match a resonance frequency driven by the application

2.1 The MEMS Approach

At the macroscale, the most widely used structure to harvest mechanical inputs with piezoelectric materials is a resonant cantilever composed of a stiff material (Si or plastic), a piezoelectric layer (typically PZT or AlN, between others), and a seismic mass (Fig. 1). The whole structure is tuned to resonate at a specific frequency driven by the application (typically between a few Hz and several 100 Hz). The aim is to increase the quality factor (Q) of the resonator, at the expense however of bandwidth narrowing [6]. Many devices have been reported using this structure, with generated energy density values ranging from 0.1 to 40 mW/cm^3 [3], for input accelerations between 1 and 10 m/s^2, corresponding to a force in the range of 30 μN–40 mN. Commercial devices can also be found from companies such as Midé (USA) proposing devices generating 9 mW/cm^3 at 15 V using a PZT film integrated into a plastic beam.

Several strategies have been proposed to increase the bandwidth of cantilever based devices, including bi-stables structures, coupled oscillators, arrays of cantilevers featuring different resonance frequencies, or the addition of amplitude limiters, between others [7].

2.2 Piezoelectricity at the Nanoscale

The reduction of beam size down to the nanoscale has been proven to improve the elastic properties, such as flexibility compared to bulk materials. Fracture strain is also increased [8]. It has also been shown that size reduction resulted in improvements of the piezoelectric coefficients (in particular d_{33}, longitudinal, along the c-axis) of semiconducting materials such as GaN and ZnO in form of nanowires (NWs) with diameters wider than 150 nm [9] or nanoribbons thicker than 500 nm [10] (see Table 1) leading to a higher voltage generated (i.e. more power and better energy conversion efficiency) for given deformation [11].

Table 1 Piezoelectric coefficient d_{33} in ZnO and GaN nanostructured materials compared to bulk (adapted from [16])

d_{33} [pm/V]			
Material	Experimental (bulk)	Experimental (nanoscale)	Theoretical (nanoscale)
ZnO	9.93 [10]	14-26.7 [10]	168.2 [12]
GaN	1.86 [12]	12.8 [9]	65.8 [12]

The experimental trends of these piezoelectric coefficients have been confirmed by theoretical studies on semiconducting NWs (GaN, ZnO and more recently AlN NWs), although at lower diameters (a few nm) (see Table 1). These calculations have used different approaches such as first principles-based density functional theory (DFT) [12] or continuum models including surface effects from ab initio calculations [13]. More recent calculations using the finite element method (FEM) have included the semiconducting properties of individual ZnO NWs (i.e. doping level and free charges) showing their impact on the reduction of the generated piezo-potential (screening effect) [14, 15] and thus on the effective piezoelectric coefficients. These results altogether show that there is still a gap between theoretical predictions and experimental measurements.

2.3 Integrated Piezoelectric Nanowires into Functional Devices

As dimensions decrease, the resonant approach becomes less appropriate for energy harvesting since most mechanical sources present in the environment are at low frequency or frequency-less, and solutions that exploit real-time deformations and impacts are thus better suited [17]. In addition, one single NW will not give enough energy for typical autonomous systems, so that practical harvesting devices must integrate large arrays of piezoelectric NWs. Several integration techniques have been reported in the literature, including lateral [18–22], vertical [20, 23, 24] and radial integration [25, 26] of ZnO, PZT, $NaNbO_3$ and PVDF NWs. To date, the most performing device integrates vertically grown ZnO NWs. It produces 0.78 W/cm^3 (estimated from output voltage and output current, measured in open and short circuit conditions, with values of 58 V and 134 μA, respectively) when a non-quantified mechanical input is applied on the device (palm impact) [27]. While most of the effort has been invested on proof of concept demonstrations and device fabrication, fewer references can be found in the literature about the theoretical analysis of the performances of such devices and the identification of optimization guidelines.

Analytical modeling is an effective way to predict general trends for device performance evolution and, especially, the influence of dimensions downscaling. Such approach has been used for vertically integrated ZnO NWs deformed by lateral or compressive forces. One example of device designed to use lateral forces to bend

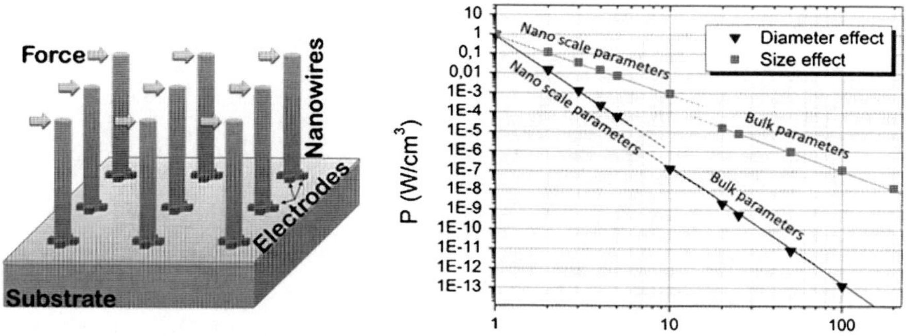

Fig. 2 Structure of a mechanical energy harvester using lateral forces applied on ZnO NWs (*left*). Effect of the NWs diameter and size (fixed aspect ratio of 20) on the power density generated from the device. The smallest scaling factor $\alpha = 1$ represents 50 nm-wide and 600 nm-long NWs (*right*) (adapted from [11])

vertically grown NWs is schematically presented in Fig. 2 (left) [11]. An analytical model has been used to study the power density harvested as a function of NWs geometrical parameters (Fig. 2 right panel) under the assumption of a constant peak lateral force (10 nN) and an average of 50 mechanical deformations per second (not necessarily periodic). Downscaling of NW size (diameter and length scaled together with fixed aspect ratio) or diameter (diameter scaled alone with fixed length) results in a large power density increase, with a maximum of 1 W/cm^3 at the minimal geometry considered (scaling factor $\alpha = 1$ representing 50 nm-wide, 600 nm-long NWs), which is compatible with the requirements of autonomous systems. Although promising, the main issue of this device would be the placement of the metallic contacts at the bottom of the device (see Fig. 2 left panel) that would be technologically challenging. These contacts could be fabricated using E-beam lithography or nanoimprint for instance, before the selective growth of the NWs.

Vertically grown ZnO NWs integrated into devices operating in compression mode have been modeled by several groups using different approaches. In 2012, Graton et al. developed a lumped circuit model of one million of NWs connected in parallel with bottom (Ohmic) and top (Schottky) metallic contacts (Fig. 3a). A maximum of 2.9 W_{rms}/cm^3 (19.4 V and 1.86 nA with an optimal load of 10 GΩ) has been reported when a compressive force of 1.2 μN at 50 Hz is applied to the device. Simulation results have shown that the reduction of NWs diameter should increase power density [15]. In 2011, Hu et al. have modeled two layers of vertical NWs integrated on a flexible polymer substrate as a continuous medium using FEM [28], predicting 80 V output voltage under bending. The fabricated devices delivered 10 V, which is quite high, even if still one step behind theoretical predictions. In 2012, Hinchet et al. have determined preliminary design guidelines for the VING (Vertically Integrated Nano Generator) architecture (Fig. 3b). This device includes a bottom electrode, a layer of ZnO NWs immersed in a polymer matrix (around and over the NWs) and a top electrode. The whole structure is fabricated typically on a Si substrate. FEM models (Fig. 4a) were developed

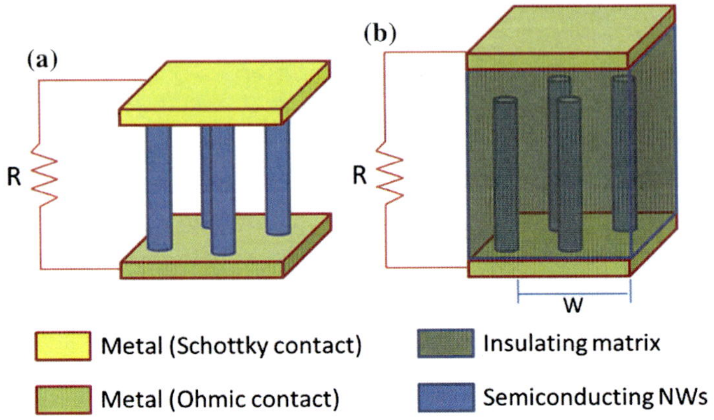

Fig. 3 **a** Structure of a device integrating vertical NWs between two metallic contacts. **b** Structure of a VING device (adapted from [16])

showing that Poly(methyl methacrylate) (PMMA) was a good matrix material and that an optimum in generated energy density was obtained for: (i) NWs separated by distances similar to their diameter (Fig. 4b) and (ii) a thinner PMMA layer over the NWs [29]. The range of parameters explored corresponded to typical experimental values for grown NWs, with 50–200 nm-wide NWs, separated by distances of 50 nm–1 μm. The maximal reported energy density was close to 10 pJ/cm^2 (65 mV) for a given peak pressure of 1 MPa. This corresponds to 20 nW/cm^3 after 50 compressions (and decompressions) per second, considering a 500 μm thick device (substrate). These results still need to be validated by experiments.

2.4 Further Improvements at the Nanoscale

Piezoelectric properties can still be further improved at the nanoscale. Recent near field (AFM) experiments on GaN NWs (25 nm-wide, 500 nm-long) including a thin AlN (8 nm) barrier along their c-axis, have shown an estimated piezoelectric coefficient 9 times higher compared to intrinsic GaN NWs with the same dimensions [30]. This would increase the efficiency of the energy conversion and the power density generated by harvesting devices [11].

3 Solar Energy Harvesting Using Semiconducting Nanostructures

Most of the world solar cell production is based on bulk (thick) silicon wafers. These solar cells are called first generation solar cells. Their main drawback is the cost of the material mainly induced by silicon purification, crystallisation and in-got and wafers

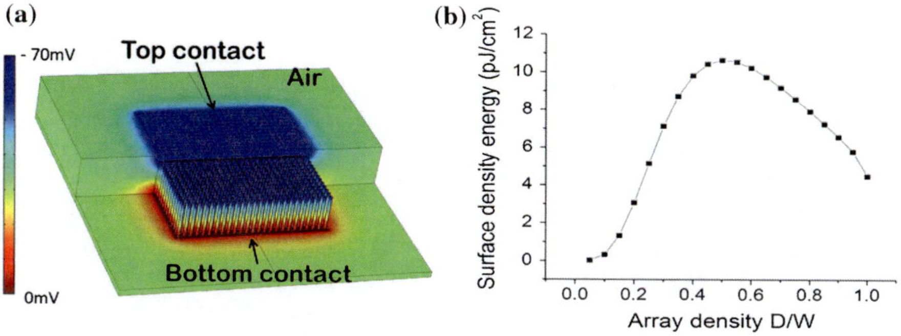

Fig. 4 **a** FEM simulation of the potential generated in a VING device surrounded by air and integrating 625 ZnO NWs in a PMMA matrix, for a pitch (defined as the ratio D/W between NW diameter and cell width) of 0.5, which corresponds to NWs separated by a distance equal to their diameter. **b** Surface density of mechanical energy stored in the VING as a function of D/W

sawing. One way to lower photovoltaic cost is therefore to reduce material consumption. However, it is then necessary to improve light trapping scheme in order to keep high absorption in the material. Silicon nanowires (NWs) based solar cells are an attractive approach to realize solar cells with an efficient light trapping scheme, potentially combined with high collection efficiency in the case of radial junctions. Indeed, the high-aspect-ratio of nanowires permits to reduce significantly solar cell thickness without loss of optical absorption while simultaneously providing effective carrier collection in the case of radial junction [31]. This structure benefits from the long optical path within the NW length and by exploiting a radial junction, a shorter path for carrier collection corresponding to the NW radius, leading to a smaller carrier recombination rate. Efficiencies similar or higher than the ones obtained with first generation solar cells (about 14–18 %) are expected for single junction nanowire solar cells with a cost reduction thanks to reduced material consumption and to low-cost growth methods.

There are two main approaches to elaborate the NWs arrays: a bottom-up approach based on the growth of the NWs and a top-down approach based on etching methods. Top-down approach has the disadvantage of wasting large quantity of matter, thus increasing the device cost. In contrast, the bottom-up approach is low-cost, technologically competitive and promising for photovoltaic energy.

Therefore, Si NWs arrays for photovoltaic applications are usually grown by Chemical Vapor Deposition (CVD) on top of silicon wafer, glass substrate or metal [31–35]. In the framework of the CVD method, the most used technique is the vapor liquid solid method which uses a metal catalyst to form a liquid eutectic with the desired NW material. Radial pn junction can be realized by diffusion of doping species from the NW surface. However, due to the small diameter of the NW, if the diffusion time is too long, there might be a complete doping resulting in a suppression of the pn junction. Another way to create the radial pn junction is to deposit a conformal and doped polysilicon layer on the NW. However, in the latter case, the interface is usually highly recombinant and should be passivated.

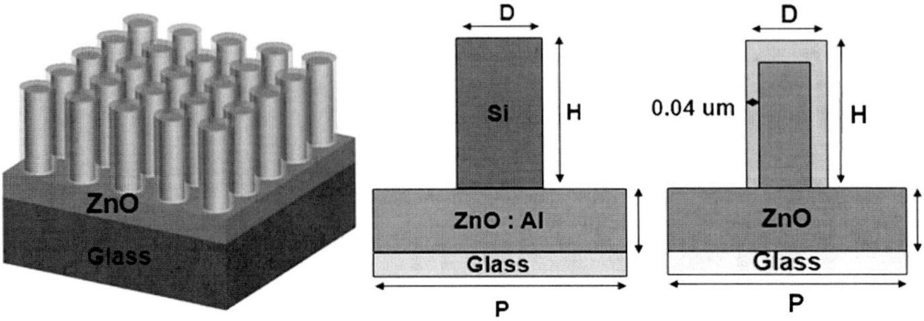

Fig. 5 NW based solar cell on ZnO on glass substrate (*left*); Si NW (*middle*); CdTe/ZnO NW (*right*), yellow shell is CdTe (adapted from Ref. [46])

The amorphous silicon / crystalline silicon (a-Si/c-Si) heterostructure is a good candidate for NW based solar cells since the heterostructure is able to efficiently separate the carriers while a-Si acts as a good surface passivation. The a-Si/c-Si heterojunction has already demonstrated high efficiency for first generation solar cells [36].

Increasing efforts have also been dedicated to the development of nanostructures based on ZnO thanks to its ability to grow within the NW morphology by a wide variety of growth methods such as CVD [37], chemical bath deposition [38] or electro-deposition [39]. Such NWs can be covered with CdSe [40], ZnS [41], ZnSe [42], or CdTe [38, 43] in order to create a type II band alignment heterojunction. The latter is a very efficient absorbing material with a bandgap energy of 1.5 eV at room temperature and will be also studied in this work in order to perform a comparison with Si NWs which are less absorbent due to the Si indirect band gap.

The relatively low experimental efficiencies obtained up to now (experimental power conversion efficiency of 7.9 and 4.74 % has already been reported for grown Si NW arrays [44] and ZnO/CdSe [45], respectively) are mainly due to high surface recombination velocity and to series and shunt resistance. There are still technological improvements needed to grow high quality NWs and to reduce surface recombination.

To optimize the absorption, it is necessary to define the best geometry by using optical simulations. Two types of materials have been compared: silicon NWs as the reference material with indirect band gap; ZnO NWs with CdTe radial heterojunction as the absorbing direct band gap material [46].

Simulation activities were performed using a Rigourous Coupled Wave Analysis (RCWA) 3D software for the optical simulations. For each structure (Fig. 5), the solar cells based on Si and ZnO/CdTe NWs have been simulated and an optimized geometry from an optical point of view has been defined [46]. The absorption versus wavelength of the incident light was deduced from the simulations of each structure (defined period and diameter) and the ideal short circuit current density (all generated electron/hole pairs are collected) corresponding to

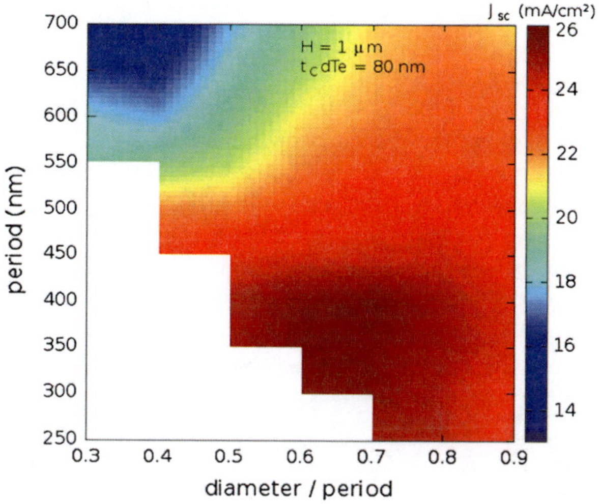

Fig. 6 Ideal short circuit current density computed with the RCWA tool for ZnO/CdTe core/shell NW arrays for different values of period and ratio between diameter and period. NW length is 1 μm. CdTe thickness is 80 nm

the standard AM1.5 incident light was then calculated [46]. An example of short circuit current density map versus period and diameter to period ratio is presented in Fig. 6 for ZnO/CdTe structure.

It was found that the NW structure significantly increases photons absorption compared to a planar structure with the same amount of material, especially in the case of indirect band gap semiconductor.

Compared to Si NW arrays, ZnO/CdTe NW array provided higher absorption with a less compact structure resulting in a smaller amount of material used thanks to the higher absorption of CdTe.

4 Thermal Energy Harvesting Using Semiconducting Nanostructures

In most of electronic and mechanical systems, a significant amount of power is wasted into heat. This power could be partially harvested by converting the resulting temperature gradients into electric power thanks to thermoelectric (TE) materials.

The efficiency of the thermoelectric devices is related to the dimensionless figure of merit $ZT = \sigma S^2 T/\kappa$, where σ is the carrier conductivity, S is the Seebeck coefficient, T is the temperature, κ is the thermal conductivity and Z is called the power factor. In order to attain large values of ZT, it is required a device/material with high carrier conductivity, large Seebeck coefficient and, at the same time, low thermal conductivity. For ordinary bulk semiconductors, ZT is far below 1. In 1990s, the thermoelectric materials regained attention as the low-dimensional systems were proposed to potentially have high thermoelectric figure of merit due to the presence of interfaces and the consequent thermal conductivity reduction below the alloy limit [47]. A largely investigated possibility to increase *ZT* is to consider bulk nanostructured materials. By using this approach, an enhancement of the figure of merit of BiTe was obtained [48] from 1 to 1.4. An alternative strategy to improve the thermoelectric efficiency is using energy filtering at the interfaces [49, 50]. In the energy-filtering technique, energy barriers are used to block the low-energy electrons and, therefore, increase the average heat transported per carrier. Hence, the Seebeck coefficient increases and could result in an enhanced power factor [51]. However, the same interfaces can also substantially reduce the mobility and, therefore, such an approach requires careful design of the nanostructures. Alternatively, the introduction of resonant impurity levels inside the conduction or valance band was proposed to create sharp features in the density of states and increase the Seebeck coefficient [52]. Another possibility is to increase the electron conductivity via modulation doping. In such an approach, charge carriers are spatially separated from their parent impurity atoms and consequently the impurity scattering is reduced [53].

Finally, much attention has been devoted to semiconductor NWs, which are particularly promising structures due to their low density of states and the high surface/volume ratio. Although this interest was initially motivated by hopes of taking advantage of electron confinement in the structures, it soon became clear that another advantage of NWs was their potentially strongly reduced thermal conductivity [54]. A large reduction in Si NW lattice thermal conductivity was experimentally reported in 2003, further stimulating research activities in this area [55]. An astonishingly low thermal conductivity has also been claimed on Si NWs due to the effect of surface roughness [56]. Recent works show that the interplay between alloy scattering and scattering by the nanostructured features can lead to interesting qualitative differences between the behavior of the thermal conductivity of alloy and non-alloy structures [57].

From a theoretical point of view, accurate models free of adjustable parameters are the most reliable way of computing fundamental phonon transport properties [58, 59]. Thermal conductivity in NWs in the presence of roughness or other spatial defects is addressed either within the semi-classical Boltzmann transport [60], which cannot take into account phase-coherent phenomena, or within non-equilibrium Green's function techniques, which usually consider only elastic transport [61, 62].

4.1 Simulation of Thermoelectric Properties of Rough Si NWs

The understanding of phonon confinement and phonon scattering effects in nanostructures is a key to any thermal transport engineering for the improvement of the thermoelectric performances. Here, we present 3D simulations of phonon properties in confined structures as semiconductor NWs in the presence of spatial

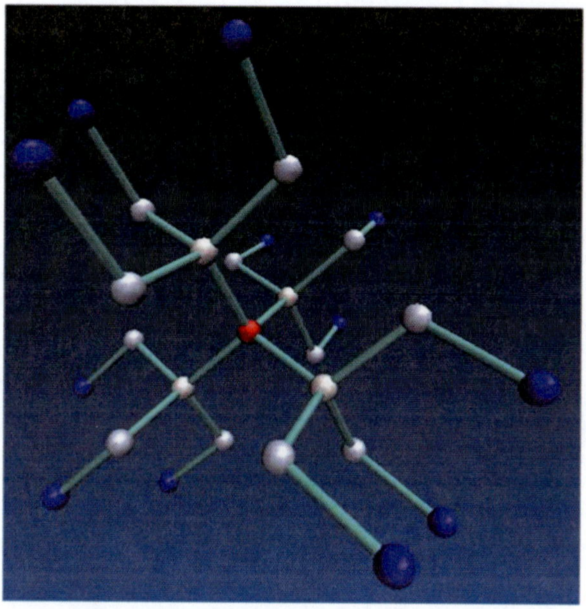

Fig. 7 Sketch of the extended valence force model showing the coupling of a single Si atom with its 28 first neighbor atoms

fluctuations. We address phonon band structures and heat flux within a full-quantum mechanical theory and further couple these results with self-consistent electron transport calculations in order to extract relevant factors of merit of thermoelectric devices.

The phonon band structure calculations were obtained by implementing an extended Keating model including four terms (bond-stretching, bond-bending, angle-angle and bond-bond interactions) for the determination of the dynamical matrices of nanosystems [63]. A scheme of the coupling of a single atom with its neighbors is shown in Fig. 7.

This model uses material constants that are chosen to reproduce the bulk phonon dispersion and then it is extended to compute the confined modes of NWs, which are assumed to be infinite and composed of identical unit cells. The NWs simulated with such an atomistic description can be naturally generalized to any crystallographic orientation and include the presence of random disorder (e.g. roughness, crystal defects). From the dynamical matrices, the NW basic phonon properties, as band structure and density of states, can be extracted. For example, Fig. 8 shows the phonon band structure and the corresponding density of states (DOS) of a square <100> oriented Si NW with a lateral cross section of $2 \times 2\ nm^2$.

Hence, starting from the dynamical matrices computed with the extended Keating model, we were able to implement a recursive algorithm based on the Sancho-Rubio iterative scheme [64] to compute the surface and bulk Green's function of Si NWs. The first application of this code was to evaluate in an alternative way the DOS of the NW in Fig. 8(left), previously computed via a

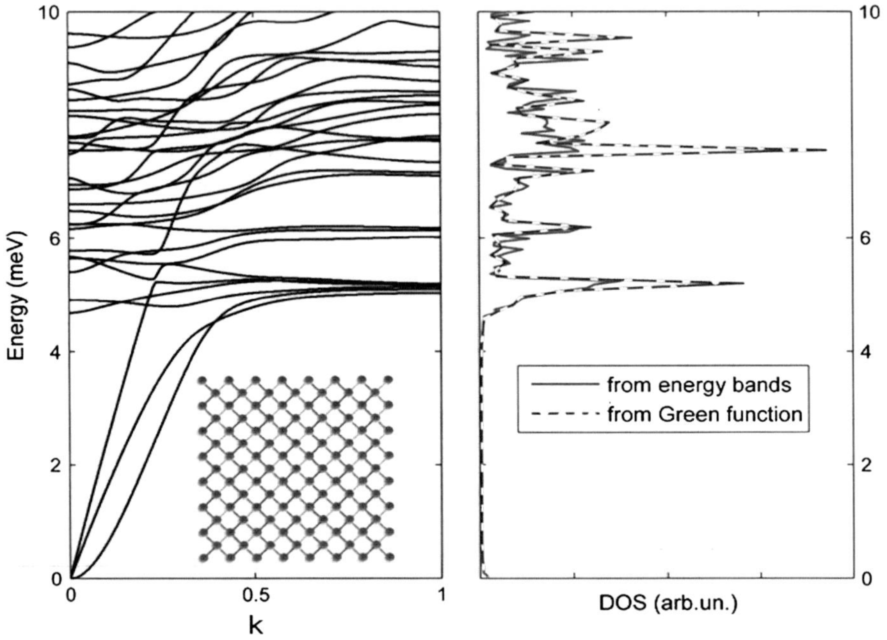

Fig. 8 Phonon band structure of a <100> Si NW with a squared cross section of 2 × 2 nm2 (*left*) and the corresponding density of states (*right*) obtained from the direct counting of energy eigenvalues in the band structure (*solid line*) and from the retarded Green's function (*dotted line*)

direct counting of the energy bands. The DOS computed via the two alternative methods presented the same features validating the methodology.

Based on the non-equilibrium Green's function formalism [65], we computed the phonon transport properties as thermal conductivity at different temperatures of silicon NWs in the presence of surface roughness. Importantly, such a kind of calculation can be easily extended to other geometries as superlattices and quantum dots and other semiconductor materials as Ge and III-V compounds.

We considered a square <100> oriented NW with an edge of 5 nm and different roughness root mean square (r.m.s.) values [66]. Surface roughness was geometrically generated with a random algorithm as described in [67]. Our results reported in Fig. 9 clearly show that surface roughness induces a strong decrease of the thermal conductance. Even a small value of roughness *r.m.s.* is able to considerably reduce the transmission in the whole frequency spectrum (left panel) and consequently strongly suppresses the thermal conductivity (right panel).

Finally, 3D atomistic simulations within the Keldysh-Green's function formalism were exploited to evaluate the increase of the factor of merit *ZT* due to the presence of surface roughness in silicon NWs. The Seebeck coefficient *S*, the electrical conductance *G* and the corresponding power factor S^2G have been computed for rectangular NWs with cross sections of 5 × 5 nm^2 and 3 × 3 nm^2 and for surface roughness *r.m.s.* 0.2 and 0.4 nm. The evolution of these parameters

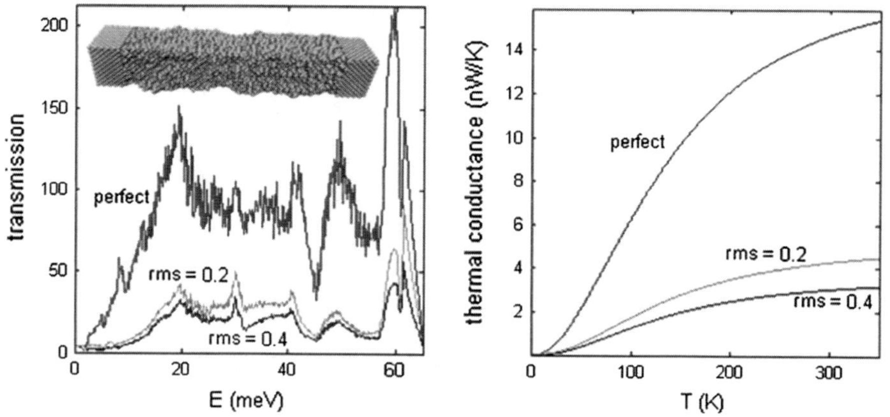

Fig. 9 Phonon transmission (*left panel*) and thermal conductivity (*right panel*) of a Si NW (*inset*) with square section of 5 × 5 nm^2 and in the absence (*blue lines*) and in the presence of surface roughness with *r.m.s.* 0.2 nm (green lines) and 0.4 nm (*red lines*)

as a function of the surface roughness *r.m.s.* is shown in Fig. 10a, where we can observe the opposite behavior of the Seebeck coefficient and of the electrical conductance as the roughness increases. The increase of the Seebeck coefficient due to surface roughness can be explained by analyzing the shape of the spectral density of the transmission probability. This implies that, as shown in Fig. 10b, the power factor, which expresses the electrical performance of the thermoelectric materials, monotonically decreases with increasing the roughness.

However, such decrease of the factor S^2G has to be compared with the corresponding decrease of the thermal conductance shown in Fig. 11a. For this, we can remark that the phonon conductance is strongly suppressed by both the lateral confinement and by the surface roughness. Phonon transmission is therefore decreased when the surface/volume ratio is as small as possible. As shown in Fig. 11b, this behavior results in an increase of the factor of merit *ZT* up to about 0.7 for very thin NWs with a 3 × 3 nm^2 lateral section and 0.2 nm of surface roughness *r.m.s.* In this configuration, *ZT* turns out to be increased, because phonon thermal conductance decreases faster than the power factor with decreasing the wire cross sections.

5 Conclusions and Perspectives

Several properties can be improved at the nanoscale compared to bulk materials: higher piezoelectric coefficients and flexibility, higher photon absorption, lower thermal conduction between others. These improvements make nanostructures promising for mechanical, solar and thermal energy harvesting but also for sensing

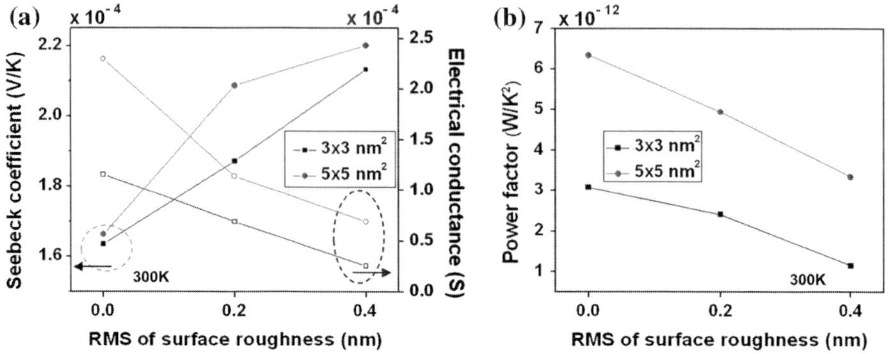

Fig. 10 **a** Seebeck coefficient and electrical conductance and **b** power factor as a function of the surface roughness rms of Si NWs with different cross sections of 3 × 3 and 5 × 5 nm^2

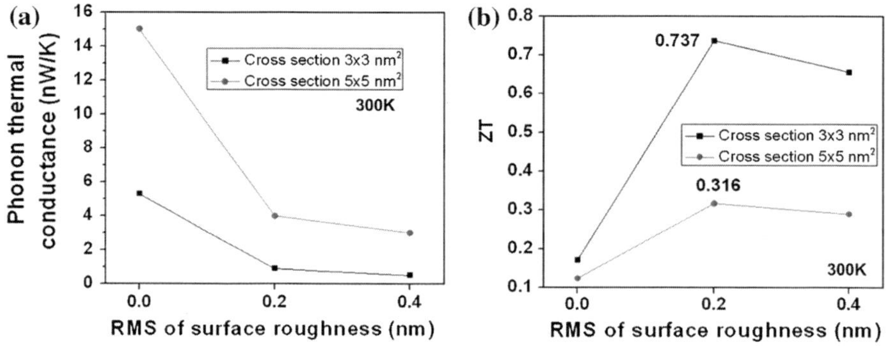

Fig. 11 **a** Phonon thermal conductance and **b** the ZT factor of merit as a function of the surface roughness rms of Si NWs with different cross sections of 3 × 3 and 5 × 5 nm^2

applications, although multiple technical issues concerning their integration into functional devices need to be solved to improve the global efficiency.

Very few models can be found in the literature concerning the performances optimization of devices based on NWs for energy conversion applications. The models reviewed in this work proposed optimization guideline rules on the choice of materials, NWs geometries (diameter, length) and roughness to improve the energy conversion efficiency for the three mentioned harvesting applications, although experimental validation is still required.

The energy conversion efficiency can be further improved at the nanoscale using axial or radial (core-shell) heterostructured NWs depending of the application. For instance, GaN NWs with thin AlN axial barriers can increase the mechanical harvesting efficiency, while ZnO (core)/CdTe (Shell) NWs can increase the optical absorption efficiency for photovoltaic applications.

Finally, the integration of several energy conversions into one single device could be a solution to increase the harvested energy density and to build truly autonomous systems working in any ambient condition.

Acknowledgments This work has been partly supported by the European Union 7th Framework Program, within the Network of Excellence NanoFunction under grant agreement FP7/ICT/NoE n° 257375.

References

1. Vullers, R.J.M., Schaijk, R.V., Visser, H.J., Penders, J., Hoof, C.V.: Energy harvesting for autonomous wireless sensor networks. IEEE Solid-State Circuits Mag. **2**, 29–38 (2010)
2. Nechibvute, A., Chawanda, A., Luhanga, P.: Piezoelectric energy harvesting devices: an alternative energy source for wireless sensors. Smart Mater. Res. **2012**, 13 (2012)
3. Cook-Chennault, K.A., Thamby, N., Sastry, A.S.: Powering MEMS portable devices—a review of non-regenerative and regenerative power supply systems with special emphasis on piezoelectric energy harvesting systems. Smart Mater. Struct. **17**, 043001 (2008)
4. Li, D., Wu, Y., Kim, P., Shi, L., Yang, P., Majumdar, A.: Thermal conductivity of individual silicon nanowires. Appl. Phys. Lett. **83**, 2934–2936 (2003)
5. Minary-Jolandan, M., Bernal, R.A., Kuljanishvili, I., Parpoli, V., Espinosa, H.D.: Individual GaN nanowires exhibit strong piezoelectricity in 3D. Nano Lett. **12**, 970–976 (2012)
6. Mitcheson, P.D., Yeatman, E.M., Rao, G.K., Holmes, A.S., Green, T.C.: Energy harvesting from human and machine motion for wireless electronic devices. Proc. IEEE **96**, 1457–1486 (2008)
7. Zhu, D., Tudor, M.J., Beeby, S.P.: Strategies for increasing the operating frequency range of vibration energy harvesters: a review. Meas. Sci. Technol. **21**, 022001 (2010)
8. Espinosa, H.D., Bernal, R.A., Minary-Jolandan, M.: A review of mechanical and electromechanical properties of piezoelectric nanowires. Adv. Mater. **24**, 4656–4675 (2012)
9. Minary-Jolandan, M., Bernal, R.A., Kuljanishvili, I., Parpoil, V., Espinosa, H.D.: Individual GaN nanowires exhibit strong piezoelectricity in 3D. Nano Lett. **12**, 970–976 (2012)
10. Zhao, M.-H., Wang, Z.-L., Mao, S.X.: Piezoelectric characterization of individual zinc oxide nanobelt probed by piezoresponse force microscope. Nano Lett. **4**, 587–590 (2004)
11. Hinchet, R., Ferreira, J., Keraudy, J., Ardila, G., Pauliac-Vaujour, E., Mouis, M., Montes, L.: Scaling rules of piezoelectric nanowires in view of sensor and energy harvester integration. In: Electron Devices Meeting (IEDM) 2012 IEEE International, pp. 1–4 (2012)
12. Agrawal, R., Espinosa, H.D.: Giant piezoelectric size effects in zinc oxide and gallium nitride nanowires. A first principles investigation. Nano Lett. **11**, 786–790 (2011)
13. Hoang, M.-T., Yvonnet, J., Mitrushchenkov, A., Chambaud, G.: First-principles based multiscale model of piezoelectric nanowires with surface effects. J. Appl. Phys. **113**, 1–9 (2013)
14. Araneo, R., Lovat, G., Burghignoli, P., Falconi, C.: Piezo-semiconductive quasi-1D nanodevices with or without anti-symmetry. Adv. Mater. **24**, 4719–4724 (2012)
15. Graton, O., Poulin-Vittrant, G., Tran Huu Hue, L.P., Lethiecq, M.: Strategy of modelling and simulation of electromechanical conversion in ZnO nanowires. Adv. Appl. Ceram. **112**, 85–90 (2013)
16. Ardila, G., Hinchet, R., Montes, L., Mouis, M.: Mechanical energy harvesting with piezoelectric nanostructures: great expectations for autonomous systems. In: Luryi, S., Xu, J., Zaslavsky, A. (eds.) Future Trends in Microelectronics: Frontiers and Innovations. Wiley, New York (2013)

17. Ardila, G., Hinchet, R., Mouis, M., Montès, L.: Scaling prospects in mechanical energy harvesting using piezoelectric nanostructures. In: ISCDG, IEEE Conference Publications, pp. 75–78 (2012)
18. Chen, X., Xu, S., Yao, N., Shi, Y.: 1.6 V nanogenerator for mechanical energy harvesting using PZT nanofibers. Nano Lett. **10**, 2133–2137 (2010)
19. Yang, R., Qin, Y., Dai, L., Wang, Z.L.: Power generation with laterally packaged piezoelectric fine wires. Nat. Nanotechnol. **4**, 34–39 (2009)
20. Xu, S., Qin, Y., Xu, C., Wei, Y., Yang, R., Wang, Z.L.: Self-powered nanowire devices. Nat. Nanotechnol. **5**, 366–373 (2010)
21. Qi, Y., McAlpine, M.C.: Nanotechnology-enabled flexible and biocompatible energy harvesting. Energy Environ. Sci. **3**, 1275–1285 (2010)
22. Chang, C., Tran, V.H., Wang, J., Fuh, Y.-K., Lin, L.: Direct-write piezoelectric polymeric nanogenerator with high energy conversion efficiency. Nano Lett. **10**, 726–731 (2010)
23. Jung, J.H., Lee, M., Hong, J.-I., Ding, Y., Chen, C.-Y., Chou, L.-J., Wang, Z.L.: Lead-free NaNbO3 nanowires for a high output piezoelectric nanogenerator. ACS Nano **5**, 10041–10046 (2011)
24. Wang, X., Song, J., Liu, J., Wang, Z.L.: Direct-current nanogenerator driven by ultrasonic waves. Science **316**, 102–105 (2007)
25. Qin, Y., Wang, X., Wang, Z.L.: Microfibre-nanowire hybrid structure for energy scavenging. Nature **451**, 809–813 (2008)
26. Lee, M., Chen, C.-Y., Wang, S., Cha, S.N., Park, Y.J., Park, Y.J., Kim, J.M., Chou, L.-J., Wang, Z.L.: A hybrid piezoelectric structure for wearable nanogenerators. Adv. Mater. **24**, 1759–1764 (2012)
27. Zhu, G., Wang, A.C., Liu, Y., Zhou, Y., Wang, Z.L.: Functional electrical stimulation by nanogenerator with 58 V output voltage. Nano Lett. **12**, 3086–3090 (2012)
28. Hu, Y., Zhang, Y., Xu, C., Lin, L., Snyder, R.L., Wang, Z.L.: Self-powered system with wireless data transmission. Nano Lett. **11**, 2572–2577 (2011)
29. Hinchet, R., Lee, S., Ardila, G., Montes, L., Mouis, M., Wang, Z.L.: Design and guideline rules for the performance improvement of vertically integrated nanogenerator. In: PowerMEMS 2012, The 12th international workshop on micro and nanotechnology for power generation and energy conversion applications (2012)
30. Xu, X., Potie, A., Songmuang, R., Lee, J.W., Bercu, B., Baron, T., Salem, B., Montès, L.: An improved AFM cross-sectional method for piezoelectric nanostructures properties investigation: application to GaN nanowires. Nanotechnology **22**, 105704 (2011)
31. Garnett, E.C., Brongersma, M.L., Cui, Y., McGehee, M.D.: Nanowire solar cells. Annu. Rev. Mater. Res. **41**, 11.1–11.27 (2011)
32. Kelzenberg, M.D., Boettcher, S.W., Petykiewicz, J.A., Turner-Evans, D.B., Putnam, M.C., Warren, E.L., Spurgeon, J.M., Briggs, R.M., Lewis, N.S., Atwater, H.A.: Enhanced absorption and carrier collection in Si wire arrays for photovoltaic applications. Nat. Mater. **9**, 239–244 (2010)
33. Fan, Z., Razavi, H., Do, J., Moriwaki, A., Ergen, O., Chueh, Y.-L., Leu, P.W., Ho, J.C., Takahashi, T., Reichertz, L.A., Neale, S., Yu, K., Wu, M., Ager, J.W., Javey, A.: Three-dimensional nanopillar-array photovoltaics on low-cost and flexible substrates. Nat. Mater. **8**, 648–653 (2009)
34. Tsakalakos, L., Balch, J., Fronheiser, J., Shih, M.-Y., LeBoeuf, S.F., Pietrzykowski, M., Codella, P.J., Korevaar, B.A., Sulima, O.V., Rand, J., Davuluru, A., Rapol, U.: Strong broadband optical absorption in silicon nanowire films. J. Nanophotonics **1**, 013552 (2007)
35. O'Donnell, B., Yu, L., Foldyna, M., Roca i Cabarrocas, P.: Silicon nanowire solar cells grown by PECVD. J. Non-Cryst. Solids **358**, 2299–2302 (2012)
36. Mishima, T., Taguchi, M., Sakata, H., Maruyama, E.: Development status of high-efficiency HIT solar cells. Sol. Energy Mater. Sol. Cells **95**, 18–21 (2011)
37. Latu-Romain, E., Gilet, P., Feuillet, G., Noel, P., Garcia, J., Levy, F., Chelnokov, A.: Optical and electrical characterizations of vertically integrated ZnO nanowires. Microelectron. J. **40**, 224–228 (2009)

38. Consonni, V., Rey, G., Bonaime, J., Karst, N., Doisneau, B., Roussel, H., Renet, S., Bellet, D.: Synthesis and physical properties of ZnO/CdTe core shell nanowires grown by low-cost deposition methods. Appl. Phys. Lett. **98**, 111906 (2011)
39. Schmidt-Mende, L., MacManus-Driscoll, J.L.: ZnO nanostructures, defects, and devices. Mater. Today **10**, 40–48 (2007)
40. Levy-Clement, C., Tena-Zaera, R., Ryan, M., Katty, A., Hodes, G.: CdSe-Sensitized p-CuSCN/Nanowire n-ZnO Heterojunctions. Adv. Mater. **17**, 1512–1515 (2005)
41. Wang, K., Chen, J.J., Zeng, Z.M., Tarr, J., Zhou, W.L., Zhang, Y., Yan, Y.F., Jiang, C.S., Pern, J., Mascarenhas, A.: Synthesis and photovoltaic effect of vertically aligned ZnO/ZnS core/shell nanowire arrays. Appl. Phys. Lett. **96**, 123105 (2010)
42. Zhang, Y., Wu, Z., Zheng, J., Lin, X., Zhan, H., Li, S., Kang, J., Bleuse, J., Mariette, H.: ZnO/ZnSe type II core/shell nanowire array solar cell. Solar Energy Mater. Solar Cells **102**, 15–18 (2012)
43. Wang, X., Zhu, H., Xu, Y., Wang, H., Tao, Y., Hark, S., Xiao, X., Li, Q.: Aligned ZnO/CdTe core/shell nanocable arrays on indium tin oxide: synthesis and photoelectrochemical properties. ACS Nano **4**, 3302–3308 (2010)
44. Putnam, M.C., Boettcher, S.W., Kelzenberg, M.D., Turner-Evans, D.B., Spurgeon, J.M., Warren, E.L., Briggs, R.M., Lewis, N.S., Atwater, H.A.: Si microwire-array solar cells. Energy Environ. Sci. **3**, 1037–1041
45. Xu, J., Yang, X., Wang, H., Chen, X., Luan, C., Xu, Z., Lu, Z., Roy, V.A.L., Zhang, W., Lee, C.S.: Arrays of ZnO/ZnxCd1−xSe nanocables: band gap engineering and photovoltaic applications. Nano Lett. **11**, 4138 (2011)
46. Michallon, J., Zanuccoli, M., Kaminski-Cachopo, A., Consonni, V., Morand, A., Bucci, D., Emieux, F., Szambolics, H., Perraud, S., Semenikhin, I.: Comparison of optical properties of Si and ZnO/CdTe core/shell nanowire arrays. Mater. Sci. Eng. B **178**, 665–669 (2013)
47. Chen, G.: Thermal conductivity and ballistic-phonon transport in the cross-plane direction of superlattices. Phys. Rev. B **57**, 14958–14973 (1998)
48. Poudel, B., Hao, Q., Ma, Y., Lan, Y., Minnich, A., Yu, B., Yan, X., Wang, D., Muto, A., Vashaee, D., Chen, X., Liu, J., Dresselhaus, M.S., Chen, G., Ren, Z.: High-thermoelectric performance of nanostructured bismuth antimony telluride bulk alloys. Science **320**, 634–638 (2008)
49. Shakouri, A., Bowers, J.E.: Heterostructure integrated thermionic coolers. Appl. Phys. Lett. **71**, 1234–1236 (1997)
50. Mahan, G.D.: Thermionic refrigeration. Semicond. Semimetals **71**, 157–174 (2001)
51. Zide, J.M.O., Vashaee, D., Bian, Z.X., Zeng, G., Bowers, J.E., Shakouri, A., Gossard, A.C.: Demonstration of electron filtering to increase the Seebeck coefficient in In0.53Ga0.47As/In0.53Ga0.28Al0.19As superlattices. Phys. Rev. B **74**, 205335(5) (2006)
52. Hicks, L.D., Dresselhaus, M.S.: Effect of quantum-well structures on the thermoelectric figure of merit. Phys. Rev. B **47**, 12727–12731 (1993)
53. Zebarjadi, M., Joshi, G., Zhu, G., Yu, B., Minnich, A., Lan, Y., Wang, X., Dresselhaus, M., Ren, Z., Chen, G.: Power factor enhancement by modulation doping in bulk nanocomposites. Nano Lett. **11**, 2225–2230 (2011)
54. Zhang, G., Zhang, Y.-W.: Thermal conductivity of silicon nanowires: From fundamentals to phononic engineering. Phys. Status Solidi RRL (2013). doi:10.1002/pssr.201307188
55. Li, D., Wu, Y., Kim, P., Shi, L., Yang, P., Majumdar, A.: Thermal conductivity of individual silicon nanowires. Appl. Phys. Lett. **83**, 2934–2936 (2003)
56. Hochbaum, A.I., Chen, R., Diaz Delgado, R., Liang, W., Garnett, E.C., Najarian, M., Majumdar, A., Yang, P.: Enhanced thermoelectric performance of rough silicon nanowires. Nature **451**, 163–167 (2008)
57. Bera, C., Mingo, N., Volz, S.: Marked effects of alloying on the thermal conductivity of nanoporous materials. Phys. Rev. Lett. **104**, 115502(4) (2010)
58. Broido, D.A., Malorny, M., Birner, G., Mingo, N., Stewart, D.A.: Intrinsic lattice thermal conductivity of semiconductors from first principles. Appl. Phys. Lett. **91**, 231922–231994 (2007)

59. Baroni, S., de Gironcoli, S., Dal Corso, A., Giannozzi, P.: Phonons and related crystal properties from density-functional perturbation theory. Rev. Mod. Phys. **73**, 515–562 (2001)
60. Martin, P.N., Aksamija, Z., Pop, E., Ravaioli, U.: Reduced thermal conductivity in nanoengineered rough Ge and GaAs nanowires. Nano Lett. **10**, 1120–1124 (2010)
61. Mingo, N., Yang, L., Li, D., Majumdar, A.: Predicting the thermal conductivity of Si and Ge nanowires. Nano Lett. **3**, 1713–1716 (2003)
62. Mingo N, Yang L (2003) Phonon transport in nanowires coated with an amorphous material: An atomistic Green's function approach. Phys. Rev. B 68: 245406(12)
63. Sui, Z., Herman, I.P.: Effect of strain on phonons in Si, Ge, and Si/Ge heterostructures. Phys. Rev. B **48**, 17938–17953 (1993)
64. Lopez Sancho, M.P., Lopez Sancho, J.M., Rubio, J.: Quick iterative scheme for the calculation of transfer matrices: application to Mo (100). J. Phys. F: Met. Phys. **14**, 1205–1215 (1984)
65. Rogdakis, K., Poli, S., Bano, E., Zekentes, K., Pala, M.G.: Phonon- and surface-roughness-limited mobility of gate-all-around 3C-SiC and Si nanowire FETs. Nanotechnology **20**, 295202(6)(2009)
66. Buran, C., Pala, M.G., Bescond, M., Dubois, M., Mouis, M.: Three-dimensional real-space simulation of surface roughness in silicon nanowire FETs. IEEE-Trans. Elec. Dev. **56**, 2186–2192 (2009)
67. Poli, S., Pala, M.G., Poiroux, T., Deleonibus, S., Baccarani, G.: Size dependence of surface-roughness-limited mobility in silicon-nanowire FETs. IEEE-Trans. Elec. Dev. **55**, 2968–2976 (2008)

Energy Harvesting Using THz Electronics

Stephen Hall, Ivona Z. Mitrovic, Naser Sedghi, Yao-chun C. Shen, Yi Huang and Jason F. Ralph

Abstract In this chapter we present a review and appraisal of energy harvesting using rectenna devices ('rectifying antennas') at THz and solar frequencies. The concept involves capturing the electromagnetic radiation in a nano-scale antenna and rectifying it to direct current. Rectennas may offer much higher efficiency than photovoltaics, in principle but there are considerable challenges in the engineering of such devices. In particular, the rectifier must provide a good match to the antenna at very low currents. Although high efficiency rectennas have been demonstrated at microwave frequencies, the device cannot simply be scaled to the higher frequencies of interest, due to the significant changes in materials properties and these are explained in the paper. Finally a design framework for one rectifier type is presented, namely the metal-insulator diode. This study serves to highlight the considerable challenges associated with the matching issue.

1 Introduction

The terahertz (10^{12} Hz) region is often described as the final unexplored region of the electromagnetic spectrum. Usually it is defined as the frequency range 0.1–10 THz (wavelengths of 3 mm–30 μm) although in this paper, the definition is extended to include the IR and visible light regions of the spectrum. The solar energy covers a very broadband of the electromagnetic spectrum, spreading over the ultraviolet (UV), visible light, and infrared (IR) regimes as shown in Fig. 1. Part of the energy (about 20 %) is attenuated by the atmosphere, but most of it reaches the surface of the Earth (at sea level) and can in principle be captured. It is

The erratum of this chapter can be found under DOI 10.1007/978-3-319-08804-4_21

S. Hall (✉) · I.Z. Mitrovic · N. Sedghi · Y.C. Shen · Y. Huang · J.F. Ralph
Department of Electrical Engineering and Electronics, The University of Liverpool, Brownlow Hill, Liverpool L69 3GJ, UK
e-mail: s.hall@liverpool.ac.uk

A. Nazarov et al. (eds.), *Functional Nanomaterials and Devices for Electronics, Sensors and Energy Harvesting*, Engineering Materials,
DOI 10.1007/978-3-319-08804-4_12

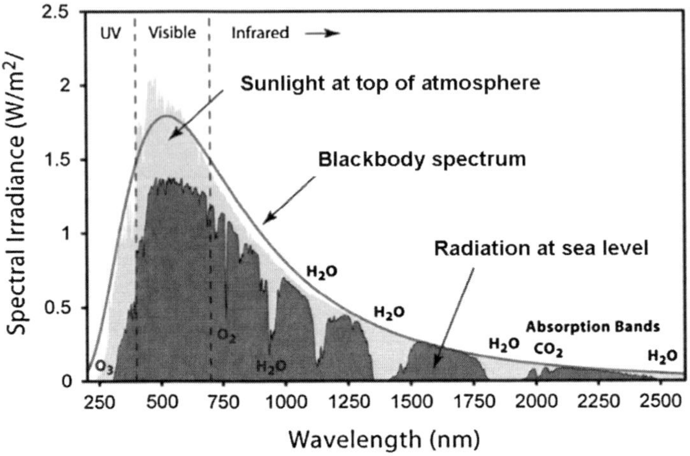

Fig. 1 The solar spectrum [1]

also interesting to note that most of the energy (>85 %) is concentrated in the spectrum between 400–1750 nm, that is, 750–170 THz. Thus the ability to harvest energy from this part of the available energy of the Sun presents a great opportunity and potential for applications and social impact. However, it presents many challenges for both electronic and photonic technologies.

Existing solid-state devices that are able to operate in the THz region have several major disadvantages, in particular they are expensive, cumbersome, or require cryogenic cooling. The energy reaching the Earth in both the visible (430–790 THz) and infra-red (IR) regions and the re-radiated IR energy ($\sim$20–40 THz) are under-utilized by current technology and so a considerable research effort with a large investment has been made in order to develop an effective way to convert this energy to usable direct current (DC) electricity. Current solid-state solar technologies, such as photovoltaics (PV), are unable to capture energy over a broad band due to the physical limitations of their operating principle of band-to-band mediated absorption. Although solutions are available, they introduce more complexity and hence cost into the technology. The average efficiency of a solar PV module is around 15 %; hence 85 % of the solar radiation is lost through heating and reflection. Furthermore, most of the world's energy supply is provided by thermoelectric power stations where a tremendous amount of heat, which radiates at tens of THz, is wasted. A 1 % increase in efficiency by converting some waste heat into electricity would have a significant economic impact.

The development of solid-state, room-temperature rectifiers operating over the THz to visible light bands could therefore constitute a significant breakthrough and benefit a number of disciplines but in particular, energy harvesting. For energy harvesting purposes, the rectifier operates in conjunction with a nano-antenna to form a so-called rectenna (rectifying antenna) and a typical configuration is shown in Fig. 2. The two key elements are the antenna and rectifier. The antenna receives solar energy (electromagnetic waves) and converts it to high frequency electricity which is then converted to DC power by the rectifier.

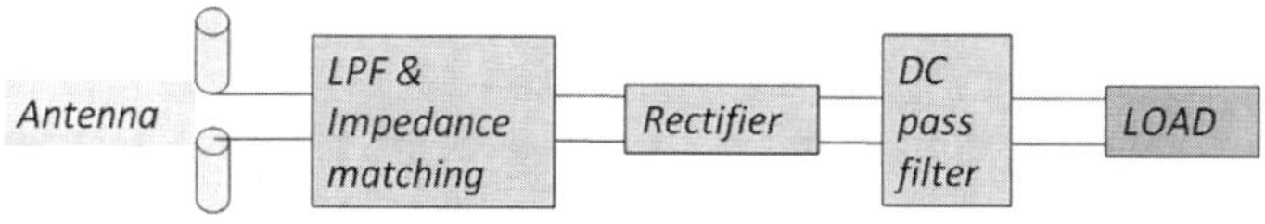

Fig. 2 Block diagram of a rectenna system

Unlike conventional photovoltaic devices, rectennas do not have to use solar power and can be designed to operate in the infrared and microwave regimes. Thus they can be used also through the night. An even more important advantage is that, unlike PVs, there is no fundamental limit on the efficiency of rectennas. In practice, it is necessary to take thermodynamic limitations into account, feasible design constraints and associated losses. Nevertheless, the achievable efficiency can in principle, still be very high, and well in excess of that of PVs. For example, efficiencies of more than 85 % have been observed at microwave frequencies in the laboratory [4].

This chapter is organized as follow. Section 2 contains a review of rectenna-related devices in the THz regime. Section 3 delivers an account of materials properties that need to be taken into account in the design of THz devices. A discussion of antenna design is presented in Sect. 4 which includes simulation and experimental results in the RF regime together with a discussion of scaling challenges to the THz regime and beyond. Section 5 contains details of a quantum mechanical model to assist the design of metal-insulator-metal based rectifiers. Conclusions are drawn in Sect. 6.

2 Review of Rectennas

The success of rectennas in the RF/microwave regions [2–6] has provided the inspiration to extend the rectenna concept to THz, infrared (IR) and even to solar optical frequencies. The motivation is to harvest solar energy at a higher conversion efficiency than the one achieved using current photovoltaic technology, which is typically around 15 %. There have been some attempts to make THz rectennas [7–13]. Bailey was the first to propose the idea of collecting solar energy using rectenna devices in 1972 [7, 8]. Dipole antennas were formed of modified pyramids which were connected to a diode, low-pass filter and load. Marks was the first to patent the use of an array of submicron cross-dipoles on an insulating sheet with full-wave rectification in 1984 [9] and Kraus proposed an array of dipole antennas for solar energy collection in his popular book published in 1988 [10]. However, the first experimental evidence for light absorption in a fabricated resonant nanostructure and rectification at optical frequency was reported by Lin et al. in 1996 [11], where a parallel dipole antenna array on a silicon substrate was employed. A relatively in-depth study was conducted by ITN Energy Systems [12] with the aim to demonstrate the feasibility of using rectennas to collect energy at a

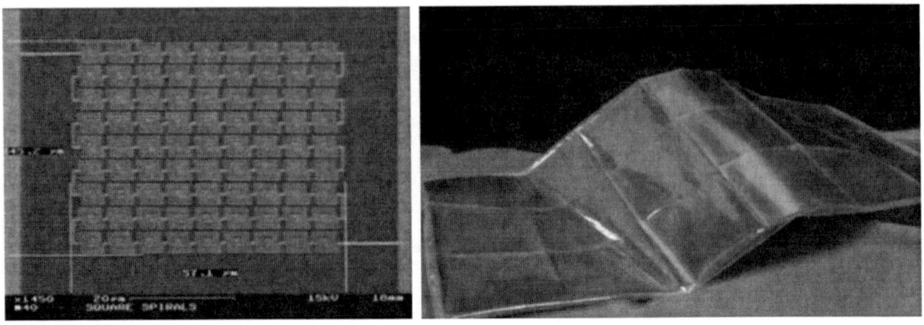

Fig. 3 A nano-antenna array image (*left*) and a nano-antenna array on plastic sheet (*right*) [13]

single wavelength in the solar spectrum. Very low efficiency (<1 %) was observed at 30 THz. More recently nanoantennas (no rectifiers) have been made on a flexible substrate over a large area [13] as shown in Fig. 3, which is cost-effective and may be applied to wider applications than the current PVs.

The rectifier however, remains perhaps the most critical part of the system and a brief account of the challenges and choices is now presented. The rectifier could be either semiconductor or dielectric based. For the semiconductor option, there are two choices, namely the pn junction or Schottky diode, both of which have an exponential dependence of current with voltage, namely $I \sim \exp(V/mV_{th})$ with the 'ideality factor', $m = 1$ for a good diode—easily achievable with a Schottky diode; V_{th} is 25 mV at room temperature and linear with temperature. Pn junction diodes operate on the principle of minority carrier injection; taking an $n^{+}p$ diode as an example, the current under forward bias is carried predominantly by electrons injected into the p-side of the junction (electrons are minority carriers in p-type material). When the bias is reversed (as in a rectifier), the minority carrier charge must be removed before the reverse 'blocking' bias can be established. There is therefore a recovery time for this process to occur during which approximately equal but opposite polarity current will flow in the diode (non-rectifying). Thus removal of electrons occurs by a reverse current of electrons across the junction and hence through the external circuit, and the internal process of recombination of minority electrons with majority carrier holes. Both of these processes are relatively slow and certainly preclude the use of such diodes in the THz regime. The Schottky diode (SD) fares better at high frequencies because the current is carried by majority carriers, electrons for a metal-n-type semiconductor system, which are injected into the metal side of the junction. Charge storage does not occur in metals; the electrons are accommodated within the so-called dielectric relaxation time of the metal, $\tau_{dM} = \sigma/\varepsilon << 1$ ps, where σ is the conductivity and ε, the permittivity. The limiting factor for the frequency response of the SD is set initially, by the capacitance of the junction, but at high THz, by the dielectric relaxation response of the n-type material which is related to the conductivity of the material which is in turn dependent on the doping density and associated mobility of the charge carriers. Thus the high frequency response is limited by physical properties of the semiconductor and the limit is insurmountable. In spite of this, Schottky diodes are currently the basic elements in THz technologies up to about 25 THz [14–16], with relatively low noise in the range <5 THz [16–20]. Several new designs have been proposed recently to alleviate the series resistance and parasitic capacitance issues [21, 22]. Their coupling with antennas and waveguides as well as the fabrication of large arrays also pose additional engineering problems [14]. A possible contender therefore, for these ultra high frequencies of interest, is the metal-insulator-metal (MIM) diode which operates on the principle of quantum mechanical tunnelling of electrons through the insulator [23, 24]. The performance of MIMs can be improved by the use of double dielectric layers (MIIM) and these will be explored later in this chapter. Typically the insulator is sufficiently thin to allow the direct tunnelling mechanism to dominate the leakage current. The transit time of charge carriers through the dielectric is given as the inverse of the probability of tunnelling and is in the femtosecond regime. Thus the device is intrinsically able to operate at 100's THz (visible light) if the parasitic capacitance can be reduced sufficiently. Modern processing techniques such as atomic layer deposition (ALD) have allowed the engineering of ultra-thin dielectric layers with monolayer control which has opened the way for such devices to be manufactured to a high degree of control and reliability. The capacitance of the device can be made very small by appropriate lithography.

Several solutions are beginning to emerge and are summarized in Table 1. The key is to develop planar diodes that are sufficiently non-linear and asymmetric (high forward-to-reverse current ratio). Promising results for MIMs based on Al/AlOx/Pt rectannas at the frequency of 30 THz [25–27] have been demonstrated by Indiana/Central Florida Universities. It has been shown that by controlling the oxidation pressure, current voltage (IV) characteristics can be altered and this leads to the proposed non-linearity of DC characteristics [27]. MIMs based on Au/Nb_2O_5/Nb showed the operating frequency at around 30 THz but the efficiency was found to be low ~0.1 % [12]. The experiments were undertaken at ITN Energy Systems, USA. MIMs based on Ni-NiO-Ni were tested in ETH, Switzerland, IBM and Columbia University, USA and they have obtained good results at 30 THz and near IR region [24, 28]. To enhance asymmetry, dissimilar metal electrodes [26, 27] or geometric [29, 30] field enhancement [31] schemes have been proposed. There has been recent investment into developing a unipolar semiconductor-based nanodiode [32–34] at Manchester University, UK. University of Colorado, USA is a key researcher in the field of rectannas as they have managed to obtain successful results on Nb/Nb_2O_5/Nb, Ni/NiO/Ni and Ta/Ta_2O_5 MIMs [35]. They have also reported encouraging results on W/Nb_2O_5/Ta_2O_5/W double insulator (MIIM) diode [36]. The MIIM diode may however struggle to perform in the visible light range due to problems in achieving low voltage operation and adequate impedance matching with the antenna, although there is considerable scope for application at lower frequencies.

Devices based on the travelling wave principle have been reported [35]. This approach can circumvent the matching limitation but losses in the metallic regions

Table 1 Emerging concepts and state-of-the-art in THz rectification: IR and visible domain

Principle	Group	Performance
MIMs based on Al/AlOx/Pt [25–27]	Indiana/Central Florida Universities, USA	~30 THz
MIMs based on $Au/Nb_2O_5/Nb$ [12]	ITN, USA	30 THz, low efficiency 0.1 %
MIMs based on Ni-NiO-Ni [24, 28]	ETH, Switzerland IBM + Columbia Uni., USA	30 THz near-IR
Geometric metal/polysilicon MIMs [31]	Maryland University, USA	high sensitivity ~14.5 V^{-1}
Geometrical metal-vacuum-metal diodes [29, 30]	Imperial College, UK	simulations for IR and visible
Metallic structures with sharp corners [37]		Visible /THz
Self-switching semiconductor nanodiode [32–34]	Manchester University, UK	~10 THz
MIM in traveling-wave configuration; $Nb/Nb_2O_5/Nb$; Ni/NiO/Ni; $Ta/Ta_2O_5/Ta$ [35]	Colorado University, USA	IR removes RC limitation and improved matching
Double insulator MIIMs; $W\text{-}Nb_2O_5\text{-}Ta_2O_5\text{-}W$ [36]		superior to MIMs (on-resistance)
Geometric diodes [40, 41]; on graphene [42, 43]		28 THz
Au nano-antennas on Si [38]	Rice University, USA	visible THz
MIM based on surface plasmon excitation and hot electron extraction; $Au/Al_2O_3/Au$ [39]	Stanford University, USA	visible THz, efficiency 2.7 %

limit efficiency in the visible light regime. A third class of device is based on plasmonic energy [35, 37, 38] whereby the light creates hot electrons in the upper metal layer [39]. These hot electrons can tunnel through the insulator or be emitted over a low barrier to form electric current. Finally, a new type of device has recently come to the attention of the community, the so-called geometric diode [40, 41]. This device is a simple arrow shaped constriction which allows current flow in one direction and makes it less likely in the other. The new material graphene looks very promising for this device due to its long mean free path for electron transport [42, 43]. Some of the latest designs of rectifiers are illustrated in Fig. 4. All of the device concepts mentioned above bring the considerable benefit of low cost, compatibility with CMOS technology and the ability to form large area structures for larger harvested currents. There have been some significant developments on the manufacturing aspects of realizing large area arrays of these devices, vital if they are to compete with PV in the energy market. Indeed rectennas have shown very strong potential for energy harvesting of radio-frequency radiation including 'on-body' applications.

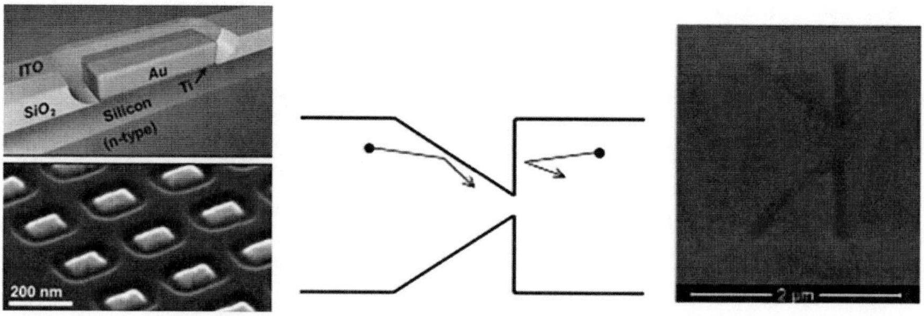

Fig. 4 (*left*) A rectifier design [44]; (*right*) geometric diode illustration and photo [40]

Scaling of such devices to the higher frequencies above is being tackled by an increasing number of teams world-wide but many challenges still remain. Rectenna technology is therefore very much a 'work in progress'.

3 Materials Properties and Design Issues for Rectennas

Conventional solar cells operate as quantum energy conversion devices, and are therefore subject to the thermodynamic efficiency limit. Photons with energy below the band gap E_g (e.g., $hv < E_g$) will not be absorbed by the solar cell thus they will not contribute to the photovoltaic process. Photons with their energy above the band gap will be absorbed, although the excess kinetic energy ($hv - E_g$) will convert to heat thus will not contribute to the photovoltaic process either. Thus the band gap E_g of a semiconductor material is the most significant factor setting the theoretical conversion efficiency limit. This also provides a simple and easy way for selecting appropriate materials for photovoltaic solar cells.

In contrast, the rectenna solar cell operates in a totally different way. A rectenna comprises two key elements: a nano-scale antenna for collecting electromagnetic waves (THz radiation and solar energy) and a high-speed diode for converting the collected electromagnetic radiation to DC power/signal. At low frequencies, a simple dipole antenna comprises two metal rods that are separated by a small gap. The metal collects the incoming electromagnetic radiation, thus inducing an electric current into the conductor. It is well-known that that the electric current flows mainly at the "skin" of the metal conductor. The skin depth, which is defined as the depth below the surface of the conductor at which the current density has fallen to 1/e of its value at the conductor surface, can be calculated as $\delta = (2\ \rho/\omega\mu)^{1/2}$ where ω, ρ and μ are the angular frequency of the incoming radiation and the resistivity and magnetic permeability of the conductor, respectively. Figure 5 shows the calculated skin depth of five common metals in the frequency range of 10^{-3}–10^{3} THz. As a rule of thumb, the thickness of the metal should be at least four skin depths. Thicker metal can be used for deposition

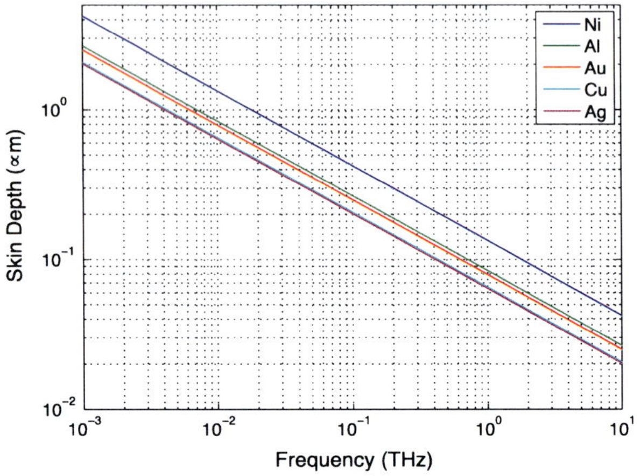

Fig. 5 Calculated skin depth of five common metals as a function of frequency. The resistivity used in the calculation is 6.99×10^{-8}, 2.8×10^{-8}, 2.44×10^{-8}, 1.68×10^{-8}, and 1.59×10^{-8} Ω-m for nickel, aluminium, gold, copper, and silver, respectively. The skin depth above 1 THz is for illustration purpose only, as the plasma effect may have to be considered when approaching high frequencies

convenience, but it will not reduce the resistance any further. As shown in Fig. 5, the skin depth of most good conductors (for example silver, gold, copper and aluminum) is less than 1 μm at 10 GHz and above. Therefore antennas operating at THz frequency range or above are expected to be made of thin metal film (on a supporting substrate), rather than bulk metals. Consequently properties of both the metal and the supporting material need to be considered in the antenna design and performance. As a general guide, the thin metal film of the antenna should be of high quality with defects much smaller than the skin depth, otherwise the electromagnetic waves will experience significant scattering losses. Metal granularity should also be considered.

Metal can be treated as a pure conductor at RF or lower frequencies but this may not be the case at high frequencies. Figure 6 shows a schematic diagram of a typical THz antenna which has been widely used for the coherent generation and detection of broadband THz radiations [45–47]. The bowtie antenna has a planar structure, and typically incorporates gold film patterns on a GaAs substrate. Usually an ultrafast laser is necessary to gate the antenna; that is, to turn on and off the receiver antenna. However, this antenna could also be connected to an ultrafast diode to enable broadband THz detection at room temperature. At 2 THz, the skin depth of gold is about 50 nm. Assuming two gold strips of 1.0 mm length and 20 μm width are used to feed the antenna output to the feed-point of an ultrafast diode, the corresponding resistance of the gold strips is of the order 50 Ω. This resistance will become even larger at higher frequencies. Therefore for the proposed THz and optical rectenna, the distance between the antenna output-point and the rectifier feed-point should be kept as small as possible. It is strongly

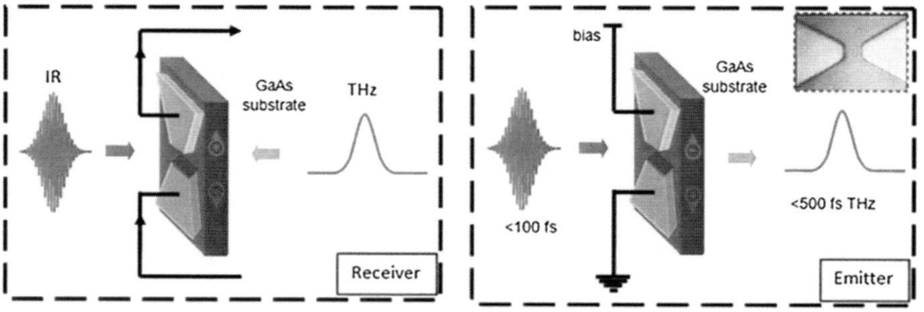

Fig. 6 Generation (*left*) and detection (*right*) of broadband THz pulses in a photoconductive antenna. Electron–hole pairs are excited in the GaAs crystal using an above–band gap femtosecond near infra-red light. For THz emitter antenna that is biased, the photo–generated carriers are accelerated by the applied electric field. This produces a transient current, which generates a pulse of electromagnetic radiation in the THz frequency range. In contrast to the THz emitter antenna, the THz detector antenna is not biased. The photo–generated carriers are actually driven by the electric field of the THz pulse, which arrives in synchronism with the NIR light. Both the amplitude and the phase of the THz electric field could be obtained by measuring the resultant photocurrent in the antenna [48]

recommended that the rectifier be integrated into the antenna, that is, the rectifier will be located at the gap between the two antenna electrodes to minimize the propagation losses.

3.1 Optical Properties of Metals

At low frequencies, electrons in a conductor follow the external electric field. As the frequency of the applied field increases, the inertia of electrons introduces a phase lag into the electron response to the electric field. Consider electrons in a conductor displaced from a uniform background of ions. Electric fields will be established in a direction so as to restore the neutrality of the plasma by pulling the electrons back to their original positions. Inertial considerations mean that electrons will overshoot and oscillate around their equilibrium positions with a characteristic 'plasma' frequency, ω_p given as

$$\omega_p = \frac{n_f q^2}{m^* \omega_\tau} \tag{1}$$

where n_f is the free electron density, m^* is the electron effective mass, q is the electronic charge and $\omega_\tau = 1/\tau$ is the damping frequency with τ being the electron relaxation time. The optical properties of metals in the infrared and visible/UV range are fairly represented using a simple Drude model. The complex dielectric constant and the complex refractive index can be written as [49]:

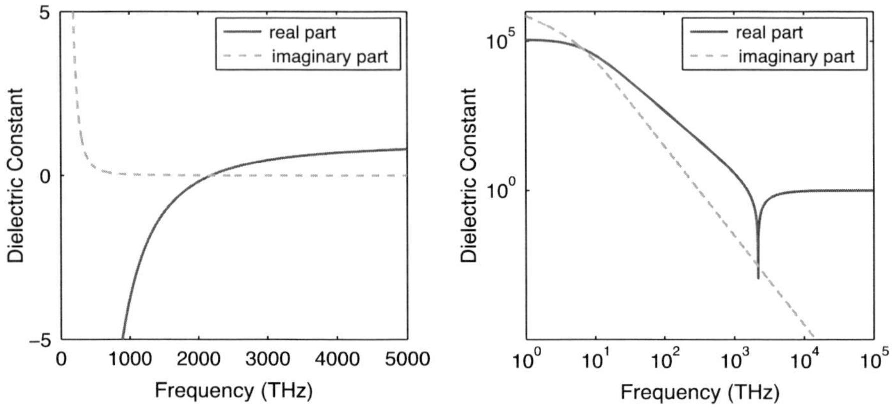

Fig. 7 A plot of complex dielectric constant of gold as a function of frequency in linear (*left*), and logarithmic (*right*) scales. The plasma frequency and the damping frequency of bulk gold is 2184 and 6.45 THz respectively

$$\frac{\varepsilon(\omega)}{\varepsilon_o} = n(\omega)^2 = 1 - \frac{\omega_p^2}{\omega^2 + j\omega\omega_\tau} \tag{2}$$

where ω_p and ω_τ are the plasma frequency and damping frequency, respectively.

As an example, Fig. 7 shows our calculations of the complex dielectric constant of gold in the frequency range of 1–10^5 THz. The real and imaginary parts are equal in magnitude at the damping frequency whilst the real part has a zero crossing at the plasma frequency. In addition, the imaginary part gets very small at high frequencies, indicating less loss at higher frequencies.

Perhaps a better way to understand the effect of the plasma on the optical properties of metal is to plot the complex refractive index, n as a function of frequency. As shown in Fig. 8, n is dominated by the real part for frequencies above the plasma frequency while the imaginary part of n is very small and can be neglected. The metal behaves like a non-absorbing dielectric medium and thus becomes transparent. This also explains why X-rays, which correspond to a frequency range of 3×10^{16}–3×10^{19} Hz, can pass through metal objects. In contrast, for frequencies below the plasma frequency such as the case of infrared and visible light whose frequency is below the plasma frequency of most metals, n is complex. The electromagnetic wave is attenuated and does not propagate very far into the metal. Note that the plasma in the conductor will also have an impact on the dynamic conductivity of metals which in turn will affect the resistance of the rectenna device. Furthermore, the complex dielectric constant of the metal will also affect the performance of the antenna as there will be a strong coupling interaction between the incoming light with surface plasma in the metal film of the antenna in a similar way as in attenuated total reflection experiments [50]. The plasmons confined to the air/film and/or film/substrate interfaces will interact with the light to form a surface plasmon polariton which propagates along the interface,

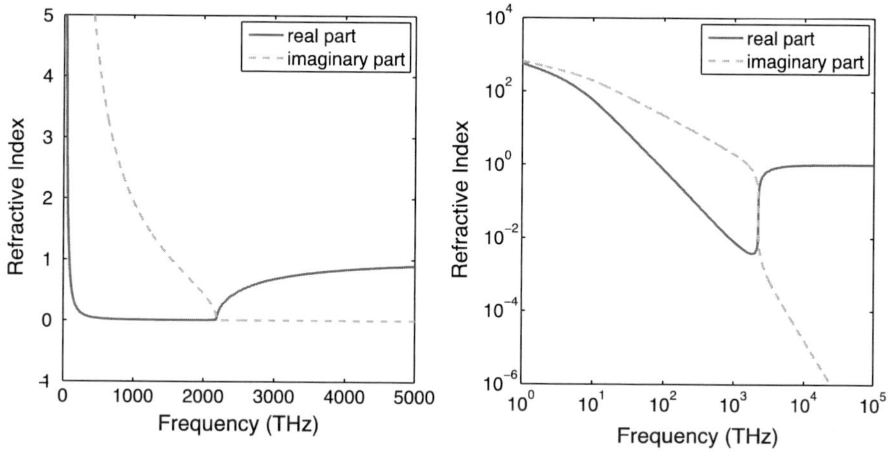

Fig. 8 A plot of the complex refractive index of gold as a function of frequency in linear (*left*) and logarithmic (*right*) scales. The plasma frequency and the damping frequency of bulk gold is 2184 and 6.45 THz respectively

and decays exponentially into both media. Some of the device concepts described in Sect. 2 make use of these plasmonic modes to increase the coupling between sunlight and the antenna, thus increasing the overall photo-electric conversion efficiency.

3.2 Substrates for Rectennas

In a rectenna device, the light is collected by the antenna and the collected radiation is converted directly to DC electricity by the rectifier. The main function of the substrate is thus to support the metal structures of the antenna. The substrate material should be chosen to be mechanically strong, thermally stable, electrically insulating, with low optical absorption loss over the frequency range of interest. The less-expensive materials such as transparent glass or plastics meet these criteria well.

Although the main function of the substrate is to support the metal structures of the antenna, it may actually have unexpected effects on the performance of the rectenna because of the above-mentioned surface plasma coupling. In an early study, we observed clear evidence of interaction between the THz radiation and the coherent optical phonons in GaAs [46, 51]. As shown in Fig. 9, there is strong coupling between the incoming THz radiation with the optical phonons of GaAs crystal, as evident from the spectral features at the GaAs phonon mode of 8 THz and a broad spectral feature at plasmon-phonon coupled modes of 20 THz. Another important issue is the substrate mode which is generated due to the fact that the thickness of the substrate may become comparable with the wavelength of

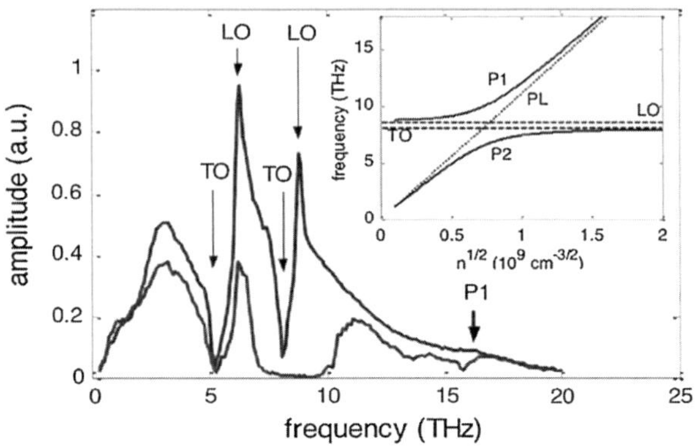

Fig. 9 THz spectra measured before and after transmission through a 0.53-mm-thick GaAs wafer. The arrows mark spectral features corresponding to TO and LO phonons of ZnTe (5.3 THz and 6.2 THz), and GaAs (8.06 THz and 8.76 THz), together with the coupled plasmon-phonon mode of GaAs (P1). *Inset* Density dependence of coupled plasmon-phonon mode frequencies (*upper branch* P1; *lower branch* P2) calculated for GaAs parameters. *Dashed* and *dotted lines* represent GaAs TO phonon, LO-phonon and plasmon (PL) frequencies, respectively [51]

the THz waves. Substrate energy loss could be considerable so there is a need for a systematic approach when selecting metals and supporting materials. The design approach should also seek to make use of the plasmonic modes to increase the overall photo-electric conversion efficiency.

In summary, the THz and optical rectenna is a new device concept for THz solar energy harvesting. It differs in key respects from the RF rectenna where the antenna simply collects the incoming RF radiation and the rectifier converts it to DC with the two devices working more or less independently, thus allowing each to be designed and optimized independently. At THz and optical frequencies however, these two core components have to be integrated into a compact design in order to minimize the propagation losses that increase at high frequencies. In addition, when approaching optical frequencies, the influence of plasma on the optical properties of metals (for example, the complex refractive index) becomes significant. It is necessary to take a systematic approach when selecting appropriate metals and its supporting materials. Most common metals can still be used for the antenna as they are still very conductive at THz frequencies that are still well below the plasma frequency of metal. However, as discussed in this section, the electrical and optical properties of metals are largely different from their low frequency counterparts because of the plasmonic influence particularly for thin metal films. Furthermore, the electrical and optical properties may change with frequency and hence it is necessary to consider the dimensions and shape of the device itself to allow tuning for plasmon confinement at particular frequencies. These issues lead to significant challenges but also exciting opportunities for the design of efficient antennas working at optical frequencies, opening up a new branch of engineering science.

4 Antennas

The antenna is a critical element of rectennas and has a major influence on the overall performance. Optimization is required to allow reception of the maximum energy from the space over the desired frequency band. The selection of antennas is based on the application. Work in [3] shows an example for RF/microwave wireless power transmission, where both the transmitter and the receiver characteristics are known, allowing a clear choice for the adoption of a dipole antenna with linear polarization for 2.45 GHz operation. This proved to be a cost-effective solution. After optimisation a very high energy conversion efficiency of over 90 %, was reported and a modified dipole design for dual band was subsequently introduced in [4]. However for solar energy harvesting, the incoming electromagnetic solar waves encompass a very broadband and have unknown polarisation. A simple dipole antenna is therefore not acceptable. The antenna has to meet a number of challenging requirements which include very broad bandwidth. The output power is approximately proportional to the bandwidth of the rectenna; the wider the bandwidth, the more power the rectenna can harvest. Dual-polarization is preferred as, unlike photovoltaic (PV) solar cells, the rectenna is sensitive to the polarization of incoming waves. The polarization of the solar waves is not well defined and could be random. A linearly polarized antenna may only receive half of the incoming energy, which results in a reduction of the efficiency by 50 %. A unidirectional radiation pattern is required since the solar energy is coming from one direction along a line of sight path. The antenna radiation pattern should be relative broad so that it is not too sensitive to the incident angle. The main potential advantage of the rectenna for energy harvesting over the conventional PV cells is higher efficiency. The antenna efficiency is part of the overall energy conversion efficiency of the rectenna. Finally, easy of integration with the rectifier is paramount; as discussed in Sect. 3, the propagation loss at THz can be very large.

Based on these considerations, a circular patch antenna array as shown in Fig. 10a has been proposed [52]. The detailed link to MIM rectifiers is shown in Fig. 10b. This initial design had the aim of demonstrating the feasibility of the concept. The important message here is that the antenna is a broadband planar structure and can be easily fabricated with state-of-the-art nano-fabrication technology, at least at the small scale.

When an array is formed by many horizontal and vertical circular dipoles, dual polarization is realized. If a ground plane is employed, a unidirectional radiation pattern can be obtained. Figure 11 shows results for an RF design implementation of the patch antenna array. The structure was simulated using CST Microwave Studio which is a tool offering full-wave electromagnetic field integration. The simulations were conducted with the aim of optimising the bandwidth and radiation pattern. As shown in Fig. 11a, the reflection coefficient of the antenna is very sensitive to the feeding gap between the circular patches. Without the ground plane, the patch antenna will not produce a uni-directional radiation pattern as

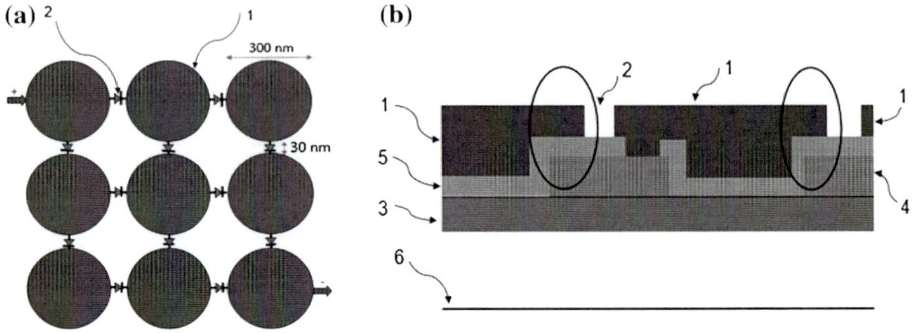

Fig. 10 A proposed rectenna array **a** Circular patch array; **b** Antenna with MIM rectifier (where "1" is for the circular patch antenna and made of Metal 1; "2" is the rectifier; "3" is the Si substrate with thermal oxide; "4" is the deposit metal; "5" is the deposit dielectric layer; "6" is the ground plane and made of metal)

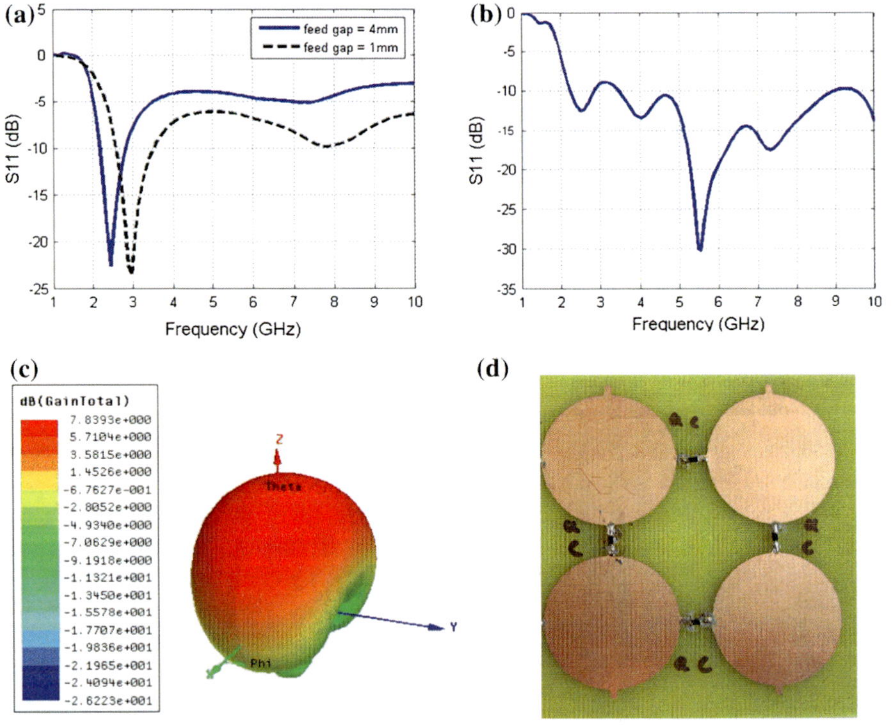

Fig. 11 Simulation results of the RF rectenna. **a** The reflection coefficients of the dipole with different feeding gaps; **b** The reflection coefficients of the dipole with different feeding gaps; **c** antenna radiation pattern; **d** fabricated rectenna

required. The ground plane serves to improve the reflection coefficient as shown in Fig. 11b and also produces a unidirectional radiation pattern as demonstrated in Fig. 11c. The optimized antenna was fabricated as indicated in Fig. 11d; measured and simulated results are in good agreement. Although this design is realized in the GHz regime, it can be scaled to THz and beyond, although the final limit depends on the increasing frequency dependence of materials properties as explained in the previous section. For example, if a gold antenna is used, the conductivity of gold is reduced from 4.1×10^7 S/m at 10 GHz–2.2×10^7 S/m at 10 THz—the conductivity loss is almost doubled, thus antenna parameters like the impedance and efficiency will be significantly affected.

There are other possible antenna designs which may be suitable for this application. For example, a broadband circularly polarized spiral rectenna array was proposed [6], the application here is for ambient wireless energy harvesting in GHz. The fabrication of such an antenna array would be more challenging due to the details required for the antenna. It should be pointed out that the selection of antenna should also be linked to the rectifier; especially the impedance match would be a major cause of reduced efficiency. The antenna performance is also heavily affected by the presence of rectifier and the supporting substrate [53].

5 Metal-Insulator-Metal Based Rectifiers

In this section we discuss design aspects of MIM based rectifiers. We also present a model for the double dielectric resonant tunnelling rectifier based on the metal-insulator-insulator-metal (MIIM) structure. The dielectric layers of different energy band gaps are chosen to ensure the creation of a potential well between the layers, which contains bound energy states. The states can provide 'stepping stones' for efficient electron transport between the electrodes, under forward bias. This study also serves to highlight challenges in the matching issues in rectenna design. It follows closely that reported in [54].

5.1 Rectifier Circuit: Design Challenges

A simple, lumped equivalent circuit for a rectenna system is shown in the Fig. 12. We now illustrate some of the design challenges by considering a single insulator MIM. The basic ideas are also appropriate for the double dielectric case. Power matching requires $r_A \sim r_D$. Physically, the incoming optical radiation is captured by the antenna (A) as an electromagnetic wave thereby inducing an 'optical voltage' which in turn causes electron flow through the diode 'detector' (D) by tunnelling. The small signal capacitor C_D' includes parasitic capacitance to substrate as well as the capacitance of the diode. A key requirement is for good matching with the antenna to allow high efficiency for the energy conversion.

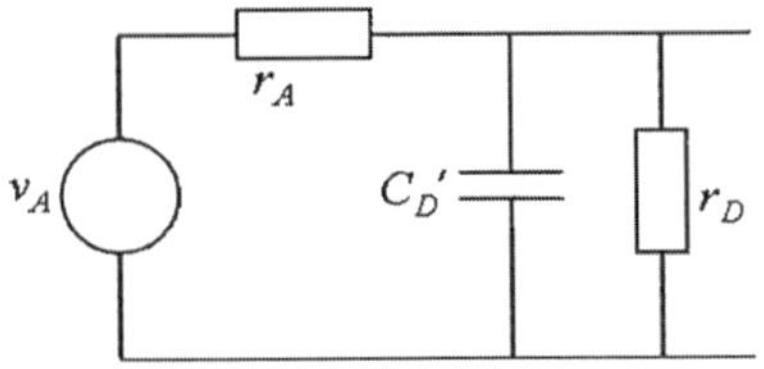

Fig. 12 Small-signal equivalent circuit of a rectenna ('A' signifies antenna and 'D' diode)

This places a very severe limitation on the diode forward characteristic. We now present a simple model for the MIM diode to illustrate the design challenges. A simple expression for the current in the diode, I with dielectric of thickness, t_{ox} under bias V, is [55]:

$$I = aA \exp\left[-\frac{B\Delta E_C t_{ox}}{V}\right] \tag{3}$$

where A, B can be considered as constants for the purposes of this discussion, a is the device area and ΔE_C is the offset in the conduction band between the injecting electrode and dielectric. As discussed in Sect. 2, ΔE_C can be engineered to provide for low and high values according to the bias requirements. In classical terms, the small-signal dynamic resistance of the diode is given as

$$r_D = \left[\frac{dI}{dV}\right]^{-1} = \frac{V^2}{aAB\Delta E_C t_{ox}} \exp\left(\frac{B\Delta E_C t_{ox}}{V}\right) \tag{4}$$

which indicates the strong degradation of r_D with increasing t_{ox} making matching with an antenna resistance of the order 100 Ω extremely difficult. The second requirement is for high responsivity which is defined as the DC current induced by the rectified incident power calculated from the ratio of the second to the first derivative of current versus voltage [56], hence write:

$$R(V) = \frac{1}{2}\frac{I''}{I'}\Big|_{V=V_{app}} = \frac{1}{2V^2}\left(1 - \frac{2V}{aAB\Delta E_C t_{ox}}\right) \tag{5}$$

This indicates that R increases with increasing t_{ox} giving a design trade-off with r_D. A third key requirement is the need to minimize the capacitance of the diode, C_D,

$$C_D = \frac{\varepsilon_o \varepsilon_d}{t_{ox}} a \tag{6}$$

Note that ε_d is the dynamic permittivity of the dielectric, appropriate for the very high frequency of operation and 'a' is the diode area. The capacitance (plus parasitics) forms a time constant, $\tau = C_D \backslash r_A$. Thus thicker oxides are favored but that compromises the matching requirement. Considering $r_A = 50\ \Omega$, and a cut-off frequency $f_c = [2\pi r_A C_D]^{-1} = 100$ THz, gives $C_D \sim 0.032$ fF. Fortuitously, the parameter ε_d is relatively low at optical frequencies as ionic polarization components are unable to respond. Taking $\varepsilon_d = 2$ (typical for oxides suitable for this application), area, a $\sim 1.3 \times 10^{-14}$ m^{-2} translating into linear dimensions of the order of 100 nm ($t_{ox} \sim 7$ nm) (diameter of a circular metal patch antenna); easily engineered using current technology. However, scaling to optical frequencies implies the need for nanoscale lithography. The oxide thickness t_{ox} needs to be as thick as possible with the constraint that it needs to be thin enough to ensure direct tunnelling, hence very rapid ~ femto-second electron transit time. These considerations suggest an optimal value for t_{ox} of the order 4 nm. This value is not optimal for R and r_D however. Finally, it should be recalled that this analysis is useful in the THz regime but less so in the IR and optical regimes as indicated in the earlier section on materials properties. Recent papers by the University of Colorado at Boulder group indicate rectenna efficiencies of the order 44 % [57] using MIM diodes. A useful model describing rectenna solar cell operation is available in [58].

5.2 *Resonant Tunnelling MIM Rectifiers: Bound States in the Potential Well*

We now consider a double dielectric (MIIM) rectifier which operates on the principle of resonant tunnelling in the forward bias direction. This serves to reduce the dynamic resistance of the diode.The energy band diagram of an MIIM device is shown in Fig. 13a.

It consists of two dielectric layers, one with a large band gap such as Al_2O_3, and another one with smaller band gap such as Ta_2O_5. The metal contacts with low work function at the left and high work function at the right provide different barrier heights and hence different tunnelling probability which results in asymmetry in tunnelling current. However, metals with similar or close work functions can also be used, since the asymmetry can be achieved by resonant tunnelling in one direction. The conduction band offset between the two dielectrics creates a potential well at the interface of two dielectrics when a large enough negative bias is applied to the left metal contact, as shown in Fig. 13b. The resonant tunnelling through the quantized energy levels in the potential well has higher probability than direct tunnelling which gives rise to current in this direction. The calculation of the properties of the device is conducted in two steps. Firstly, the system parameters and voltage bias are studied to identify regimes where bound states may occur. Secondly, a transfer matrix method is used to estimate the current-voltage characteristics.

The quantized energy levels in the potential well can be found from the numerical solution of the time-independent Schrödinger equation within the potential well. The number and energy levels of bound states are obtained by defining a Hamiltonian matrix comprising a set of localized basis states in the gate

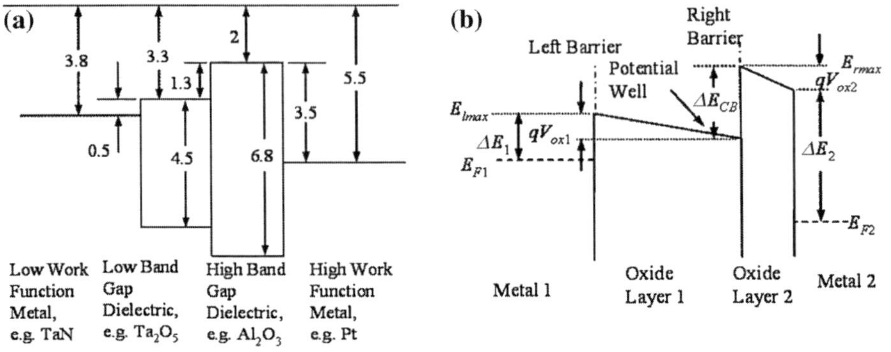

Fig. 13 **a** Energy band diagrams of a representative materials system for an MIIM rectifier (all units are in eV), **b** Conduction energy band diagram showing the potential well

stack and diagonalization of the matrix. The energy eigenstates are calculated by taking a general superposition state, formed from sinusoidal basis states $\psi_j(x)$:

$$\varphi_n(x) = \sum_{j=1}^{M} A_{nj}\psi_j(x) = \sqrt{\frac{2}{t_{HK} + t_{IL}}} \sum_{j=1}^{M} A_{nj} \sin\left(\frac{\pi j x}{t_{HK} + t_{IL}}\right) \tag{7}$$

where M is the number of basis states used in the numerical expansion, and A_{nj} is the coefficient of the nth eigenstate associated with the jth basis state. The diagonalization of the Hamiltonian matrix is equivalent to solving the time-independent Schrödinger equation for the potential in the insulator layers,

$$H(x)\varphi_n(x) = E_n^{(x)}\varphi_n(x). \tag{8}$$

Only the states with energy levels lower than $E_{l\max}$ and $E_{r\max}$ in Fig. 13b are localized and are considered as bound states. The states with energy level above these limits are unstable and can leak to metal 1 or the conduction band of oxide layer 2. Furthermore, those states with energy levels close (within about kT, where k is the Boltzmann constant and T is temperature) to these boundaries are not thermally stable and can leak by thermal activation.

For each bound state in the one dimensional (1D) well, there are also a set of transverse excitations, corresponding to the same 1D excitation, but in two transverse directions. The lateral or transverse dimension of the device is significantly larger than the IL thickness; therefore, each 1D state forms a band of closely spaced three dimensional (3D) energy states. The transverse dimensions are modeled as a rectangular box, where excitations are independent of the longitudinal dimension. These energy levels are given by

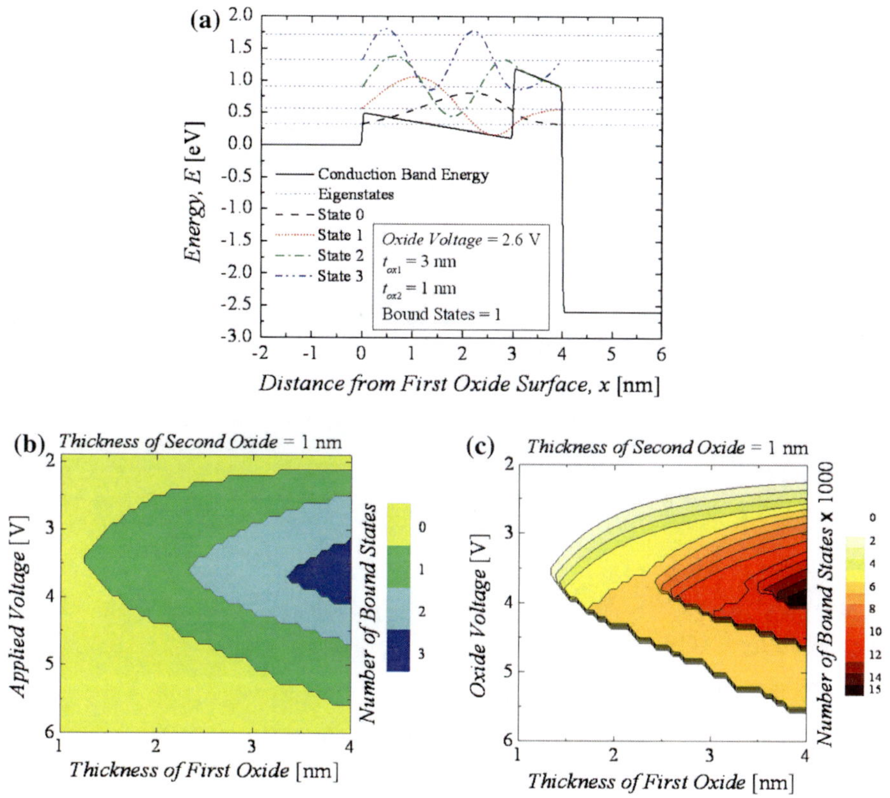

Fig. 14 **a** Energy band diagram of the MIIM device with one bound state, number of bound state for **b** 1D and **c** 3D structures

$$E^{(y,z)} = \frac{h\pi}{2m^*}\left(\frac{n_y^2}{L_y^2} + \frac{n_z^2}{L_z^2}\right) \tag{9}$$

where h is Plank's constant, m^* is the effective electron mass, n_y and n_z are transverse excitations, and L_y and L_z are the transverse dimensions of the device. The number and energy levels of bound states in the 1D and 3D well can be calculated using the above method. The thickness of the second oxide layer has been fixed to 1 nm and the thickness of the first oxide layer is changed between 1 and 4 nm. The energy band diagram with one bound state at applied voltage of 2.6 V is shown in Fig. 14a. The number of bound states as a function of applied voltage and thickness of the first oxide layer is shown in Fig. 14b for 1D and in Fig. 14c for 3D structures with lateral dimensions of 100 nm. The maximum number of bound states occurs with the thickest first oxide layer at an applied voltage of circa 4 V.

5.3 Calculation of Current Density

The current in the MIIM diode is established by tunnelling of electrons on the left metal contact through the oxide, transmitted via the available quantum states. There is also transmission in the opposite direction for electrons on the right metal contact. The net current at each applied voltage then can be calculated by deducting the right to left current from the left to right current. The current depends on the density of available states at each energy level and also on the average occupancy of each state, based on the Fermi-Dirac distribution. Assuming zero potential on the left electrode and potential on the right electrode equal to the applied voltage the total current density at applied voltage of V_{app} can be expressed as [36]:

$$\begin{aligned} J &= J_{L\to R} - J_{R\to L} \\ &= \frac{m^* q}{2\pi^2\hbar^3}\int_0^\infty T_{coeff}(E_x)dE_x \int_{E_x}^\infty \left[f_L(E) - f_R\left(E + qV_{app}\right)\right]dE \end{aligned} \tag{10}$$

where f_L and f_R are Fermi-Dirac distribution functions at left and right electrodes described as

$$f_L = \frac{1}{1+\exp\left(\frac{E-E_{FL}}{kT}\right)} \tag{11}$$

$$f_R = \frac{1}{1+\exp\left[\frac{E-\left(E_{FR}-qV_{app}\right)}{kT}\right]} \tag{12}$$

q is the electronic charge, $\hbar$ is the reduced Plank constant, T_{coeff} is the transmission probability, E is the total energy of electron, E_x is the energy component of electron in transmission direction, E_{FL} and E_{FR} are the Fermi levels on left and right electrodes, k is Boltzmann constant, and T is temperature. The inner integral in Eq. 10 is solved to obtain,

$$J = \frac{m^* qkT}{2\pi^2\hbar^3}\int_0^\infty T_{coeff}(E_x)\ln\left\{\frac{1+\exp[(E_x - E_{FL})/kT]}{1+\exp\left[\left(E_x - E_{FR} - qV_{app}\right)/kT\right]}\right\}dE_x. \tag{13}$$

The transmission probability T_{coeff} is calculated from the plane-wave solution for the Schrödinger equation. The most common methods used for calculations are the transfer matrix method (TMM) [59] which is based on the Tsu-Esaki method [60] and the quantum transmitting boundary method (QTBM) [61]. The Tsu-Esaki model is based on the WKB approximation for the wavefunction at each 'slice' through a potential barrier by constructing a piecewise constant transfer matrix for

Fig. 15 The current-voltage characteristics of the MIIM device with the first oxide thickness as a parameter

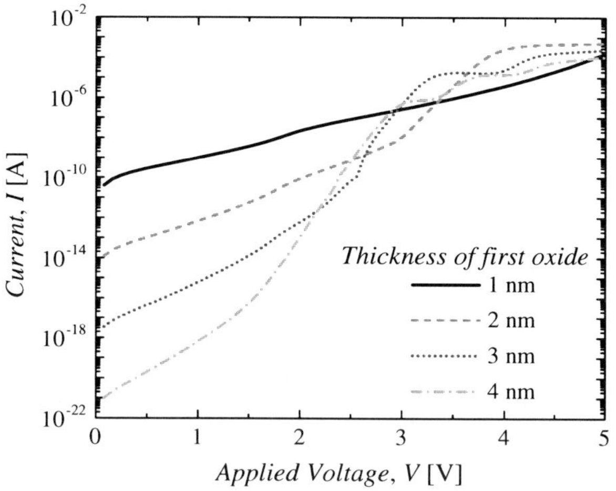

each slice. The model has been used for tunneling through multiple barriers, containing resonant states. The WKB tunneling probability for each slice can be expressed as

$$P_{Tj} = \exp\left\{-2\left[m^*\left(q\phi_{Bj} - E_{xj}\right)^{1/2}\right]d_j\right\} \tag{14}$$

where ϕ_{Bj} is the barrier height and d_j is the thickness of the piecewise rectangular barrier of the slice j.

The IV characteristics of a device with $d_2 = 1$ nm and $1 < d_1 < 4$ nm, where d_1 and d_2 are the thicknesses of dielectric layers, with lateral dimensions of 100 nm are shown in Fig. 15. At lower voltages, the current reduces with increasing d_1. However, at voltages above 2 V, the current in the diodes with thicker first oxide increases because of resonant tunneling of electrons to the states in quantum well. The diode with $d_1 = 2$ nm has the highest on-current.

Variation of the device dynamic resistance with applied voltage with d_1 as a parameter is shown in Fig. 16a. The on-state dynamic resistance of the diode is decreased significantly due to resonant tunneling. In particular, the diode with $d_1 = 4$ nm has the largest on-off ratio of dynamic resistance, hence the highest rectification. Figure 16b shows the responsivity of the MIIM devices. A significant increase in responsivity can be observed at positive voltages when the resonant tunneling via the states in the quantum well starts. The responsivity also shows an increase with d_1 due to increase in number of bound states.

The MIIM results indicate that the dynamic resistance of the diode configuration presented is still too high for good matching with the antenna and further work is required on materials selection and device architecture to address this key roadblock. To keep the dynamic resistance low, low barrier height metals can be used on both sides, with a cost of decrease in device asymmetry and hence responsivity. The MIIM offers a good solution for integration and there is still

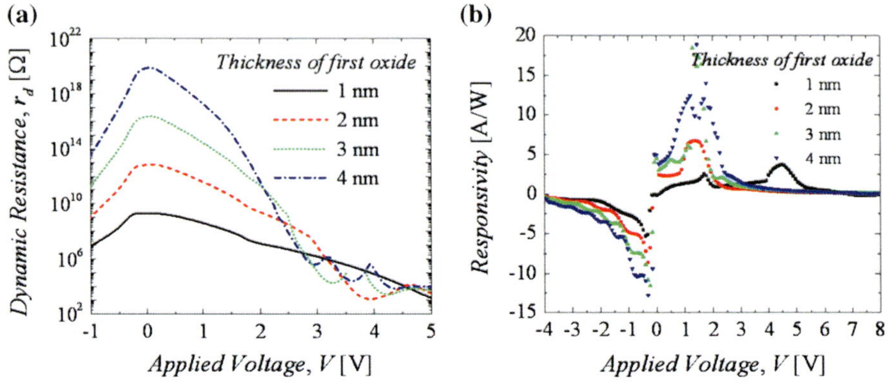

Fig. 16 Variation of dynamic resistivity (**a**) and responsivity (**b**) of MIIM device with voltage at various thicknesses of the first oxide

scope for further research to assess its potential. Parasitic RC considerations are also key requirements in the layout and materials choice, dependent on the particular architecture and fabrication process.

6 Conclusions

The current status of rectifying attennas (rectennas) has been reviewed. The issues around materials, rectifier and antenna design have been outlined and discussed, illustrated by experiment and simulation results. An important conclusion is that the antenna and rectifier may be considered as separate parts at GHz frequencies but there is a need to consider the rectenna as an integrated device in its own right, as frequency increases through the THz to visible parts of the spectrum. There are considerable scientific and engineering challenges to address if viable solutions are to be realized and this may open up an exciting new branch of engineering with considerable potential.

Acknowledgments Most of this work has been supported by the UK Engineering and Physical Sciences Research Council. The antenna computer simulations were performed by Ms Jingwei Zhang at the University of Liverpool.

References

1. http://en.wikipedia.org/wiki/Sunlight
2. Stressner II, B., Chang, K.: Microwave power transmission: historical milestones and system components. Proc. IEEE **101**(6), 1379–1396 (2013)
3. Brown, W.C.: Optimization of the efficiency and other properties of the rectenna element. IEEE MTT-S Int. Microwave Symp. Dig. **76**, 142–144 (1976)

4. Suh, Y.H., Chang, K.: A high efficiency dual frequency rectenna for 2.45 and 5.8 GHz wireless power transmission. IEEE Trans. MTT. **50**(7), 1784–1789 (2002)
5. Hagerty, J., Popovic, Z.: An experimental and theoretical characterization of a broadband arbitrarily-polarized rectenna array. IEEE MTT-S Int. Microwave Symp. Dig. **3**, 1855–1858 (2001)
6. Hagerty, J.A., Helmbrecht, F.B., McCalpin, W.H., Zane, R., Popovic, Z.B.: Recycycling ambient microwave energy with board-band rectenna arrays. IEEE Trans. MTT **52**(3), 1014–1024 (2004)
7. Bailey, R.L.: A proposed new concept for a solar energy converter. J. Eng. Power. **94**(2), 73–77 (1972)
8. Fletcher, J.C., Bailey, R.L.: Electromagnetic waver energy converter. US Patent No. 3760257 (1973)
9. Marks, A.M.: Device for conversion of light power to electric power. US Patent No. 4445050 (1984)
10. Kraus, J.D.: Antennas, 2nd edn. McGraw-Hill, New York (1988)
11. Lin, G.H., Abdu, R., Bockris, J.: Investigation of resonance light absorption and rectification by subnanostructures. J. Appl. Phys. **80**(1), 565–568 (1996)
12. Berland, B.: PV technologies beyond the horizon. Final Report, NREL/SR-520-33263. ITN Energy System, 1–16 (2002)
13. Kotter, D.K., Novack, S., Slafer, W., Pinhero, P.: Solar nantenna electromagnetic collectors. In: Proceedings of Energy Sustainability 2008, pp. 1–7. Florida (2008)
14. Sizov, F., Rogalski, A.: THz detectors. Prog. Quantum Electron. **34**, 278–347 (2010)
15. Hubers, H.W.: Terahertz heterodyne receivers. IEEE J. Sel. Top. Quantum Electron. **14**(2), 378–391 (2008)
16. Crowe, T.W., Bishop, W.L., Porterfield, D.W., Hesler, J.L., Weikle, R.M.: Opening the terahertz window with integrated diode circuits. IEEE J. Solid-State Circuits **40**, 2104–2110 (2005)
17. Ito, H., Nakajim, F., Ohno, T., Furuta, T., Nagatsuma, T., Ishibashi, T.: InP-based planar-antenna-integrated Schottky-barrier diode for millimeter- and sub-millimeter-wave detection. Jpn. J. Appl. Phys. **47**, 6256–6261 (2008)
18. Shiktorov, P., Starikov, E., Gruzinskis, V., Perez, S., Gonzalez, T., Reggiani, L., Varani, L., Vaissiere, J.C.: Theoretical investigation of Schottky-barrier diode noise performance in external resonant circuits. Semicond. Sci. Technol. **21**, 550–557 (2006)
19. Brown, E.R.: A system-level analysis of Schottky diodes for incoherent THz imaging arrays. Solid-State Electron. **48**, 2051–2053 (2004)
20. Kazemi, H., Shinohara, K., Nagy, G., Ha, W., Lail, B., Grossman, E., Zummo, G., Folks, W.R., Alda, J., Boreman, G.: First THz and IR characterization of nanometer-scaled antenna-coupled InGaAs/InP Schottky diode detectors for room temperature infrared imaging. Proc. Infrared Technol. Appl. XXXIII. **6542**(1), 65421 J-1–6542 J-4 (2007)
21. Manohara, H.M., Wong, E.W., Schelcht, E., Hunt, B.D., Siegel, P.H.: Carbon nanotube Schottky diodes using Ti-Schottky and Pt-ohmic contacts for high frequency applications. Nano Lett. **5**(7), 1469–1474 (2005)
22. Pavelev, D.G., Demarina, N.V., Koshurinov, Y.I., Vasilev, A.P., Semenova, E.S., Zhukov, A.E., Ustinov, V.M.: Semiconductors **38**, 1105–1110 (2004)
23. Hartman, T.E.: Tunneling of a wave packet. J. Appl. Phys. **33**, 3427–3433 (1962)
24. Nagae, M.: Response time of metal-insulator-metal tunnel junctions. Jpn. J. Appl. Phys. **11**, 1611–1621 (1972)
25. Bean, J.A., Tiwari, B., Bernstein, G.H., Fay, P., Porod, W.: Thermal infrared detection using dipole antenna-coupled metal-oxide-metal diodes. J. Vac. Sci. Technol. B **27**(1), 11–14 (2009)
26. Tiwari, B., Bean, J.A., Szakmany, G., Bernstein, G.H., Fay, P., Porod, W.: Controlled etching and regrowth of tunnel oxide for antenna-coupled metal-oxide-metal diodes. J. Vac. Sci. Technol. B **27**, 2153–2160 (2009)

27. Bean, J.A., Weeks, A., Boreman, G.D.: Performance optimization of antenna-coupled Al/AlOx/Pt tunnel diode infrared detectors. IEEE J. Quantum Electron. **47**, 126–135 (2011)
28. Choi, K., Yesilkoy, F., Ryu, G., Cho, S.H., Goldsman, N., Dagenais, M., Peckerar, M.: A focused asymmetric metal-insulator-metal tunneling diode: fabrication, DC characteristics and RF rectification analysis. IEEE Trans. Electron. Dev. **58**(10), 3519–3528 (2011)
29. Mayer, A., Chung, M.S., Weiss, B.L., Miskovsky, N.M., Cutler, P.H.: Simulations of infrared and optical rectification by geometrically asymmetric metal-vacuum-metal junctions for applications in energy conversion devices. Nanotechnology **21**, 145204 (8 pp) (2010)
30. Mayer, A., Chung, M.S., Weiss, B.L., Miskovsky, N.M., Cutler, P.H.: Three-dimensional analysis of the rectifying properties of geometrically asymmetric metal-vacuum-metal junctions treated as an oscillating barrier. Phys. Rev. B **78**, 205404 (13 pp) (2008)
31. Dagenais, M., Choi, K., Yesilkoy, F., Chryssis, A.N., Peckerar, M.C.: Solar spectrum rectification using nano-antennas and tunneling diodes. Proc. SPIE **7605**, 76050E-2–76050E-12 (2010)
32. Balocco, C., Kasjoo, S.R., Lu, X.F., Zhang, L.Q., Alimi, Y., Winnerl, S., Song, A.M.: Room-temperature operation of a unipolar nanodiode at terahertz frequencies. Appl. Phys. Lett. **98**, 223501 (3 pp) (2011)
33. Xu, K.Y., Lu, X.F., Song, A.M., Wang, G.: Terahertz harmonic generation using a planar nanoscale unipolar diode at zero bias. Appl. Phys. Lett. **92**, 163503 (3 pp) (2008)
34. Song, A.M., Missous, M., Omling, P., Maximov, I., Seifert, W., Samuelson, L.: Nanometer-scale two-terminal semiconductor memory operating at room temperature. Appl. Phys. Lett. **86**, 042106 (3 pp) (2005)
35. Grover, S., Dmitriyeva, O., Estes, M.J., Moddel, G.: Traveling-wave metal/insulator/metal diodes for improved infrared bandwidth and efficiency of antenna-coupled rectifiers. IEEE Trans. Nanotechn. **9**(6), 716–722 (2010)
36. Grover, S., Moddel, G.: Engineering the current-voltage characteristics of metal-insulator-metal diodes using double-insulator tunnel barriers. Solid-State Electron. **67**, 94–99 (2012)
37. Luo, Y., Lei, D.Y., Maier, S.A., Pendry, J.B.: Broadband light harvesting nanostructures robust to edge bluntness. Phys. Rev. Lett. **108**, 023901 (5 pp) (2012)
38. Knight, M.W., Sobhani, H., Nordlander, P., Halas, N.J.: Photodetection with active optical antennas. Science **332**, 702–704 (2011)
39. Wang, F., Melosh, N.A.: Plasmonic energy collection through hot carrier extraction. Nano Lett. **11**, 5426–5430 (2011)
40. Moddel, G., Zhu, Z., Grover, S.: Solar power conversions using diodes coupled to antennas. SPIE Newsroom. (2011). doi:10.1117/2.1201108.003807 (http://spie.org/x51851.xml)
41. Joshi, S., Zhu, Z., Grover, S., Moddel, G.: Infrared optical response of geometric diode rectenna solar cells. In: IEEE PVSC Conference, pp. 002976–002978 (2011)
42. Zhu, Z., Grover, S., Krueger, K., Moddel, G.: Optical rectenna solar cells using graphene geometric diodes. In: IEEE PVSC Conference, pp. 002120–002122 (2011)
43. Zhu, Z., Joshi, S., Grover, S., Moddel, G.: Graphene geometric diodes for terahertz rectennas. J. Phys. D: Appl. Phys. **46**, 185101 (6 pp) (2013)
44. Nano-antenna fashions charge from light (2011) http://physicsworld.com/cws/article/news/45921
45. Auston, D.H., Cheung, K.P., Smith, P.R.: Picosecond Photoconducting Hertzian Dipoles. Appl. Phys. Lett. **45**, 284–286 (1984)
46. Shen, Y.C., Upadhya, P.C., Linfield, E.H., Beere, H.E., Davies, A.G.: Ultrabroadband terahertz radiations from GaAs photoconductive emitters. Appl. Phys. Lett. **83**, 3117–19 (2003)
47. Shen, Y.C., Upadhya, P.C., Beere, H.E., Linfield, E.H., Davies, A.G., Gregory, I.S., Baker, C., Tribe, W.R., Evans, M.J.: Generation and detection of ultrabroadband terahertz radiation with photoconductive emitter and receiver antennas. Appl. Phys. Lett. **85**, 164–166 (2004)
48. Shen, Y.C., Jin, B.B.: Terahertz applications in pharmaceutical industry and science. In: Saeedkia, D. (ed.) Handbook of terahertz technology for imaging, sensing and communications, pp. 579–614. Woodhead Publishing Ltd, Cambridge (2013) (ISBN-10: 0857092359)

49. Ordal, M.A., Bell, R.J., Alexander, R.W., Long, L.L., Querry, M.R.: Optical properties of fourteen metals in the infrared and far infrared: Al Co, Cu, Au, Fe, Pb, Mo, Ni, Pd, Pt, Ag, Ti, V, and W. Appl. Opt. **24**, 4493–4499 (1985)
50. Shen, Y.C., Zhang, S.Y., Jiang, Y.S., Zhu, R., Wei, Y.: Angular resonance absorption spectra of Langmuir-Blodgett films studied by the photoacoustic technique. Thin Solid Films **248**, 36–40 (1994)
51. Shen, Y.C., Upadhya, P.C., Linfield, E.H., Beere, H.E., Davies, A.G.: Terahertz generation from coherent optical phonons in a biased GaAs photoconductive emitter. Phys. Rev. B **69**, 235325 (2004)
52. Huang, Y., Hall, S., Shen, Y.C.: A novel rectenna for solar energy harvesting British Patent, GB1017401.9 (2010)
53. Li, D., Huang, Y., Shen, Y.C.: Effects of Substrate on Photoconductive THz Antennas. IEEE iWAT2010, Lisbon (2010)
54. Sedghi, N., Zhang, J. W., Ralph, J. F., Huang, Y., Mitrovic, I. Z., and Hall, S.: Towards rectennas for solar energy harvesting. Proc. 43rd ESSDERC, 131–134, Bucharest, Romania (2013)
55. Engström, O., Raeissi, B., Hall, S., Buiu, O., Lemme, M.C., Gottlob, H.D.B., Hurley, P.K., Cherkaoui, K.: Navigation aids in the search for future high-k dielectrics: physical and electrical trends. Solid-State Electron. **51**, 622–626 (2007)
56. Tucker, J.R., Millea, M.F.: Photon detection in nonlinear tunneling devices. Appl. Phys. Lett. **33**, 611–613 (1978)
57. Joshi, S., Moddel, G.: Efficiency limits of rectenna solar cells: theory of broadband photon-assisted tunnelling. Appl. Phys. Lett. **102**, 083901-1–5 (2013)
58. Grover, S., Joshi, S., Moddel, G.: Quantum theory of operation for rectenna solar cells. J. Phys. D Appl. Phys. **46**, 135106-1–7 (2013)
59. Ghatak, A.K., Thyagarajan, K., Shenoy, M.R.: A novel numerical technique for solving the one-dimensional Schroedinger equation using matrix approach-application to quantum well structures. IEEE J. Quant. Electron. **24**, 1524–1531 (1988)
60. Tsu, R., Esaki, L.: Tunneling in a finite superlattice. Appl. Phys. Lett. **22**, 562–564 (1973)
61. Lent, C.S., Kirkner, D.J.: The quantum transmitting boundary method. J. Appl. Phys. **67**, 6353–6359 (1990)

Uncooled Detector Challenges for mm/sub-mm Range

Fedor Sizov, Mykola Sakhno and Alexandr Golenkov

Abstract Responsivity *R* and noise equivalent power *NEP* of long channel unbiased silicon field effect transistors (FETs) as mm-wave/THz detectors, accounting for resistive and capacitive parasitics, are compared with those of contemporary Schottky barrier diode (SBD) mm/sub-mm detectors. The ultimate performance limits of such detectors are estimated. It is shown that with account of the parasitics and detector-antenna matching one can describe these FET and SBD detector parameters. As compared to SBD detectors, the FET ones seem to be preferable in future applications for active imaging, especially in the radiation region above the frequency range of 1 THz or a little bit lower. They should overcome SBD ones because of possible better adjustment of FET parameters to antenna impedance due to nowadays better developed silicon technologies and the possibility of proper integrated detector design/fabrication compared to technologies of III-V ternary compounds applied to SBD detectors.

1 Introduction

Detectors belong to the critical components of the mm/sub-mm (these spectral regions are also denoted as sub-THz/THz) imaging systems. To make such systems cost-effective, it is desirable to use uncooled detectors and arrays on their basis. Estimations of ultimate possible noise equivalent power (*NEP*) of the detector or array that define their applicability in active or passive imaging systems are important when comparing the estimated *NEP*s with experimental ones for a given radiation frequency range. Exact identification of *NEP* from experimental signal data in mm or THz spectral region is rather challenging as estimation of the

F. Sizov (✉) · M. Sakhno · A. Golenkov
Institute of Semiconductor Physics of the National Academy of Sciences of Ukraine, Nauki Av., 4, Kiev 03028, Ukraine
e-mail: sizov@isp.kiev.ua

A. Nazarov et al. (eds.), *Functional Nanomaterials and Devices for Electronics, Sensors and Energy Harvesting*, Engineering Materials,
DOI 10.1007/978-3-319-08804-4_13

incoming power in the detector is error prone and can be mistakable leading to responsivity, and thus *NEP*, overestimations. This problem is mainly due to the use antenna on dielectric substrate for which the effective area is difficult to determine experimentally.

Here the comparison of parameters of different types (Golay cells, bolometers, pyroelectric detectors, SBDs, FETs and some others) of un-cooled mm/sub-mm detectors is made. Golay cells and pyroelectric detectors are rather sensitive ($NEP \sim 5 \times 10^{-10}$–$10^{-9}$ W/Hz$^{1/2}$) and are widely used in IR, THz and mm-wave regions (see e.g. [1]) but are slow ones (response time $\tau \sim 5 \times 10^{-2}$–$10^{-1}$ s). Moreover Golay cells are difficult to be assembled into arrays. Pyroelectric detectors are relatively slow and sensitive (see Table 1) but can be assembled into arrays for active imaging. Metallic bolometers and thermocouples can be effective in use with relatively high sensitivity and also can be assembled into arrays (see e.g. [2, 3]). Conventional VO_x microbolometer matrix arrays, well developed for IR imaging, can be used in the radiation frequency range $\nu > \sim 3$ THz [4] with $NEP \sim 10^{-10}$ W/Hz$^{1/2}$. For α-Si microbolometers with thick dielectric layer in the pixel cavity the very favorable for applications $NEP \sim 10^{-12}$ W/Hz$^{1/2}$ ($\nu \sim 0.9$ THz) was reported [5] though, perhaps, this needs to be confirmed.

Signal rectification by uncooled mm/sub-mm detectors seems to be an advantageous technique for fast radiation direct detection allowing for combining such detectors into focal plane arrays by integrated technologies for applications in real time active or passive imaging. From this point of view FETs (see e.g. [19, 21]) at zero bias source-drain condition and zero bias SBDs (see e.g. [22, 23]) are favorable for uncooled vision systems. They are square-law rectification detectors in which the response is proportional to the incoming power over the wide range. Together with antennas they can be suitable for the spectral range of $\nu \sim 100$ GHz …1 THz and both of them can be assembled into arrays. Main advantages of FET detectors and arrays are related to CMOS mature integrated technologies.

The interest to FET as THz detectors has begun with the Dyakonov–Shur [24] publication, which paid attention to signals registration possibility in the range of frequencies much higher the cut-off frequency ν_{co}. Within the small signal approximation of FET operation there is a cut-off frequency ν_{co} at which the input current exceeds the output one, and the unit gain frequency at which the input power exceeds the output one. Below these frequencies FET is commonly characterized by linear response but not as the non-linear device. Non-linear response is important at higher frequencies compared to ν_{co}. This phenomenon in [24] was proposed for THz radiation detection.

In fact THz detection is related to generation of static component in distortion analysis with generation of higher order harmonics included. So, [24] and distortion analysis investigate the same phenomenon from the different points of view. Distortion analysis of FET at high frequencies with detailed explanation of common assumptions is given e.g. in [25].

Table 1 Parameters of uncooled FET and SBD mm-wave/THz detectors, together with parameters of some other uncooled detectors

Detector type	Modulation frequency, Hz	Radiation frequency, GHz	NEP, W/Hz$^{1/2}$	Refs. #
Golay cell	$\leq$20	$\leq$30	$\sim 10^{-9}$–5×10^{10}	Commercial
Pyroelectric	$\leq 10^2$	$\leq$30	$\approx(1\ldots3) \times 10^{-9}$	Commercial
Nb microbolometer	–	$\leq$30	$^{3}5 \times 10^{-11}$	[6]
SiN membrane	$\leq$200	$\approx$1.6… 4.3	10^{-9}	[7]
VO_x microbolom	$\leq 10^2$	4.3	$\geq 3 \times 10^{-10}$	[8]
VO_x microbolom	$\leq 10^2$	4.25; 2.54	7.9×10^{-11}; 9×10^{-11}	[4]
BiSb/Sb thermocouples	$\leq 4 \times 10^4$	812	1.7×10^{-10}	[3]
Zero bias SBDs	Up to $\sim 10^{10}$	150, 300, 400	$\sim(5\ldots20) \times 10^{-12}$	[9]
Zero bias SBDs, ErAs/InAlGaAs/InP	–	104	1.2×10^{-12}	[10]
Zero bias SBDs, InGaAs/InP	–	$\sim$(300…700)	$\approx 5 \times 10^{-10}$	[11]
Zero bias SBDs, InGaAs	–	$\sim$<300	$\approx 3 \times 10^{-11}$ (estimated)	[12]
Zero bias SBDs, AlAs/InGaAs/InAs	–	$\leq$100, $\sim$400	$\approx 4.5 \times 10^{-12}$, 8×10^{-12}	[13]
SBD in 0.13 digital CMOS	–	860	3.2×10^{-11}	[14]
Si FET	–	295	$\approx 10^{-11}$	[15]
65 nm SiGe CMOS and BiCMOS	–	1027	$\approx 6.6 \times 10^{-11}$	[16]
Si FET	3×10^4	650	3×10^{-10}	[17]
Si n-MOS FET	–	$\sim$320	3.2×10^{-10}	[18]
Si CMOS FET	–	595; 2.91 THz	4.2×10^{-11}; 4.87×10^{-10}	[19]
MCT hot electron bolometer	$\sim 10^6$	$\sim$75–150	$\sim(1\ldots.3) \times 10^{-10}$	[20]

Up to date FET detectors have demonstrated *NEP* in the spectral range $\nu \sim$ 100–600 GHz appropriate for active imaging $\sim(10^{-10}-10^{-11})$ W/Hz$^{1/2}$ (see e.g. [15, 17, 19]) that is about one order of magnitude worse or comparable to the best SBDs (see Table 1). The latter ones are successfully used in pre-amplified systems for active and passive imaging (see e.g. [26–28]). In mm-wave region it seems that FETs (GaAs HEMTs) as mm/sub-mm detectors were first applied for imaging in [29].

Here the comparison of MOSFET and SBD detectors based on known models is carried out and available up to date experimental data are also compared. The purpose of this investigation is to consider the limits of electrical (NEP^{el}) and optical (NEP^{opt}) noise equivalent power of FET and SBD detectors for mm and THz spectral regions with the aim of their applicability as arrays in direct detection vision systems. FET detector theory similar to that known for SBD one [30, 31] was considered to compare the known data for these types of detectors. Since there are some confusion in determining the responsivity R and NEP of detectors with antennas, applications of these parameters to certain cases are specified.

Comparison of MOSFET and SBD detectors based on models known and used for data analysis (for SBDs, see e.g. [10, 30–32]) is carried out. Another objective of the research is to relate these data to measured quantities with the aim of their applicability as arrays in uncooled direct detection active or passive vision systems at radiation frequencies $\nu > 100$ GHz (but not the preamplified direct detection systems having a lot of additional elements such as Dicke switches, LNAs, demodulators, etc. (see e.g. [27, 33, 34]) and operating mainly at $\nu \sim <100$ GHz).

A nonlinear current-voltage characteristic is the base of operation SBDs as well as FETs in THz region. The feature of these both types of detectors as compared to very sensitive but deeply cooled ones (NEP up to $\sim 10^{-18}$–10^{-19} W/Hz$^{1/2}$ [35]) is their much lower sensitivity (NEP $\sim 10^{-10}$–10^{-12} W/Hz$^{1/2}$, see Table 1). But when compared to cooled detectors the latter ones still have enough sensitivity to be used in many active imaging applications and spectroscopy. Room-temperature operation, possible relatively low noise (when zero biased), high measurement rates, small dimensions and availability of technologies make these detectors, among others, favorable for THz/sub-THz wave detection in low-cost imaging or spectroscopic systems.

Direct detection of THz/sub-THz wave signals allows a system simplification compared to frequency down-conversion (coherent) or pre-amplified direct detection systems by eliminating expensive and complicated components such as local oscillators for mixers and GHz signal amplification (LNAs, LPFs, Dicke switches, etc.).

2 General Considerations for FET and SBD Detector Models

Usually FET is divided into intrinsic and extrinsic parts [36]. The intrinsic part is the channel. The extrinsic one is the rest of FET. It includes parasitic capacitances, gate resistance, source and drain resistance, etc. Parasitic effects become more important at high frequencies.

Simplified FET representation is shown in Fig. 1. We consider the "long-channel" FETs, in which the channel length L is larger than the "detection effective channel length" L_{eff}, which is short (L_{eff} $\sim$ <100 nm) in Si FETs at room temperature [37, 38]. L_{eff} is the length of spacing where a.c. signal rectification occurs.

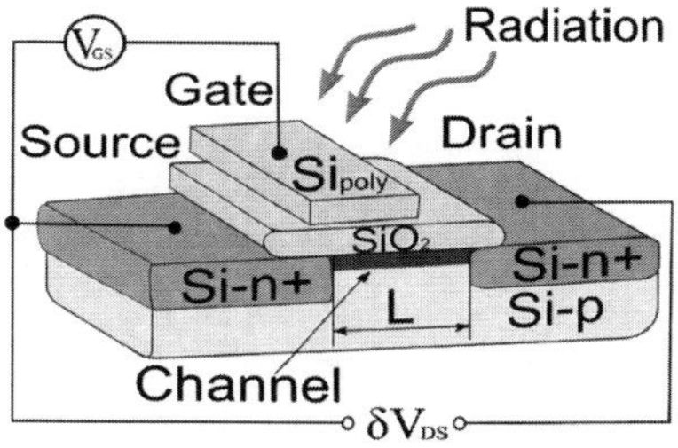

Fig. 1 Simplified FET representation. V_{GS} is a gate-source voltage (e.g., from antenna and external voltage), δV_{DS} is a d.c. drain voltage appearing under irradiation

The reason of the mm-wave/THz radiation detection in long-channel FET is a rectification of the high frequency ν signal at a short distance near the source (if the antenna receiving a signal is connected to the gate-source). This rectification was experimentally observed in *Si*-FET detectors up to radiation frequencies of $\nu \sim 4.3$ THz [19], though these FETs can not operate, e.g. as signal amplifiers in this frequency range. Simple explanation why they operate as a square-law mm/sub-mm detector lies in the fact that channel parasitic capacitances do not shunt the rectified signal at the remaining channel length, and thus the rectified signal can be reliably registered.

Rectification in FETs at THz/sub-THz frequencies is closely related to distortion (non-linear) phenomenon that was considered as a negative one, and manifests itself at higher frequencies, where FET operation is nonquasistatic. This nonquasistatic mode of FET operation was investigated a long time ago (see e.g. [25, 39, 40]).

To describe FET THz sensitivity there exist two known basic models. One is based on Dyakonov-Shur theory [24]. The second one is so called distributive resistive self-mixing model (see e.g. [17]). Both these models, concerning MOS-FETs, are based on drift current equation and strong inversion layer assumptions. The Dyakonov-Shur model is based on hydrodynamic approximation and does not seem to be applicable for inverse layer charge dynamic for MOSFET for weak and moderate inversion regions (were the maximum MOSFET sensitivity to THz/sub-THz radiation is observed) due to the dominant diffusion current in these regions. From starting expressions of [24] the theory of only *I–V*-dependence for the strong inversion region (where the drift current dominates) can be obtained. In [19] the drift equation with some form of inverse layer charge dependence approximation was used to describe rectification in all inversion regions and extension of the model [24] was used to describe the FET impedance, responsivity and *NEP*.

Here the so called "transmission line" model is used for long-channel ($L \gg L_{\mathrm{eff}}$) FET characterization as a THz detector in which the channel and gate form a transmission line for a.c. voltage signal (with impedance per unit length $Z = R_{CH}/L$, and admittance per unit length $Y = j\omega C/L$ (see e.g. Ref. [41]) with typical parameters of constant propagation $k = [(R_{CH}/L)\cdot(j\omega C/L)]^{1/2}$ ($k \sim >10^5$ cm^{-1}). Here C is the capacitance between the gate and channel (it is dependent on the gate-source voltage V_{GS} and in the region of strong inversion $C = C_{ox}$). For accepted estimations $C \sim C_{ox}$ for any V_{GS} biases, $\omega = 2\pi\nu$ is a radiation circular frequency.

In the model developed for *Si*-MOSFET sensitivity it is assumed that a short part of FET channel near the source, where the rectification of THz signal occurs, operates similar to the whole FET. To describe *I–V*-dependences (drain-source current) of this part the dependence for the whole FET channel can be used with the proper coefficient (L_{eff}/L). Since this part is rather short (<100 nm) the quasi-static approximation can be applied.

2.1 I–V All-Region FET Static Dependence

There exist several different titles for FET operation modes (e.g., weak inversion, moderate inversion, strong inversion, sub-threshold, above threshold). In this chapter we use widely accepted notifications from [36]. The FET channel inversion region, that is controlled by the gate-source voltage V_{GS}, can be separated into three regions:

(i) The weak inversion region, where the channel (drain-source) current I_{DS} of diffusion character can be described by expression

$$I_{DS} = I_0 \times \left(1 - e^{-V_{DS}/\varphi_t}\right) \times e^{(V_{GS}-V_t)/n\,\varphi_t} \tag{1}$$

The factor I_0 includes the channel width W to length L ratio W/L. The current in this region is basically of the diffusion type, $V_{FB} + \phi_F \leq V_G \leq V_M = V_{FB} + 2\phi_F + \gamma\sqrt{2\phi_F + V_{SB}}$. The lower boundary is approximately defined at the point where the inversion begins or where the leakage current becomes dominant. The upper one is treated as the point where the charge of inversion layer is comparable with that of ionized acceptors in the depletion region. Here, I_0 is a constant specific to technology, V_{DS} is the drain-source voltage, $\varphi_t = (k_B{\cdot}T)/q$ is the thermal potential, k_B is the Boltzmann constant, $n \sim 1.5$ is the sub-threshold turn-on swing factor, which is a function of the gate voltage V_{GS} and other FET parameters, V_{FB} is the flat-band voltage, ϕ_F is the Fermi level relatively to the middle of the band gap, V_M is the voltage of moderate inversion, γ is the transistor body effect factor, V_{SB} is the voltage between the source and bulk.

(ii) The moderate inversion region. In this region, the simple analytical expressions do not exist. Diffusion and drift components can be comparable. Several approximate formulas exist [36].

(iii) The strong inversion region, where the channel current is of the drift character

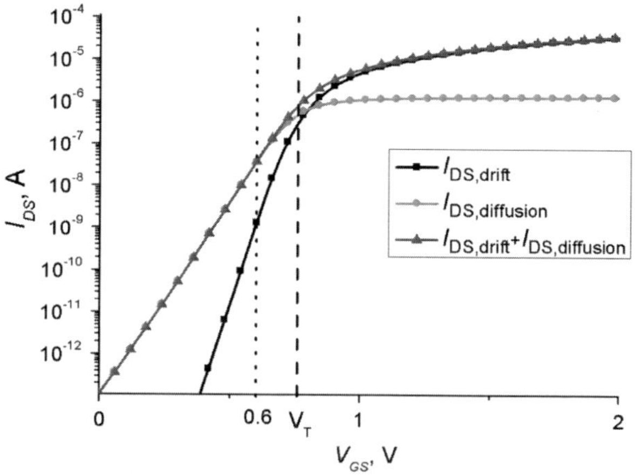

Fig. 2 FET drift $I_{DS,drift}$ and diffusion $I_{DS,diffusion}$ currents. $W/L = 20/2$ (μm) at $V_{GS} = 0.6$ V (at signal maximum), $V_{DS} = 0.05$ V. $I_{DS,diffusion}/I_{DS,drif} \approx 20$ according to the universal model [36]. $V_T = 0.758$ V

$$I_{DS} = \frac{W}{L}\mu_n C'_{ox} \times \left[(V_{GS} - V_T) \times V_{DS} - \frac{1}{2}(1+\delta) \times V_{DS}^2\right] \tag{2}$$

In this region the current is basically the drift one and $V_{GS} \geq V_H = V_M + V_Z$. Here, $V_Z \sim 0.5\ldots0.6$ V is some constant that is dependent on technology and accuracy of calculations, C'_{ox} is the specific capacitance of the gate insulator, δ is the bulk-charge factor accounting for bulk effects, V_H is the strong inversion voltage, V_M is the moderate inversion voltage.

Figure 2 shows different components of the drain-source current in FET channel at various V_{GS}. It demonstrates the importance to consider the different inversion regions. As the signal responses have a maximum at $\sim V_{GS} \sim 0.6$ V for transistors investigated (see also [15, 17, 37, 42]), from Fig. 2 it is clear that the static current is mainly a diffusion current at this V_{GS}. Thus the signal response process is mainly governed by the diffusion current component.

THz/sub-THz signals can be detected by FET detectors in all these three regions. As the maximal responsivity R_V and minimal *NEP* in most cases are experimentally observed in the range of V_{GS} corresponding to the moderate inversion ones then more general expressions should be used.

There exist the analytical models that describe FET current-voltage dependences for all V_{GS}, however, they require numerical computation [36] or numerical integration (see e.g. [43]), or are rather complex (e.g. BSIM, EKV, PSP). So, here the "balanced" model that satisfies the demands of moderate complexity, accuracy and absence of numerical computations was chosen. The semi-empirical expression that is applicable to all regions (see, e.g., [36, 44]) of drain-source current I_{DS} was used for experimental data description and *NEP* evaluations

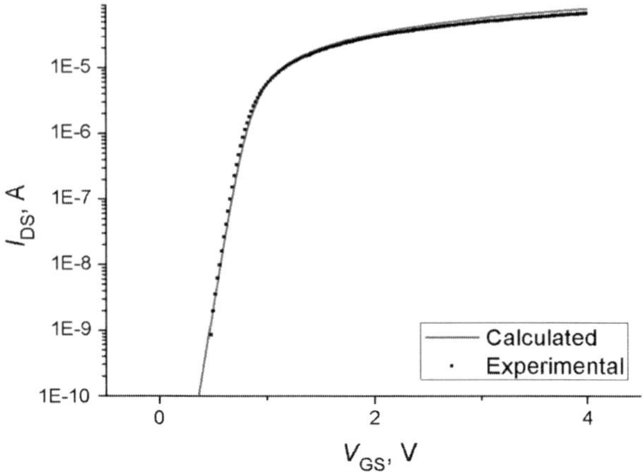

Fig. 3 Comparison of experimental (*points*) and calculated (*solid line*) from Eq. 3 *I*–*V*-characteristics of *Si*-MOSFET with $W/L = 20/2$ (μm), $V_T = 0.758$ V, $n = 1.625$

$$I_{DS}(V_{DS},V_{GS}) = \frac{W}{L}\frac{\mu_n}{1+u_a\left(\frac{V_{GS}+V_T}{T_{ox}}\right)+u_b\left(\frac{V_{GS}+V_T}{T_{ox}}\right)^2} \times C'_{ox}(2n)\varphi_t^2 \times \left[\ln\left(1+\exp\left(\frac{V_{GS}-V_T}{2n\varphi_t}\right)\right)^2 - \ln\left(1+\exp\left(\frac{V_{GS}-V_T-nV_{DS}}{2n\varphi_t}\right)\right)^2\right] \tag{3}$$

The electron mobility dependence [36, 45] on V_{GS} was taken into account (the effective mobility from BSIM3 model was used). In Eq. 3 μ_n is the electron mobility in the channel (given by the parameter u_0 in BSIM file) and u_a, u_b are the parameters of BSIM model [45], T_{ox} is the oxide thickness and C'_{ox} is the oxide layer capacitance between the gate and channel. Basically, the term with mobility degradation does not play an important role at $V_{GS} < 1$ V, and if it is not used the result will not change noticeably (in the region of maximal sensitivity).

When comparing with the experimental data, the parameters *W* and *L* were taken without any corrections. The transistor sub-threshold turn-on swing factor *n* was chosen from the region of weak inversion and V_T was chosen from the region of strong inversion for the better fitting with experimental data for static current-voltage dependences. In Fig. 3 the experimental and calculated *I*–*V*-characteristics for one of the investigated *Si*-MOSFETs are shown. As the signal responses have max at $V_{GS} \sim 0.6$ V for transistors investigated, it is seen from Fig. 2 that the static current is mainly the diffusion current at these V_{GS} values, and thus, the signal response is mainly governed by the diffusion current component.

FET or SBD THz/sub-THz detector response is conditioned by signal rectification due to current-voltage nonlinearities in these devices. The series expansion of FET current in Eq. 1 near the V_{GS}, $V_{DS} = 0$ point gives

$$I_{DS}(V_{GS}+\Delta V_{GS},V_{DS}+\Delta V_{DS}) = I_{DS}(V_{GS},V_{DS}) + \frac{\partial I_{DS}}{\partial V_{GS}}\Delta V_{GS} + \frac{\partial I_{DS}}{\partial V_{DS}}\Delta V_{DS} + \frac{1}{2}\frac{\partial^2 I_{DS}}{\partial V_{GS}^2}\Delta V_{GS}^2 + \frac{\partial^2 I_{DS}}{\partial V_{GS}\partial V_{DS}}\Delta V_{GS}\Delta V_{DS} + \frac{1}{2}\frac{\partial^2 I_{DS}}{\partial V_{DS}^2}\Delta V_{DS}^2 + \ldots \quad (4)$$

Taking for internal (channel) part of FET $V_{GS} = V_{G0} + \Delta V_{GS}$, $\Delta V_{GS} = \Delta V_{GS,int}\cdot cos(\omega t)$, $\Delta V_{DS} = \Delta V_{DS,int}\cdot cos(\omega t + \Delta\phi)$, where $\Delta\phi$ is a phase shift, and using time-averaging from Eq. 3 in general case for rectified signal current we get for relatively small signals level $\Delta V_{GS,\text{int}} < \varphi_t$ at zero drain-source bias $V_{DS} = 0$ V:

$$\langle I_{DS}\rangle = \frac{1}{T}\int_0^T I_{DS}(V_{DS,\text{int}},V_{GS,\text{int}})dt \approx \left(\frac{1}{2}\frac{\partial^2 I_{DS}}{\partial V_{GS}\partial V_{DS}} + \frac{1}{4}\frac{\partial^2 I_{DS}}{\partial V_{DS}^2}\right)|_{V_{DS}=0} \times \Delta V_{GS,\text{int}}^2 \quad (5)$$

Here $V_{DS,int}$, $V_{GS,int}$ mean the voltages on the internal (channel) part of FET.

In some models, in which the threshold voltage is counted from the source (source referenced model) (see e.g. [17, 46–49]), the derivative $\partial^2 I_{DS}/\partial V_{DS}^2$ does not exist at $V_{DS} = 0$ point as it has different values astride this point. If, for example, one takes a model (see e.g. [17, 47, 48])

$$I_D = \begin{cases} A\left((V_{GS}-V_T)V_{DS}-\frac{1}{2}V_{DS}^2\right), & V_{DS}\geq 0 \\ -A\left((V_{GS}-V_T)(-V_{DS})-\frac{1}{2}(-V_{DS})^2\right), & V_{DS}<0 \end{cases} \quad (6)$$

that is applicable in strong inversion region, then

$$\left.\frac{\partial^2 I_{DS}}{\partial V_{DS}^2}\right|_{V_{DS}\to 0+} = -A, \quad \left.\frac{\partial^2 I_{DS}}{\partial V_{DS}^2}\right|_{V_{DS}\to 0-} = A \quad (7)$$

where A is a constant. It is seen that $\partial^2 I_{\text{D}}/\partial V_{\text{DS}}^2$ doesn't exist at $V_{\text{DS}} = 0$. This is due to referencing the threshold voltage V_{T} from the source [36, 46]. Let's assume that (i) no V_{DS} bias is applied (zero bias approximation) so, $I_{DS}(V_{DS} = 0, V_{GS}) = 0$ and thus $\partial^2 I_{\text{DS}}/\partial V_{\text{GS}}^2 = 0$ and (ii) transistor is symmetrical $I_{DS}(V_{DS}, V_{GS}) = -I_{DS}(-V_{DS}, V_{GS})$ and thus $\partial^2 I_{\text{DS}}/\partial V_{DS}^2 = 0$ at $V_{DS} = 0$.

Then from Eq. 5

$$\langle I_{DS}\rangle = \frac{1}{T}\int_0^T I_{DS}\left(V_{DS,\text{int}},V_{GS,\text{int}}\right)dt \approx \left.\frac{1}{2}\frac{\partial^2 I_{DS}}{\partial V_{DS}\partial V_{GS}}\right|_{V_{DS}=0}\Delta V_{GS,\text{int}}^2 \quad (8)$$

and response at drain-source contacts of FET appears as signal mixing at source and drain. If one applies THz/sub-THz voltage at source-drain of the symmetrical FET, when connections and parasitics are the same at both ends of the channel, then no signal should appear. Practically the response can appear, as at high signal frequencies there exist various parasitic capacitances through which signals penetrate to the gate (in the most cases through connection wires and busses that are asymmetrical).

The FET response for internal part of transistor is $V_{\text{det,int}} = \left(\frac{\partial I_{DS}}{\partial V_{DS}}\right)^{-1}_{V_{GS}} \cdot \langle I_{DS} \rangle$, and if one takes into account that at $V_{DS} \to 0$ the differential resistance is close to the ordinary one

$$\frac{\partial I_{DS}}{\partial V_{DS}}\Big|_{V_{DS}=0} \approx \frac{I_{DS}}{V_{DS}}\Big|_{V_{DS}\to 0} = \sigma_{CH} \tag{9}$$

Then the detector internal voltage response can be written as

$$V_{\text{det}} = \frac{1}{2} \cdot \frac{1}{\sigma_{CH}} \cdot \frac{d\sigma_{CH}}{dV_{GS}} \cdot \Delta V^2_{GS,\text{int}}. \tag{10}$$

The similar expression was obtained earlier [24, 47, 50–52] but with a coefficient "1/4" due to the fact that the derivative here is taken as mixing derivative on V_G and V_D. The presence of "1/2" instead of "1/4" means that FETs would have better responsivity over SBDs that was mentioned long time ago [53].

Equation 10 using Eq. 3 can be written in more explicit form giving the possibility to compare FET detector signal with that one for SBD detector. The channel conductivity at $V_{DS} = 0$ can be presented as

$$\sigma_{CH} = \sigma_{CH0} \cdot f_\sigma(x), \tag{11}$$

where $\sigma_{CH0} = (W/L) \cdot \mu_{\text{n}} \cdot C'_{\text{ox}} \cdot n \cdot \varphi_{\text{t}}, f_\sigma(x) = 2 \cdot \ln\left(1 + e^{x/2}\right) \cdot e^{x/2} \cdot \left(1 + e^{x/2}\right)^{-1}$. This is the first term in Taylor expansion of Eq. 3. Mobility reduction term wasn't accounted in this expression, and $x = (V_{GS} - V_T)/n \cdot \varphi_{\text{t}}$. Then

$$V_{\text{det}} = \frac{\Delta V^2_{GS,\text{int}}}{2} \cdot \frac{1}{n \cdot \varphi_t} \cdot \frac{1}{f_\sigma} \cdot \frac{\partial f_\sigma}{\partial x} \tag{12}$$

Equations 10 and 12 are different from SBD detectors by factor "1/2" (see e.g. [30] and Eq. 42 in which 1/4 is present), but FET voltage response depends also on the function describing the channel conductivity (f_σ) and its derivative, which are the functions of gate-source bias, sub-threshold turn-off swing factor, etc. In SBD Eq. 42 describes the response that is independent on bias.

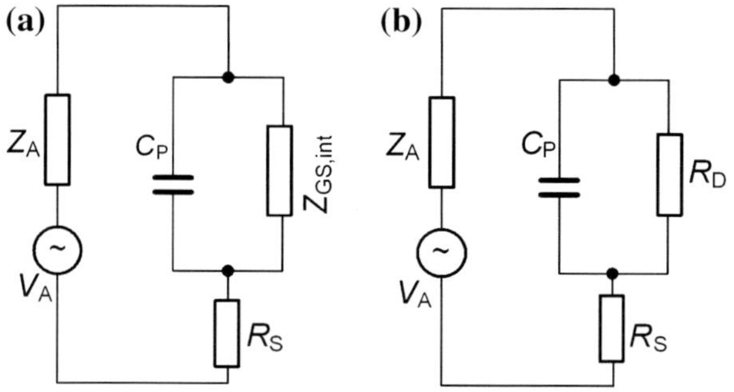

Fig. 4 FET (**a**) and SBD (**b**) simplified schematic representations taking into account the basic parasitic components. Z_A is the antenna impedance; V_A is the antenna voltage amplitude; $R_S = R_G + R_{source}$ in FET is the active series (parasitic) resistance of FET, where R_G is the gate active resistance; R_S in SBD is series parasitic active resistance; R_D is SBD differential active resistance; C_P is the parasitic reactance (usually capacitive), R_S is the active serial resistance; $Z_{GS,int}$ is internal source-gate impedance

2.2 FET Responsivity with Account of Rarasitics

To characterize FET as mm-wave/THz detector one should find the relationship between the output signal and absorbed radiation power, thus, voltage R_V or current R_I responsivity. In order to much such estimation one should know the FET equivalent circuit. High-frequency FET equivalent circuit is rather complex (see e.g. [36, 54]). However, in order to find R_V or R_I one is interested in FET part between source and gate. This simplified FET circuit connected to the antenna can be represented by the circuit shown in Fig. 4a [36, 54]. SBD schematic [10, 30–32] is shown in Fig. 4b. These schematics are presented in the way to be similar to each other, thus allowing direct comparison of FETs and SBDs characteristics as mm-wave/THz detectors.

FET can be divided into two parts: (i) internal part (where rectification occurs), $Z_{GS,int}$, and (ii) external part (where parasitic components influence on signal characteristics can be substantial), C_P, R_S. R_S includes the source series and gate resistances.

The input impedance of the internal transistor part $Z_{GS,int}$ can be defined as an input impedance of transmission line with length $d = L$, propagation constant k, characteristic impedance Z_0, and loaded by impedance Z_L by expression [41]:

$$Z(d) = Z_0 \cdot \frac{Z_L + Z_0 \cdot \tanh(kd)}{Z_0 + Z_L \cdot \tanh(kd)} \tag{13}$$

and since (L•Rek) $\gg$ 1, then $Z = Z_0$ due to $\tanh(Lk) \approx 1$.

Thus, the internal source-gate impedance is equal to transmission line characteristic impedance (for the case of long channel transistor):

$$Z_{GS,\text{int}} = Z_0 = \sqrt{\frac{R_{CH}}{j\omega C}} = \frac{\alpha(1-j)}{\sqrt{\omega}}, \tag{14}$$

where $\alpha = (R_{CH}/2C)^{1/2}$. Here C is a function of V_{GS} bias. For estimations it is accepted $C \approx C_{ox}$. For example with $C \approx C_{ox}$ for $\nu \sim 100$ GHz the internal impedance $Z_{GS,int} \approx 5.3\bullet(1\text{–}j)\bullet 10^3\ \Omega$.

It is important to find the voltage R_V (or current R_I) responsivity and noise equivalent power of FET detector. For electrical responsivity it can be written $R_V^{el} = \eta \cdot R_{V,int}$, where the responsivity for internal part of transistor is $R_{V,int}$, and η is the ratio of power absorbed in FET internal part to all power absorbed in FET.

2.2.1 Power Transfer Coefficient

For mm-wave/THz radiation power absorbed at the internal transistor part

$$P_{in,\text{int}} = \frac{1}{T}\int_0^T I_{GS,\text{int}}(t)V_{GS,\text{int}}(t)dt = \frac{1}{2}\frac{\text{Re}Z_{GS,\text{int}}}{|Z_{GS,\text{int}}|^2}|\Delta V_{GS,\text{int}}|^2, \tag{15}$$

where $\Delta V_{GS,int}$ is the high frequency voltage between the source and drain in internal part of the transistor and coefficient 1/2 is due to the use of amplitude value.

The average current arising at $V_{DS} = 0$, V_{GS} is defined by Eq. 8. Then from Eqs. 8 and 15 for current responsivity $R_{I,int}$ at the internal part of FET it follows

$$R_{I,\text{int}} = \frac{\langle I_{DS}\rangle}{\langle P_{in,\text{int}}\rangle} = \frac{(\partial^2 I_{DS}/\partial V_{DS}\partial V_{GS})|_{V_{DS}=0}}{\left(\text{Re}Z_{GS,\text{int}}/|Z_{GS,\text{int}}|^2\right)} \tag{16}$$

It also can be written using Eqs. 9, 11 and 14:

$$R_{I,\text{int}}^{FET} = \frac{1}{n\cdot\varphi_t}\cdot\sqrt{\frac{2\cdot\sigma_{CH0}}{\omega\cdot C_{CH}}}\cdot\frac{1}{\sqrt{f_\sigma}}\cdot\frac{\partial f_\sigma}{\partial x}. \tag{17}$$

In the case of low input power level the current and voltage responsivities are related by $R_{V,int} = R_{I,int}\bullet R_{CH}$. Finally, using Eqs. 14 and 16, one can obtain:

$$R_{V,\text{int}} = 2\frac{\alpha}{\sqrt{\omega}}R_{CH}\frac{\partial^2 I_{DS}}{\partial V_{DS}\partial V_{GS}}\bigg|_{V_{DS}=0}, \tag{18}$$

$$R_{V,\text{int}} = \frac{1}{n\cdot\varphi_t}\cdot\sqrt{\frac{2}{\sigma_{CH0}\cdot\omega\cdot C_{CH}}}\cdot\frac{1}{f_\sigma\cdot\sqrt{f_\sigma}}\cdot\frac{\partial f_\sigma}{\partial x} \tag{19}$$

The voltage responsivity $R_{V,int}$ can be very large as the channel resistance is typically large in the weak inversion region. This voltage responsivity was calculated [48] with account of the parasitic resistances but without parasitic capacitances, which, however, do not change the result sufficiently.

One should find the ratio of power that is absorbed in internal part of the transistor $P_{in,int}$ to power Pin that is absorbed in the transistor as a whole (in its internal and external parts). That is (for circuit in Fig. 4a)

$$\eta = \frac{P_{in,\mathrm{int}}}{P_{in}} = \left|\frac{Z_1}{Z_{GS}}\right|^2 \frac{\mathrm{Re}Z_{GS,\mathrm{int}}/\left|Z_{GS,\mathrm{int}}\right|^2}{\mathrm{Re}Z_{GS}/\left|Z_{GS}\right|^2}, \tag{20}$$

where $Z_1 = Z_{GS,\mathrm{int}}\|X_P$, $Z_{GS} = Z_1 + R_S$, $X_P = (j\omega C_P)^{-1}$. Here η is the power transmission coefficient. Thus, electrical responsivity can be written as

$$R_V^{el} = \eta \cdot R_{V,\mathrm{int}} \tag{21}$$

It should be pointed out that in the weak inversion region the input transistor impedance $Z_{GS,int}$ increases with V_{GS} decrease and thus the internal voltage sensitivity $R_{V,int}$ seems to increase due to the decrease of power that is absorbed in internal part $P_{in,int}$. At $V_{DS} \to 0$ the voltage sensitivity $R_{V,int}$ can reach large-scale values as was pointed out in [48].

This behavior of $R_{V,int}$ can be clearly seen from Eq. 19, as $f_\sigma(x) \approx 2e^x$ and for $x \leq -5$ it is $(f_\sigma)^{-3/2}\cdot(\partial f_\sigma/\partial x) \approx 2^{1/2}\cdot e^{-x/2}$ which increases exponentially with decreasing V_{GS} (and x). Thus, at $V_{GS} \to 0$ the voltage sensitivity $R_{V,int}$ can reach large-scale values. However, as the power transmission coefficient η is going down with transistor internal input impedance $Z_{GS,int}$ increase ($Z_{GS,int} \sim (R_{CH})^{1/2}$) (Eq. 14), the resulting electrical responsivity $R_V^{el} = \eta R_{V,\mathrm{int}}$ of FET remains practically the same (see Fig. 5). This indicates that it is necessary to take into account the external part of FET.

To see the detected signal dependence on radiation frequency ν let's present η in the following form. From Eqs. 14 and 20 it can be found

$$\eta = \frac{1}{1 + (\omega/\omega_{c1})^{1/2} + \omega/\omega_{c2} + (\omega/\omega_{c3})^{3/2}}, \tag{22}$$

where the coefficients $\omega_{c1} = (R_S/\alpha)^{-2}$, $\omega_{c2} = (2 \cdot C_P \cdot R_S)^{-1}$, $\omega_{c3} = (2C_P^2 \cdot R_S \cdot \alpha)^{-2/3}$, $\alpha = (R_{CH}/2C_{ox})^{1/2}$ as $C \approx C_{ox}$. Here $\omega_{ci} = 2\pi\nu_{ci}$ are certain coefficients.

The estimated voltage responsivities R_V^{el} for different devices are shown in Fig. 6. At radiation frequency $\nu = 77$ GHz the FET detector responsivity, estimated for 0.35 μm design rule technology with the channel width-to-length ratio $W/L = 1/1$ (μm), should be approximately 20 times larger as compared to FET detector responsivity manufactured using 1 μm design rules technology with $W/L = 20/2$ (μm). With an increase of radiation frequency ν this difference increases.

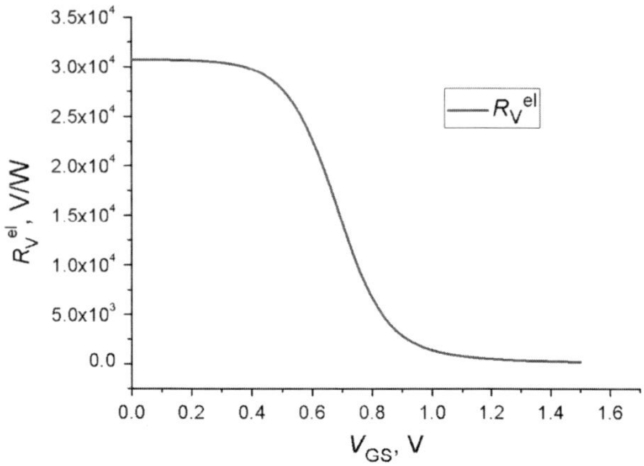

Fig. 5 Dependence of calculated electrical voltage responsivity $R_{V,el} = R_{V,int} \bullet \eta$ for MOSFET investigated with 1 μm design rules technology $W/L = 20/2$ (μm)

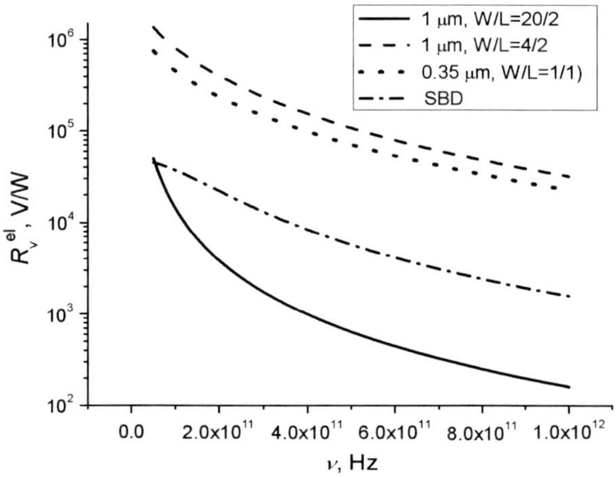

Fig. 6 Comparison of calculated electrical responsivity R_V^{el} for different devices

For estimations the following devices were compared:

1. Experimentally investigated FET manufactured by 1 μm technology, $W/L = 20/2$ (μm), $C_P = 4$ fF, $R_S \approx 200\ \Omega$, $V_{GS} \approx 0.6$ V, $R_{CH} \approx 1.1 \times 10^6\ \Omega$ (at $V_{GS} \approx 0.6$ V), $C_{ox} \approx 35 \bullet 10^{-15}$ F, $\alpha \approx 4.0 \bullet 10^9\ \Omega/\text{s}^{1/2}$, $\nu_{c1} \approx 6.0 \times 10^{13}$ Hz, $\nu_{c2} \approx 9.8 \times 10^{10}$ Hz, $\nu_{c3} \approx 1.8 \times 10^{10}$ Hz.
2. FET detector based on 1 μm technology with low parasitic resistances: $W/L = 4/2$ (μm), $C_P = 1$ fF, $R_S \approx 15\ \Omega$, $R_{CH} \approx 5.5 \times 10^6\ \Omega$ (at $V_{GS} \approx 0.6$ V), $C_{ox} \approx 7 \times 10^{-15}$ F, $\alpha \approx 2 \times 10^{10}\ \Omega/\text{s}^{1/2}$, $\nu_{c1} \approx 2.8 \times 10^{17}$ Hz, $\nu_{c2} \approx 5.4 \times 10^{12}$ Hz, $\nu_{c3} \approx 2.25 \times 10^{11}$ Hz.
3. Transistors based on 0.35 μm technology: $W/L = 1/1$ (μm), $C_P = 0.2$ fF, $R_S \approx 426\ \Omega$, $R_{CH} \approx 1.26 \times 10^6\ \Omega$ (at $V_{GS} \approx 0.5$ V), $C_{ox} \approx 4.6 \times 10^{-15}$ F, $\alpha \approx 1.2 \times 10^{10}\ \Omega/\text{s}^{1/2}$, $\nu_{c1} \approx 1.2 \times 10^{14}$ Hz, $\nu_{c2} \approx 9.3 \times 10^{11}$ Hz, $\nu_{c3} \approx 2.93 \times 10^{11}$ Hz. The value $V_{GS} = 0.5$ V instead of 0.6 V was chosen due to different

threshold voltage. Higher C_{ox} value is due to thinner oxide. The parasitics per unit width are almost the same.

4. SBD (see below): $R_S = 10\ \Omega$, $R_D = 3000\ \Omega$, $C_P = 5$ fF, $n = 1.2$.

Parasitic parameters for FET were taken from BSIM3 model file provided by manufacturer. Parameters for gate resistance were chosen from the technology data-sheet.

Usually $\nu_{c1} > \nu_{c2} > \nu_{c3}$. Thus, let's consider FET operation at sufficiently large frequencies $\nu > \nu_{c3}$. Then

$$R_V^{el} = 2\alpha R_{CH} \frac{\partial^2 I_{DS}}{\partial V_{DS} \partial V_{GS}}\bigg|_{V_{DS}=0} \omega_{c3}^{3/2} \omega^{-2} = \frac{1}{C_p^2 R_S} R_{CH} \frac{\partial^2 I_{DS}}{\partial V_{DS} \partial V_{GS}}\bigg|_{V_{DS}=0} \omega^{-2} \tag{23}$$

doesn't depend on parameter $\alpha = (R_{CH}/2C_{ox})^{1/2}$ (from the physical point of view the parasitic capacitance C_P shunts the internal part $Z_{GS,int}$, and thus the power is absorbed mainly by R_S).

This formula can be rewritten in the form similar to one for SBDs

$$R_V^{el} \approx R_{V0}^{el} \left(\frac{\omega_c}{\omega}\right)^2, \tag{24}$$

where for FETs $R_{V0}^{el} = R_{CH}^2 \frac{\partial^2 I_{DS}}{\partial V_{DS} \partial V_{GS}}\Big|_{V_{DS}=0}$, $\omega_c = \frac{1}{C_p \sqrt{R_S R_{CH}}}$.

For example, for 1-μm technology FET ($W = 20$ μm, $L = 2$ μm), $\nu_c = 2.7 \times 10^9$ Hz, $\frac{\partial^2 I_{DS}}{\partial V_{DS} \partial V_{GS}}\Big|_{V_{DS}=0} = 2 \times 10^{-5} \frac{A}{V^2}$, $R_{CH} = 1.1 \times 10^6\ \Omega$, $R_{V0}^{el} = 2 \cdot 10^7 \mathrm{V/W}$, the "zero" current responsivity $R_{I0}^{el} = R_{CH} \frac{\partial^2 I_{DS}}{\partial V_{DS} \partial V_{GS}}\Big|_{V_{DS}=0} \approx 21 \cdot \mathrm{A/W}$, which is larger than the theoretical ultimate performance R_I for SBD detector (at $T = 300$ K, $R_I = 19.3$ A/W). This is due to the lack of factor "1/2" in the approach used (FET responsivity is based on mixed derivatives, see Eq. 8 and thus can be about twice higher compared to SBD detectors though one should take into account the other existing factors for FETs which level the situation).

Also, the Eq. 23 for R_V^{el} can be rewritten in a form

$$R_V^{el} = (C_p^2 \cdot R_S)^{-1} \times R_{I0}^{el} \cdot \omega^{-2} \tag{25}$$

This formula is useful for estimations as R_{I0}^{el} is within ~10…40 A/W for FET. For some estimations the next expressions can be used

$$NEP^{el} = \frac{\sqrt{4kTR_{CH}}}{R_V^{el}} \approx \sqrt{4kT} \frac{(R_{CH})^{1/2} C_p^2 R_S}{R_{I0}^{el}} \omega^2, \tag{26}$$

where the thermal noise, which is important for NEP ultimate performance estimations at $V_{DS} = 0$ for FET detectors, is included.

Using $C_p = C_x W, R_S = \frac{\rho_1}{W} + \rho_2 \frac{W}{L}, R_{CH} = \frac{L}{W} R_{CH0}$ (model used in BSIM3 valid for $W > L$) the next equation can be written

$$NEP^{el} = \frac{\sqrt{4kTR_{CH0}}}{R_{I0}^{el}} \omega^2 \left(C_x^2 \rho_1 \sqrt{LW} + C_x^2 \rho_2 W^{5/2} L^{-1/2} \right) \tag{27}$$

Here R_{CH0} (Ω) is the channel resistance, C_x (F/m), ρ_1 (Ω·m), ρ_2 (Ω/□) are the technological parameters for given design rules. For example for 1-μm and 0.35-μm design rules $R_{CH0} = 1.1 \times 10^7$ Ω ($V_{GS} = 0.6$ V) and $R_{CH0} = 1.3 \times 10^6$ Ω ($V_{GS} = 0.5$ V), respectively. The parameter C_x is the same $C_x = 2 \times 10^{-10}$ F/m, $\rho_2 = 16$ Ω/□ for 1-μm technology, and $\rho_2 = 12$ Ω/□ for 0.35-μm technology, and $\rho_1 = 9 \times 10^{-4}$ and 4.1×10^{-4} Ω•m, respectively. From Eq. 27 the electrical noise equivalent power decreases with W reduction. So, for a given L value the transistor with smaller W has lower (better) NEP^{el}.

Optimizing this expression with technological constraints (1 μm $\leq L \leq$ 5 μm, 1 μm $\leq W \leq$ 20 μm) the optimal FET dimensions can be estimated to be $L \approx 1$ μm and $W \approx 1$ μm. For example for 1.0-μm CMOS technology one can evaluate the optimal responsivity $R_{I0}^{el} \approx 21$ A/W and electrical $NEP^{el} \sim 2.8 \times 10^{-12}$ W/Hz$^{1/2}$ at $\nu \sim 77$ GHz. For 0.35-μm CMOS technology these values are $R_{I0}^{el} \approx 21$ A/W and $NEP^{el} \approx 2.5 \times 10^{-13}$ W/Hz$^{1/2}$. However, better NEP for 0.35-μm technology may be due to lower value of parasitic capacitance, different V_{GS} and different oxide thickness which lead to lower channel resistance. So, NEP^{el} (that is only valid for FET channel itself) is smaller for FET with lower W. However the estimations within this model should be used only as rough ones.

2.2.2 FET Antenna and Voltage Divider Matching

When using FET as a detector with antenna, to estimate the detector sensitivity it is important to find the part of power that is absorbed in the transistor to maximal possible power that can be taken off from antenna. Here we consider the case when transistor is connected to antenna assuming antenna connection to gate-source (antenna impedance Z_{ant}, transistor impedance Z_{GS}). In the case when transmission line is used, Z_{ant} means the impedance at the end of the transmission line. Then for antenna transfer coefficient η_a one can write

$$\eta_a = \frac{4\mathrm{Re}Z_{ant}\mathrm{Re}Z_{GS}}{|Z_{ant} + Z_{GS}|^2}. \tag{28}$$

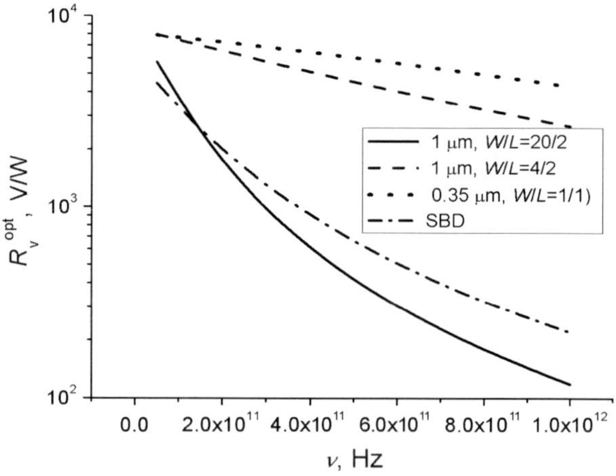

Fig. 7 Comparison of calculated R_V^{opt} for SBD, 1-μm design rules, 0.35-μm FET technologies and "optimized" MOSFET based on 1-μm technology. $Z_{ant} = (100 - j100)\ \Omega$

Thus, the optical responsivity can be defined as

$$R_V^{opt} = R_V^{el} \times \eta_a. \tag{29}$$

Obviously, R_V^{opt} is always lower than $R_V^{el}\left(R_V^{opt} < R_V^{el}\right)$ due to the impedances antenna-detector mismatch. Comparison of optical responsivity for different devices is shown in Fig. 7 assuming the frequency independent antenna impedance $Z_{ant} = (100 - j100)\ \Omega$. Comparing the results of Figs. 6 and 7 one can see that the difference between R_V^{opt} and R_V^{el} can be significant.

When measuring the voltage at the transistor gate-source by voltmeter, the FET channel and voltmeter input circuit form the voltage divider. The voltage drop part at the voltmeter (the voltage transfer coefficient from transistor to voltmeter) is [47, 55]

$$\eta_l = (|1 + R_{CH}/Z_{load}|)^{-1}, \tag{30}$$

where $Z_{load} = R_{load} \| \frac{1}{j2\pi f_{\text{mod}} C_{load}}$ is the impedance of the measuring device. For example, for one of FETs investigated the voltmeter is represented by parallel connected resistor $R_{load} = 10\ \text{M}\Omega$, capacitor $C_{load} = 90$ pF, and modulation frequency $f_{mod} = 172$ Hz. Thus, $Z_{load} \approx (5.0 - j5.1) \times 10^6\ \Omega$.

The resulting voltage measured by the voltmeter at FET as a detector is

$$V_{det} = P_{ant,max} \cdot R_{V,int} \cdot \eta \eta_a \eta_L, \tag{31}$$

where η_L is the loading transmission coefficient.

The maximum power $P_{ant,max}$ that can be supplied to FET channel from antenna at plane wave illumination [56, 57] is

$$P_{ant,\max} = G\frac{\lambda^2}{4\pi}I_0. \tag{32}$$

It is for the case of ideal impedance matching between the antenna and loading when $Z_{GS} = Z^*_{ant}$. Here Z_{GS} is the transistor input impedance and Z_{ant} is the antenna impedance, G is the antenna gain, λ is the wavelength in vacuum (air), and I_0 is the radiation intensity that is falling down on antenna from air.

In high dielectric permittivity and thick (d ~ >100 μm) substrates, the different substrate modes can arise strongly modifying G [56, 57] and can unpredictably change the detector responsivity and *NEP*.

One can obtain the Eq. 28 for power transferred from antenna to detector $P_{ant} = P_{ant,\max}\eta_a$ in the case of impedance mismatch $Z_{ant} \neq Z^*_{GS}$.

Setting $P_{ant,max}$ equal to power in term of voltage on antenna terminals for the case of impedance matching, one obtains

$$I_0\frac{\lambda^2}{4\pi}G = \frac{1}{2}\frac{|U_{ant}|^2}{4\mathrm{Re}Z_{ant}}, \tag{33}$$

and the antenna voltage

$$|U_{ant}|^2 = 2\frac{\lambda^2 G}{\pi}\mathrm{Re}Z_{ant}I_0 \tag{34}$$

In the case $Z_{SG} \neq Z^*_{ant}$, the power at the antenna output

$$P = \frac{1}{2}|U_{ant}|^2\left|\frac{Z_{GS}}{Z_{GS}+Z_{ant}}\right|^2\frac{\mathrm{Re}Z_{GS}}{|Z_{GS}|^2} = \frac{\lambda^2 G I_0}{\pi}\frac{\mathrm{Re}Z_{GS}\mathrm{Re}Z_{ant}}{|Z_{GS}+Z_{ant}|^2} \tag{35}$$

Then, the power transfer coefficient from the antenna to load is

$$\eta_a = \frac{P}{P_{\max}} = 4\frac{\mathrm{Re}Z_{GS}\mathrm{Re}Z_{ant}}{|Z_{GS}+Z_{ant}|^2}. \tag{36}$$

The resulting responsivity measured for one of the FETs investigated as a detector in linear and semi-log coordinates is shown in Fig. 8. One can see that in logarithmic coordinates the exponential signal decay is observed at $V_{GS} < 0.4$ V which is related to R_{CH} dependence on V_{GS} (R_{CH} ~ $\exp(-V_{GS}/n\varphi_t)$) and thus channel resistance reducing the matching coefficient η_L with a constant voltmeter input resistance.

Using Eq. 31 one can estimate the frequency dependence of the signal detected on the value of $R_{V,int}$. When comparing the estimated value of V_{det} with the experimental one, the method of experimental data acquisition should be taken into account. In experiments, usually the detected voltage V_{det} and irradiation intensity I_0 are known values. However, to obtain the responsivity of the detector R_V^{meas}, the radiation power P' supplied to detector should be known, and that one is related to irradiation

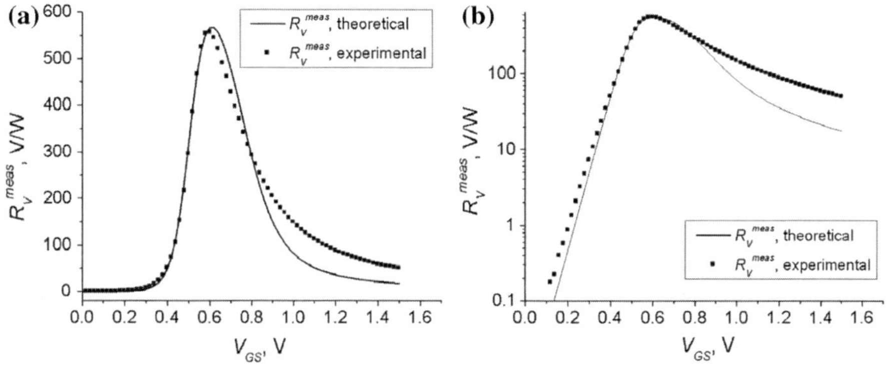

Fig. 8 Comparison of experimental (*points*) and calculated (*solid lines*) voltage responsivities versus the gate voltage V_{GS} (**a** for linear coordinates and **b** for semi-log coordinates). To fit calculations and experimental data, the latter ones were multiplied by certain coefficient as the power introduced to FET channel is uncertain because of an unknown antenna gain G and antenna input impedance. (Normalization is equivalent of use the following antenna parameters: $G = 0.4$, $Z_{ant} = 10 - j30\ \Omega$)

intensity I_0 and some effective area S' that absorbs the radiation. This area usually is uncertain: it can be the diffraction limited area, pixel area, antenna area, etc. That's why it is difficult to compare the responsivities cited in various papers. Really, it is a "hazy" procedure to obtain the performance of detector itself R_V^{el} from the performance of a system R_V^{meas} without knowing the antenna parameters.

The influence of coefficients $\eta{\cdot}\eta_a{\cdot}\eta_L$ product, that determines the power absorbed by FET detector, is crucial for determination of registered detector signals and thus, for responsivity and *NEP* evaluations. For example, taking for estimations $\eta \sim 0.1$ from Eq. 22 ($\nu \approx 77$ GHz), $\eta_a \sim 0.2$ at $Z_{ant} \sim (100 - j100)\ \Omega$ from Eq. 28, and FET impedance $Z_{FET} \sim (200 - j{\cdot}500)\ \Omega$. From Eq. 30 $\eta_L \sim 1$ at voltmeter $R_{input} \sim 10$ MΩ and at antenna gain $G \sim 1$ one has with $R_{V,int}$ at $V_{GS} \approx 0.6$ V (see Fig. 8) the value of $R_V^{meas} \sim 4 \times 10^3 \cdot \text{V/W}$. In [15], the voltage sensitivities observed in CMOS detectors is $R_V^{meas} \approx 6 \times 10^3 \cdot \text{V/W}$ at higher radiation frequencies $\nu \approx 300$ GHz, which should lead to worse *NEP* values. This could be a result of larger gain $G > 1$ due to the resonances in thick substrates used and other FET parameters that were not pointed out.

3 SBD Detectors

SBD THz/sub-THz detectors are mainly manufactured on the basis of III-V semiconductors. For quasi-optical SBD detectors over a frequency range from ~100 to ~500 GHz the typical responsivity values are from ~300 to 1000 V/W. Figure 4b shows the SBD equivalent circuit representation which is frequently used in estimations of SBD parameters (see e.g. [10, 30, 31]).

The dominant current transport mechanisms in SBDs are the emission of electrons over the barrier and the quantum mechanical tunneling of electrons through the barrier. A widely recognized model that takes into account both of these effects is the thermionic-field emission model [58].

Generalized *I–V*-characteristics of an ideal junction can be written as

$$I(V) = I_s\left(\exp\left(\frac{V}{n\phi_t}\right) - 1\right), \tag{37}$$

where $I(V)$ is the current through the SBD, I_S is the reverse bias saturation current, n is the ideality factor ($n \sim 1\ldots2$, depending on the fabrication process), $\phi_t = k_B{\cdot}T_{jun}/q \approx 25.85$ mV (at $T = 300$ K) is the thermal voltage, where k_B is the Boltzmann constant, and q is the electron charge.

The reverse bias temperature dependence of the saturation current can be expressed as

$$I_S = S \cdot A^* \cdot T_j^2 \cdot \exp\left(-\frac{\varphi_b(0)}{n \cdot \varphi_t}\right), \tag{38}$$

where $\varphi_b(0)$ is the equilibrium Schottky potential barrier, S is the junction area, T_{jun} is the temperature of the junction, the Richardson constant $A^* = 4\pi \cdot q \cdot k_B^2 \cdot m_0/h^3 = 120\ \mathrm{A}/(\mathrm{cm}^2 \cdot \mathrm{K}^2)$, where m_0 is the free electron mass. In semiconductors m_0 should be changed to effective electron mass m^*, that will decrease the Richardson constant by a factor of m^*/m_0 (this means that thermoionic current decreases by the same factor). The modified Richardson constant in dependence of reverse bias can change from $A^* = 30$ A/(cm^2·K^2) for moderate electric fields to $A^* \approx 7$ A/(cm^2·K^2) at biases $U \sim 1$ V [59]. In GaAs, as a rule, it is accepted $A^* \approx 8.2$ A/(cm^2·K^2).

Equation 37 describes the current-voltage characteristics of metal-semiconductor junction but does not do that in the SBD case since the voltage in the exponent is the voltage applied to junction, but not the voltage applied to SBD, and it does not take into account the series resistance R_S and junction capacitance C_P.

With $\varphi_b(0) \sim 0.7$ V at $T = 300$ K ($\varphi_b(0)$ changes within 0.6–1.0 V for more than 40 metals in GaAs SBDs [60]). From Eq. 38, $I_S \sim 1.5 \times 10^{-14}$ A for junction area $S \sim 1\ \mu\mathrm{m}^2$, and its resistance at zero bias $R_0 \sim 1.8 \times 10^{12}\Omega$ ($A^* = 8A/\ \mathrm{cm}^2 \cdot \mathrm{K}^2, R_0 = \left(\frac{dI}{dV}\Big|_{V=0}\right)^{-1} = \frac{n\cdot\varphi_t}{A^*\cdot T^2} \cdot \exp(\frac{q\phi_b(0)}{n\cdot k_B T_{jun}}) = \frac{n\cdot\phi}{I_S}, n = 1$). Thus, for SBD (with $\varphi_b(0) \sim 0.7$ V) operating as sub-THz detector, it is seen that it should be forward biased, which leads to additional noises up to several times and even orders of magnitude depending on modulation frequency, primarily because of the 1/*f* noise component [30–32, 61, 62]. So, it is important to lower the barrier height to obtain SBD operation at zero bias in low noise regime.

By lowering the effective barrier of SBD junction, the barrier differential resistance can be significantly reduced. With $\varphi_b(0) \approx 0.2$ V (e.g. Al/GaAs δ-doped SBDs [63], InGaAs/InP [62, 64]) I_S equals to $(1.5\ldots10) \times 10^{-6}$ A, for similar junction area and zero bias resistance ($R_O \sim 8\times10^3$ Ω). Thus, such kind of SBDs as detectors having $\varphi_b(0) \sim (0.2\ldots0.3)$ V can be used at zero biases.

One of the disadvantages of zero-bias III-V SBDs based mainly on ternary alloys is difficulty of their monolithic integration into arrays with read-out electronics (which are usually based on silicon). Moreover, III-V SBDs are difficult to produce with repeatable performance due to the difficulties in control of the doping and structure required because of a need of high-quality interface between metal and semiconductor. Another limitation of SBDs is a strong temperature dependence of their performance. As the current flow through the barrier is the thermionic, I_S changes approximately by a factor of 2 every 20 K.

SBD can be divided [30], like a FET, into two parts: internal (where rectification occurs) and external ones (where parasitic resistances and capacitances are present). Here the junction capacitance is considered as the external part.

Contrary to FET detector the SBD responsivity can be found in analytical form in the case of no bias (without Taylor expansion use). The SBD current signal I_d is an averaged quantity over the period T

$$I_d = \langle I \rangle = \frac{1}{T}\int_0^T I(t)dt = \frac{1}{T}\int_0^T I_s\left(e^{\frac{\Delta V\cos(\omega t)}{n\phi_t}} - 1\right)dt = I_s\left(I_0\left(\frac{\Delta V}{n\phi_t}\right) - 1\right) \quad (39)$$

The power that is absorbed in SBD internal part is

$$P_{in,\text{int}} = \frac{1}{T}\int_0^T I(t)V(t)dt = \frac{1}{T}\int_0^T I_s\left(e^{\frac{\Delta V\cos(\omega t)}{n\phi_t}} - 1\right)\Delta V\cos(\omega t)dt = I_s\Delta V I_1\left(\frac{\Delta V}{n\phi_t}\right) \quad (40)$$

Then for internal current responsivity $R_{I,int}$ it follows

$$R_{I,\text{int}} = \frac{I_d}{P_{in,\text{int}}} = \frac{I_s\left(I_0\left(\frac{\Delta V}{n\phi_t}\right) - 1\right)}{I_s\Delta V I_1\left(\frac{\Delta V}{n\phi_t}\right)} \Rightarrow R_{I,\text{int}} = \frac{\left.\frac{\partial^2 I}{\partial V^2}\right|_{V=0}}{2\left.\frac{\partial I}{\partial V}\right|_{V=0}} = \frac{1}{2n\phi_t} \quad (41)$$

Here I_O, I_I are the modified Bessel functions of the first kind and the right part is valid for voltage $\Delta V < \varphi_t$, that is realized in square-law region. The detected voltage is given by

$$V_d^{SBD} = \frac{\Delta V^2}{4}\cdot\left(\frac{dI}{dV}\right)^{-1}\cdot\frac{d^2I}{dV^2} = \frac{\Delta V^2}{4}\cdot\frac{1}{n\cdot\phi_t}. \quad (42)$$

Equations 41 and 42 differ from the similar expressions for FET detector (Eqs. 8 and 10) by factor "1/2" as FET is a 3 terminal device and the signal detected in the model accepted is connected to mixed derivatives.

At $T = 300$ K and $n = 1$, $R_{I,int} = 19.3$ A/W is the maximum possible internal current responsivity of zero-bias SBD. And for the voltage responsivity R_V measured at low frequencies (when SBD is connected to a high impedance load such as e.g. oscilloscope) $R_{V,int} \approx R_{I,int} \bullet R_D$, where R_D is the zero-bias barrier differential resistance.

In the square-law region the SBD current responsivity R_I (defined by ratio of current through the SBD to power absorbed in SBD) under the power with radiation frequency ν for circuit shown in Fig. 4b can be written as [30]

$$R_I^{el} = R_{I,\mathrm{int}} \cdot \eta, \tag{43}$$

where η is the power transfer coefficient (the ratio of power dissipated in internal part to total power)

$$\eta = \frac{1}{(1 + R_S/R_D) \cdot \left(1 + (\nu/\nu_c)^2\right)}, \tag{44}$$

where the cut-off frequency

$$\nu_c = \frac{(1 + R_S/R_D)^{1/2}}{2\pi \cdot C_P \cdot (R_S \cdot R_D)^{1/2}}. \tag{45}$$

With "good" values of different parameters $R_S \approx 10\ \Omega$, $R_D \approx 3 \bullet 10^3\ \Omega$ and typical zero-bias capacitance for small area SBDs $C_P \approx 5$ fF, ν_c equals 184 GHz, and at frequency $\nu \approx 100$ GHz the estimated sensitivity is $R_I \approx 15$ A/W ($n = 1$).

At $R_D \approx 6.5$ kΩ and $n = 1.6$ [22] the zero-bias SBD internal responsivity can be estimated as $R_{I,int} \approx 11.7$ A/W giving for internal voltage responsivity $R_{V,int} \approx R_{\mathrm{I,int}} \bullet R_D \approx 75 \bullet 10^3$ V/W. For comparison in one of the investigated FETs (*Si*-MOSFET, 1.0-μm design rules) the estimated internal responsivity is several times higher: $R_{V,int} \approx 240 \times 10^3$ V/W and $Z_{GS,int} \approx 5.7(1\text{–}j) \times 10^3\ \Omega$ at $V_{GS} = 0.6$ V.

For high-frequency SBDs, several characteristic frequencies are mentioned (see e.g. [10]), and they all are called as the "cut-off" frequency.

(i) The shunt cutoff frequency ν_{shunt} is the frequency at which the real part of the intrinsic diode impedance drops below the magnitude of the imaginary part. When $R_D \gg R_S$, as it is true for most of zero-bias SBDs, $\nu_{shunt} \approx (2\pi R_D C_P)^{-1}$, which is equal to $\nu_{shunt} \approx 11$ GHz for above pointed parameters.
(ii) The series cut-off frequency, well known in Schottky diode mixers and multiplier technology, is the frequency at which the real part of the junction impedance equals the series resistance and is given by $\nu_{series} \approx (2\pi R_S C_P)^{-1} \approx 3.2$ THz.
(iii) The cut-off frequency given by Eq. 45 is $\nu_c \approx 184$ GHz.

As a rule, in papers (where the cut-off frequency exceeds ~200 GHz) frequently pointed out for cut-off frequency is the upper frequency ν_{series}, but for estimations of the frequency dependence for sensitivity, the lower frequency ν_c should be used. It is worth to emphasize that the cut-off frequencies are the functions of SBD detector resistance and because of it they depends on bias.

4 NEP of FET and SBD Detectors

It is important to find the upper limit of *NEP* at a given frequency and its radiation frequency dependence for both FET and SBD detectors and try to make a conclusions which one is better e.g. for vision system applications.

The detector noise equivalent power *NEP* can be determined as $NEP = N/R_V$, where *N* is the noise level. One should discriminate the electrical, measured and optical responsivities which, in turn, define the electrical (NEP^{el}), measured (NEP^{meas}) and optical (NEP^{opt}) noise equivalent power, respectively, that are calculated from the noise *N* and responsivity R_V levels.

The electrical responsivity R_V^{el} characterizes the detector itself, the optical responsivity R_V^{opt} characterizes the detector + antenna circuit, and the measured responsivity R_V^{meas} characterizes the detector + antenna + method of measurement (e.g. lenses used arrangement). From these parameters different *NEP* values can be estimated. Thus, for different devices similar quantities should be compared (R_V^{el} with R_V^{el}, but not R_V^{el} with R_V^{opt}). The same refers to NEP^{el} and NEP^{opt}. The quantity R_V^{el} is the maximum responsivity that can be gained at optimal detector matching with antenna and optical system.

Accounting for the case of no voltage applied to FET source-drain contacts, the noise is only the thermal one [37]. For FET electrical $NEP^{FET,el}$, one has

$$NEP_V^{FET,el} = (4k_BTR_{CH})^{1/2}/R_V^{FET,el}, \tag{46}$$

The similar expression can be written for SBD detector with changing the channel resistance R_{CH} for FET by SBD resistance R_D and SBD electrical responsivity.

In high radiation frequency ν limit (where $\nu > \nu_{c3}$), from Eqs. 24 and 25 follows $R_V^{FET,el} \sim \nu^{-2}$. In the case of weak frequency ν dependence for $\frac{\partial^2 I_D^{FET}}{\partial V_{GS}\partial V_{DS}}$, R_S, Z_{ant}, one has from Eq. 46:

$$NEP_V^{FET,el} \sim \nu^2. \tag{47}$$

Taking into account that $P_{ant,max} \sim \nu^{-2}$ (when antenna gain *G* is not dependent sufficiently on radiation frequency, e.g., when only wire or printed antennas without lenses or focusing optics are used), the optical $NEP^{FET,meas}$ should have the following dependence on the radiation frequency ν

$$NEP_V^{FET,meas} \sim \nu^4. \tag{48}$$

For typical *Si*-FET parameters, the estimations for gain $G \sim 1$ with $R_{V,int} = 2.4 \times 10^5$ V/W at $V_{GS} \approx 0.6$ V (see Fig. 8) give $R_{\mathrm{V}}^{\mathrm{meas}} \sim 570$ V/W. Thus, for only thermal noise *N*th observed in *Si*-FETs [37] at $V_{DS} = 0$ and $\nu \sim 77$ GHz, $NEP^{FET,meas}$ is $\sim 2.4 \times 10^{-10}$ W/Hz$^{1/2}$, that corresponds to experimental *NEP* estimations (at $R_{CH} \sim 1.1$ MΩ, $V_{\mathrm{GS}} \sim 0.6$ V and $T \approx 300$ K, $N_{th} = (4k_BTR_{CH}\Delta f)^{1/2} \approx 1.3 \times 10^{-7}$ V/Hz$^{1/2}$, $\Delta f = 1$ Hz) and $NEP^{FET,el} \sim 5.8 \times 10^{-12}$ W/Hz$^{1/2}$. In the case when $Z_{ant} = (100 - j100)$ Ω, the responsivity R_V^{meas} is 4.1×10^3 V/W, and thus $NEP^{FET,meas}$ is $\sim 3.3 \times 10^{-11}$ W/Hz$^{1/2}$.

In the case of ideal antenna matching with the FET η_{ant} equals to 1 ($Z_{GS} = Z*_{ant}$). Then $NEP^{FET,opt}$ (ideal) equals to $NEP^{FET,el} \approx 5.8 \times 10^{-12}$ W/Hz$^{1/2}$. Optimizing η and gain G in some radiation frequency range, one can improve (decrease) $NEP^{FET,opt}$ by several times. But still these uncooled FET detectors and arrays seem hardly to be used in direct detection passive vision systems (but not preamplified direct detection systems having a lot of additional elements such as Dicke switches, one or several high gain low-noise amplifiers (LNAs), demodulators, etc. that sufficiently raise the imaging system and the overall system cost). These preamplified direct detection systems are operating, as a rule, at $\nu < \sim 100$ GHz allowing passive vision regime using Dicke switches, one or several LNAs, demodulators, etc. that allows to decrease the uncooled system *NEP* to $NEP^{sys} \sim 10^{-13}$ W/Hz$^{1/2}$. But FET detectors studied up to now ($NEP^{det} > \sim 10^{-11}$ W/Hz$^{1/2}$) have not sufficiently low *NEP* (mostly due to parasitic influence and noises) needed even for broad-band ($\Delta\nu/\nu > \sim 0.1$) mm-wave/THz passive vision systems for diffraction limited beams.

For SBDs in R_V and *NEP* ultimate performance estimations, also only thermal noise is taken into account, which defines the upper values of *NEP*, though e.g. 1/*f* noise in a broad frequency range under forward bias can change the spectral density noise level within the amplitude modulation frequency range by several orders of magnitude (see, e.g. [30–32, 61, 62]) and is, as a rule, important for estimations of these characteristics. For example, in [65] the good figure of $NEP = 33$ pW/Hz$^{1/2}$ at the radiation frequency $\nu = 280$ GHz was obtained for the amplitude modulation frequency $f = 1$ MHz. 1/*f*-noise can play an important role in the dynamic operation of SBD detectors due to self-biasing. In the presence of 1/*f*-noise, the *NEP* not only depends on the bias level but also on the video bandwidth (output measurement).

For SBDs, the internal voltage responsivity $R_{V,int} = R_{I,int} \bullet R_D \approx 16$ A/W•3•10^3 Ω $\approx 4.8 \times 10^4$ V/W ($n \approx 1.2$). The coefficients $\eta = \frac{1}{(1+R_S/R_D)\cdot(1+(\nu/\nu_c)^2)} = 0.85$, $\eta_a = \frac{P}{P_{\max}} = 4\frac{\mathrm{Re}Z_{SBD}\mathrm{Re}Z_{ant}}{|Z_{SBD}+Z_{ant}|^2} \approx 0.093$. Then, $NEP^{SBD,opt} = (4k_{\mathrm{B}}TR_D\Delta f)^{1/2}/R_V^{opt} \approx 1.85 \times 10^{-12}W/Hz^{1/2}$. For ideal antenna matching $NEP^{SBD,opt}$(ideal) $= NEP^{SBD,el} = (4k_{\mathrm{B}}TR_D\Delta f)^{1/2}/R_V^{el} = 1.7 \times 10^{-13}$ W/Hz$^{1/2}$ at $\nu \approx 77$ GHz, ($R_V^{opt} = R_{V,int} \cdot \eta \cdot \eta_a = 3810 \cdot$ V/W).

To do meaningful comparison of some parameters, the formulas should be similar or their form should be "identical" for different devices. However, for FET we can write $R_V^{FET,el} \approx A^{FET} \cdot (\omega)^{-1/2} \cdot [1 + (\omega/\omega_{c3})^{3/2}]^{-1}$, and for SBD $R_V^{SBD,el} \approx A^{SBD} \cdot [1 + (\omega/\omega_c)^2]^{-1}$, where A^{FET}, A^{SBD} are some frequency independent constants. For SBD, the electrical responsivity is almost constant for $\omega < \omega_c$ and then decreases as ω^{-2}. For FET, the electrical responsivity decreases as $\omega^{-1/2}$ at frequency region $\omega < \omega_{c3}$, and for frequencies $\omega > \omega_{c3}$ it decreases as ω^2. So, direct comparison of ω_c and ω_{c3} is meaningless because they enter to different formulas. In the higher frequency case, when $\omega > \omega_{c3}$ and $\omega > \omega_c$, the formulas can be written in a similar manner $R_V^{el,FET} \approx A^{FET} \frac{\omega_{c3}^{3/2}}{\omega^2}$ and $R_V^{el,SBD} \approx A^{SBD} \frac{\omega_c^2}{\omega^2}$. However, for 0.35-μm FET design rule technology with $W/L = 1/1$ μm, it will be valid only for $\nu > 600$ GHz. Thus, for SBD and FET direct detection detectors, the values of $A^{FET}\omega_{c3}^{3/2}$ and $A^{SBD}\omega_c^2$ can be compared. The formulas pointed out can be rewritten in the following general form:

$$R_V^{el} \approx \frac{1}{C_p^2 R_S} R_{I0} \omega^{-2}, \tag{49}$$

where $R_{I0} = R_{I,\text{int}}$ for SBD and $R_{I0} = R_{CH} \frac{\partial^2 I_{DS}}{\partial V_{DS} \partial V_{GS}}\Big|_{V_{DS}=0}$ for FET. They have rather similar values, $R_{I0} \approx 10\ldots19$ A/W for SBD detectors (depending on the ideality factor n), and $R_{I0} \approx 10\ldots40$ A/W for FET detectors (depending on the subthreshold turn-on swing factor n), and C_P, R_S have their conventional meaning.

Which detector is better (for frequencies larger than cut-off frequency, for typical values of parameters at $\nu > 400$ GHz) depends mainly on the factor $P = (C_p^2 \cdot R_S)^{-1}$. For parameters mentioned for studied FET detectors P is about 3.1×10^{26} F^{-2}•Ω^{-1}. For improved "optimized" FET, the parameter P is about 6.7×10^{28} F^{-2} Ω^{-1}, and for SBD detector used for estimations $P \approx 4 \times 10^{27}$ F^{-2} Ω^{-1}. So, improved "optimized" FET in high radiation frequency range can feature better performance.

The results of NEP^{el} calculation are shown in Fig. 9. The performance increases with an advance of technology due to the thinner oxide at better technology and thus, the lower channel resistance at the same level of parasitics.

The results for NEP^{opt}, that can be compared with experimental data, assuming constant $Z_{\text{ant}} = (100 - j100)$ Ω for both types of detectors, are shown in Fig. 10.

As one can see the model used for SBD detectors describes well the known experimental data for NEP^{opt}. However, for FET ones rather strong scatter in experimental data is seen. The main reason seems to be the fact that FET technologies are non-optimized for FETs as detectors for THz/sub-THz radiation. One can also point out that with improving technology for MOSFET design and manufacture the FETs performance in high radiation frequency range of 1 THz or a little bit lower should exceed SBD ones. Comparing Figs. 9 and 10 one can point out the appreciable difference between NEP^{el} and NEP^{opt} that is due to the

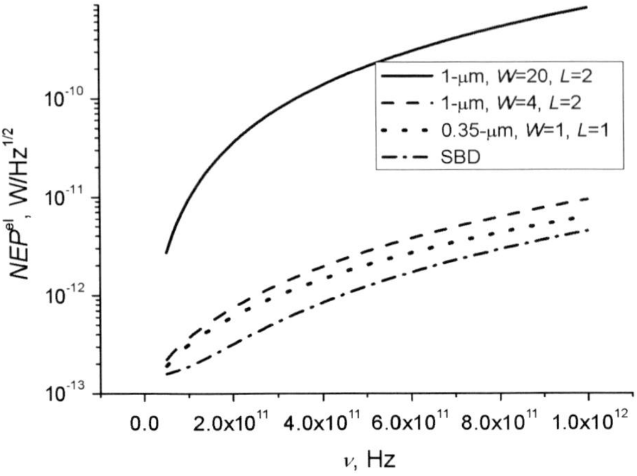

Fig. 9 Comparison of calculated electrical NEP^{el} for different devices. MOSFET *NEP* improvement performance when going from 1-μm technology ($W = 20$ μm, $L = 2$ μm) to 0.35-μm technology ($W = 1$ μm, $L = 1$ μm) is seen

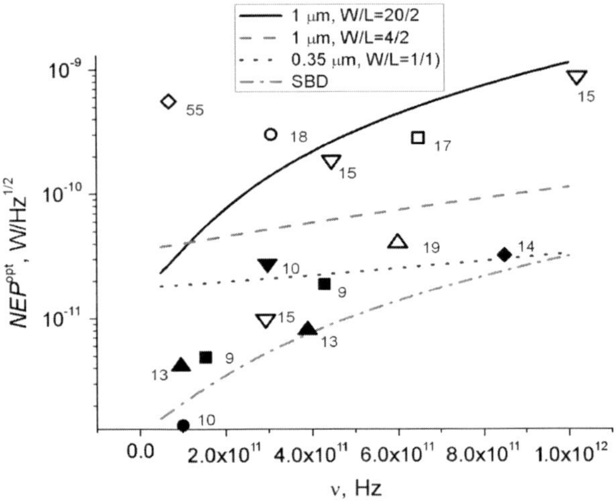

Fig. 10 Comparison of calculated optical NEP^{opt} for a few devices with the antenna impedance $Z_{ant} = (100 - j \cdot 100)$ Ω. Open marks are data for silicon FET detectors and filled marks are for SBD detectors. Numbers at experimental data signs mean the numbers of Refs. Data of [55] were obtained without antennas use

influence of parasitics in both types of detectors that suppress the *NEP* values by about an order.

Analyzing data on *Si*-FET used as mm-wave/THz detectors and also different SBD detectors, it is difficult to make a conclusion about their advantages over each other as detectors of mm-wave/THz radiation due to lack of standardized measurement procedures and interpretation of measured data (which can lead to comparison e.g. of electrical with optical responsivity).

Zero-bias SBD direct detection detectors based on III-V ternary semiconductor compounds are well developed and perhaps reached their limit *NEP* $\sim 10^{-11}....10^{-12}$ W/Hz$^{1/2}$ within the frequency range of $v \sim 100....500$ GHz (see Table 1). They seem to be preferable now as single detectors over FET ones in spectroscopic or preamplified direct detection systems.

Mm-wave/THz silicon FET direct detection detectors now have approximately an order of magnitude worse *NEP* $\sim 10^{-10}....10^{-11}$ W/Hz$^{1/2}$. These still worse *NEP* values seem to be mainly because of non-optimized FET impedance matching with antennas ($Z_{FET} \gg Z_{ant}$) and not optimized for high frequency operation (e.g., no multifinger design) FETs earlier designed frequently for other aims and not for detectors. If these issues are solved, then FETs should have better performance.

5 Conclusions

The analysis of silicon FET and SBD detectors at zero drain-source ($V_{DS} = 0$) or SBD zero bias ($V_D = 0$) was considered taking into account both the drift and diffusion current components for *Si*-MOSFETs. It is shown that the FET and SBD detectors performance is mainly limited by parasitic effects though only active and capacitive ones were taken into account. The considerations have shown that with progress in FET technology the FET detector performance can be improved with advancing of the design rules due to lowering the parasitic effects.

In spite of the fact that at the moment *NEP* of *Si*-MOSFET THz/sub-THz detectors in the range of radiation frequencies considered are about an order of magnitude worse compared to SBD ones, due to more advanced technologies applied in *Si*-CMOS area, these detectors can have some benefits because of well-developed silicon technologies and the possibility of proper integrated detector design and fabrication compared to technologies of III-V ternary compounds applied to SBD detectors. Both of these detectors can be applied mostly for active imaging at frequencies $\nu > 100$ GHz.

References

1. Rogalski, A., Sizov, F.: Terahertz detectors and focal plane arrays. Opto-Electron. Rev. **19**, 346–404 (2011)
2. Miller, A.J., Luukanen, A., Grossman, E.N.: Micromachined antenna-coupled uncooled microbolometers for terahertz imaging arrays. Proc. SPIE **5411**, 8–24 (2004)
3. Huhn, A.K., Spickermann, G., Ihring, A., et al.: Uncooled antenna-coupled THz detectors with 22 μs response time based on BiSb/Sb thermocouples. Appl. Phys. Lett. **102**, 121102 (2013)

4. Bolduc, M., Terroux, M., Marchese, L., et al.: THz imaging and radiometric measurements using a microbolometer-based camera. In: IRMMW-THz IEEE 36th International Conference, Houston, July 2011
5. Nguyen, D.-T., Simoens, F., Ouvrier-Buffet, J.-L., et al.: NIKON 2012. In: 19th International Conference on Microwaves Radar Wireless Communications, pp. 116–121. Warsaw, May 2012
6. Grossman, E.N., Miller, A.J.: Active millimeter-wave imaging for concealed weapons detection. Proc. SPIE **5077**, 62–70 (2003)
7. Kasalynas, I., Adam, A.J.L., Klaassen, T.O., et al.: Design and performance of a room-temperature terahertz detection array for real-time imaging. IEEE J. Sel. Top. Quant. Electron. **14**, 363–369 (2008)
8. Lee, A.W.M., Williams, B.S., Kumar, S., et al.: Real-time imaging using a 4.3-THz quantum cascade laser and a 320x240 microbolometer focal-plane array. IEEE Photonics Technol. Lett. **18**, 1415–1417 (2006)
9. Liu, L., Hessler, J.L., Hu, H., et al.: A broadband quasi-optical terahertz detector utilizing a zero-bias Schottky diode. IEEE Microwave Wirel. Compon. Lett. **20**, 504–506 (2010)
10. Brown, E.R., Young, A.C., Zimmerman, J.D., Gossard, A.C.: Advances in Schottky rectifier performance. IEEE Microwave Mag. **8**(3), 54–59 (2007)
11. Semenov, A., Cojocari, O., Hubers, H.-W., et al.: Application of zero-bias quasi-optical Schottky-diode detectors for monitoring short-pulse and weak terahertz radiation. IEEE Electron Device Lett. **31**, 674–676 (2010)
12. Schoenherr, D., Bleasdale, C., Goebel, T. et al.: Extremely broadband characterization of a Schottky diode based THz detector. In: 35th International Conference on Infrared and Millimeter Waves—IRMMW-THz, pp. 1–2. IEEE, 2010
13. Chahal, P., Morris, F., Frazier, G.: Zero bias resonant tunnel Schottky contact diode for wide-band direct detection. IEEE Electron Device Lett. **26**, 894–896 (2005)
14. Han, R., Zhang, Y., Kim, Y. et al.: 280 GHz and 860 GHz image sensors using Schottky-barrier diodes in 0.13 μm digital CMOS. In: 2012 IEEE International Solid State Circuits Conference, pp. 254–256. IEEE, 2012
15. Schuster, F., Coquillat, D., Videlier, H., et al.: Broadband terahertz imaging with highly sensitive silicon CMOS detectors. Opt. Express **19**, 7827–7832 (2011)
16. Hadi, R.A., Sherry, H., Grzyb, J., et al.: A Broadband 0.6 to 1 THz CMOS imaging detector with and integrated lens. In: 2011 36th International Conference on Infrared and Millimeter Waves IRMMW-THz, pp. 1–4. IEEE, 2011
17. Öjefors, E., Pfeiffer, U.R., Lisauskas, A., Roskos, H.G.: A 0.65 THz focal-plane array in a quarter-micron cmos process technology. IEEE J. Solid-State Circuits **44**, 1968–1976 (2009)
18. Pleteršek, A., Trontelj, J.: A self-mixing n-MOS channel-detector optimized for mm-wave and THz signals. J. Infrared Millimeter Terahertz Waves **33**, 615–626 (2012)
19. Boppel, S., Lisauskas, A., Mundt, M., et al.: CMOS integrated antenna-coupled field-effect transistors for the detection of radiation from 0.2 to 4.3 THz. IEEE Trans. Microwave Theory Tech. **60**, 3834–3843 (2012)
20. Sizov, F., Petriakov, V., Zabudsky, V., et al.: Millimeter-wave hybrid un-cooled narrow-gap hot-carrier and Schottky diodes direct detectors. Appl. Phys. Lett. **101**, 082108 (2012)
21. Knap, W., Dyakonov, M., Coquillat, D., et al.: Field effect transistors for terahertz detection: physics and first imaging applications. J. Infrared Millimeter Terahertz Waves **30**, 1319–1337 (2009)
22. Kazemi, H., Nagy, G., Tran, L., et al.: Ultra sensitive ErAs/InAlGaAs direct detectors for millimeter wave and THz imaging applications. In: IEEE/MTT-S International Microwave Symposium, pp. 1367–1370. Honolulu, Jun 2007
23. Liu, L., Hessler, J.L., Hu, H., et al.: A broadband quasi-optical terahertz detector utilizing a zero-bias Schottky diode. IEEE Microwave Wirel. Compon. Lett. **20**, 504–506 (2010)
24. Dyakonov, M.I., Shur, M.S.: Plasma wave electronics: novel terahertz devices using two dimensional electron fluid. IEEE Trans. Electron Devices **43**, 1640–1645 (1996)

25. Pu, L.-J., Tsividis, Y.P.: Harmonic distortion of the four-terminal MOSFET in non-quasistatic operation. IEE Proc. G (Circuits, Devices Syst) **137**, 325–332 (1990)
26. Kenneth, K.O.: Sub-millimeter wave CMOS integrated circuits and systems. In: RFIT2011-IEEE International Symposium on Radio-Frequency Integration Technology, pp. 1–8. Beijing, Nov 2011
27. Chen, Z., Wang, C.C., Yao, H.C., Heydari, P.: A BiCMOS W-Band 2 × 2 focal-plane array with on-chip antenna. IEEE J. Solid-Stat e Circuits **47**, 2355–2371 (2012)
28. Gu, Q.J., Xu, Zh., Jian, H.Yu., Chang, M.F.: A CMOS fully differential W-band passive imager with <2 K NETD. In: Radio Frequency Integrated Circuit IEEE Symposium: RFIC, pp. 1–4 (2011)
29. Chew, W., Fetterman, H.R.: Millimeter-wave imaging using FET detectors integrated with printed circuit antennas. Int. J. Infrared Millimeter Waves **10**, 565–578 (1989)
30. Cowley, A.M., Sorensen, H.O.: Quantitative comparison of solid-state microwave detectors. IEEE Trans. Microwave Theory Tech. **14**, 588–602 (1966)
31. Brown, E.R.: A system-level analysis of Schottky diodes for incoherent THz imaging arrays. Solid State Electron. **48**, S2051–S2053 (2004)
32. Yu, C., Wu, C.-L., Kshattry, S., et al.: Compact, high impedance and wide bandwidth detectors for characterization of millimeter wave performance. IEEE J. Solid-State Circuits **47**, 2335–2343 (2012)
33. Tomkins, A., Garcia, P., Voinigescu, S.P.: A passive W-band imaging receiver in 65-nm bulk CMOS. IEEE J. Solid-State Circuits **45**, 1981–1991 (2010)
34. Gilreath, L., Jain, V., Heydari, P.: Design and analysis of a W-Band SiGe direct-detection-based passive imaging receiver. IEEE J. Solid-State Circuits **46**, 2240–2252 (2011)
35. Sizov, F., Rogalski, A.: THz detectors. Progr. Quantum Electron. **34**, 278–347 (2010)
36. Tsividis, Y., McAndrew, C.: Operation and modeling of the MOS transistor. Oxford University Press, New York (2011)
37. Tauk, R., Teppe, F., Boubanga, S., et al.: Plasma wave detection of terahertz radiation by silicon field effects transistors: responsivity and noise equivalent power. Appl. Phys. Lett. **89**, 253511 (2006)
38. Knap, W., Coquillat, D., Dyakonov, M., et al.: Field effect transistors for terahertz detection: physics and first imaging applications. J. Infrared Millimeter Terahertz Waves **30**, 1319–1337 (2009)
39. Turchetti, C., Mancini, P., Masetti, G.: A CAD-oriented non-quasi-static approach for the transient analysis of MOS ICs. IEEE J. Solid-State Circuits **21**, 827–883 (1986)
40. Mancini, P., Turchetti, C., Masetti, G.: A non-quasi-static analysis of the transient behavior of the long-channel MOST valid in all regions of operation. IEEE Trans. Electron Devices **34**, 325–334 (1987)
41. Pozar, D.M.: Microwave Engineering. Wiley, USA (2011)
42. Lisauskas, A., Pfeiffer, U., Öjefors, E., et al.: Rational design of high-responsivity detectors of terahertz radiation based on distributed self-mixing in silicon field-effect transistors. J. Appl. Phys. **105**, 114511 (2009)
43. Pao, H.C., Sah, C.T.: Effects of diffusion current on characteristics of metal-oxide (insulator)-semiconductor transistors. Solid State Electron. **9**, 927–937 (1966)
44. Tsividis, Y., Suyama, K., Vavelidis, K.: Simple "reconciliation" MOSFET model valid in all regions. Electron. Lett. **31**, 506–508 (1995)
45. Liu, W.: MOSFET models for SPICE simulation, including BSIM3v3 and BSIM4. Wiley, New York (2001)
46. Gildenblat, G.: Compact Modeling: Principles, Techniques and Applications. Springer, Netherlands (2010)
47. Sakowicz, M., Lifshits, M.B., Klimenko, O.A., et al.: Terahertz responsivity of field effect transistors versus their static channel conductivity and loading effects. J. Appl. Phys. **110**, 054512 (2011)
48. Preu, S., Kim, S., Verma, R., et al.: An improved model for non-resonant terahertz detection in field-effect transistors. J. Appl. Phys. **111**, 024502 (2012)

49. Gutin, A., Kachorovski, V., Muraviev, A., Shur, M.: Plasmonic terahertz detector response at high intensities. J. Appl. Phys. **112**, 014508 (2012)
50. Knap, W., Kachorovski, V., Deng, Y., et al.: Nonresonant detection of terahertz radiation in field effect transistors. J. Appl. Phys. **91**, 9346–9353 (2002)
51. Lisauskas, A., Glaab, D., Roskos, H.G., et al.: Terahertz imaging with Si MOSFET focal-plane arrays. Proc. SPIE **7215**, 72150J (2009)
52. Lisauskas, A., Pfeiffer, U., Öjefors, E., et al.: Rational design of high-responsivity detectors of terahertz radiation based on distributed self-mixing in silicon field-effect transistors. J. Appl. Phys. **105**, 114511 (2009)
53. Ratni, M., Huyart, B., Bergeaut, E., Jallet, L.: RF power detector using a silicon a silicon MOSFET, IEEE MTTT-Symposium Digest, pp. 1139–1142 (1998)
54. Enz, C., Cheng, Y.: MOS transistor modeling for RF IC design. IEEE J. Solid-State Circuits **35**, 186–201 (2000)
55. Sizov, F., Golenkov, A., But, D., et al.: Sub-THz room-temperature sensitivity of long channel field effect transistors. Opto-Electron. Rev. **20**, 194–199 (2012)
56. Balanis, C.A.: Antenna theory analysis and design. Wiley, New Jersey (2005)
57. Volakis, H.L.: Antenna Engineering Handbook. McGraw–Hill, New York (2007)
58. Sze, S.M., Ng, K.K.: Physics of Semiconductor Devices. Wiley-Interscience, New York (2006)
59. Shepherd, F.D.: Infrared Detectors: State of the Art. Proc. SPIE **1735**, 250–261 (1992)
60. Myburg, G., Auret, F.D., Meyer, W.E., et al.: Summary of Schottky barrier height data on epitaxially grown n- and p-GaAs. Thin Solid Films **325**, 181–186 (1998)
61. Lynch, J.J., Moyer, H.P., Schaffner, J.H., et al.: Passive millimeter-wave imaging module with preamplified zero-bias detection. IEEE Trans. Microwave Theory Tech. **56**, 1592–1600 (2008)
62. Hesler, J.L., Crowe, T.W.: Responsivity and noise measurements of zero-bias Schottky diode detectors. In: 2007 18th International Symposium on Space Terahertz Technology, pp. 89–92. Pasadena, CA, March (2007)
63. Shashkin, V.I., Murel, A.V., Daniltsev, V.M., Khrykin, O.I.: Control of charge transport mode in the Schottky barrier by δ-doping: Calculation and experiment for Al/GaAs. Semiconductors **36**, 505–510 (2002)
64. Hesler, J.L., Liu, L., Xu, H., et al.: The development of quasi-optical THz detectors. In: IRMMW-THz 2008 33rd International Conference Infrared Millimeter Terahertz Waves, Pasadena, U.S. 15–19 Sept 2008
65. Han, R., Zhang, Y., Coquillat, D., et al.: A 280-GHz Schottky diode detector in 130-nm digital CMOS. IEEE J. Solid-State Circuits **46**, 2602–2611 (2011)

Structural and Luminescent Properties of Carbonized Silicon Oxide Thin Layers

Andrii V. Vasin

Abstract A critical review of structural and white light emission properties of carbonized silicon oxide thin layers (SiOC) is presented. Methods of synthesis of the SiOC layers, effects of thermal treatments, spectral properties and degradation of photoluminescence are analyzed. SiOC layers were synthesized by two methods: (1) magnetron deposition of a-SiC:H thin films on Si wafer followed by oxidation in flow oxygen or water vapor at temperature of 450–600 °C; (2) thermal treatment of porous silicon layer (grown on p-type Si wafer by anodization in HF/ C_2H_5OH solution) in flow of acetylene (850–1050 °C) followed by oxidation in flow of water vapor (600–850 °C). Working hypothesis is that main contribution in broad band photoluminescence (400–700 nm) of the materials originates from carbon nano-clusters. Hence, original experimental data and other published data are discussed in frame of this working hypothesis.

1 Introduction

Light emission of SiOC materials Light emitting materials are of great importance in modern optoelectronics, lighting, and light indication technologies. Development of white-light emitting materials with high luminous efficiency and good color properties attracts a particular interest in the field of artificial lighting. Large class of visible light emitting materials is phosphors that are used in white light sources for converting of ultraviolet (UV) or blue/violet narrow emission bands of plasma (luminescent lamps) or semiconductor (LEDs) irradiation into broad band of white light.

Currently, all commercially available white light emitting phosphors contain heavy metals including transition and rare-earth metals that are expensive and

A.V. Vasin (✉)
Lashkaryov Institute of Semiconductor Physics, Pr, Nauki 41,
Kyiv 03028, Ukraine
e-mail: av966@yahoo.com

A. Nazarov et al. (eds.), *Functional Nanomaterials and Devices for Electronics, Sensors and Energy Harvesting*, Engineering Materials,
DOI 10.1007/978-3-319-08804-4_14

toxic for human body. Hence, the development of new white light emitting materials free from heavy metals is an important task. During last decade, two alternatives to the phosphors based on metal activators were suggested and studied: (1) organic materials and (2) semiconductor nanoparticles ("quantum dots", QD). Despite the optimistic progress in the efficiency of organic white-light sources, thermal stability and long-term reliability issues of organic materials are still to be solved. QDs exhibit relatively narrow emission band, so the color-rendering index of these white light sources is low.

The general approach for obtaining white light with best spectral properties (color temperature, high color rendering index (CRI)) is a combination of materials with complementary spectral characteristics. However, single material phosphor converting blue-and/or-UV into white light is more desirable because application of multicolor phosphors reduces the light emission efficiency. Moreover, the emission spectrum of such multicomponent phosphors usually changes with operation time due to different degradation properties of different phosphor components.

One of the promising alternatives to modern phosphors is a new class of white-light emitting Si:O:C composite materials that does not contain heavy metal activators. Earliest reports on the white light emission of SiOC fabricated by sol-gel polymerization techniques are refereed to [1] and [2].

Recently, it was demonstrated that light-emitting SiOC material can be synthesized by variety of methods in form of thin layers: ion implantation of C^+ in SiO_2 thin layer [3–5], sol–gel-derived SiOC thin films [6], a-SiO(x)C(y):H thin films deposited by chemical vapor deposition techniques [7, 8], magnetron sputtering [9], a-SiO(x)C(y):H thin films synthesized by low temperature oxidation of carbon-rich a-Si(x)C(y):H thin films [10], porous layers por-SiO_2:C on silicon wafer [11–13]. Recently, it was demonstrated that the spectral properties of SiOC-based sol-gel thin films [6] and porous SiO_2:C layers [12, 13] can be tuned over the visible range by changing of preparation conditions. However, the origin of the photoluminescence is still a matter of discussion among the scientific community. The present report is a first attempt to summarize experimental data on structure and PL properties of light emitting SiOC material in form of thin layers.

Structure models of SiOC composites At the beginning, some definitions concerning SiOC ternary compound are necessary to introduce to avoid misunderstandings with terms. Carbon is insoluble in silicon oxide so that there is no thermodynamically stable solid state SiO_2:C solution (alloy) at room temperature. Also there are no known (at least to date) thermodynamically stable SiOC chemical compound with corresponding crystalline structure. Hence there are two possible structural forms of SiOC compound: (1)—phase separated SiO_2/C composites i.e. carbon particles dispersed in silicon oxide, and (2)—metastable SiOC alloys with mixed Si–O, Si–C and C–O bonds (silicon oxycarbide).

Silicon oxycarbide SiOC is a term used to denote the chemical structure in which silicon is simultaneously bonded with carbon and oxygen. These tetrahedral network species can be generally described as $[C_xSiO_{4-x}]$ where x-1, 2, or 3. The incorporation of carbon in silicate glasses presents the possibility of replacing some oxygen, which is only two-coordinated, with carbon which can be four-coordinated. This increased bonding per anion is expected to strengthen the molecular structure of the glass network, and thereby, to improve the thermal and mechanical properties. In fact, silicon oxycarbide amorphous ceramics have better mechanical and physical properties than silicon oxide. For example, silicon oxycarbide has a 40 % higher elastic modulus, 20 % higher hardness, and 160 degree higher glass transition temperature than silicon oxide [14, 15].

Incorporation of carbon in silica using high temperature processes is strongly limited due to oxidation, volatilization and decomposition as well as SiC nucleation that accompany high temperature melting. Alternative ways most commonly used for incorporation of carbon in silicon oxide are sol-gel polymerization processes, direct carbon ion implantation into silicon oxide thin layers and plasma deposition at low temperature (PECVD or magnetron sputtering).

Amorphous hydrogenated silicon oxycarbide thin films (a-SiOC:H) deposited on silicon substrates using radio frequency plasma enhanced chemical vapor deposition (PECVD) have electrically very low dielectric constants ($k \leq 2.8$) and are considered as low-k interlayer dielectrics [16, 17] to replace SiO_2 ($k = 4.1$) in modern integrated circuits with Cu metallization.

Incorporation of carbon in silicon oxide results in formation of Si–C bonding in form of $Si–CH_3$ with enhanced ionic polarizability and formation of nano-pores caused the film density to be reduced [18]. The structure of amorphous hydrogenated silicon oxycarbide is suggested to be similar to that of tetrahedral basic structure of SiO_2, but structural block now is SiO(x)C(y)H(z) tetrahons (Fig. 1a). Corresponding disordered structural network is illustrated in Fig. 1b. Silica has a molecular structure in which each Si atom is bonded to four oxygen atoms, and each oxygen atom to two silicon atoms ($SiO_{4/2}$). Amorphous SiOC:H network is suggested to be built of SiO(x)C(y)H(z) tetrahons.

Carbon–carbon bonding was demonstrated to be prevalent in these materials [19, 20], so that carbon in SiOC system tends to cluster. Despite the evidence for the presence of free carbon, direct identification of carbon segregation has proven

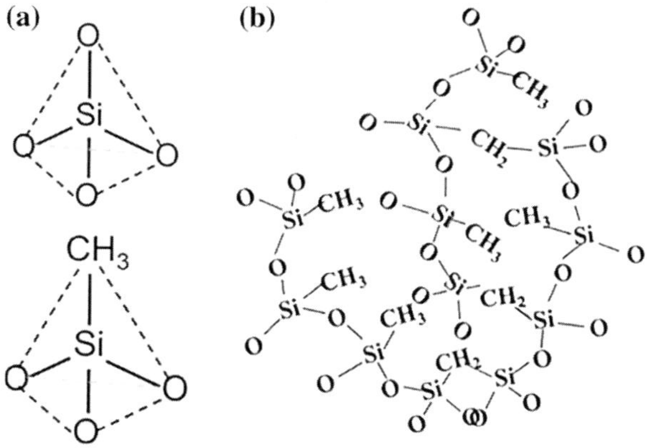

Fig. 1 **a** structural blocks in SiO_2 and SiOC:H; **b** structural model of a-SiOC:H amorphous network

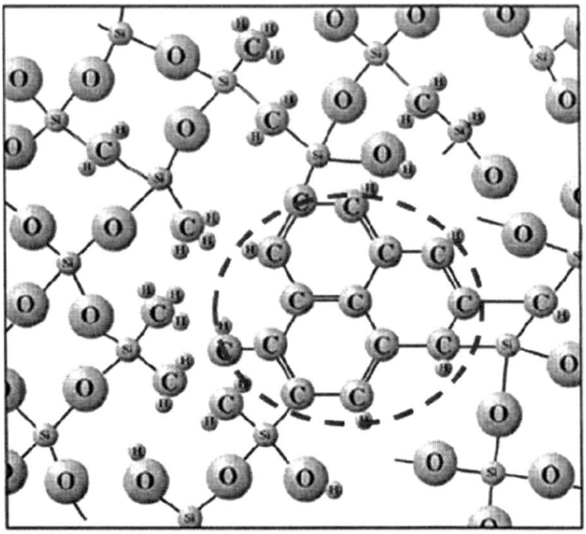

Fig. 2 Graphene-like structural realization of carbon precipitate imbedded in a-SiOC network

very difficult due to extremely small size attributed to the domains (less than 3 nm) and highly disordered local structure [20–22]. Even energy-filtered electron scattering analysis of silicon oxycarbide materials has shown only limited segregation of carbon within the structure of the material [21]. Topology of carbon precipitates have not been identified clearly. It was proposed that the carbon–carbon bonding takes the form of roughly spherical nano-domains, dispersed graphene sheet segments [23] (see Fig. 2), or interconnected net of free carbon phase [24].

2 SiOC Thin Films Fabricated by Oxidation of $Si_{1-x}C_x$:H Films

Amorphous silicon-carbon alloy thin films (a-$Si_{1-x}C_x$:H) is an excellent solid-state precursor for fabrication of SiOC layers by oxidation treatment. Bulk silicon carbide is well known as material extremely resistant to oxidation but low density amorphous hydrogenated silicon-carbon alloy films deposited at low substrate temperature can be easily oxidized by oxygen at temperature below 700 °C. Deposition of a-$Si_{1-x}C_x$:H was performed by two variations of magnetron sputtering technique: direct current (DC) magnetron sputtering of crystalline silicon target and radio frequency (RF) magnetron sputtering of SiC polycrystalline target in argon/methane gas mixture. These methods are quite suitable for variation of density and composition of the films by varying of deposition conditions. Deposition by RF-magnetron sputtering of SiC polycrystalline target in pure argon results in formation of dense and corrosion resistant near-stochiometric a-SiC layers while adding of methane to argon strongly reduces film density and oxidation resistance [25]. Thus we can easily vary density and carbon incorporation varying partial pressure of methane, total working pressure and discharge power.

2.1 *Thermal Treatments of a-$Si_{1-x}C_x$:H Films Deposited by DC-Magnetron Sputtering*

The series of near-stochiometric (x = 0.5) and carbon-rich (x = 0.7) silicon carbon alloy films have been deposited by DC-magnetron sputtering and studied to analyze the effect of carbon-enrichment on oxidation effects. After the deposition the samples were annealed at atmospheric pressure in the flow of dry Ar, wet-Ar and dry O_2 flow at 450 °C for 30 min.

2.1.1 PL Properties of Annealed and Oxidized Samples

a-$Si_{0.5}C_{0.5}$:H series Weak blue PL band centered at about 420 nm is detected in as-deposited near-stochiometric a-$Si_{0.5}C_{0.5}$:H film (Fig. 3a, spectrum 1). Annealing at 450 °C for 30 min in dry argon causes development of broad emission band with maximum intensity at about of 510 nm while blue shoulder remains unchanged (Fig. 3a, spectrum 2). Oxidation by dry wet argon and oxygen results in further development of "green" band (Fig. 3a, spectra 3 and 4) but integral intensity of the PL sample annealed in wet argon is five times larger (Fig. 3c). Narrow peaks at 700 and 727 nm present in the spectra are the laser satellites. Amplitude of these laser lines can be used for rough estimation of relative PL intensity in different samples.

a-$Si_{0.3}C_{0.7}$:H series As-deposited carbon-rich a-$Si_{0.3}C_{0.7}$:H film emits broad PL band with spectrum strongly modulated by interference effect and maximum intensity in the range of 500–550 nm (Fig. 3b, spectrum 1). PL intensity strongly increases after annealing in dry Ar while spectral shape of PL is almost unchanged (Fig. 3b, spectrum 2). Annealing in wet argon and dry oxygen results in further increase of PL intensity and blue shift of the intensity distribution (Fig. 3b, spectra 3 and 4). Integrated PL intensity of a-$Si_{0.5}C_{0.5}$:H series is presented in Fig. 3d.

2.1.2 Fourier-Transform Infrared Transmission Spectroscopy

a-$Si_{0.5}C_{0.5}$:H series Spectrum from as-deposited a-$Si_{0.5}C_{0.5}$:H (Fig. 4a) is composed by strong absorption bands at 780 cm^{-1} (Si–C stretching) and 1010 cm^{-1} (combination of C–H rocking/waging and Si–O stretching), and weaker absorption bands at 2100 cm^{-1} (Si–H_n stretching) and 2690–3000 cm^{-1} (C–H_n stretching). Traces of absorption are found at about 1260 cm^{-1} (bending vibration modes of

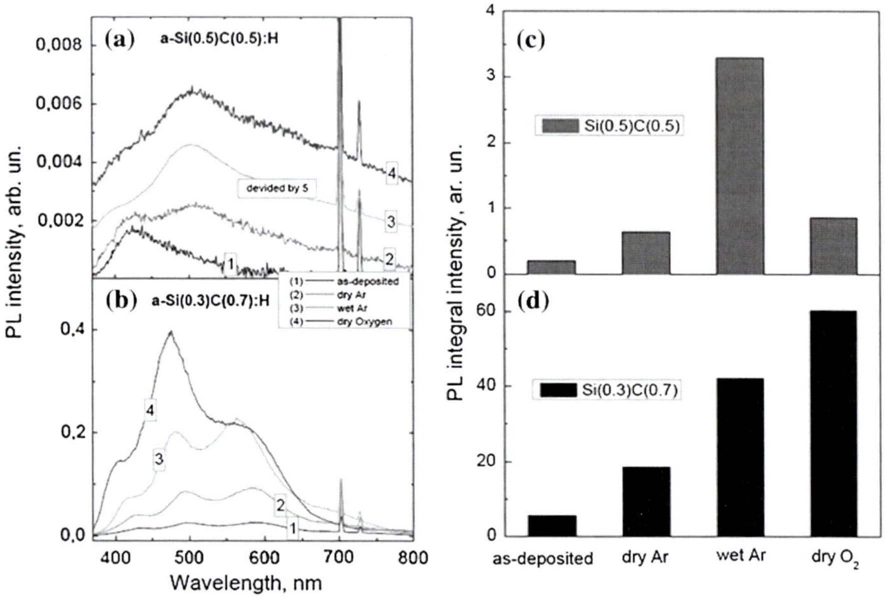

Fig. 3 **a**, **b** PL spectra of as-deposited (spectrum 1) and annealed films; **c**, **d** corresponding integrated PL intensities

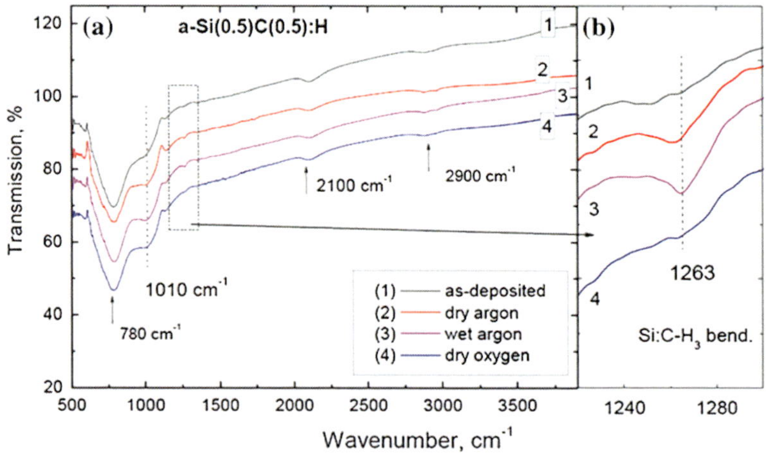

Fig. 4 FTIR spectra as-deposited and annealed of a-$Si_{0.5}H_{0.5}$:H series

Si:C–H_3 bonds) and in range of 3000–3500 cm^{-1} (O–H stretching). Weak narrow spectral features at 608 and 1100 cm^{-1} are due to imperfect subtraction of the absorption of Si substrate.

Absorption band at 1010 cm^{-1} in a-$Si_{1-x}C_x$:H films is commonly ascribed to rocking/waging vibration modes of CH_2 radicals attached to silicon atoms [26]. But taking into account significant residual oxygen contamination in our films (up to 7–10 at.%), the contribution of Si–O (x < 2) stretching vibrations in silicon suboxide (SiO_x, x < 2) structural fragments should be taken into account as well.

From Fig. 4a one can see that relative intensity of this band increases after thermal treatments regardless of annealing atmosphere (spectra 2, 3, 4). Annealing in dry and wet Ar results in appearance of Si:C–H_3 related band at 1260 cm^{-1} (Fig. 4b, spectra 2, 3) but this band is not detected after annealing in oxygen (Fig. 4b, spectrum 4). The intensity of absorption bands of Si–H_n (2100 cm^{-1}), C–H_n (2690–3000 cm^{-1}) and O–H (3000–3500 cm^{-1}) are decreased after annealing.

a-$Si_{0.3}C_{0.7}$:H series FTIR spectra of a-$Si_{0.3}C_{0.7}$:H films are shown in Fig. 5a. Relative intensity of 1010 cm^{-1} band as well as Si:C–H_3 bending (1256 cm^{-1}) and C–H_n stretching bands (2800–3000 cm^{-1}) are much stronger in carbon-rich sample than those in a-$Si_{0.5}C_{0.5}$:H, obviously indicating larger contribution of carbon-hydrogen bonds. The OH-related absorption band at 3500 cm^{-1} is also much stronger probably due to enhanced nanoporosity of carbon-rich structure.

Intensity of 1010 and 1256 cm^{-1} bands is increased after annealing in dry and wet argon. By comparing Fig. 5a and b one can find the direct correlation of intensity of these bands confirming hydrogen origin of 1010 cm^{-1} absorption band. Absorption intensity of Si–H_n and C–H_n stretching vibration bands decreases after annealing in dry and wet Ar, and completely vanishes after thermal treatment in oxygen. Obvious absorption at about 1700 cm^{-1} due to C = O bonds is observed after annealing in dry and wet argon. Only oxygen related bands are present in the spectrum of a-$Si_{0.3}C_{0.7}$:H after annealing in oxygen indicating strong oxidation of SiC matrix.

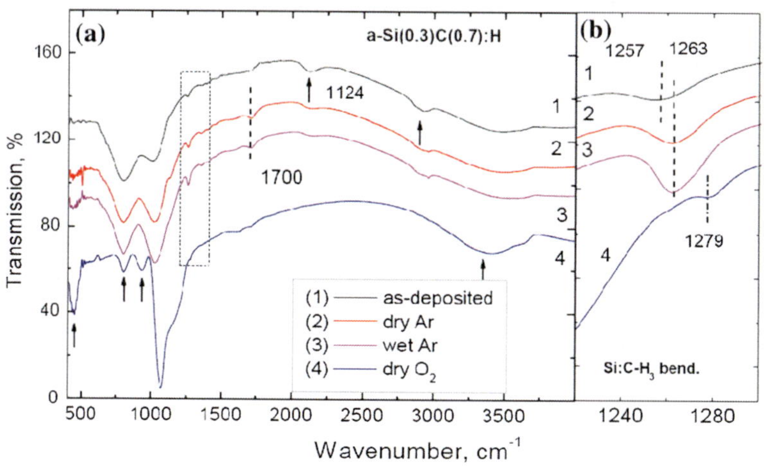

Fig. 5 FTIR spectra as-deposited and annealed of a-$Si_{0.3}H_{0.7}$:H series

2.1.3 EPR

Electron paramagnetic resonance measurements (EPR) were performed at room temperature in the X band employing 100 kHz field modulation. Single EPR-line with g-factor of 2.0026 (±0.0002) is detected in as deposited and annealed samples. Concentration of paramagnetic centers N_S in as deposited a-$Si_{0.5}C_{0.5}$:H film is as high as $2{\cdot}10^{20}$ cm^{-3} while N_S in as-deposited a-$Si_{0.3}C_{0.7}$:H sample is found to be several times smaller (Fig. 6). General tendencies of variation of spin concentration in a-$Si_{0.3}C_{0.7}$:H and a-$Si_{0.3}C_{0.7}$:H films after annealing are similar, i.e. gradual decrease in sequence (dry Ar) > (wet Ar) > (dry oxygen). No EPR signal is detected in $Si_{0.5}C_{0.5}$:H the samples after annealing in wet argon and dry oxygen probably due to small thickness of the film.

2.1.4 Presumed Nature of PL

Photoluminescence in amorphous nearstochiometric a-SiC:H films is commonly attributed (by analogy with a-Si:H) to tail-to-tail (TTT) recombination [26] i.e. radiative recombination of photo-excited electron-hole pairs from localized state in conduction band tail to localized state in valence band tail. Stocks shift of fundamental absorption edge and emission band should be observed in frame of such "classical" model. As-deposited a-$Si_{0.5}C_{0.5}$:H films exhibit blue light emission with maximum intensity at about 430 nm (2.9 eV).

Optical band gap ("Tauc" band gap [27]) of this film is about 2.5 eV, so that light emission cannot be attributed to TTT recombination in amorphous SiC matrix. It is reasonable to assign blue emission to optically active defects (oxygen deficiency centers, ODC) in SiO_x present in as-deposited films due to residual oxygen contaminations. Maximum of PL band appeared after annealing is at about

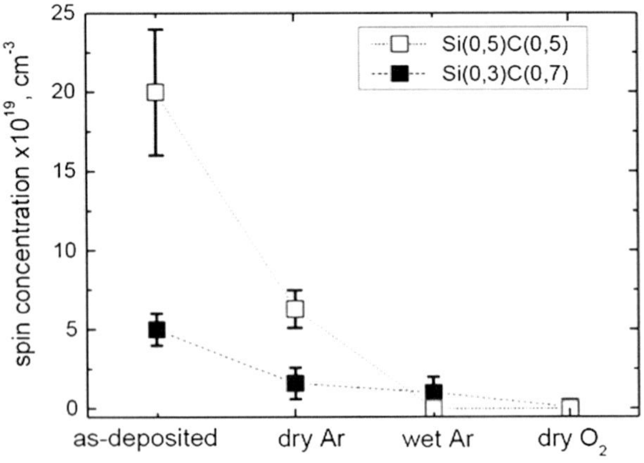

Fig. 6 Concentration of paramagnetic centers N_S in as-deposited and annealed a-$Si_{0.5}H_{0.5}$:H and a-$Si_{0.3}H_{0.7}$:H series

510 nm (2.4 eV) that is close to Tauc band gap of this sample, so that it can be assigned to SiC network providing Stocks shift is very small (in order of 0.1 eV).

Interference features in PL spectra of carbon-rich samples complicate the spectral properties analysis but some general assumptions are possible to suggest. Optical band gap in as-deposited carbon-rich films is about 2.7 eV and maximum of PL intensity is about 550 nm (2.3 eV), so that light emission can be assigned to TTT transitions in amorphous silicon–carbon network. Large Stocks shift of about 0.4 eV is quite common for disordered structures with strongly localized electron states. Annealing in dry and wet argon does not change PL spectrum so that increase of PL intensity is associated with decrease of the concentration of non-radiative recombination centers. After annealing in oxygen, PL band is obviously shifted to 460–500 nm (2.5–2.8 eV). XPS data (not presented here) and FTIR spectra indicate only Si–O bonding and no signs of Si–C in this sample so that light emission could hardly be assigned to TTT transition in amorphous SiC. For the first sight it is reasonable to assign blue light emission to defects in silicon oxide network. However, time resolved measurements show PL decay time of about 10 ns that is too short for oxygen deficiency defects in silicon oxide. Light emission in visible spectral range in silicon oxide is usually associated with recombination from triplet state to ground state ($T_1 \rightarrow S_0$) of ODC [28]. Such recombination (i.e. "phosphorescence") is natively slow process with characteristic decay time of micro- and milliseconds. To solve such discrepancies it is quite reasonable to assign light emission to carbon nano-clusters embedded in silicon oxide matrix. Recently, it was demonstrated that carbon nano-clusters with passivated surface exhibit strong visible luminescence with emission photon energy depending on the size of the cluster [29, 30].

From classical view point PL efficiency in amorphous semiconductors is mainly determined by competition of radiative recombination path through TTT and non-radiative recombination path through mid-gap defect states associated commonly with dangling bonds. In the a-$Si_{0.3}H_{0.7}$:H, PL intensity development after annealing directly correlates with decrease of the concentration of paramagnetic center (compare Figs. 3d and 6). But in case of a-$Si_{0.5}C_{0.5}$:H such correlation is not evident (compare Figs. 3c and 6). Comparing PL intensity and concentration of paramagnetic centers N_s of a-$Si_{0.5}C_{0.5}$:H sample annealed in dry argon (N_s is about $6 \cdot 10^{19}$ cm^{-3}) and as-deposited a-$Si_{0.3}H_{0.7}$:H sample (N_s is about $4 \cdot 10^{19}$ cm^{-3}) one can see that photoluminescence intensity differs by 10 times, while the concentration of paramagnetic centers is almost the same. From these observations it is obvious that concentration of paramagnetic defects as non-radiative recombination centers is not a single parameter determining PL efficiency in a-$Si_{1-x}C_x$:H. In [31] it has been suggested that enhancement of the PL in carbon-rich a-$Si_{1-x}C_x$:H films after vacuum annealing at 450 °C is associated not only with extra-passivation of paramagnetic centers but also with enhancement of localization of photo-excited electron-hole pairs due to increase of the concentration of C–H bonds.

2.2 *Thermal Treatment of a-$Si_{1-X}C_x$:H Films Deposited by RF-Magnetron Sputtering*

RF-sputtering of SiC target in argon-methane mixture is very suitable method for deposition of carbon-rich a-SiC:H films. In fact at high working pressure it is possible do deposit pure a-C:H films. If working pressure is high enough all silicon atoms sputtered from SiC target are scattered back and only plasma excited hydrocarbon radical participate in deposition on the substrate. In this case FTIR spectra indicate only C = C/C–H related absorption bands and no signs of absorption by Si–H or Si–C bonds.

Figure 7a represents PL spectra of a-SiC:H films (about 65 at.% of carbon) deposited in Si wafer before and after oxidation at 700 °C for 30 min. Measurements were performed using excitation by 480 nm argon laser radiation. Thickness of the film is as small as 180 nm so that there is almost no interference distortion of the spectra. The absence of interference distortion allows to observe low-frequency shift after oxidation. Quite similar shift is observed in the a-C:H after thermal treatments (Fig. 7b). a-C:H sample was deposited on glass and quartz substrates at high pressure of argon/methane gas mixture. Advantage of transparent substrate is minimizing of interference effects through minimizing reflection from substrate/film interface.

It is worth noting that PL intensity of as-deposited a-C:H films is much stronger than that of as-deposited a-SiC:H and can be well seen in sunny day (Fig. 8) under 40 mW radiation of semiconductor laser (370 nm). The main effect of thermal treatment is reduction of PL intensity by factor of about 2 and red-shift (about 30 nm) of PL band.

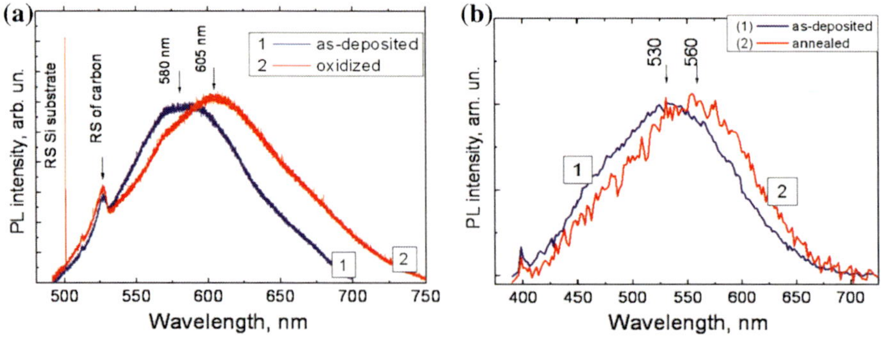

Fig. 7 **a** Normalized PL spectra of as-deposited a-SiC:H film (spectrum 1) and after oxidation in oxygen at 700 °C for 30 min (spectrum 2); **b** Normalized PL spectra of a-C:H films deposited at high working pressure before thermal treatment (spectrum 1) and after annealing in wet nitrogen

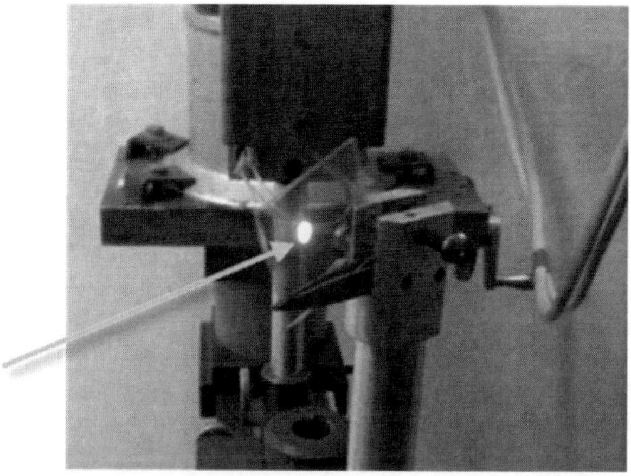

Fig. 8 Light emission of as-deposited a-C(Si):H on glass substrate (excitation by 408 nm LED, 40 mW)

2.3 The Effect of Oxidation on Morphology Reconstruction

SEM cross section image of as-deposited a-SiC:H film is presented in (Fig. 9a). It is seen that surface of as deposited a-SiC:H film is smooth and homogeneous. Oxidation treatments result in strong roughening of surface and creation of pit-like inhomogeneities with lateral size of about 50–100 nm and deepness of about 40 % of initial film thickness (Fig. 9b). This observation clearly indicates shrinkage of the film material caused by oxidation. It is most reasonable to assign material shrinkage to collapsing of nanopores in SiO_2 network.

It is important to note that oxidation treatment causes conversion of compressive stress present in as-deposited a-SiC:H films into tensile stresses. Such conversion is observed even in weakly oxidized a-SiC:H samples with small

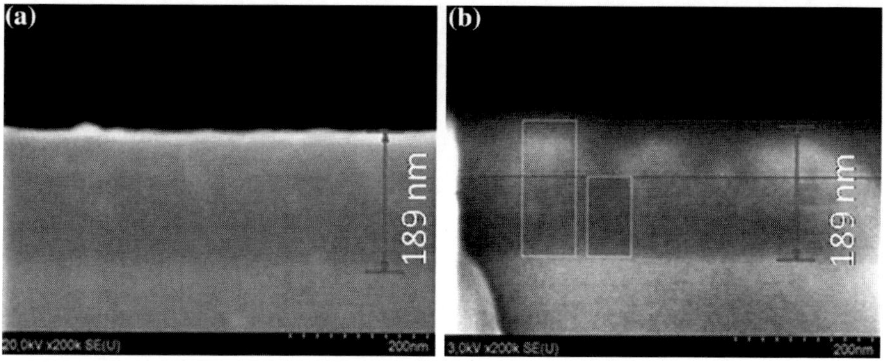

Fig. 9 SEM of the cross section of as-deposited **a** and oxidized **b** a-SiC:H film

fraction of SiO_x phase. Obviously such transformation is caused by shrinkage of the material. Important conclusion from this observation is that shrinkage of the material can lead to encapsulation of carbon clusters released in course of oxidation of SiC network.

3 Porous SiO_2:C Layers

Light-emitting porous SiO_2:C thin layers were synthesized using porous Si layer (por-Si) as a precursor. Freshly anodized por-Si subjected to thermal treatment in acetylene flow ("carbonization" stage) followed by oxidation by water vapor at temperature 600–800 °C. At such temperature water molecules readily oxidize silicon skeleton but do not react with carbon that can only migrate within porous layer.

3.1 Preparation Procedure and General Properties

Porous silicon precursors (por-Si) were prepared by anodizing of p^+-type Si(100) wafers (0.015–0.025 Ω · cm) in a hydrogen fluoride (40 %):ethanol mixture 1:1. As-anodized por-Si layers exhibit a red photoluminescence which is well known and typical for such a material. The por-Si samples were carbonized in N_2 (1.5 l/min)/C_2H_2 (1 l/min) flow in temperature range 850–1050 °C for 5–30 min resulting in the formation of carbonized porous silicon layer (por-Si:C). No detectable PL was observed after carbonization step. The next sample preparation step was the thermal treatment at atmospheric pressure in moisturized argon (Ar/H_2O) flow at temperature 600–800 °C for 1–3 h resulting in the formation of carbon-incorporated porous silicon oxide (por-SiO_2) layer. Typical cross section images of carbonized porous silicon layer before and after oxidation are presented in Fig. 10a and b. Electron diffraction pattern from the sample subjected to wet oxidation shows diffused halo (inset in Fig. 10b) indicating full amorphization of the porous Si but morphology of the layer remains porous after oxidation treatment (see Fig. 10c, d and e). Oxidation of the carbonized porous silicon layer results in strong white-light emission under excitation by UV irradiation.

Spectral properties of PL of por-SiO_2 Typical emission spectra of por-SiO_2:C layer measured in steady state and time resolved regimes are presented in Fig. 11a. Intensity of PL is normalized for pictorial demonstration. One can see that relative contribution of light emission in blue region is significantly reduced in time-resolved spectrum ("pulsed" regime of measurement) indicating that emission time of electron states responsible for high energy part of spectrum is signifi-

Fig. 10 TEM cross section image of carbonized porous Si layer before (**a**) and after oxidation (**b**). Insets represent corresponding electron diffraction pattern. **c** elastic scattering TEM image of near-substrate region, **d** corresponding image of carbon, and **e** silicon distribution (electron energy loss filtration) in SiO_2:C layer

cantly smaller than that of responsible for low energy part of spectrum. It means the spectrum of por-SiO_2:C is composed by at least two bands originated from emission sites of different structure.

Excitation PL measurements were performed with the excitation wavelength λ_{EXC} ranging from 250 to 350 nm. PL intensity at 420 and 500 nm as a function of λ_{EXC} ($I_{420}(\lambda)$ and $I_{500}(\lambda)$ respectively) is presented in Fig. 11b. The intensity of PL in the blue and green–yellow spectral range strongly increases with the decrease of excitation wavelength down to 250 nm. Such development can be attributed to the increase of the energy absorbed by defect states in silicon oxide matrix followed by transferring of excitation energy to light-emitting sites. $I_{42}0(\lambda)$ and $I_{500}(\lambda)$ are almost identical at excitation wavelength in the range of 250–300 nm while in range of 300–350 nm one can clearly see that excitation efficiency of the green–yellow component is enhanced in comparison with that of blue component. This range of excitation energy (320 nm corresponds to 3.9 eV) coincides with absorption energy of peroxy bonds (–O–O– bridges) reported in [32].

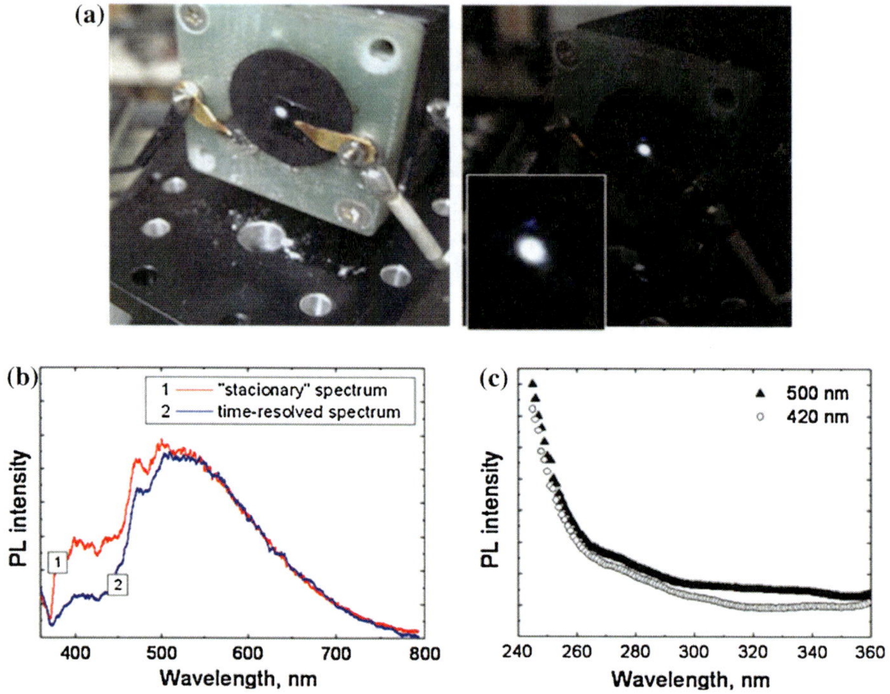

Fig. 11 **a** photo-image of light emission of por-SiO_2:C under 357 nm laser excitation; **b** typical PL spectra of por-SiO_2:C layer measured in steady state and time resolved regimes. Intensity of PL is normalized for pictorial demonstration; **c** amplitude of PL with wavelength of 500 nm and 420 nm as function of excitation wavelength

3.2 Structure Characterization of Light Emitting Porous SiO_2:C Layers by Infra-red and Raman Scattering Spectroscopy

PL spectra of por-SiO_2:C is quite strong and very broad, covering all visible spectral range. It makes hard or even impossible application of RS spectroscopy for characterization of carbon (first-order Raman scattering bands of diamond, graphite, amorphous carbon and other carbon allotropes are in range of 1000–1800 cm^{-1}) due to strong background of light emission excited by laser probe. That is why RS spectroscopy can be used if PL of the sample is weak.

Carbon precipitation during oxidations Series of por-SiO_2:C fabricated using 850 °C carbonization process were annealed in pure argon, wet argon and oxygen flow for 30 min at 600 and 800 °C after carbonization. RS spectrum of as carbonized sample (Fig. 12) is a typical spectrum of amorphous carbon composed by two broad bands: D-band (centered about 1340 cm^{-1} "disordered") and G-band (centered about 1590 cm^{-1} "graphitic"). Evolution of integral intensity of

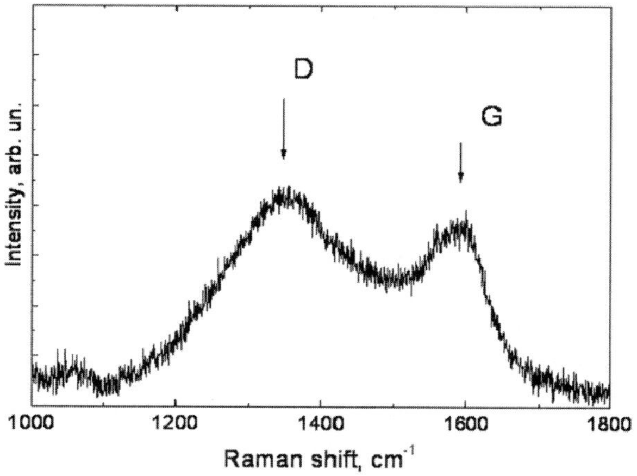

Fig. 12 Raman scattering spectrum of porous silicon layer after carbonization at 850 °C for 30 min

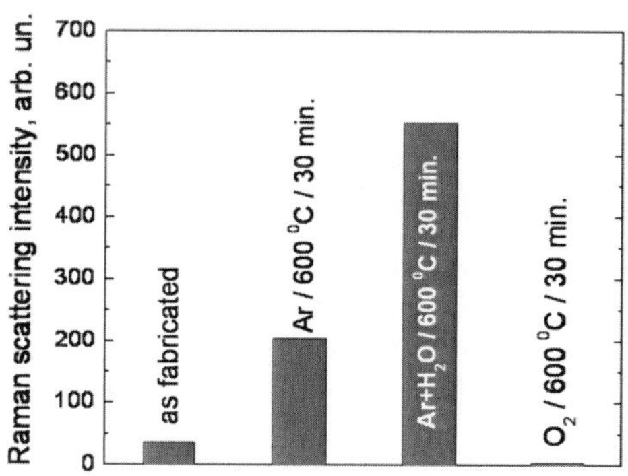

Fig. 13 Integral intensities of carbon related Raman scattering band in as-fabricated and thermally treated samples

carbon-related RS band is illustrated in Fig. 13. No carbon related RS spectrum is detected after annealing in oxygen at 600 and 850 °C indicating removing amorphous carbon in form of volatile oxides. In contrast, integrated intensity of carbon RS band is several times increased after the annealing in pure and wet argon. During the thermal treatment in dry or humid argon migration and coalescence of carbon precipitates take place resulting in increase of size and development of graphite-like local structure in carbon clusters.

Most interesting is that RS signal after wet argon treatment is 2–3 times stronger than after annealing in dry argon. Assuming that this observation is associated with larger amount of detectable amorphous carbon material in the porous layer we have to admit that after wet oxidation some "additional" carbon appears. Such "additional" carbon most likely comes after oxidation of SiC surface layer in carbonized pores. By such consideration we can estimate that amount of carbon atoms chemically bonded to silicon in as-carbonized sample is comparable with amount of free carbon atoms in free C-precipitates.

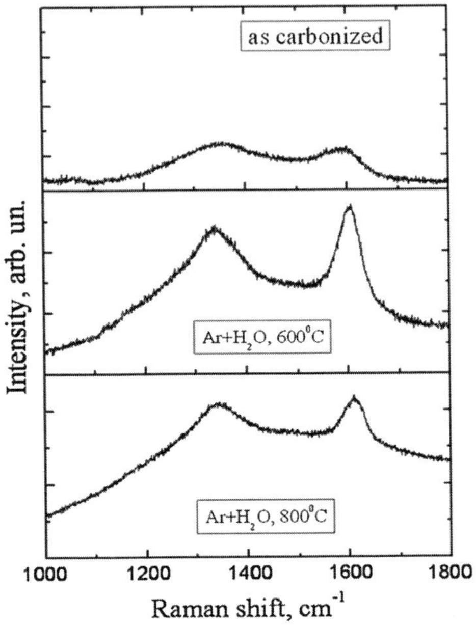

Fig. 14 Carbon-related Raman scattering of the porous silicon, carbonized at the temperature of 850 °C during 20 min and subjected to wet oxidation at 600 and 800 °C for 30 min

Figure 14 represents typical evolution of carbon related RS spectra after annealing in wet argon at 600 and 800 °C. It is seen that increase of the annealing temperature is accompanied by (1) narrowing of G- and D-bands, (2) shift of G-band from 1580 to 1600 cm^{-1}) and (3) decrease of relative intensity of D-band indicating development of local graphite-like structure and increase of the size of graphite-like carbon precipitates.

Obvious background of photoluminescence appears after annealing at 800 °C. Oxidation duration is too short (30 min) and results in weak photoluminescence, so that we can still detect RS peaks but appearance of luminescence background results in the RS signal amplitude reduction. If oxidation duration is increased up to 2–3 h PL becomes much stronger and it is impossible to detect carbon related RS band.

Effect of carbonization temperature Carbonization process was performed at 850, 950 and 1050 °C followed by wet oxidation at 600 °C for 2 h. Labeling of the samples is presented in Table 1.

RS spectra of as-carbonized samples are presented in Fig. 15. It is obvious that increase of carbonization temperature results in increase of carbon related RS band.

Spectral shift of PL band PL spectra of the samples PSO850, PSO950 and PSO1050 measured in steady state and time resolved regimes are presented in Fig. 16a and b, respectively.

It is clearly seen that relative contribution of the blue shoulder in steady state spectra is increasing with decreasing of carbonization temperature so that color of the PL changes from orange in PSO1050 to white in PSO850. "Pulse regime" strongly reduces blue shoulder indicating that decay time of blue band is significantly longer

Table 1 Labeling of the samples

Carbonization temperature, °C	por-Si:C	por-SiO_2:C
850	PS850	PSO850
950	PS950	PSO950
1050	PS1050	PSO1050

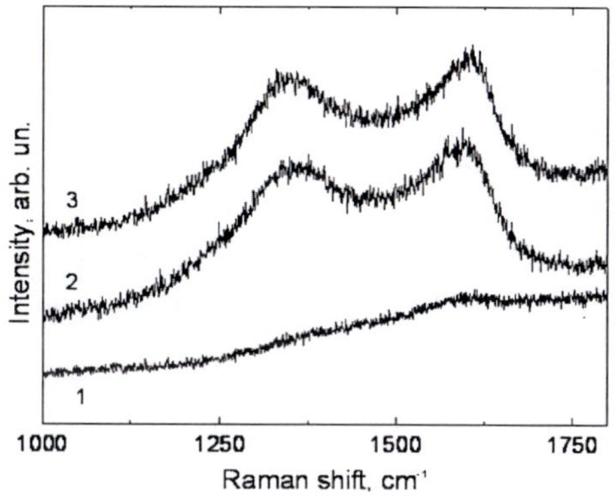

Fig. 15 Raman scattering spectra of: 1-PS850; 2-PS950; 3-PS1050

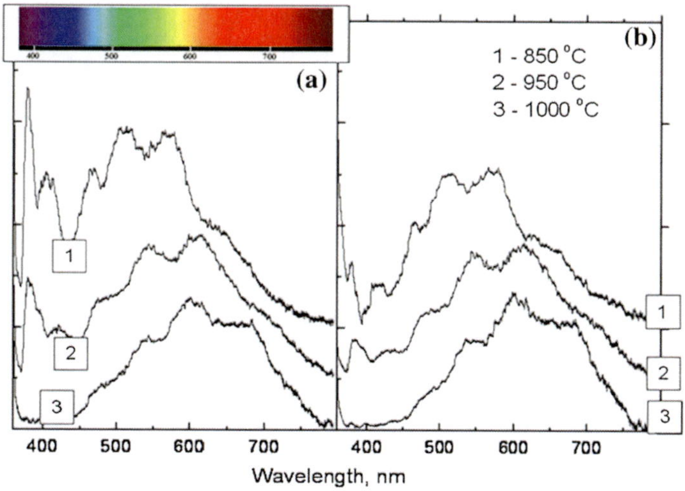

Fig. 16 PL spectra of the samples carbonized at different temperatures and measured in steady-state (**a**) and pulse (**b**) regimes: 1-PSO850; 2-PSO950; 3-PSO1050

than that of carbon-related band (Fig. 16b). Elimination of the blue shoulder by applying "pulse regime" allows for demonstrating that carbon-related band is gradually shifted to shorter wavelength with the decrease of carbonization temperature.

3.3 Thermally and Photo-Induced Degradation Effects

Thermal annealing in oxygen Figure 17a represents PL spectra of por-SiO_2:C layers before and after annealing in flow of pure oxygen at 600 °C for 30 min. The narrow band at 715 nm is an instrumental feature that arises due to unfiltered scattering of excitation source (Xe-lamp). Thermal treatment of as-prepared por-SiO_2:C sample results in strong reduction of the green–yellow component while the blue shoulder remains almost unchanged (spectrum 2). It is important to note that annealing of as-prepared por-SiO_2:C sample at the same conditions but in dry argon does not change PL spectrum and intensity. Electron energy loss spectroscopy indicates that thermal treatment in oxygen results in a strong reduction of the carbon content (Fig. 17b). The carbon reduction can be explained by converting of surface carbon into volatile carbon oxides (i.e. "burning"). Hence it is quite reasonable to attribute PL degradation to reduction of carbon-related light-emission centers.

Photo-induced PL effects in atmospheric air and vacuum PL degradation is also observed under intense UV radiation. Curve 1 in Fig. 18 represents evolution of green PL component during measurements at atmospheric conditions using high-power laser regime (325 nm). A rapid degradation is clearly observed. But if we perform measurements in evacuated volume (at pressure 10^{-3} Pa) we observe quite opposite effect, i.e. small increase of PL intensity with radiation time (curve 2, dashed horizontal straight line is just for eye guidance). Origin of this phenomenon is currently under discussion and we have no ready explanation at present time but in general it can be summarized that: the contact with atmospheric

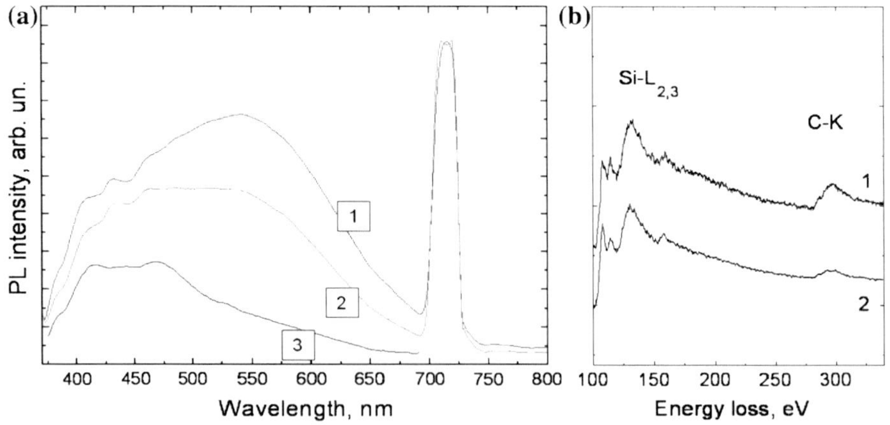

Fig. 17 **a** PL spectra of por-SiO2:C layer prepared by 3 and 4 h wet oxidation before annealing in oxygen (spectra 1 and 2 respectively) and after thermal treatment in oxygen at 600 °C for 30 min (spectrum 3); **b** electron energy loss spectra of SiO_2:C layer before (1) and after (2) thermal treatment in oxygen

Fig. 18 PL amplitude at 500 nm as function of laser radiation time (*1*) atmospheric conditions; (*2*) evolution of amplitude at 512 nm as function of laser radiation time in vacuum

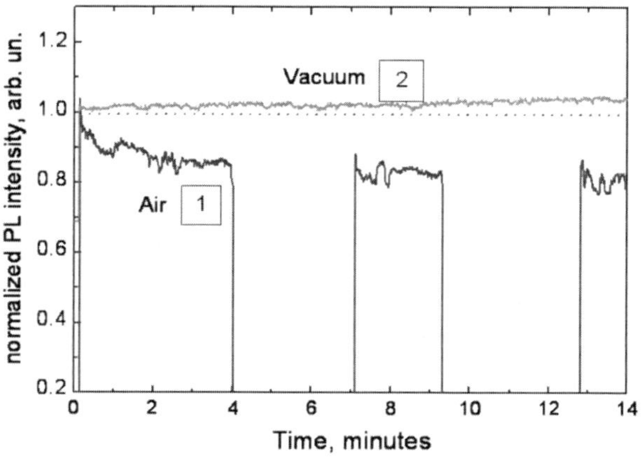

air causes photo-induced PL degradation. Following the hypothesis on carbon nano-cluster origin of the light emission centers, it is then reasonable to suggest photo-stimulated reduction (oxidation) of light emission carbon clusters.

Reversible degradation To analyze photo-induced PL degradation under intense UV radiation in more details we have performed the study of time dependent PL evolution under cyclic Xe lamp radiation (on/off cycles). PL amplitude at 500 nm emission wavelength was measured as a function of time in virgin por-SiO_2:C layer and por-SiO_2:C layer coated with protective (encapsulation) a-SiC thin film deposited by magnetron sputtering (Fig. 19a). One can see that the degradation rate in unprotected material is faster. On the background of gradual PL quenching one can see step-like features at the moments of Xe lamp switching on/off.

After Xe lamp switching on, the intensity immediately drops step-like for about several percents followed by continuous monotonic reduction. After the lamp switching off, PL amplitude raises step-like.

Measurements have been performed in time resolved regime i.e. light detection system is synchronized with laser pulses so that only pulsed light is detected. It is obvious that additional UV radiation from Xe lamp results in increase of total emission intensity, but intensity of pulsed component decreases. It means that quantum efficiency of PL is reduced by intense UV radiation and this reduction is reversible.

Typical evolution of PL amplitude within single "on/off" cycle in SiC coated sample is illustrated in Fig. 19b. One can see that under laser radiation PL intensity is almost constant but with Xe lamp being switched on the emission intensity drops fast. Continuous exponential reduction of PL intensity within "on" cycle now can be attributed to irreversible degradation associated with photo stimulated interaction of carbon with atmospheric oxygen. Height of the step in "on" and "off" edge is approximately the same indicating full recovery of PL efficiency.

Fig. 19 **a** evolution of PL amplitude at 500 nm under constant N_2 laser irradiation and cyclic on/off switching of Xe-lamp; **b** typical PL evolution within one cycle

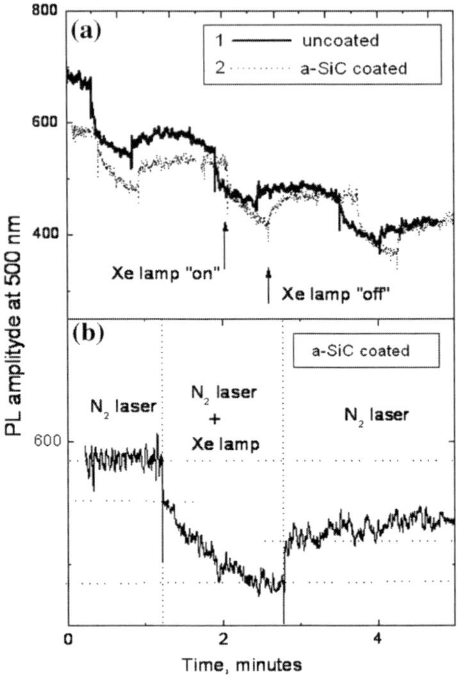

Thermal dependence of PL efficiency PL spectra of selected samples were measured in temperature range of 20–300(350) K. Typical evolution of integral PL intensity of por-SiO_2:C as function of temperature $I_{PL}(T)$ is presented in Fig. 20a. Measurements were performed and HeCd (325 nm) laser applying reduced laser radiation power (3 mW) to minimized photo-induced transformations.

Two temperature regions can be identified: (I)—low-temperature region characterized by weak variation of PL, and (II) higher temperature region characterized by increase of PL intensity. Spectral distribution of PL at low temperature and high temperature are shown in (Fig. 20b). It is seen that contribution of "green" component is slightly increased at room temperature i.e. temperature enhancement of "green" component is more efficient.

Independence of $I_{PL}(T)$ on temperature in range (I) indicates strong localization of photo exited carriers. It is more problematic to identify the reason of $I_{PL}(T)$ increase in range (II). In fact, in amorphous semiconductors (for example a-Si:H or a-SiC:H thin films) PL intensity is strongly reduced with temperature due to thermally activated diffusion of photo-excited electron-hole pairs to non-radiative recombination centers. In our case we observe quite reverse behavior. At present time there is no ready model for explanation of such evolution of $I_{PL}(T)$, but it is reasonable to suggest a general hypothesis that temperature enhanced PL efficiency is associated with thermally enhancement of energy transfer from light absorption sites (for example defects in SiO_2 matrix) to light emission sites (in frame of our working hypothesis—carbon nano-clusters).

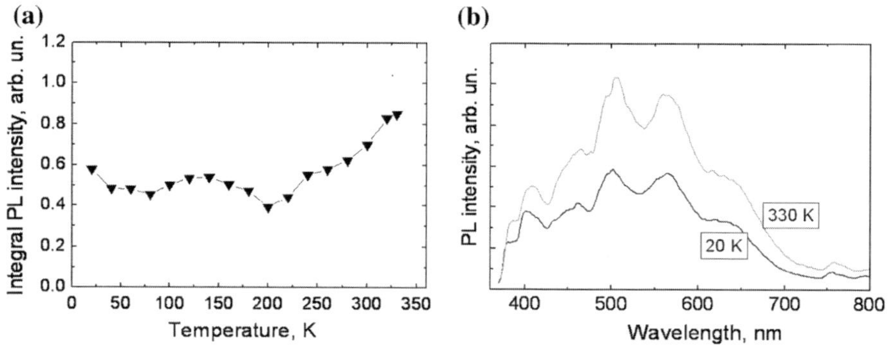

Fig. 20 **a** Evolution of integral PL intensity ($I_{PL}(T)$) of por-SiO_2:C as function of temperature measured low power regime of Cd laser (325 nm, 3 mW); **b** PL spectra of por-SiO_2:C layer measured at 20 and 330 K

4 Summarizing Discussion

SiO(x):C(y) compounds that exhibited broad visible PL can be divided into three general groups: (1) carbon implanted silicon oxide layers; (2) silicon oxycarbide thin films and (3) nanostructured SiO_2:C composites.

(1) In [5, 33], it was suggested that PL of carbon implanted silicon oxide in blue–yellow range can be associated with the formation of carbon-related clusters in form of graphitic or SiC clusters. However in [3, 4], light emitting SiO_2:C layers were fabricated by similar ion implantation method but no Si–C related bands in FTIR spectra were detected that makes SiC formation an unlikely assumption. They also argued that Si–C formation is energetically unfavorable.

(2) Broad visible light emission band was observed in hydrogenated and non-hydrogenated SiO(x)C(y) thin films. Thin films a-SiCO:H are intensively investigated as a material for low-k barrier dielectric for Cu-based multilevel interconnection in ultra-large-scale integrated circuits. Hence there are many reports on optical, electrical, and chemical properties of this material, but there are only few publications reporting strong white PL. This may be an indication that effective luminescence is not inherent to hydrogenated silicon oxycarbide. Structure of an ''ideal'' a-SiCO:H material is represented by amorphous SiO_x network terminated by $-CH_3$ or –OH groups with possible substitution of Si–O–Si bridges by $-CH_2-$ crosslinking (Fig. 1b). Such ''ideal'' structure contains "no defects'' like homonuclear Si–Si and C–C bonds or Si–O–C bridges, but real materials contain variety of structural ''defects''. It is logical to suggest that the reported light emission of silicon oxycarbide films results from structural fluctuations associated with the ''defects''.

The authors of [7] have found correlation of PL intensity and position of the 437 cm^{-1} IR absorption band in a-SiO(x)C(y):H films deposited by CVD method. This band is associated with bending (rocking) vibrations of Si–O–Si bridges, and in thermally grown dense silicon oxide layers this band is located at 457 cm^{-1} [34]. Small shift of this band in CVD a-SiO(x)C(y):H films was ascribed to the presence of Si–O–C bonds. Based on the correlation of the IR band intensity and PL intensity, it was concluded that light emission originates from localized electron states associated with Si–O–C bonds. Unfortunately, no clear proof of this interpretation has been presented. In fact, shift of the vibration frequency can be caused by several factors, such as mechanical stresses, local Si–O–Si bonding distortions etc.

In [35], white luminescence of a-SiOC:H films deposited by atmospheric pressure micro-plasma jet was ascribed to neutral oxygen vacancy (NOV) defects in SiO_x matrix. This interpretation was based on the PL intensity and hysteresis of C–V characteristics associated with trapped positive space charges. Authors assumed that all positive charges in a-SiOC:H/Si heterostructure are trapped by bulk NOV defects. They excluded from consideration a large number of other bulk and interface defects that are also able to trap positive charge and, thus, to contribute to C–V hysteresis.

In some studies white light emission observed in *sol–gel-derived* SiOC layers has been associated with the formation of dangling bonds and sp^2 C atoms during pyrolysis [36], others researchers explained the PL origin by precipitation of sp^2 C clusters into an amorphous $Si_xC_yO_zH_w$ [37]. Combination of emission bands associated with Si, SiC, and C related precipitates was also suggested to explain variable spectral properties of the material [5, 6].

(3) Another possible origin of light emission in SiOC compound is radiative recombination in carbon nanoparticles. In fact, the first report on luminescent properties of individual carbon nanoparticles was published in 2006 [38]. Thereafter, chemists have discovered and developed several methods of fabrication of light emitting nanoparticles, so called ''carbon nanodots'' (reviewed in [29, 30]). It was shown that carbon particles with a size of several nanometers after an appropriate surface passivation become highly luminescent in broad visible spectral range. Unfortunately physical mechanism of light emission has not been studied properly. Nevertheless, the most important conclusion of the studies is that amorphous carbon nanoparticles can effectively emit visible light and spectral position of PL maximum depends on the size of carbon dots because in carbon-related molecular-like clusters the gap between HOMO and LOMO of *p*-orbitals decreases with increasing of the number of atoms. It is also important to note that light emitting carbon dots usually contain very large amount of oxygen (tens of at.%).

Let us consider light emitting properties of SiOC layers in frame of "carbon dot" (CD) hypothesis. *First*, CD-hypothesis naturally explains spectral shifts of PL bands in thermally treated SiOC thin films (see Sect. 2.2) and effect of carbonization temperature on PL band position in por-SiO_2:C layers (see Sect. 3.3). Maximum intensity of PL was observed to be shifted from about 2.1 to 2.3 eV with decreasing of carbon incorporation in SiOC films deposited by RF-magnetron sputtering [9]. Spectral shift was also observed in SiOC films synthesized by polymerization processes. At low pyrolysis temperatures (800–1000 °C), a blue emission band was observed, while at higher temperatures (1000–1250 °C), an orange emission dominates the luminescence [6].

Secondly, on the basis of the carbon cluster hypothesis, it is quite easy to explain irreversible degradation induced by thermal annealing in oxygen and UV radiation in air conditions. Such degradation is related with thermally or photo-induced oxidation of carbon nano-clusters.

Thirdly, one of the reasonable explanations of reversible photo-induced PL quenching/recovery under cyclic UV exposure is the mechanism of fluorescence intermittency (or ''blinking''), phenomenon, which is well known in light emitting semiconductor nanoparticles [39]. Under irradiation by energetic photons, some of nanoparticles can be charged by electron jumping from core state to neighboring energy state at the interface or in the surrounding wide band gap matrix. In the charged state, the non-radiative Auger recombination becomes the dominating recombination path making the nanocrystal ''dark'' until neutrality will be recovered. Obviously, the higher relative fraction of charged particles the lower is the integral PL intensity. No blinking phenomenon was observed in colloidal carbon nanodots in water solution [29, 30, 38] but there is no any data on the "blinking" of CD in solid state.

5 Summary

It is demonstrated that all experimental data on PL properties of carbonized silicon oxide layers (both in form of thin films and porous layers) are self consistent in frame of the hypothesis that broad PL band in por-SiO_2:C layers are associated with carbon nano-clusters. Formation of carbon nano-clusters takes place through competition of thermally activated precipitation of small light-emitting clusters and coalescence/growth of non light-emitting large carbon precipitates with disordered graphite like structure.

References

1. Green, W.H., Le, K.P., Grey, J., Au, T.T., Sailor, M.J.: White phosphors from a silicate carboxylate sol-gel precursor that lack metal activator ions. Science **276**, 1826–1828 (1997)
2. Yamada, N.: Photoluminescence in carbon/silica gel nanocomposites. In: Supercarbon: Synthesis, Properties and Applications, pp. 211–225. Springer, New York (1998)
3. Yu, Y.H., Wong, S.P., Wilson, I.H.: Visible photoluminescence in carbon-implanted thermal SiO_2 films. Phys. Status Solidi A **168**, 531–534 (1998)
4. Zhao, J., Mao, D.S., Lin, Z.X., Jiang, B.Y., Yu, Y.H., Liu, X.H., Wang, H.Z., Yang, G.Q.: Intense short-wavelength photoluminescence from thermal SiO2 films co-implanted with Si and C ions. Appl. Phys. Lett. **73**, 1838–1840 (1998)

5. Perez-Rodrigues, A., Gonzales-Varona, O., Garrido, B., Pellegrino, P., Moprante, J.R., Bonafos, C., Carrada, M., Claverie, A.: White luminescence from Si^+ and C^+ ion-implanted SiO_2 films. J. Appl. Phys. **94**, 254–262 (2003)
6. Karakuscu, A., Guider, R., Pavesi, L., Soraru, G.D.: Broad-band tunable visible emission of sol–gel derived SiBOC ceramic thin films. Thin Solid Films **519**, 3822–3826 (2011)
7. Gallis, S., Nikas, V., Suhag, H., Huang, M., Kaloyeros, A.E.: White light emission from amorphous silicon oxycarbide a-SiC_xO_y thin films: Role of composition and postdeposition annealing. Appl. Phys. Lett. **97**, 081905-1–081905-3 (2010)
8. Seo, S.Y., Cho, K.S., Shin, J.H.: Intense blue–white luminescence from carbon-doped silicon-rich silicon oxide Appl. Phys. Lett. **84**, 717–719 (2004)
9. Hayashi, S., Kataoka, M., Yamamoto, K.: Photoluminescence Spectra of Carbon Clusters Embedded in SiO_2. Jpn. J. Appl. Phys. **2**(32), L274–L276 (1993)
10. Vasin, A.V., Ishikawa, Y., Kolesnik, S.P., Konchits, A.A., Lysenko, V.S., Nazarov, A.N., Rudko, GYu.: Light-emitting properties of amorphous Si:C:O: H layers fabricated by oxidation of carbon-rich a-Si:C: H films. Solid State Sci. **11**, 1833–1837 (2009)
11. Vasin, A.V., Ishikawa, Y., Shibata, N., Salonen, J., Lehto, V.P.: Strong white photoluminescence from carbon-incorporated silicon oxide fabricated by preferential oxidation of silicon in nano-structured Si: C layer Jpn. J. Appl. Phys. **19**, L465–L467 (2007)
12. Ishikawa, Y., Vasin, A.V., Salonen, J., Muto, S., Lysenko, V.S., Nazarov, A.N., Shibata, N., Lehto, V.-P.: Color control of white photoluminescence from carbon-incorporated silicon oxide. J. Appl. Phys. **104**, 083522-1–083522-6 (2008)
13. Vasin, A., Rusavsky, A., Nazarov, A., Lysenko, V., Rudko, G., Piryatinski, Y., Blonsky, I., Salonen, J., Makila, E., Starik, S.: Excitation effects and luminescence stability in porous SiO2: C layers. Phys. Status Solidi A **209**, 1015–1021 (2012)
14. Pantano, C.G., Singh, A.K., Zhang, H.: Silicon oxycarbide glasses. J. Sol-Gel. Sci. Technol. **14**, 7–25 (1999)
15. Rouxel, T., Massouras, G.: High Temperature behavior of a gel-derived SiOC glass: elasticity and viscosity. J. Sol-Gel. Sci. Technol. **14**, 87–94 (1999)
16. Grill, A., Patel, V.: Low dielectric constant SiCOH films as potential candidates for interconnect dielectrics. Mat. Res. Soc. Symp. Proc. **565**, 107–110 (1999)
17. Grill, A., Perraud, L., Patel, V., Jahnes, C., Cohen, S.: Ultralow-*k* dielectrics prepared by plasma-enhanced chemical vapor deposition. App. Phys. Let. **79**, 803–805 (2001)
18. Kim, Y.H., Lee, S.K., Kim, H.J.: Low-k Si–O–C–H composite films prepared by plasma-enhanced chemical vapor deposition using bis-trimethylsilylmethane precursor. J. Vac. Sci. Tech. A **18**, 1216–1219 (2000)
19. Soraru, G.D., Dandrea, G., Campostrini, R., Babonneau, F., Mariotto, G.: Structural characterization and high-temperature behavior of silicon oxycarbide glasses prepared from sol-gel precursors containing Si-H bonds. J. Am. Ceram. Soc. **78**, 379–387 (1995)
20. Turquat, C., Kleebe, H.J., Gregori, G., Walter, S., Soraru, G.D.: Transmission electron microscopy and electron energy-loss spectroscopy study of nonstoichiometric silicon-carbon-oxygen glasses. J. Am. Ceram. Soc. **84**, 2189–2196 (2001)
21. Gregori, G., Kleebe, H.J., Blum, Y.D., Babonneau, F. Evolution of C-rich SiOC ceramics: part II. Characterization by high lateral resolution techniques: electron energy-loss spectroscopy, high-resolution TEM and energy-filtered TEM. Int. J. Mater. Res. **97**, 710–720 (2006)
22. Kleebe, H.J., Gregori, G., Babonneau, F., Blum, Y.D., MacQueen, D.B., Masse, S.: Evolution of C-rich SiOC ceramics: Part I. Characterization by integral spectroscopic techniques: Solid-state NMR and Raman spectroscopy. Int. J. Mater. Res. **97**, 699–709 (2006)
23. Brus, J., Kolar, F., Machovic, V., Svitilova, J.: Structure of silicon oxycarbide glasses derived from poly(methylsiloxane) and poly[methyl(phenyl)siloxane] precursors. J. Non-Cryst. Solids **289**, 62–74 (2001)
24. Saha, A., Raj, R., Williamson, D.L.: A model for the nanodomains in polymer-derived SiCO. J. Am. Ceram. Soc. **89**, 2188–2195 (2006)

25. Vasin, A.V., Muto, Sh, Ishikawa, Yu., Rusavsky, A.V., Kimura, T., Lysenko, V.S., Nazarov, A.N.: Comparative study of annealing and oxidation effects in a-SiC: H and a-SiC thin films deposited by radio-frequency magnetron sputtering. Thin Solid Films **519**, 2218–2222 (2011)
26. Bullot, J., Schmidt, M.P.: Physics of silicon-carbon alloys. Phys. Status Solidi B **143**, 345–418 (1987)
27. Tauc, J., Grigorovici, R., Vancu, A.: Optical properties and electronic structure of amorphous germanium. Physica Status Solidi **15**, 627–637 (1966)
28. Skuja, L.: Optically active oxygen-de®ciency-related centers in amorphous silicon dioxide. J. Non-Cryst. Solids **239**, 16–48 (1998)
29. Baker, ShN, Baker, G.A.: Luminescent carbon nanodots: emergent nanolights angew. Chem. Int. Ed. **49**, 6726–6744 (2010)
30. Kang, H., Li, Zh., Liu, Y., Lee, Sh.-T.: Carbon nanodots: synthesis, properties and applications. J. Mater. Chem. **22**, 24230–24253 (2012)
31. Vasin, A., Kolesnik, S.P., Konchits, A.A., Rusavsky, A.V., Lysenko, V.S., Nazarov, A.N., Ishikawa, Y., Koshka, Y.: Paramagnetic defects and light-emission of carbon-rich a-SiC:H films. J. Appl. Phys. **103**, 123710-1–123710-7 (2008)
32. Nishikawa, H., Tohmon, R., Ohki, Y.: Defects and optical absorption bands induced by surplus oxygen in high-purity synthetic silica. J. Appl. Phys. **65**, 4672–4678 (1989)
33. Rebohle, L., Gebel, T., Froöb, H., Reuther, H., Skorupa, W.: Ion beam processing for Si/C-rich thermally grown SiO_2 layers: photoluminescence and microstructure. Appl. Surf. Sci. **184**, 156–160 (2001)
34. Kirk, C.T.: Quantitative analysis of the effect of disorder-induced mode coupling on infrared absorption in silica. Phys. Rev. B. **38**, 1255–1273 (1988)
35. Ding, Y., Shirai, H.: White light emission from silicon oxycarbide films prepared by using atmospheric pressure microplasma jet. J. Appl. Phys. **105**, 043515-1–043515-4 (2009)
36. Menapace, I., Mera, G., Riedel, R., Ermen, E., Eichel, R.-A., Pauletti, A., Appleby, G.A.: Luminescence of heat-treated silicon-based polymers: promising materials for LED applications. J. Mater. Sci. **43**, 5790–5796 (2008)
37. Pivin, C., Sendova-Vassileva, M., Colombo, P., Martucci, A.: Photoluminescence of composite ceramics derived from polysiloxanes and polycarbosilanes by ion irradiation. Mater. Sci. Eng. B **69–70**, 574–577 (2000)
38. Sun, Y.P., Zhou, B., Lin, Y., Wang, W., Fernando, K.A.S., Pathak, P., Meziani, M.J., Harruff, B.A., Wang, X., Wang, H.F., Luo, P.J.G., Yang, H., Kose, M.E., Chen, B.L., Veca, L.M., Xie, S.Y.: Quantum-Sized Carbon Dots for Bright and Colorful Photoluminescence. J. Am. Chem. Soc. **128**, 7756–7757 (2006)
39. Nirmal, M., Dabbousi, B.O., Bawendi, M.G., Macklin, J.J., Trautman, J.K., Harris, T.D., Brus, L.E.: Fluorescence intermittency in single cadmium selenide nanocrystals. Nature **383**, 802–804 (1996)

Preparation, Luminescent Properties and Bioimaging Application of Quantum Dots Based on Si and SiC

Valeriy A. Skryshevsky, Tetiana Serdiuk, Yuriy E. Zakharko, Sergei A. Alekseev, Alain Géloën and Vladimir Lysenko

Abstract Well-known, the interest to the colloidal solution with quantum dots (QDs) lies in their fluorescence properties. Among the advantages of QDs are the high resistance to photooxidation, the size and composition variation allowing to obtain the narrow emission spectra with high quantum yield from the ultraviolet to the near infrared region. In this chapter we present the last achievements in forming and bio-medical applications of luminescent Si and SiC QDs. It is shown that a broad size distribution of Si QDs are obtained at electrochemical etching. The dimensions of the Si QDs undergone filtering in colloidal solution vary discretely with a radius quantum equal to 0.12 nm. Existing of this quantum may correspond to step-like increasing of Si QDs radius on one new shell at the surface of Si QDs. The formed QDs show intense luminescent in visual region. However, one of the major drawbacks of Si QDs for bio-medical application is instability over time in water or buffer solutions. To overcome this drawback the several methods of surface functionalization are discussed. The SiC QDs are stable in water solutions and do not require supplementary surface functionalisation for bioimaging. A strong fluorescence from the SiC QDs, which undoubtedly penetrate into the cell, has been

V.A. Skryshevsky (✉) · T. Serdiuk
Institute of High Technologies, Taras Shevchenko National University of Kyiv, 64 Volodymyrska Str, Kyiv 01601, Ukraine
e-mail: skryshevsky@yandex.ru

T. Serdiuk · Y.E. Zakharko · V. Lysenko
Nanotechnology Institute of Lyon (INL), University of Lyon, UMR-5270, CNRS, INSA de Lyon, 69621 Villeurbanne, France

S.A. Alekseev
Chemistry Faculty, Taras Shevchenko National University of Kyiv, 64 Volodymyrska Str, Kyiv 01601, Ukraine

T. Serdiuk · A. Géloën
CarMeN Laboratory, U 1060 INSERM, INSA de Lyon, 69621, Villeurbanne, France

A. Nazarov et al. (eds.), *Functional Nanomaterials and Devices for Electronics, Sensors and Energy Harvesting*, Engineering Materials,
DOI 10.1007/978-3-319-08804-4_15

observed. The studying of health and cancer cells using SiC QDs shows that simple modification of surface charge of QDs gives strong opportunity to target the same QDs in intracellular space with their preferential localisation inside or outside the cell nucleus.

1 Introduction

Quantum dots (QDs) are semiconductor nanoparticles in which "quantum confinement" effect is observed [1]. Quantum confinement of electrons in semiconductor nanoparticles is observed when a crystallite radius less than the Bohr exciton radius (5 nm for silicon) and results in specific optical and electronic properties—size dependent bandgaps and thus a Stokes shift with high quantum yield photoluminescence.

Structurally, the central part of QDs consists of a core composed of a variety of substances such as A^{IV}, A^2B^6, A^3B^5 and more complex semiconductors which can be amorphous or crystalline, doped and undoped [2]. The core can be covered by other semiconductor shell and stabilizing layer. Depending on functional application of QDs this layer can be hydrophilic or hydrophobic. The linkers and ligands can be attached to these layers allowing formation of different complex. The examples of possible materials for core, doping ions, shell, stabilizing layer, linkers and ligands are presented in Fig. 1.

Due to their specific photophysical, magnetic, electronic and biological properties quantum dots are widely applied in electronics and optoelectronics (light emission diodes, memory devices, displays), photochemistry (hydrogen generation, sensitizes, photoelectrodes), analytical chemistry (chemical analysis), molecular biology and medicine (biosensors, fluorescent labels, photodynamic therapy, drug biotracking), etc. [1–4].

In the last few decades, the emergence of nanotechnology gave rise to exciting developments in the field of cell biology. Indeed, the ability to control matter at the nanoscale length paves the way for the sensing of biological systems at the level of single molecules by optical or electrical detection. The implementation of QDs for use in cell labeling has been one of the fastest growing areas in this field. The reason for this intense research activity, in particular fluorescence labeling, lies in the considerable advantages of semiconductor QDs over the classically used organic dye molecules or fluorescent proteins. Numerous reviews have been devoted to these QDs [1–9] including the method of QDs formation and possible application to bioimaging. Here we review the last achievements in the branch of Si and SiC QDs formation by electrochemical etching, their surface functionalization, analysis of luminescent properties of QDs colloidal solutions, and different bio-medical application.

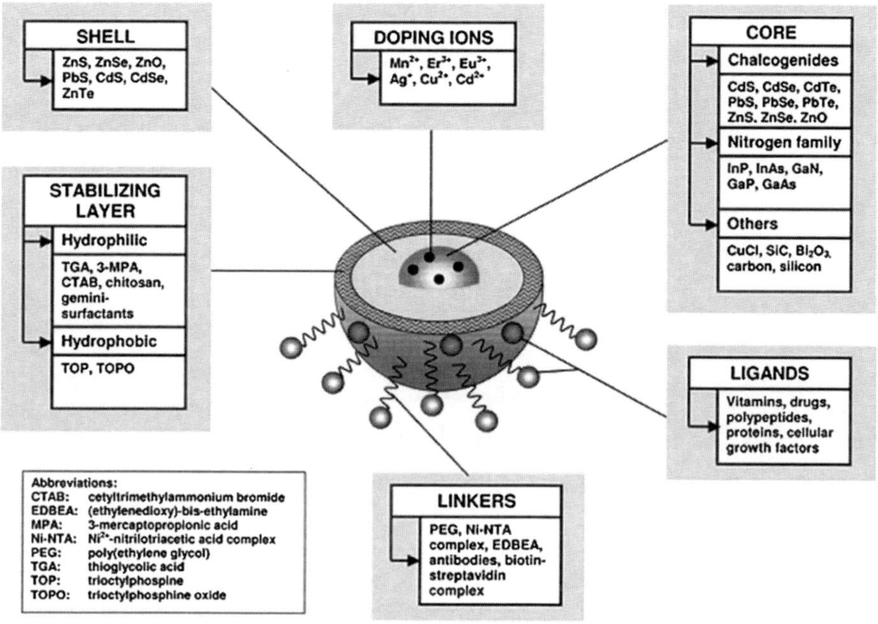

Fig. 1 Scheme of quantum dot structure (from [2])

2 Properties and Advantages of Colloidal Solutions with QDs

The interest in colloidal solution with QDs lies in their fluorescence properties. Among the advantages of QDs, the principal one is the higher resistance of QDs to light-induced reactions such as photooxidation. This photochemical inertness known as the photobleaching phenomenon prevents the degradation of luminescence properties with time. Another important feature of QDs is their size and composition-induced tunable emission spanning from the ultraviolet to the near infrared (UV to NIR) with a narrow emission spectra and high quantum yield. The conjugation of a narrow emission with a large effective Stokes shift [8, 10], a constant emission wavelength no matter what excitation wavelength is used [11], and a broad absorption spectrum are key parameters to achieve multiplexed imaging as the fluorescence signal of each QD can be readily separated and individually analyzed using a single excitation source. This multicolor imaging potentiality is particularly interesting for the monitoring of intracellular processes [3, 12]. As the last advantage, it is also important to note that QDs have orders of magnitude larger cross sections for two-photon excitation compared to the organic chromophore. Thanks to this property, infrared excitation in the transparency window of tissues (700–900 nm) can be used with visible radiation to perform fluorescence detection.

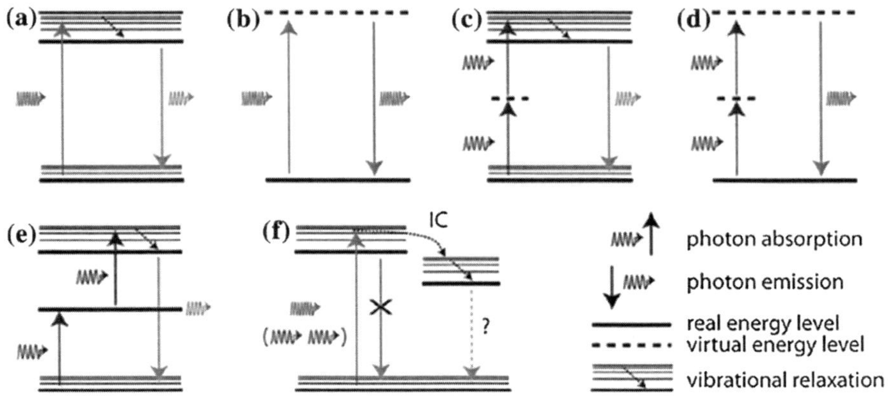

Fig. 2 **a** One-photon excited fluorescence. **b** Linear scattering. **c** Two-photon-excited fluorescence. **d** Second harmonic generation. **e** Sequential absorption and upconverted emission. **f** Principle of molecular bleaching showing intersystem crossing upon one- or two-photon absorption (from [7])

Besides fluorescence (one photon and two-photon excited) we can note other effects like linear scattering, second harmonic generation, sequential absorption and upconverted emission. The schemes in Fig. 2 summarize emission effects in QDs that can be applied to bio-medical applications (see detailed explanation in [7]).

QDs present also two important drawbacks. The first one is that most QDs, as synthesized, have hydrophobic organic ligands coating their surface [13]. To make the QDs water soluble, their organophilic surface species have to be exchanged with more polar species. A complex multilayer [14] ligand shell has then to be realized, which leads in some cases of nonspecific optical absorption. The second and the major drawback concerns cytotoxicity. For example, the widely used II–VI semiconductor QDs (e.g. CdSe) were found to be cytotoxic through the release of free cadmium ions [15, 16]. As the result, these QDs strongly influence biological cell function and their application in efficient cell imaging is limited. Therefore, a protective shell (ZnS or a polymer, for example) must be systematically added to prevent cell death when using these types of QDs. To completely avoid this problem, one has to use QDs on the basis of more benign materials such as group IV semiconductors, first of all Si and SiC QDs.

3 Technology of Colloidal Solution of Si QDs

Different approaches have been proposed and described in literature to create Si nanoparticles and QD_S. The isolated particles were obtained from a gas-phase preparation using silanes or other Si sources via thermal decomposition, thermal annealing, reactive sputtering, thermal vaporization, microwave plasma, gas-evaporation, chemical vapour deposition [1, 9, 17–20].

"Wet chemistry" route to Si QDs typically involves the reaction of Zintl salts (KSi/NaSi) or reductive organometallic Si compounds with $SiCl_4$ or NH_4Br [21, 22]. Chemical reduction of silicon halogenides (SiX_4; X = Cl or Br) by $LiAlH_4$ in water-free reverse-micelle solutions gives Si QDs with narrow size distribution [23]. However, comparing to chemical route, electrochemical anodization is much more reproducible and simple technique for the fabrication of Si QDs emitting light in a large spectral range at room temperature under photoexcitation. Visible light emitting colloidal Si QDs are extensively studied [1]. Otherwise UV emitting Si QDs were only mentioned in a few papers [24–26], because it is relatively difficult to control fabrication of stable and homogeneously distributed Si QDs with dimensions less than 1.5 nm having the energy band gaps larger than 3 eV.

Commonly the anodization to prepare Si QDs as well as porous silicon (PS) takes place in the mixtures of HF (able to transform silicon into soluble H_2SiF_6 under anodization) and organic solvent (commonly ethanol) decreasing the wetting angle and thus allowing the penetration of electrolyte inside pores of PS. An experimental protocol for the preparation of Si QDs with desired emission wavelength (from NIR to UV) was described in details in the works [26, 27]. At the first stage, highly porous (>90 %) 300 μm thick PS layer was prepared by anodization (current density $j = 55$ mA/cm^2) of p-type (1–10 Ω cm) boron-doped (100)-oriented Si wafers with a backside Al ohmic contact in 1:1 (v:v) mixture of 48 % aqueous HF and absolute ethanol during 2 h. To remove all HF traces the layer was washed several times in absolute ethanol and naturally dried, which resulted in its self-transformation into a strongly hydrogenated PS micropowder with about 6.5 % mass content of H [27]. Finally the micropowder was mechanically grinded and dispersed in absolute ethanol, giving Si QDs saturated colloidal solution with 5 mg/ml concentration.

Transmission Electron Microscopy (TEM) pictures shown in Fig. 3 give a general view of quasi-spherical Si particles constituting the nanopowder suspension into the initial colloidal solution. As one can see, the general size distribution of the shown particles is relatively large. In particular, the initial nanopowder consists of large (>10 nm) porous particles (Fig. 3 b) and numerous small denser

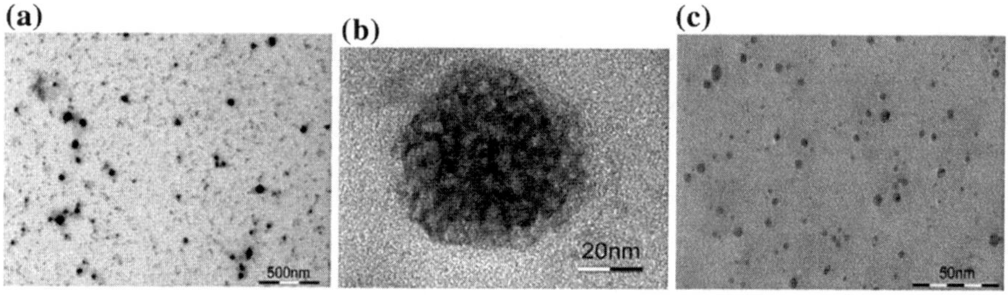

Fig. 3 TEM pictures of Si NPs constituting the initial strongly hydrogenated PS nanopowder: **a** general view; **b** large porous NPs; **c** small dense Si NPs with dimensions less than 5 nm (from [26])

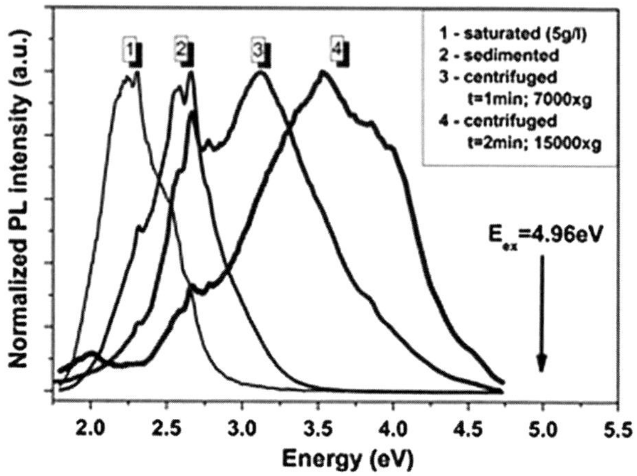

Fig. 4 PL spectrum of the initial PS nanopowder in ethanol solution and its evolution upon sedimentation and moderate centrifugation (from [26])

(<10 nm) particles (Fig. 3c). Total absence of clearly defined electron diffraction features indicates an amorphous-like state of the studied particles [27]. Clear and dark spots of the large porous particles (Fig. 3b) correspond to the pits and solid Si phase, respectively. In general, their shape was already described in terms of fractal geometry [28]. As for the small (<10 nm) particles (Fig. 3c), they are well known to exhibit highly efficient luminescence in a large spectral range due to quantum and spatial confinements of photogenerated charge carriers and these particles can be considered as SiQDs.

Figure 4 shows normalized photoluminescence (PL) spectra of the Si QDs in ethanol solutions [26]. First, the PL signal of the initial saturated solution is represented by spectrum 1. It is very similar to a usual PL spectrum of a red emitting PS layer. The spectra blueshifting in comparison to the initial PL is observed after sedimentation in the dark during 1 day (curve 2). Associating both PL spectra to particular size distributions of SiQDs and taking into account that the emission energy for the smaller QDs is higher, the observed sedimentation-induced evolution of the PL signal can be explained by the removal of the largest submicron PS QDs from the ethanol solution. Thus, the observed green-blue PL maxima of spectrum 2 may correspond to luminescence of the smaller porous QDs as those shown in Fig. 3b, for example. Further centrifugation leads to almost complete elimination of the QDs with dimensions larger than 2 nm from the colloidal solution. The PL spectrum 4 of this doubly centrifuged solution has a main maximum at 3.5 eV and a set of other UV peaks at the high energy wing situated in the spectral region between 3.5 and 4.5 eV.

In [29, 30] we proposed a new method allowing more precise size selection of colloidal Si QDs. This approach is based on the use of meso-porous Si (meso-PS) free standing membranes as QDs filters (Fig. 5). In particular, due to auto-filtration phenomenon an efficient filtering was achieved and size quantization of the amorphous Si QDs was brought to the fore. Via variation of the porosity and thickness of meso-porous layer it is possible to obtain PL spectra with maxima

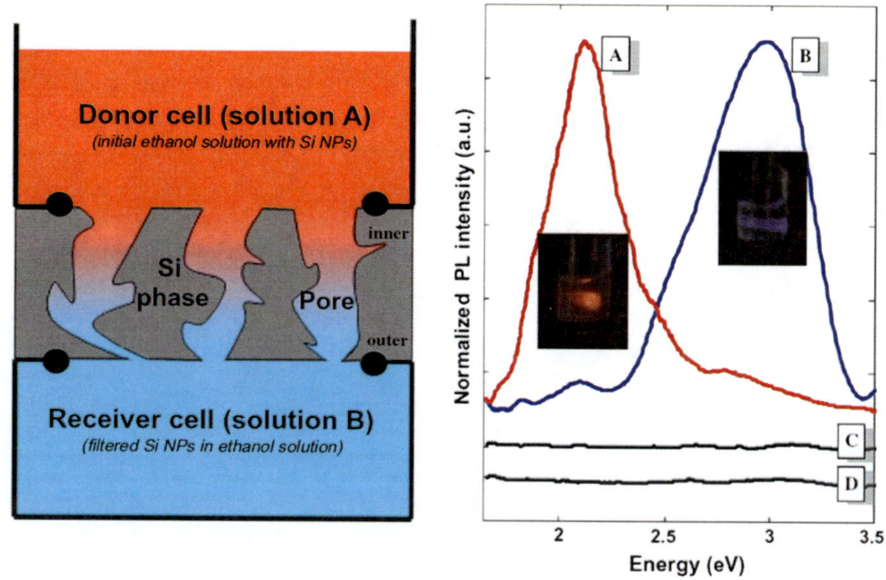

Fig. 5 *Left* Scheme of a double-tank homemade hermetic filtering set-up with upper (donor) cell and *bottom* (receiver) cell and *Right* PL spectra with corresponding photos of initial (*curve A*) and filtered (*curve B*) ethanol solutions with dispersed Si QDs. PL signals from pure absolute ethanol and from pure ethanol without Si QDs passed through the meso-PS membrane are represented by non-normalized spectra *C* and *D*, respectively (from [29])

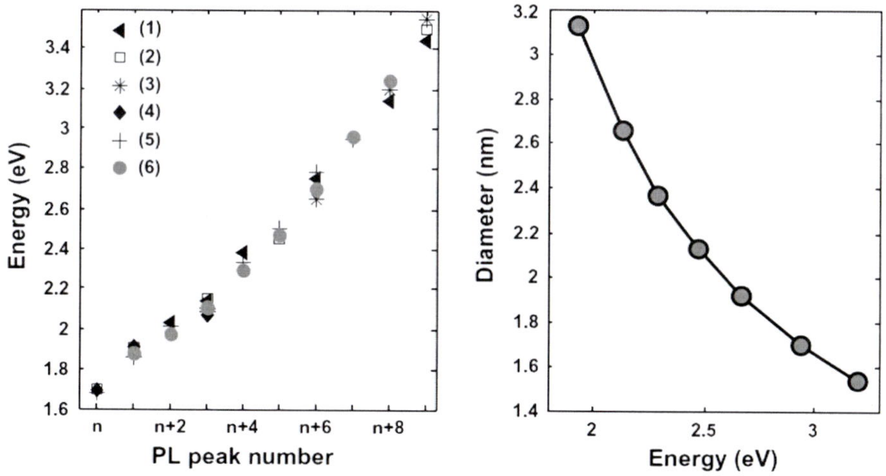

Fig. 6 *Left* spectral positions of PL maxima/shoulders reported in (*1*) [31], (*2*) [32], (*3*) [26], (*4*) [33], (*5*) [34], and (*6*) [29]; *Right* dependence of Si QDs size calculated according to the model described in [35] on the PL peak energies reported in [29]

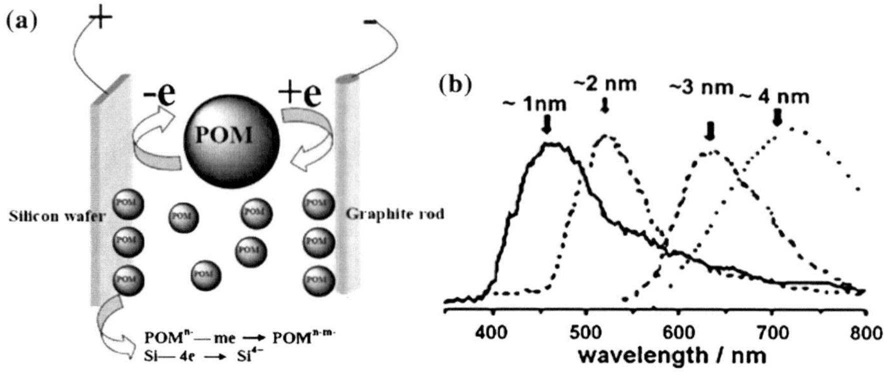

Fig. 7 Scheme for the POM-assisted electrochemical etching process (**a**). Typical PL spectra of SiQDs with sizes ranging from 1 to 4 nm (**b**) from [9]

from 1.7 to 3.4 eV. Energy positions of these features are summarized in Fig. 6. The lowest (in terms of energy) feature is designated as n and all others as $n + i$, where i is the index number. A remarkable coincidence of the data for each PL peak number can be stated. In particular, one can deduce from Fig. 6 that the PL energy values seem to vary discretely which is especially interesting taking into account total independence of the picked out data.

Since each spectral energy value assigned to a specific QD dimension, Fig. 6b shows this correlation according to the model of Delerue et al. [35] for which:

$$\mathrm{Eg\ (d)} = 1.167 + 3.73 * \mathrm{d}^{-1.39}, \quad (1)$$

where $E_g(d)$ is the energy band gap of a spherical-like QD with diameter d. The QD diameter was estimated from Eq. (1) for each spectral PL feature n + i characterized by its proper energy value $E_g(d)$. Thus, in frame of experimental errors, Fig. 6b shows that dimensions of the Si QDs undergone filtering vary discretely with a radius quantum equal to 0.12 nm. Existing of this quantum may correspond to step-like increasing of Si QDs radius on one new shell at the surface of Si QDs.

The chromatography and other complex separation methods were applied to separate electrochemically etched Si QDs into monodisperse fractions [1, 9, 36–38].

For particle size control, Kang et al. have developed a polyoxometalate (POM)-assisted electrochemical method, which yielded highly monodisperse Si dots (1, 2, 3, and 4 nm) with a narrow size distribution requiring no further separation (Fig. 7) [36]. It represents a convenient and rational synthesis of a variety of Si nanostructures via the simple control of current density and choice of catalyst POMs. POMs clearly play a vital role incontrolling the size of Si dots and formation of Si nanostructures. The size control of Si dots may be attributed to the unique electronic characteristics of POMs in gaining and donating electrons simultaneously while keeping their structure unchanged.

4 Grafting of Si QDs and Their Application for Bioimaging

One of the major drawbacks of SiQDs for application for bioimaging is their instability over time in water or buffer solutions. It is well-known that PS surface slowly oxidizes in air which is resulting in degradation of optical and luminescent properties. Several methods have been used to achieve a change in the surface composition for improving the stability of materials. The PS passivation can be achieved by replacing the Si–H_x bonds with stronger S–C and Si–OC bonds [39, 40]. Fortunately, the hydride species which cover the PS surface after Si etching provide a versatile starting point for various reactions, allowing the attachment of different species. An application of carbon grafting is prospective due to the fact that carbon directly bonded to silicon stabilizes porous Si against dissolution in aqueous solutions [41].

Si–C bonds are usually formed on hydride-terminated porous Si surfaces by hydrosilylation, i.e. an addition of Si–H fragments to unsaturated bonds of alkene or alkyne according to Scheme 1 [39, 41–46].

Aryl and alkyl radicals of different length, aminogroups, thiols, carboxylic acid and other variable functionalities were successfully attached on the surface of PS as well as Si QDs by this approach. Differently from organosilicon compounds, the hydrosilylation on the surface of Si nanostructures does not require metalocomplex catalysts such as H_2PtCl_6; the $Si_{4-x}SiH_x$ groups can efficiently be activated to hydrosilylation by temperature (approx. 130 °C), visible light or Lewis acid catalysts [39, 44–46]. As an alternative to hydrosilylation, covalently attached organic layers can be formed on the porous Si surfaces under treatment with Grignard and alkyl- or aryllithium reagents, which is considered in the reviews [41, 47]. An interaction of surface SiH_x groups with alcohols at elevated temperature resulted in alkoxy-coated Si particles [1, 48]. In general, alkyl coated Si QDs demonstrate superior qualities, such as better oxidation and hydrolytic stability as well as PL efficiency, comparing to unmodified SiH_x coated QDs [9].

In [49] the Si QDs with alkyl passivation of the surface were obtained through mechanical milling of Si wafers in a reactive liquid medium like alkynes and alkenes under an inert atmosphere. The reaction of the terminal triple or double bond with the reactive Si–Si and silicon surface radicals results in the formation of covalent Si–C bonds, which are preventing further oxidation of the silicon surface. This cycloaddition of unsaturated hydrocarbons resulted in organic passivated Si QDs. The milling in monofunctionalized organic liquids forms Si QDs that are soluble exclusively in organic solvents.

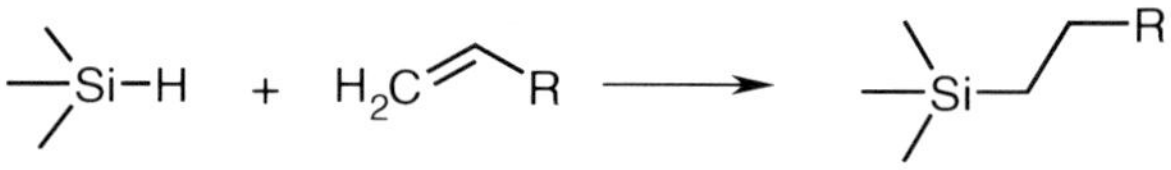

Scheme 1 Hydrosilylation on the surface of Si nanostructures

Fig. 8 3 nm H–SiQDs are oxidized in EtOH–H_2O_2; emission of the oxidized SiQDs ranges from salmon pink to blue (from [51])

Very stable and bright emitting amine-terminated Si QDs with different alkyl chain lengths between the Si core and amine end-group were synthesized in [50]. The obtained Si QDs have a spherical shape and homogeneous size distribution (1.6 nm). Their emission can be tuned from the UV to the blue spectral region, in a controllable fashion, by only changing the alkyl spacer length. The emission quantum yields are 12 % for all synthesized Si QDs. Excited state lifetimes are in the ns range and point to a direct band gap excitation. NH_2-terminated Si QDs exhibit an exceptional stability over a wide pH range (1–13) and high temperatures (120 °C).

Lee and co-workers recently presented an EtOH/H_2O_2—assisted oxidation method to synthesize water-dispersed Si/SiO_xH_y core/shell QDs with a Si core of different controlled diameters [51]. Significantly, this method allows for fine tuning emission wavelengths of QDs, producing seven luminescent colors from blue to red, the fluorescent silicon nanospheres (Si NSs) containing several hundreds of SiQDs (Fig. 8). The as-prepared nanospheres possess excellent aqueous dispersibility, strong fluorescence (quantum yield: ~15 %), robust photo-stability and favorable biocompatibility. They developed a new kind of water-dispersed oxidized SiNSs (O–SiNSs) prepared via thermal oxidation of the precursor SiNSs [52, 53]. The quantum yield of the O–SiNSs was dramatically increased to as high as 25 %. More significantly, O–SiNSs are stable under high-power UV irradiation and in acidic-to-basic environments covering pH = 2–12. This extremely high pH stability leads to facile conjugation of nanospheres with antibodies, resulting in brightly luminescent silicon bioconjugates for immunofluorescent bioimaging (Fig. 9).

A production of stable silicon QDs with various surface functionalities through microwave-assisted hydrosilylation of hydride-terminated silicon QDs in the presence of various reactive compounds (decene, undecylenyl alcohol, and undecenoic acid) are considered in [54]. Reagents were selected from bifunctional compounds. One functional group should be the C=C which is involved in the hydrosilylation process and the other group could bear any other functionality to enable widespread applications of SiQDs.

A significant enhancement of the PL efficiency is observed for aqueous suspensions of PS particles coated by bioresorbable polymers, i.e., polylactic-co-glycolic acid (PLGA) and polyvinyl alcohol (PVA). The inner hydrophobic

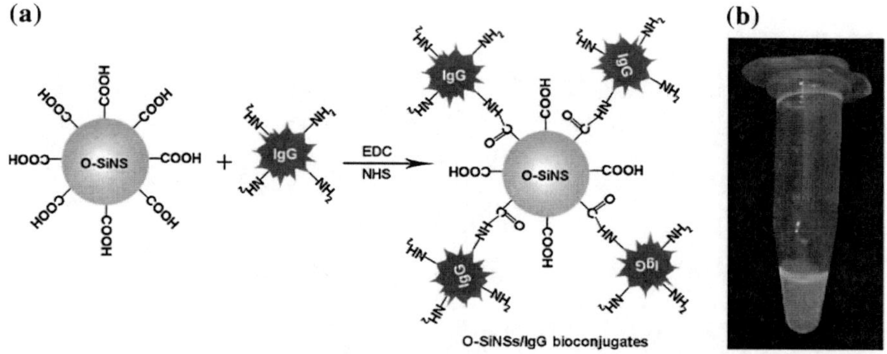

Fig. 9 **a** Schematic illustration of silicon nanospheres conjugating with goat-anti-mouse IgG. The carboxylic acid groups of nanospheres readily reacted with the amino groups of IgG by using EDC and NHS as zero-length cross-linkers (Figure is not to scale). **b** Optical micrographs of the prepared silicon bioconjugates under UV (365 nm) irradiation. The prepared bioconjugates preserve stable and bright fluorescent intensity for over 1 month (*right*) (from [53])

PLGA layer protects the QDs from the reaction with water, while the outer PVA layer makes the QDs hydrophilic [55]. UV-induced graft polymerization of acrylic acid on the surface of silicon nanoparticles was used to prepare a stable aqueous luminescent silicon QDs solution. By grafting a water-soluble polymer on the particle surface, the dispersions in water of the silicon nanoparticles became very stable and clear aqueous solutions could be obtained as in [56]. Hydrogen-capped silicon QDs with strong blue PL were synthesized by the metathesis reaction of sodium silicide, NaSi, with NH_4Br. The hydrogen-capped Si QDs were further terminated with octyl groups and then coated with a polymer to render them water-soluble [57]. Silicon QDs were prepared and functionalized with alkene-terminated poly (ethylene oxide) to impart amphiphilic solution properties to the particles [58]. The alkyl-passivated Si nanocolloids were obtained via thermally-initiated hydrosilylation of the hydrogenated Si nanoparticles, formed by electrochemical etching of Si wafer, with 1-undecene [59].

The carboxylic acid functionalized Si QDs (with the using of short alkyl chain organic peroxide, succinic acid peroxide) were applied for HeLa cells imaging [60] in several emission ranges. HeLa or heLa cell, is an immortalized cell type used in scientific research in vitro. The HeLa cells treated the carboxylic acid terminated Si QDs demonstrate the fluorescence with wide emission spectra arising from Si QDs (Fig. 10b–d). The bright blue fluorescence (450 nm) from the Si QDs is distributed uniformly and shows that the Si QDs were taken up into the cytoplasm (Fig. 10b). Remarkably, when the image for the 515 nm channel was collected, a bright green fluorescence from the internalized Si–QDs was clearly observed from the robust optical signal of the cells (Fig. 10c). Red fluorescence from the internalized Si QDs was also observed for the 605 nm channel (Fig. 10d). This behavior indicates that it might be possible to use hydrophilic Si QDs as chromophores for biological fluorescence imaging.

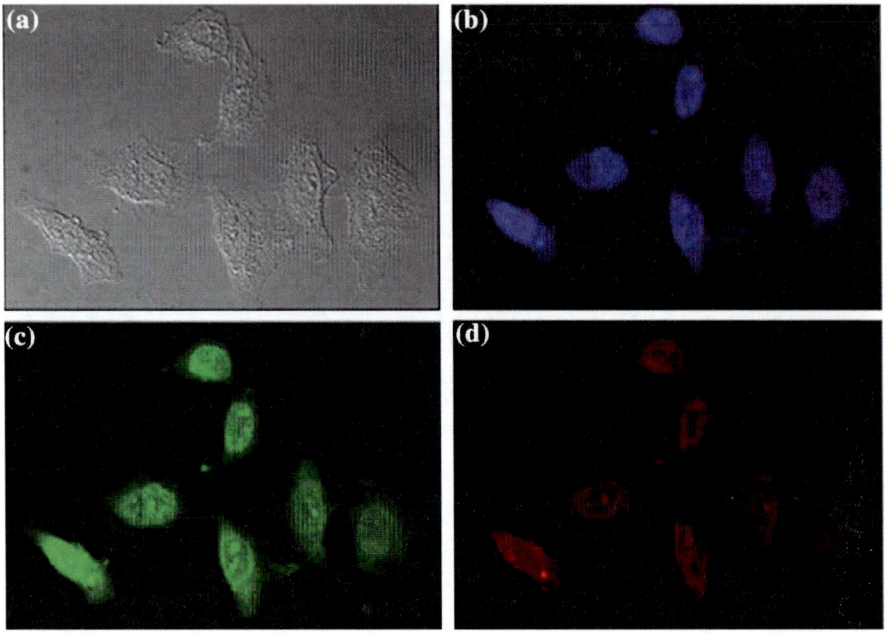

Fig. 10 Confocal microscopic visualisation of HeLa cells treated with carboxylic acid terminated Si QDs collected at different channels: **a** bright field image; and luminescence images collected at: **b** 450 nm, **c** 515 nm, and **d** 605 nm (from [60])

Application of functionalized luminescent Si QDs for bioimaging of tumors are considered in [18, 61, 62]. The micelle encapsulated SiQDs were injected intravenously into a Balb C mouse which was then sacrificed after 24 h. The mouse was dissected and the organs were imaged to observe where the SiQDs were concentrating in the body (Fig. 11). The QDs concentrated preferentially in the spleen with some seen in the liver. The particles are not seen in the heart, lungs, kidneys, or the brain [62].

5 Fabrication and PL Properties of Colloidal Solutions with SiCQDs

Luminescent properties of silicon carbide (SiC) nanostructures have been extensively studied during about last 15 years. In comparison to extremely weak PL of bulk SiC substrates at room temperature, PL intensity of the SiC nanostructures with dimensions <10 nm is significantly enhanced [63]. Particularly, room temperature PL of porous SiC nanostructures obtained by electrochemical etching has received a special attention from scientists. Radiative recombinations via surface states and impurity levels have been discussed in literature as main mechanisms of the observed PL signals coming from the SiC nanostructures [64, 65]. In [66] it was

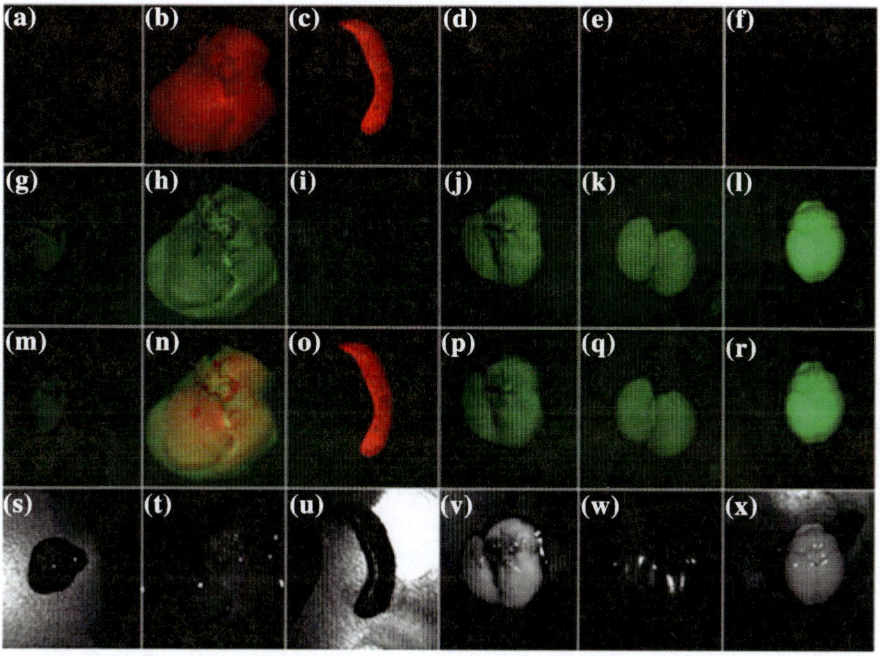

Fig. 11 Images of organs of a Balb C mouse 24 h after injection of the Si QDs. A, G, M, S: *Heart* B, H, NN, T: *Liver* C, I, O, U: *Spleen* D, J, P, V: *Lungs* E, K, Q, W: *Kidneys* F, L, R, X: *Brain*. *Top Row* Fluorescence; *Second Row* Autofluorescence; *Third Row* Overlay; *Fourth Row* White light picture (from [62])

reported about ultraviolet PL from SiC nanoparticles with dimensions below 3 nm, which were extracted from a porous layer formed by electrochemical etching. In particular, quantum confinement effect of photogenerated charge carriers was discussed by the authors to explain an important PL signal detected at energies higher than 3 eV. In addition, it has been also shown that when the non-radiative surface states were efficiently passivated by solvent molecules and concentration of the nanoparticles in the solutions was sufficiently low reducing considerably physical interactions between the nanoparticles, significant above-gap PL spectral tails due to recombination of quantum confined photoexcited carriers could be clearly seen [67]. Several years ago, it has been reported on experimental evidence of quantum confinement effect in the PL spectra of colloidal suspensions of SiC nanoparticles with cubic crystalline symmetry (3C-polytype) obtained by an ultrasonic treatment of porous 3C–SiC layer in various solvents [68, 69]. As the excitation wavelength was increased, the colloidal nanocrystals were found to give a strong above-gap emission band whose maximum was redshifted in the range from 2.8 to 2.2 eV. At the same time, as expected by the quantum confinement model, the emission intensity rises up to a maximum before decreasing which correlates to the size distribution supported by TEM measurements.

The method of forming SiC QDs obtained by electrochemical etching of bulk substrates of 3C polytype was described in [70, 71]. Firstly, 3C–SiC nano-porous layer was formed by means of electrochemical anodization of a low resistivity grade (<1 Ω cm) bulk 3C–SiC polycrystalline wafer. The etching process took place for 2–3 h at a current density of 25 mA/cm^2 using a 1:1 HF(50 %)/ethanol electrolyte. After the etching, a highly porous network constituted by numerous interconnected 3C–SiC nanocrystals was formed. The ultra-porous layer was washed with water, naturally dried in ambient air and then removed from the wafer. An intense mechanical dry grinding of the formed free nano-porous layer transformed it into a nano-powder state. The as-prepared 3C–SiC nanopowder can be dispersed in various polar solutions. To change surface charges, the suspended QDs are treated in different surfactants like cetyl-tri-methyl-ammonium bromide (CTMA) and sodium lauryl sulphate (SLS) which are added to the colloidal solutions to communicate positive and negative surface charges to the QDs, respectively.

FTIR spectra of the SiC QDs recorded in transmittance (pellet in KBr) and diffuse reflectance modes, are presented in Fig. 12. The assignment of the bands coincides in general with results obtained for porous 6H–SiC and 3C–SiC [70–72]. Wide ν (OH) band centred at about 3,500 cm^{-1} corresponds to the adsorbed water molecules as well as to C–OH and Si–OH groups. Multiple bands at 2,985, 2,943 and 2,914 cm^{-1} (ν(CH_x)) and 1,469, 1,450 and 1,375 δ (CH_x) indicate on presence of different types of aliphatic CH_x groups in the SiC QDs. Intense ν (C=O) band at 1,743 cm^{-1} which is clearly seen in the spectrum of SiC sample, can be assigned not only to the presence of carboxylic acid groups, but also to their esters such as –$COOC_2H_5$, because the position of the band is shifted to higher wavenumbers comparing to common spectra of carboxylic acids (usually the C=O band of hydrogen bonded –COOH groups is observed at 1,700–1,725 cm^{-1}, however for –COOH group without H-bonds it can be found at 1,770 cm^{-1}). Wide band at centred at 1,213 cm^{-1} probably appeared due to the combination of ν (C–O) band of –COOH groups, ν (Si–O) band and some lattice bands of SiC.

Fig. 12 FTIR spectra (transmittance and diffuse reflectance) of palletized and ground 3C–SiC samples (from [70])

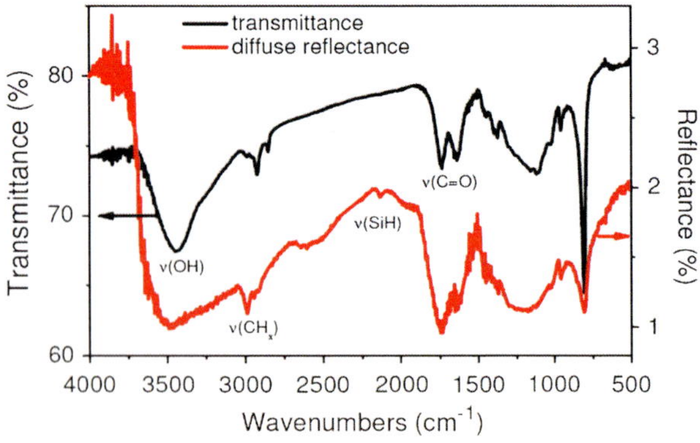

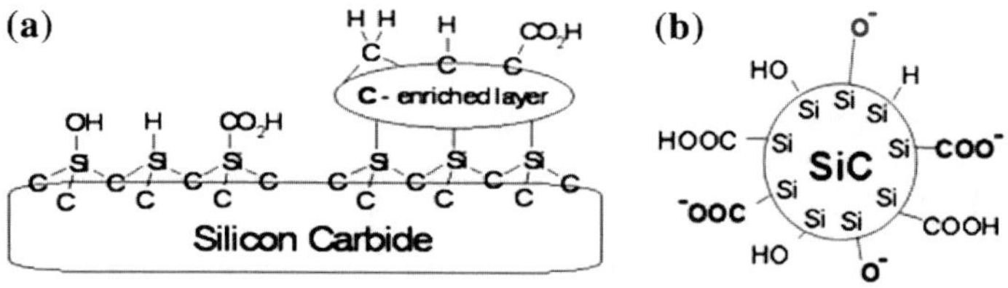

Fig. 13 Schematic representation of: **a** general surface chemistry and **b** surface charges of 3C–SiC nanoparticles [70]

The surface groups of the 3C–SiC nano-powder are very similar to those of recently studied 6H–SiC porous nanostructures and schematically represented by a scheme in Fig. 13a. According to [73], the 3C–SiC QDs surface contains carbon-enriched layer appeared mainly due to the preferential removing of Si atoms during the electrochemically assisted nano-powder formation. Dissociation of the carboxylic (Si–COOH) and/or silanol (Si–OH) acid groups in polar solvents leads to appearance of negative electric charges at the nanoparticle surface as it is illustrated by scheme in Fig. 13b. In particular, these charges are responsible for extremely stable and optically homogeneous colloidal solutions obtained with the 3C–SiC QDs dispersed in solvents.

Electrostatic screening of the radiative band gap states by highly polar solvent media leads to a blue shift and a decrease in the width at half maximum of the PL spectra of the QDs. As for the surface charges, they govern band bending slope and thus influence strongly the radiative transitions via energy states in the band gap.

Transmission electron microscopy (TEM) image of the spherical-like SiC QDs is given in Fig. 14a. An example of a daylight photo of the centrifuged optically homogeneous colloidal suspension of the 3C–SiC is shown by the inset in Fig. 14a. Typical size distribution of the 3C–SiC QDs is presented in Fig. 14b. As one can see, the majority of the QDs dimensions are found to be below 5 nm with the most probable size value being around 2.5 nm. The inset in Fig. 14b shows typical PL spectrum of the 3C–SiC QDs dispersed in aqueous solutions under UV excitation. Since the mean size value of the SiC QDs are smaller then the Bohr's diameter of the exciton in bulk SiC substrate ($\sim$5.4 nm), they exhibit highly efficient luminescence with energies higher than bandgap energy of bulk SiC due to spatial and quantum confinement effects [71].

Well defined spherical colloidal cubic SiC QDs with average diameter of 5 nm were synthesized using the reactive bonding and wet chemical etching method in [74, 75]. These QDs show strong violet-blue PL emission. ATR–IR measurements revealed the surface structure of the SiC QDs which consists of Si–O–Si, C–O–C, CH, COOH, and COO-surface terminations as for electrochemically prepared QDs.

There are only few reports on surface functionalization of SiC nanoparticles [4, 75–79]. Polymerization of the conducting polyaniline in the presence of camphorsulfonic acid and SiC nanoparticles results in the formation of polyaniline-camphorsulfonic acid shell with a thickness in the range from 0.5 nm to a few

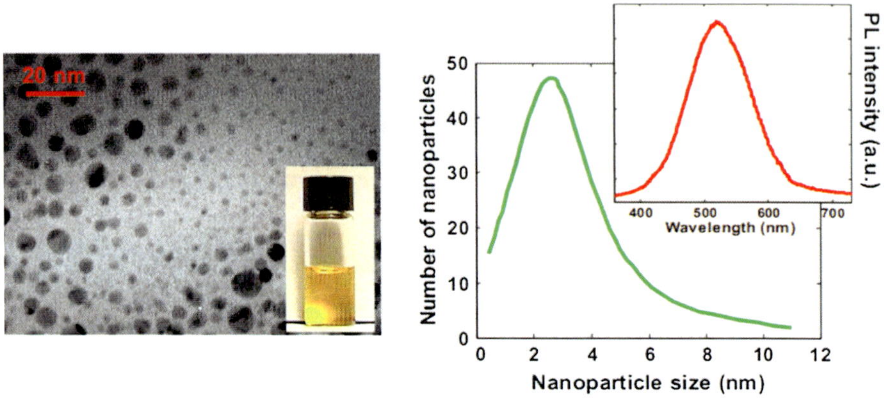

Fig. 14 *Left* The general TEM image of the SiC QDs. Inset shows a colloidal solution with the dispersed SiC QDs; Right size distribution of the SiC QDs. Inset shows a typical PL spectrum of the colloidal 3C–SiC QDs (from [81])

nanometers at the surface of SiC nanocrystals [79]. Zhang et al. [76] reports on PL and light reabsorption in the SiC QDs embedded in the binary poly(allylamine hydrochloride)-sodium poly(styrene sulfonate) polyelectrolytes solid matrix. The emission maximum wavelength can be tuned over a wide blue-violet spectral region indicating the quantum confinement as the origin of light emission.

6 SiC QDs as Fluorescent Bio-Labels

The highly luminescent 3C–SiC QDs was successfully explored as fluorescent agents for living cell imaging. By using fluorescent microscope we studied the cells of different plants, mouse fibroblast, healthy and cancer epithelial human cells with SiC QDs [8, 71, 80, 81]. Covalent grafting of amino groups onto the carboxylic acid functionalities, naturally covering the surface of fluorescent nanoparticles produced from silicon carbide, allowed tuning of their surface charge from negative to highly positive (Fig. 15). Grafting of aminogroups was performed by the reaction of SiC QDs surface carboxylic acid groups with ethylenediamine (En, $H_2N–CH_2CH_2–NH_2$) (Fig. 15).

Incubating 3T3-L1 fibroblast cells with differently charged SiC QDs demonstrates the crucial role of the charge in cell fluorescent targeting [71]. Negatively charged SiC QDs concentrate inside the cell nuclei. Close to neutrally charged SiC QDs are present in both cytoplasm and nuclei while positively charged SiC QDs are present only in the cytoplasm and are not able to move inside the nuclei (Fig. 16). This effect opens the door for the use of SiC QDs for easy and fast visualization of long-lasting biological processes taking place in the cell cytosol or nucleus as well as providing a new long-term cell imaging tool. Moreover, here we

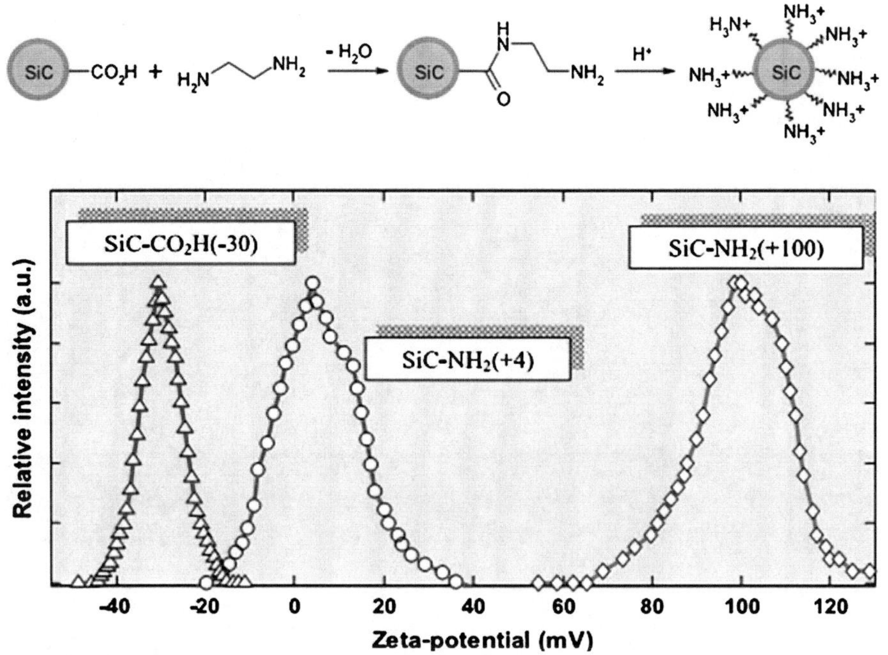

Fig. 15 *Top* Grafting of amino groups on the surface of as-prepared SiC QDs; *Bottom* Zeta-potential measurements performed on colloidal solutions of SiC–CO_2 H(−30), SiC–NH_2 (+4) and SiC–NH_2 (+100) (from [71])

have shown that the interaction between charged QDs and nuclear pore complex plays an essential role in their penetration into the nuclei. The importance of electrostatic interactions was more recently strengthened by authors studying the molecular structure of the yeast nuclear pore complex (NPC) and the translocation of model particles [82].

The uptake of QDs and its intra-nuclei concentration is dependant to cell proliferation (Fig. 17). This has been shown by two different ways, first in healthy cells and second in cancer cells [81]. In healthy SG cells (a human immortalized gingival epithelioid cell line), the labeling is altered by the state of confluence, at confluence the labeling is low and absent in the nuclei (c) while it is strong and present in the nuclei when cells divide (a and b). In cancer cells HSC-2 (a human oral squamous carcinoma line), that proliferate even at confluence, the labeling is not influence by confluence (d, e and f). Although the mechanism linking intra-nuclear accumulation of SiC QDs and cell proliferation has not been yet identified, this observation is important for at least two reasons. Firstly, it brings evidence of a specificity of SiC QDs to be uptaken by proliferating cells that are promising to give specificity for cancer treatment or drug delivery with SiC QDs.

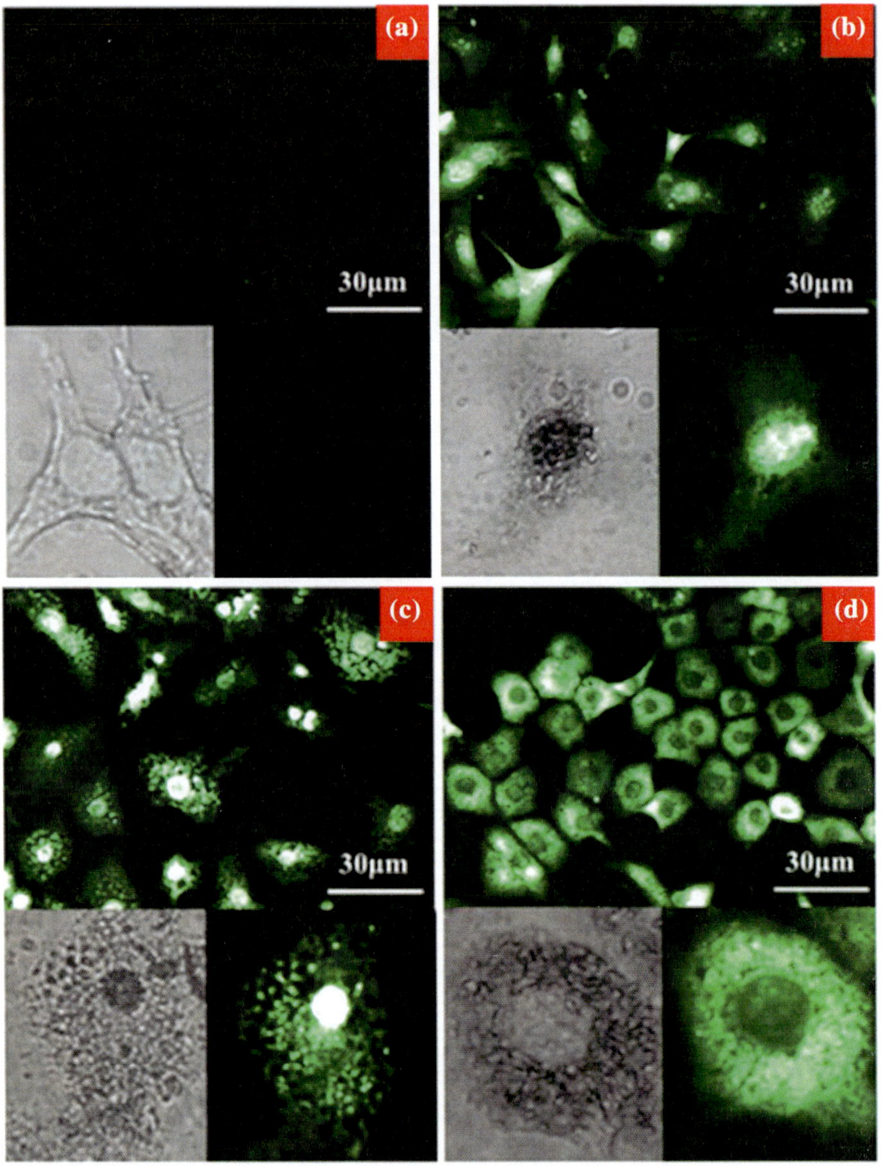

Fig. 16 Fluorescence and visible microscopy images of 3T3-L1 cells: **a** without SiC QDs; **b** labeled with negatively charged SiC–CO_2H(–30) QDs; **c** labeled with quasineutral SiC–NH_2 (+4) QDs and **d** labeled with positively charged SiC–NH_2 (+100) QDs (from [71])

Secondly, a majority of studies on QDs are done on proliferating cells, according to the present results, the behaviour of QDs may be not the same for non confluent (proliferating) and confluent (non proliferating) cells.

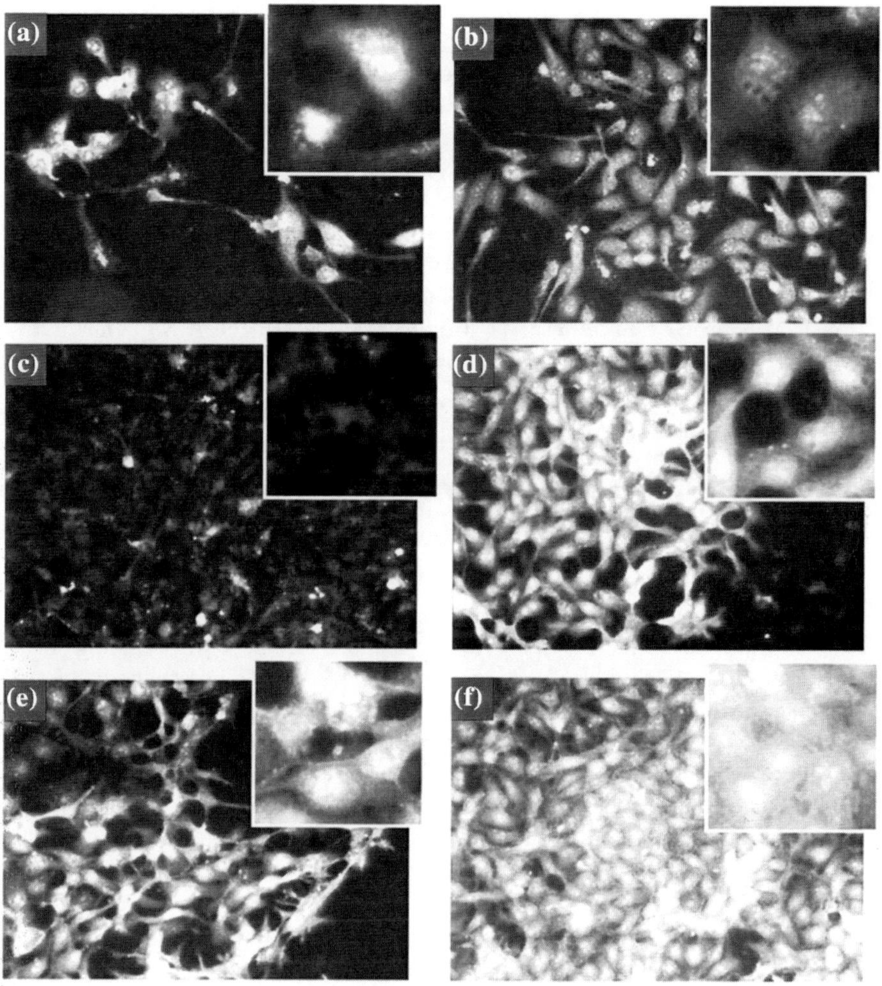

Fig. 17 Fluorescence microscopy images of SG (**a**, **b**, **c**) and HSC-2 (**d**, **e**, **f**) cells labeled by SiC QDs: (**a**), (**d**) low confluence (**b**), (**e**) intermediate confluence, (**c**), (**f**) total confluence (from [81])

To test the cytotoxic potential of SiC QDs, we chose two oral squamous carcinoma and one immortalized oral epithelioid lines. Like the two cancer lines, the control cells we used possess a high growth rate and are used as a control in comparison to HSC-2 cells [83]. Higher concentrations of SiC QDs and longer exposure time are required to reduce S–G proliferation when compared to both human and murine tumoral cells. The differential toxicity of SiC QDs against rapidly dividing cancer cells suggests their potential use as antitumoral agents. The precise mechanism(s) of action of SiC QDs in inhibiting cell proliferation remains to be investigated. We considered the possibility of using QDs for antitumor therapy. When SiC QDs concentration of 2 mg/mL kills 100 % of the cancer cells HSC-2 is already at 6:00 incubation and almost no effect on the survival of the culture of healthy cells that have stopped cell division through a mechanism of contact inhibition. This means that toxic effects, as well as the total amount of nanoparticles in the cell decreases dramatically when the cell stops proliferation (division) [8].

SiC QDs have, however, few disadvantages preventing their use in complex bio-imaging applications. This includes low quantum yield and rather broad emission band, which makes multicolour imaging difficult. Low quantum yield also influences concentration at which labeling effect is detectable. Therefore, to compete with usual fluorophores, a strong PL enhancement of SiC QDs may be ensured by their nearfield coupling with multipolar plasmons [84]. In order to achieve a high PL enhancement factors, an approach based on: (1) localization of SiC QDs in the vicinity of randomly arranged silver (Ag) nanoparticles (NPs) chemically formed on silicon nitride (SiN_X)/glass substrates and (2) careful tuning of the multipolar plasmon bands of the nanoparticles to overlap the excitation and emission spectral bands of SiC QDs, has been developed.

Fluorescence images of fibroblast cells on glass and plasmonic substrates with or without SiC QDs can be seen in Fig. 18. All images were obtained under the same observation conditions in terms of excitation (UV/violet excitation band ($\lambda = 350$–460 nm) and an observation spectral range corresponding to the wavelengths $\lambda > 470$ nm) and acquisition time. Natural fluorescence of the cells grown on the usual glass cover slip substrates is too weak to be observed in such experimental conditions (Fig. 18a). On the other hand, the cells grown on the glass substrates and labeled with the SiC QDs can be better seen in Fig. 18b. Strong auto-fluorescence enhancement of the fibroblast cells grown onto the nano-Ag/SiN_X/glass substrates is shown in Fig. 18c. This label-free natural green cell auto-fluorescence is significantly increased in comparison with the cells grown on a simple glass substrate (Fig. 18a). The cells with the up-taken SiC QDs grown onto the nano-Ag/SiN_X/glass substrates show a drastic fluorescence enhancement (Fig. 18d) in comparison to the cells containing the same quantity of the SiC QDs and grown on a usually-used glass cover slips (Fig. 18b).

Different mechanisms can be responsible for the observed fluorescence enhancement of the cells. Firstly, the localized plasmons appearing in the Ag NPs leads to an increase of excitation/absorption and photostimulated auto-emission of the cell membranes due to their close proximity to the Ag NPs. The fluorescence enhancement can be only observable when the plasmons are localized in very close vicinity to the cell membrane. Moreover, the cell membrane thickness is estimated to be around a few nanometers which allows plasmon-induced PL enhancement of the NPs up-taken by the cells and localized near the membrane after the dehydration and the subsequent fixation of the cells on the plasmonic nano-Ag/SiN_X/glass substrate. All these factors can explain the detected enhancement of cell auto-fluorescence and PL of the up-taken SiC QDs (see Fig. 18c and d, respectively).

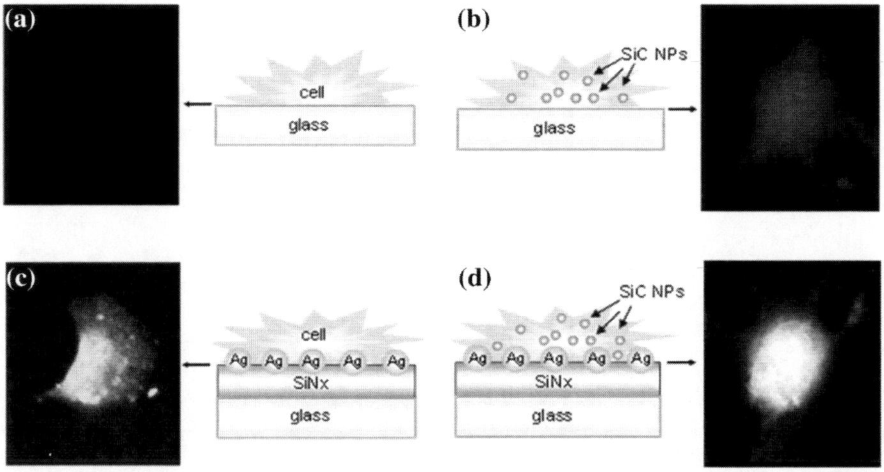

Fig. 18 Sample schematics and corresponding fluorescent images of fibroblast cells: **a** on glass substrate, **b** labelled with the SiC QDs on glass substrate, **c** on nano-Ag/SiN_X/glass substrate, **d** labelled with the SiC NPs on nano-Ag/SiNX/glass substrate

Secondly, the cell labeling with the SiC QDs can induce modification of the local effective dielectric constant which will lead to a better near-filed coupling (in terms of spectral overlap) with the localized plasmon modes under the same excitation conditions. In addition, the higher values of the local dielectric constant will also ensure the higher intensities of the plasmon-induced local electric field penetrating deeper inside the labelled cells. Finally, a certain degree of the excitation scattering within the nanostructured area and inside the cells may explain general PL enhancement taking place in the whole cell volume. Moreover, the use of a UV/violet filter selecting the exciting photons from the relatively large spectral range ($\lambda = 350$–460 nm) of a white lamp leads to a simultaneous photo-excitation of a wide absorptive part of the extinction spectra, provoking significant increase of the absorption cross-sections of numerous cell fluorescent agents as well as of the big number of the SiC QDs with various sizes.

In addition, the auto-fluorescence cell signal and PL of the SiC QDs incorporated inside the cells (detected in the spectral range $\lambda > 470$ nm) are strongly enhanced by near-field interaction with the coupled dipole plasmon modes, as it is discussed in details in [84].

7 Conclusion

This chapter focuses on the preparation of colloidal solutions of Si and SiC QDs by the method of electrochemical etching of bulk Si and 3C–SiC. It was shown that a broad size distribution of crystalline Si and SiC QDs are obtained. It is worth noting that the dimensions the dimensions of the Si QDs undergone filtering in colloidal solution vary discretely with a radius quantum equal to 0.12 nm. Existing of this quantum may correspond to step-like increasing of Si QDs radius on one new shell at the surface of Si QDs. The formed QDs show intense luminescence in visual region that is promising for fluorescent labeling of biomaterials. However, one of the major drawbacks of SiQDs for application for bioimaging is instability over time in water or buffer solutions. To overcome this drawback the several methods of surface functionalization are proposed and discussed. From the other side, the SiC QDs are stable in water solutions and do not require supplementary surface functionalisation for bioimaging. A strong fluorescence from the SiC QDs, which undoubtedly penetrate into the cell, has been observed. The studying of health and cancer cells using SiC QDs show that simple modification of surface charge of QDs give strong opportunity to target the same QDs in intracellular space with their preferential localisation inside or outside the cell nucleus. Electrostatic interaction between QDs and positively charged chromatin gives platform to completely switch-open penetration or it blocking corresponding to the surface charge of QDs that plays role of key to open/close nucleus membrane doors. These effects open wide perspectives for understanding and discovering nucleus transport mechanisms and give possibility for their bright imaging. Moreover, the uptake of QDs and its intra-nuclei concentration is strongly dependant on cell proliferation.

References

1. Kumar, V. (ed.): Nanosilicon. Elsevier Ltd, Amsterdam (2007)
2. Geszke-Moritz, M., Moritz, M.: Quantum dots as versatile probes in medical sciences: Synthesis, modification and properties. Mater. Sci. Eng., C **33**, 1008–1021 (2013)
3. Michalet, X., Pinaud, F.F., Bentolila, L.A., Tsay, J.M., Doose, S., Li, J.J., Sundaresan, G., Wu, A.M., Gambhir, S.S., Weiss, S.: Quantum dots for live cells, in vivo imaging, and diagnostics. Science **307**, 538–544 (2005)
4. Fan, J., Chu, P.K.: Group IV nanoparticles: synthesis, properties, and biological applications. Small **6**, 2080–2098 (2010)
5. Alivisatos, A.P.: The use of nanocrystals in biological detection. Nat. Biotechnol. **22**, 47–52 (2004)
6. Parak, W.J., Pellegrino, T., Plank, C.: Labelling of cells with quantum dots. Nanotechnology **16**, R9–R25 (2005)
7. Bonacina, L.: Nonlinear nanomedecine: harmonic nanoparticles toward targeted diagnosis and therapy. Mol. Pharm. **10**, 783–792 (2013)
8. Bluet, J.-M., Botsoa, J., Zakharko, Y., Géloën, A., Alekseev, S., Marty, O., Mognetti, B., Patskovsky, S., Rioux, D., Lysenko, V.: SiC as a biocompatible marker for cell labeling, Chapter 11. In: Silicon Carbide Biotechnology, Saddow, S (ed) Elsevier Inc., pp. 327–429. 2012
9. Kang, Z., Liu, Y., Lee, S.-T.: Small-sized silicon nanoparticles: new nanolights and nanocatalysts. Nanoscale **3**, 777–791 (2011)
10. Dabbousi, B.O., Rodriguez-Viejo, J., Mikulec, F.V., Heine, J.R., Mattoussi, H., Ober, R., Jensen, K.F., Bawendi, M.G.: CdSe-ZnS core-shell quantum dots: synthesis and characterization of a size series of highly luminescent nanocrystallites. J. Phys. Chem. B. **101**, 9463–9475 (1997)

11. Nozik, A.J.: Spectroscopy and hot electron relaxation dynamics in semiconductor quantum wells and quantum dots. Annu. Rev. Phys. Chem. **52**, 193–231 (2001)
12. Giljohann, D.A., Mirkin, C.A.: Drivers of biodiagnostic development. Nature **462**, 461–464 (2009)
13. Hines, M.A., Guyot-Sionnest, P.: Synthesis and characterization of strongly luminescing ZnS-capped CdSe nanocrystals. J. Phys. Chem. **100**, 468–471 (1996)
14. Bruchez Jr, M., Moronne, M., Gin, P., Weiss, S., Alivisatos, A.P.: Semiconductor nanocrystals as fluorescent biological labels. Science **281**, 2013–2016 (1998)
15. Derfus, A.M., Chan, W.C.W.: Bhatia SN, Probing the cytotoxicity of semiconductor quantum dots. Nano Lett. **4**, 11–18 (2004)
16. Kirchner, C., Liedl, T., Kudera, S., Pellegrino, T., Munoz Javier, A., Gaub, H.E., Stölzle, S., Fertig, N., Parak, W.J.: Cytotoxicity of colloidal CdSe and CdSe/ZnS nanoparticles. Nano Lett. **5**, 331–338 (2005)
17. Fojtik, A., Henglein, A.: Luminescent colloidal silicon particles. Chem. Phys. Lett. **221**, 363–367 (1994)
18. Erogbogbo, F., Yong, K.-T., Roy, I., Hu, R., Law, W.-C., Zhao, W., Ding, H., Wu, F., Kumar, R., Swihart, M.T., Prasad, P.N. In Vivo Targeted Cancer Imaging, Sentinel Lymph Node Mapping and Multi-Channel Imaging with Biocompatible Silicon Nanocrystals, ACS NANO. **5**, 413–423 (2011)
19. Knipping, J., Wiggers, H., Rellinghaus, B., Roth, P., Konjhodzic, D., Meier, C.: Synthesis of high purity silicon nanoparticles in a low pressure microwave reactor. J. Nanosci. Nanotechn. **4**, 1039–1044 (2004)
20. Sankaran, R.M., Holunga, D., Flagan, R.C., Giapis, K.P.: Synthesis of blue luminescent si nanoparticles using atmospheric-pressure microdischarges. Nano Lett. **5**, 537–541 (2005)
21. Rowsell, B.D., Veinot, J.G.C.: Reductive thermolysis of a heterocyclic precursor: a convenient method for preparing luminescent, surfactant-stabilized silicon nanoparticles. Nanotechnology **16**, 732–736 (2005)
22. Baldwin, R.K., Pettigrew, K.A., Ratai, E., Augustine, M.P. Kauzlarich, S.M.: Solution reduction synthesis of surface stabilized silicon nanoparticles, Chem. Commun. **17**, 1822–1823 (2002)
23. Zou, J., Baldwin, R.K., Pettigrew, K.A., Kauzlarich, S.M.: Solution synthesis of ultrastable luminescent siloxane-coated silicon nanoparticles. Nano Lett. **4**, 1181–1186 (2004)
24. Rao, S., Sutin, J., Clegg, R., Gratton, E., Nayfeh, M.H., Habbal, S., Tsolakidis, A., Martin, R.M. Excited states of tetrahedral single-core Si29 nanoparticles. Phys. Rev. B . **69**, art.205319 (2004)
25. Tanaka, A., Saito, R., Kamikake, T., Imamura, M., Yasuda, H.: Electronic structures and optical properties of butyl-passivated Si nanoparticles. Solid State Commun. **140**, 400–403 (2006)
26. Lysenko, V., Onyskevych, V., Marty, O., Skryshevsky, V.A., Chevolot, Y., Bru-Chevallier, C.: Extraction of ultraviolet emitting silicon species from strongly hydrogenated nanoporous silicon, App. Phys. Let. **92**, art.251910 (2008)
27. Lysenko, V., Bidault, F., Alekseev, S., Turpin, C., Geobaldo, F., Rivolo, P., Garrone, E., Zaitsev, V., Barbier, D.: Study of porous silicon nanostructures as hydrogen reservoirs. J. Phys. Chem. B. **109**, 19711–19718 (2005)
28. Nychyporuk, T., Lysenko, V., Barbier, D. Fractal nature of porous silicon nanocrystallites, Phys. Rev. B. **71**, art. 115402 (2005)
29. Serdiuk, T., Lysenko, V., Alekseev, S., Skryshevsky, V.A.: Size tuning of luminescent silicon nanoparticles with meso-porous silicon membranes. J. Colloid Interface Sci. **364**, 65–70 (2011)
30. Serdiuk, T., Skryshevsky, V.A., Ivanov, I.I., Lysenko, V.: Storage of luminescent nanoparticles in porous silicon: Toward a solid state golden fleece. Mater. Lett. **65**, 2514–2517 (2011)

31. Belomoin, G., Therrien, J., Smith, A., Rao, S., Twesten, R., Chaieb, S., Nayfeh, M.H., Wagner, L., Mitas, L.: Observation of a magic discrete family of ultrabright Si nanoparticles, Appl. Phys. Lett. **80**, art. 841 (2002)
32. Patrone, L., Nelson, D., Safarov, V.I., Sentis, M., Marine, W.: Photoluminescence of silicon nanoclusters with reduced size dispersion produced by laser ablation, J. Appl. Phys. **87**, art. 3829 (2000)
33. Ledoux, G., Guillois, O., Porterat, D., Reynaud, C.: Photoluminescence properties of silicon nanocrystals as a function of their size. Phys. Rev. B. **62**, 15942–15951 (2000)
34. Feng, Z.C., Tsu, R. (eds.): Porous Silicon. World Scientific, Singapore (1994)
35. Delerue, C., Allan, G., Lannoo, M.: Theoretical aspects of the luminescence of porous silicon. Phys. Rev. B. **48**, 11024–11036 (1993)
36. Kang, Z.H., Tsang, C.H.A., Zhang, Z.D., Zhang, M.L., Wong, N.B., Zapien, J.A., Shan, Y.Y., Lee, S.T.: A polyoxometalate-assisted electrochemical method for silicon nanostructures preparation: from quantum dots to nanowires. J. Am. Chem. Soc. **129**, 5326–5327 (2007)
37. Kang, Z.H., Tsang, C.H.A., Wong, N.B., Zhang, Z.D., Lee, S.T.: Silicon quantum dots: a general photocatalyst for reduction, decomposition, and selective oxidation reactions. J. Am. Chem. Soc. **129**, 12090–12091 (2007)
38. Heinrich, J.L., Curtis, C.L., Credo, G.M., Kavanagh, K.L., Saylor, M.J.: Luminescent colloidal silicon suspensions from porous silicon. Science **255**, 66–68 (1992)
39. Buriak, J.M., Allen, M.J.: Lewis acid mediated functionalization of porous silicon with substituted alkenes and alkynes. J. Am. Chem. Soc. **120**, 1339–1340 (1998)
40. Manilov, A.I., Skryshevsky, V.A.: Hydrogen in porous silicon—a review. Mater. Sci. Eng., B **178**, 942–955 (2013)
41. Anglin, E.J., Cheng, L., Freeman, W.R., Sailor, M.J.: Porous silicon in drug delivery devices and materials. Adv. Drug Deliv. Rev. **60**, 1266–1277 (2008)
42. Boukherroub, R., Petit, A., Loupy, A., Chazalviel, J.N., Ozanam, F.: Microwave-assisted chemical functionalization of hydrogen-terminated porous silicon surfaces. J. Phys. Chem. B. **107**, 13459–13462 (2003)
43. Wu, E.C., Park, J.-H., Park, J., Segal, E., Cunin, F., Sailor, M.J.: Oxidation-triggered release of fluorescent molecules or drugs from mesoporous Si microparticles. ACS Nano. **2**, 2401–2409 (2008)
44. Stewart, M.P., Buriak, J.M.: Photopatterned hydrosilylation on porous silicon. Angew. Chem. Int. Ed. **37**, 3257–3260 (1998)
45. Sam, S., Chazalviel, J.N., Gouget-Laemmel, A.C., Ozanam, F., Etcheberry, A., Gabouze, N.: Peptide immobilisation on porous silicon surface for metal ions detection. Nanoscale Res. Lett. **6**, art.412 (2011)
46. Hua, F., Swihart, M.T., Ruckenstein, E.: Efficient surface grafting of luminescent silicon quantum dots by photoinitiated hydrosilylation. Langmuir **21**, 6054–6062 (2005)
47. Ruizendaal, L., Pujari, S.P., Gevaerts, V., Paulusse, J.M.J., Zuilhof, H.: Biofunctional silicon nanoparticles by means of thiol-ene click chemistry. Chem. Asian J. **6**, 2776–2786 (2011)
48. Sweryda-Krawiec, B., Cassagneau, T., Fendler, J.H.: Surface modification of silicon nanocrystallites by alcohols. J. Phys. Chem. B. **103**, 9524–9529 (1999)
49. Hallmann, S., Fink, M.J.: Mechanochemical synthesis of functionalized silicon nanoparticles with terminal chlorine groups. J. Mater. Res. **26**, 1052–1060 (2011)
50. Rosso-Vasic, M., Spruijt, E., Popovic, Z., Overgaag, K., Lagen, B., Grandidier, B., Vanmaekelbergh, D., Dominguez-Gutierrez, D., De Cola, L., Zuilhof, H.: Amine-terminated silicon nanoparticles: synthesis, optical properties and their use in bioimaging. J. Mater. Chem. **19**, 5926–5933 (2009)
51. Kang, Z.H., Liu, Y., Tsang, C.H.A., Ma, D.D.D., Fan, X., Wong, N.B., Lee, S.T.: Water-soluble silicon quantum dots with wavelength-tunable photoluminescence. Adv. Mater. **21**, 661–664 (2009)
52. He, Y., Su, Y.Y., Yang, X.B., Kang, Z.H., Xu, T.T., Zhang, R.Q., Fan, C., Lee, S.T.: Photo and pH stable, highly-luminescent silicon nanospheres and their bioconjugates for immunofluorescent cell imaging. J. Am. Chem. Soc. **131**, 4434–4438 (2009)

53. He, Y., Fan, C., Lee, S.-T.: Silicon nanostructures for bioapplications. Nano Today **5**, 282–295 (2010)
54. Wang, J., Liu, Y., Peng, F., Chen, C., He, Y., Ma, H., Cao, L., Sun, S.: A general route to effi cient functionalization of silicon quantum dots for high-performance fluorescent probes. Small **8**, 2430–2435 (2012)
55. Gongalsky, M.B., Yu, K.A., Osminkina, L.A., Yu, T.V., Jeong, J., Lee, H. Chung, B.H.: Enhanced photoluminescence of porous silicon nanoparticles coated by bioresorbable polymers, Nanoscale Res. Lett. **7**, art.446 (2012)
56. Li, Z.F., Ruckenstein, E.: Water-soluble poly (acrylic acid) grafted luminescent silicon nanoparticles and their use as fluorescent biological staining labels. Nano Lett. **4**, art. 14631467 (2004)
57. Zhang, X., Neiner, D., Wang, S., Louie, A.Y., Kauzlarich, S.M.: A new solution route to hydrogen-terminated silicon nanoparticles: synthesis, functionalization and water stability. Nanotechnology. **18**, art. 095601 (2007)
58. Sudeep, P.K., Page, Z., Emrick, T.: PEGylated silicon nanoparticles: synthesis and characterization. Chem. Commun. **46**, 6126–6127 (2008)
59. Wang, K., Wang, Q., Chao, Y.: Simple fabrication of nanostructured silicon and photoluminescence. Appl. Phys. A **109**, 437–440 (2012)
60. Ge, J., Liu, W., Zhao, W., Zhang, H., Zhuang, X., Lan, M., Wang, P., Li, H., Ran, G., Lee, S.-T.: Preparation of highly stable and water-dispersible silicon quantum dots by using an organic peroxide. Chem. Eur. J. **17**, 12872–12876 (2011)
61. Erogbogbo, F., Chang, C.-W., May, J.L., Liu, L., Kumar, R., Law, W.-C., Ding, H., Yong, K.T., Roy, I., Sheshadri, M., Swihart, M.T., Prasad, P.N.: Bioconjugation of luminescent silicon quantum dots to gadolinium ions for bioimaging applications. Nanoscale **4**, 5483–5489 (2012)
62. May, J.L., Erogbogbo, F., Yong, K.-T., Ding, H., Law, W.-C., Swihart, M.T., Prasad, P.N.: Enhancing silicon quantum dot uptake by pancreatic cancer cells via pluronic® encapsulation and antibody targeting. J. Solid Tumors **2**, 24–37 (2012)
63. Fan, J.Y., Wu, X.L., Chu, P.K.: Low-dimensional SiC nanostructures: fabrication, luminescence, and electrical properties. Prog. Mat. Sci. **51**, 983–1031 (2006)
64. Matsumoto, T., Takahashi, J., Tamaki, T., Futagi, T,. Mimura, H., Kanemitsu, Y.A.: Bluegreen luminescence from porous silicon carbide. Appl. Phys. Lett. **64**, art. 226 (1994)
65. Petrova-Koch, V., Sreseli, O., Polisski, G., Kovalev, D., Muschik, T., Koch, F.: Luminescence enhancement by electrochemical etching of SiC(6H). Thin Solid Films **255**, 107–110 (1995)
66. Rossi, A.M., Murphy, Th.E., Reipa, V.: Ultraviolet photoluminescence from 6H silicon carbide nanoparticles. Appl. Phys. Lett. **92**, art.253112 (2008)
67. Botsoa, J., Bluet, J.-M., Lysenko, V., Marty, O., Barbier, D., Guillot, G.: Photoluminescence of 6H–SiC nanostructures fabricated by electrochemical etching. J. Appl. Phys. **102**, art. 083526 (2007)
68. Wu, X.L., Fan, J.Y., Qiu, T., Yang, X., Siu, G.G., Chu, P.K.: Experimental evidence for the quantum confinement effect in 3C-SiC Nanocrystallites. Phys. Rev. Lett. **94**, art. 026102 (2005)
69. Fan, J.Y., Wu, X.L., Li, H.X., Liu, H.W., Siu, G.G., Chu, P.K.: Luminescence from colloidal 3C-SiC nanocrystals in different solvents. Appl. Phys. Lett. **88**, art. 041909 (2006)
70. Zakharko, Y., Botsoa, J., Alekseev, S., Lysenko, V., Bluet, J.-M., Marty, O., Skryshevsky, V.A., Guillot, G.: Influence of the interfacial chemical environment on the luminescence of 3C- SiC nanoparticles. J. Appl. Phys. **107**, art. 013503 (2010)
71. Serdiuk, T., Alekseev, S.A., Lysenko, V., Skryshevsky, V.A., Géloën, A.: Charge-driven selective localization of fluorescent nanoparticles in live cells. Nanotechnology **23**, art.315101 (2012)
72. Alekseev, S., Botsoa, J., Zaitsev, V.N., Barbier, D.: Fourier transform infrared spectroscopy and temperature programmed desorption mass spectrometry study of surface chemistry of porous 6H-SiC. Chem. Mater. **19**, 2189–2194 (2007)

73. Shin, W., Seo, W., Takai, O., Koumoto, K.: Surface chemistry of porous silicon carbide. J. Electron. Mater. **27**, 304–307 (1998)
74. Beke, D., Szekre'nyes, Z., Balogh, I., Veres, M., Fazakas, E., Varga, L.K., Kamara's, K., Cziga'ny, Z., Gali, A.: Characterization of luminescent silicon carbide nanocrystals prepared by reactive bonding and subsequent wet chemical etching, Appl. Phys. Lett. **99**, art. 213108 (2011)
75. Li, Y., Chen, C., Li, J.-T., Yang, Y., Lin, Z.-M.: Surface charges and optical characteristic of colloidal cubic SiC nanocrystals. Nanoscale Res. Lett. **6**, art. 454 (2011)
76. Zhang, N., Dai, D., Zhang, W., Fan, J.: Photoluminescence and light reabsorption in SiC quantum dots embedded in binary-polyelectrolyte solid matrix. J. Appl. Phys. **112**, art. 094315 (2012)
77. Che, J., Wang, X., Xiao, Y., Wu, X., Zhou, L., Yuan, W.: Effect of inorganic–organic composite coating on the dispersion of silicon carbide nanoparticles in non-aqueous medium. Nanotechnology **18**, art.135706 (2007)
78. Iijima, M., Kamiya, H.: Surface modification of silicon carbide nanoparticles by azo radical initiators. J. Phys. Chem. C **112**, 11786–11790 (2008)
79. Pud, A.A., Noskov, Y.V., Kassiba, A., Fatyeyeva, K.Y., Ogurtsov, N.A., Makowska-Janusik, M., Bednarski, W., Tabellout, M., Shapoval, G.S.: New aspects of the low-concentrated aniline Polymerization in the solution and in SiC nanocrystals dispersion. J. Phys. Chem. B. **111**, 2174–2180 (2007)
80. Serdiuk, T., Lysenko, V., Skryshevsky, V.A., Géloën, A.: Vapor phase mediated cellular uptake of sub 5 nm nanoparticles. Nanoscale Res. Lett. 7: art. 212 (2012)
81. Serdiuk, T., Lysenko, V., Mognetti, B., Skryshevsky, V., Géloën, A.: Impact of cell division on intracellular uptake and nuclear targeting with fluorescent SiC-based nanoparticles. J. Biophotonics **6**, 291–297 (2013)
82. Tagliazucchi, M., Peleg, O., Kroger, M., Rabin, Y., Szleifer, I.: Effect of charge, hydrophobicity, and sequence of nucleoporins on the translocation of model particles through the nuclear pore complex. PNAS **110**, 3363–3368 (2013)
83. Weisburg, J.H., Weissman, D.B., Sedaghat, T., Babich, H.: In vitro cytotoxicity of epigallocat echin gallate and tea extracts to cancerous and normal cells from the human oral cavity, Basic Clin. Pharmacol. Toxicol. **95**, 191–200 (2004)
84. Zakharko, Yu., Serdiuk, T., Nychyporuk, T., Geloen, A., Lemiti, M., Lysenko, V.: Plasmon-enhanced photoluminescence of SiC quantum dots for cell imaging applications. Plasmonics **7**, 725–732 (2012)

Rare Earth Implanted MOS Structures: Advantages and Drawbacks for Optoelectronic Applications

Lars Rebohle

Abstract In this paper the advantages and drawbacks of rare earth-implanted MOS structures for optoelectronic applications are investigated. The discussion starts with a short overview of the electroluminescence properties and highlights the ambivalent role of hot electrons in these devises, namely the efficient excitation of rare earths ions and the efficient creation of defects. In addition, the defect shell model is addressed which explains the slight but continuous fading of the electroluminescence during device operation. Based on this, strategies for improving critical device parameters are discussed. The potential for voltage downscaling is not yet fully exploited but is generally limited by the extension of a dark zone in which the hot electrons do not yet have enough kinetic energy to excite rare earth ions. The most frequent strategies for enhancing the power efficiency comprise the increase of the excitation cross section by pumping via Si nanoclusters or via other rare earth ions. Finally, the discussion closes with the different possibilities to improve the operation lifetime, followed by a few remarks about potential applications.

1 Introduction

Integrated photonics is a key technology of the 21st century, as it combines optical and electrical functions at chip level. Si-based photonic components are of special interest as they propose an easy integration into the CMOS platform. An essential building block is the electrically driven light emitter which, however, is difficult to realize. However, despite an intense research for more than 20 years in the field of Si-based light emission, the results are quite mixed. On one hand, there exists an

L. Rebohle (✉)
Institute of Ion Beam Physics and Materials Research, Helmholtz-Zentrum Dresden, Rossendorf, Bautzner Landstraße 400, 01328 Dresden, Germany
e-mail: l.rebohle@hzdr.de

A. Nazarov et al. (eds.), *Functional Nanomaterials and Devices for Electronics, Sensors and Energy Harvesting*, Engineering Materials,
DOI 10.1007/978-3-319-08804-4_16

enormous bandwidth of Si-based material systems showing electroluminescence (EL), ranging from crystalline Si with different types of luminescence centers over Si-rich oxides and nitrides to rare earth (RE) based materials. On the other hand, these light emitters are not yet able to compete with their counterparts based on compound or organic semiconductors with respect to important key parameters like power efficiency or operation lifetime.

In this chapter, these issues are discussed at the example of RE implanted, MOS-based light emitting devices (MOSLED), which feature a high conformity with standard CMOS processes combined with the excellent optical properties of RE elements. Within the group of RE elements, Er is the most often investigated element [1–3] due to its main emission wavelength of 1.54 μm fitting well with the global absorption minimum of SiO_2 [4]. However, if implanted into a MOS structure, other RE elements also exhibit strong EL in the UV (Gd [5]), in the visible (Ce [6], Tb [7–9], Eu [10]) and in the near IR (Nd [11]) spectral region. Thus, the following discussion includes MOSLEDs implanted with different RE elements with a slight focus on Er.

2 Experimental

In most cases the MOS-based light emitting devices (MOSLED) consist of a Si substrate, a dielectric stack and a transparent or semi-transparent top electrode. In the present case, a thermally grown gate oxide was produced by LOCOS (LOCal Oxidation of Silicon) technology on 4 inch (100) Si wafers, followed by rare earth ion implantation, SiON layer deposition using plasma-enhanced chemical vapour deposition, and furnace annealing (FA), rapid thermal annealing (RTA) or flash lamp annealing (FLA). If not otherwise stated, both the SiO_2 and SiON layer thickness amount to 100 nm. The implantation energy was chosen in such a way that the projected range of implantation is roughly in the middle of the gate oxide layer. The atomic peak concentration of the rare earth elements is about 1.5 %. Whereas FA was applied in the temperature range of 700 and 1,000 °C for 30 min in nitrogen ambient, FLA temperatures between 900 and 1,200 °C (pulse duration 3 or 20 ms, nitrogen ambient) were used. Finally, a transparent front electrode made of indium tin oxide (ITO) and a back electrode made of aluminium were processed. The front electrode was lithographically patterned into circular dots with a diameter of 300 μm. The composition of the light emitter is schematically shown in Fig. 1.

The EL spectra were recorded at room temperature under constant current injection using a Keithley 2411 source measurement unit and a monochromator in combination with a spectroscopic CCD camera or an uncooled InGaAs detector for Er-implanted MOSLEDs. Current-Voltage (IV) characteristics were measured by using a Keithley 237 source measurement unit. In order to determine the EL power efficiency, the EL output was monitored simultaneously with the IV curves by a calibrated Si photodiode (visible and near IR spectral range) or a calibrated Ge

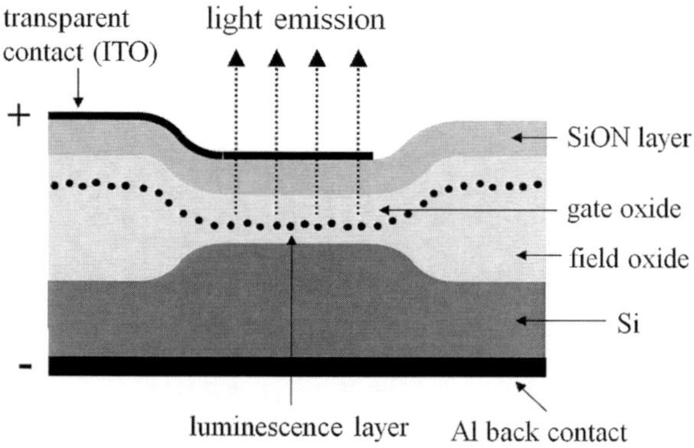

Fig. 1 Basic view of the rare earth-implanted MOSLEDs

photodiode (IR range) in a dark box. The operation lifetime of individual devices was measured by applying a constant injection current and by recording the applied voltage and the EL output as a function of time. The technical equipment is the same as for the IV measurements.

3 Electroluminescence Properties of RE-Implanted MOSLEDs

3.1 Spectrum and Efficiency

Figure 2 displays the normalized EL spectrum of various rare earth-implanted MOSLEDs covering the UV, visible and NIR spectral range depending of the used rare earth element. Most of the emission lines are due to electronic, optical dipole forbidden transitions within the 4f shell of trivalent rare earth ions with decay times in the ms range. Comprehensive surveys about the energy level systems can be found in [12, 13]. Fortunately, these selection rules do not apply in case of electric excitation which results in relatively high excitation cross sections. In case of Tb- and Er-implanted devices power efficiencies (the ratio of optical output to electrical input power) up to 0.3 % were reached [9].

Generally, maximum EL efficiency is obtained for an optimum choice of the implantation and annealing conditions. Annealing is performed in order to anneal out implantation induced defects and to activate the RE luminescence centers. However, annealing also triggers processes driven by diffusion, namely the formation and growth of RE oxide clusters as well as the diffusion of the RE ions towards the interfaces of the SiO_2 layer. In most cases the compromise is a medium thermal budget with relatively high temperatures but short annealing times. As seen in Fig. 3a at the example of Gd-implanted MOSLEDs, maximum EL efficiency is achieved for FLA at temperatures around 1,000 °C [15]. In some cases, namely for Er, FA in the temperature range between 800 and 900 °C can lead to similarly good results as FLA.

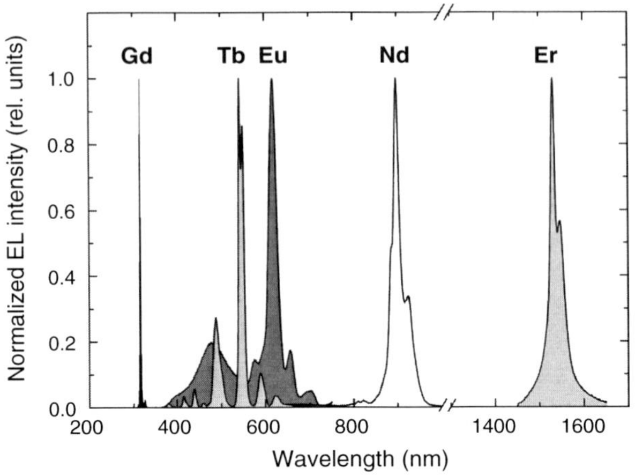

Fig. 2 Normalized EL spectrum of various rare earth-implanted MOSLEDs (after [14])

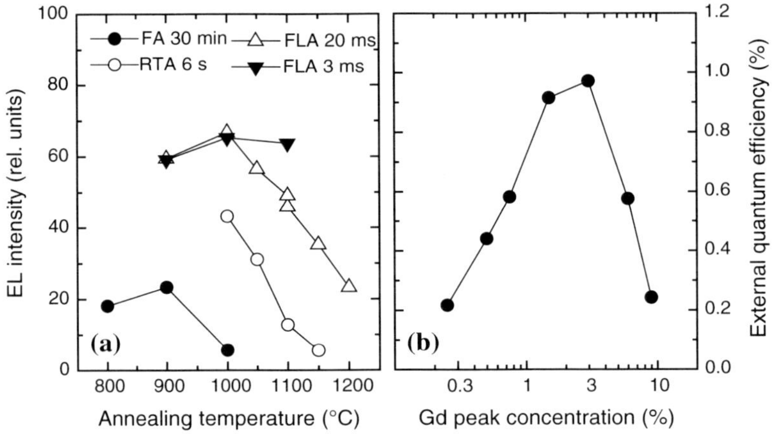

Fig. 3 Annealing temperature dependence for MOSLEDs with 2 % Gd and annealed by FA, RTA and FLA (**a**) and concentration dependence (**b**) of Gd-implanted MOSLEDs annealed by FA 900 °C (data taken from [15, 16])

There is also an optimum RE concentration for maximum EL output. Starting at low RE ion concentration, the EL efficiency will increase because of the increasing number of RE ions that can be excited. At higher concentrations quenching and saturation effects will dominate the EL behavior. According to Fig. 3b the external quantum efficiency (the ratio of the number of emitted photons to the injected electrons per time interval) becomes maximal for a Gd concentration between 1 and 3 at% [16]. This also applies to most of the other RE elements. However, the change of the matrix material may change the concentration of maximum EL efficiency. Much more details can be found in [17].

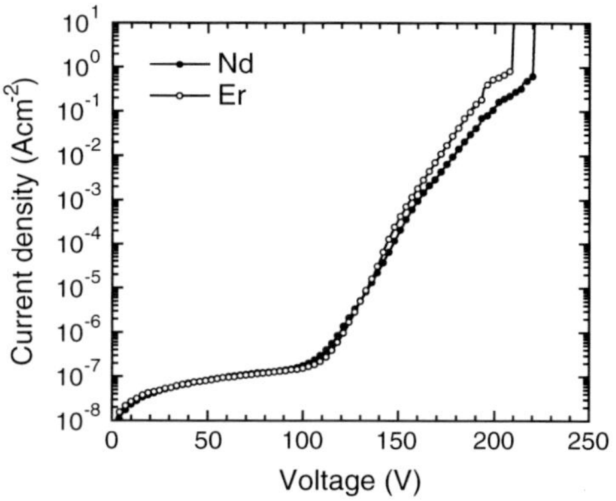

Fig. 4 Typical *I–V-curves* of Nd- and Er-implanted MOSLEDs

As shown at the example of Nd- and Er-implanted devices in Fig. 4, the *I–V* curves are governed by Fowler-Nordheim tunneling: after very low values the current density increases approximately by an exponential law, followed by a dielectric breakdown at very high voltages. After injection into the conduction band of SiO_2 by Fowler-Nordheim tunneling, there are several possibilities for charge transportation. A part of the electrons may move from trap to trap by Poole-Frenkel or Hopping conduction. However, these electrons will not gain enough kinetic energy to excite the luminescence centers. Another part of the electrons will be accelerated and gain an equilibrium energy distribution depending on the electric field in the SiO_2 layer. If they have enough kinetic energy it can be transferred by inelastic scattering to rare earth ions. Finally, the radiative de-excitation of these excited rare earth ions will cause the observed EL.

3.2 The Number of Excited Luminescence Centers

For low and medium level injection currents the EL intensity shows a linear dependence on the injection current density over several orders of magnitude, before it saturates at high injection currents (not shown). This translates into an approximately constant EL power efficiency (the ratio between optical output power and the electrical input power) in the linear region as shown for the Nd-implanted MOSLED in Fig. 5a. For Er-implanted MOSLEDs saturation starts at lower injection current densities, but the exact value strongly depends on the annealing conditions.

According to [18] the fraction of excited Nd or Er ions f is given by

$$f = \frac{N_{exc}}{N_{tot}} = \frac{\tau V_0 \eta j}{E_\gamma D} \tag{1}$$

where N_{exc} is the number of excited RE ions, N_{tot} is the total number of RE ions in the active layer, τ is the EL decay time, V_0 is the applied voltage across the whole device, η is the measured power efficiency, j is the injection current density, E_γ is

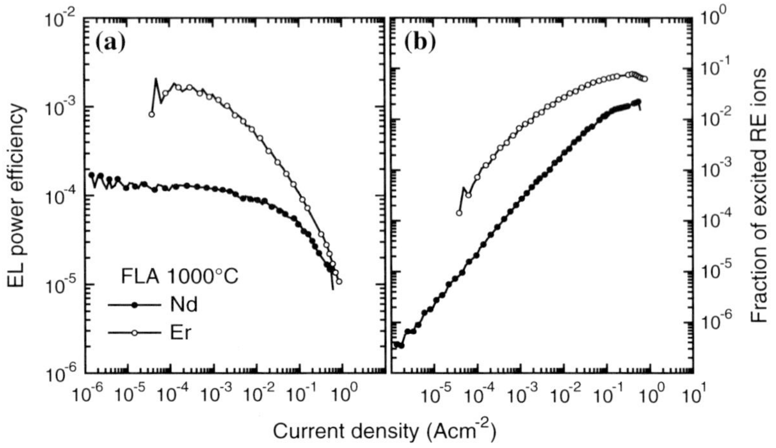

Fig. 5 EL power efficiency (**a**) and the fraction of excited RE ions (**b**) for Nd- and Er-implanted devices

the energy of the emitted photon and D is the RE implantation dose. This basic estimation simply assumes that an excited RE ion stays in the excited state for the timespan τ, and it neglects excited ions which de-excite via a non-radiative decay channel. In addition, the out-coupling efficiency is not considered, for which reason Eq. (1) is an underestimation or lower limit for the true value of f.

In the present case the EL decay constant is quite insensitive to the annealing conditions (except the as-implanted case) and amounts to 8.6 and 12.3 ms for Nd- and Er-implanted samples, respectively [19]. Using these values the fraction of excited Nd or Er ions as a function of the injection current density is obtained (Fig. 5b). Similar to the EL intensity f linearly increases with increasing injection current density and saturates at high currents. In case of Er and for high injection currents f reaches values in the order of 10 %, whereas the power efficiency has significantly fallen. This leads to the assumption that the falling power efficiency is most probably caused by an increasing number of hot electrons passing the device without exciting Er ions. Furthermore, both f and the power efficiency of Nd are roughly one order of magnitude lower than the values of Er. This difference can be either due to a lower excitation cross section of Nd or due to a lower out-coupling efficiency of Nd caused by the different refraction index of ITO at 900 and 1,550 nm.

3.3 Operation Lifetime

The operation lifetime of these devices is determined by the electric charge which can be injected into the SiO_2 layer up to the breakdown. As shown in Fig. 6, the EL intensity is fairly constant under constant current injection for a certain period of time, followed by EL quenching and the dielectric breakdown of the device. As

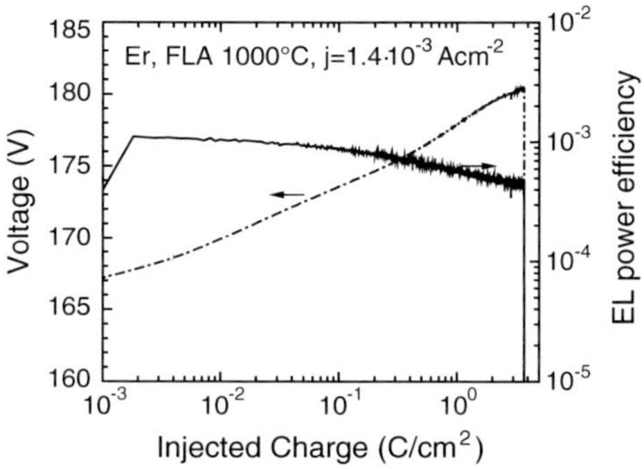

Fig. 6 Voltage (*dashed line*) and EL power efficiency (*solid line*) of an Er-implanted device under constant current injection

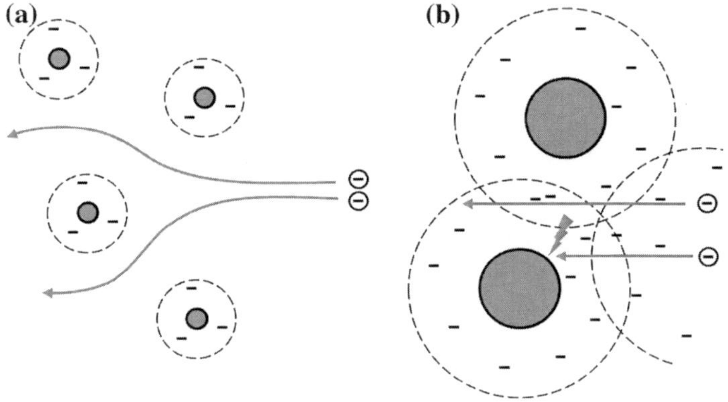

Fig. 7 Basic view of the defect shell model according to [23, 24]

known for decades, hot electrons in SiO_2 continuously create defects by anode hole injection [20], hydrogen release [21], band-to-trap ionization and band-to-band ionization. These defects accumulate and lead to a breakdown if a critical defect density is exceeded [22].

The specific quenching behavior depends on the implantation and annealing conditions, but generally it is believed that the EL quenching is most probably due the Coulomb screening of luminescence centers by trapped electrons. This scenario is supported by the continuous increase of the voltage needed to hold the constant injection current (Fig. 6) which indicates the accumulation of negative charges within the dielectric stack.

According to the defect shell model [23, 24] RE ions tend to cluster in form of RE oxide clusters with RE luminescence centers in the interface region of such clusters. As such clusters do not fit perfectly into the SiO_2 network, a shell of structural defects will form around the clusters. During injection, electrons will be trapped in these defects and prevent the further excitation of the nearby RE luminescence centers by Coulomb repulsion (Fig. 7a). An increase of the annealing temperature usually leads to an increase of the RE oxide cluster size, which has two consequences. At first, a larger fraction of RE ions is located within the RE cluster, for which reason the number of potentially available luminescence centers and thus the EL efficiency will decrease. Secondly, the defect shells around the clusters increase and start to overlap. However, if the trapped negative charges are homogeneously distributed in the bulk, the electron injection into the dielectric is still hindered, but the Coulomb repulsion itself is lowered (Fig. 7b). As a consequence, MOSLEDs with small clusters feature a high EL efficiency and a high EL quenching, whereas MOSLEDs with large clusters are characterized by both lower EL efficiencies and lower EL quenching cross sections.

There is further experimental evidence for this model. In [17, 25] it was shown that the EL quenching cross section correlates well with the electron trapping cross section. Furthermore, a low temperature FLA treatment (400 °C) was applied to a previously stressed device with an already strongly quenched EL and led to a nearly complete recovery of the EL intensity [26]. This behavior was explained by the photon-mediated release of trapped electrons due to the high energetic part (UV and blue spectral region) of the flash lamp spectrum.

In summary, the three main drawbacks of rare earth-implanted MOSLEDs are the high operation voltage, the low power efficiency compared to compound semiconductors with a direct bandgap, and the insufficient operation lifetime. In the following, these issues will be discussed in more detail.

4 Challenges for Critical Performance Parameters

4.1 Voltage Downscaling

It is known that electrons injected into SiO_2 under high electric fields need a certain acceleration distance to gain their equilibrium energy distribution, for which reason there is a thin zone with low defect concentration next to the injecting interface [27]. Recently, this model was advanced and applied to rare earth-implanted MOSLEDs [18]. According to this there is a dark zone close to the injecting interface where the electrons do not yet have enough energy to excite rare earth ions, and rare earth ions located in this zone are regarded as lost material.

The experimental data are given in Fig. 8 showing the fraction of excited Tb ions f as a function of the active gate oxide layer thickness. The layer sequence corresponds to Fig. 1 with electron injection from the Si substrate. In order to avoid saturation effects a fairly low injection level was chosen. As shown, the fraction of excited Tb ions is on its original level for a layer thickness of 75 nm and larger, but strongly decreases for smaller thicknesses. According to the dark zone model, in the case of a gate oxide layer thickness of 50 and 30 nm approximately 42 and 87 % of the Tb ions are located in the dark zone, respectively. The extension of the dark zone was estimated to be in the order of 20 nm [18].

Considering this, it seems possible to scale the layer thickness down to values around 20 nm resulting in an operation voltage of about 20 V. However, further downscaling without losing efficiency will become more difficult. This limit might be circumvented by shifting the excitation mechanism to charge carrier recombination or by changing the matrix. However, regarding the latter point it must be kept in mind that SiO_2 is already superior to most of the alternative materials with respect to defect and trap concentration. At least for SiON and Si_3N_4, which are characterized by a much higher defect and trap density, the fraction of hot electrons is much lower than in case of SiO_2 [28]. On the other hand, a higher intrinsic defect concentration may lead to a higher robustness of the material to additional defect creation by hot electrons which potentially prolongs the operation lifetime.

4.2 Enhancement of the EL Power Efficiency

Although EL power efficiencies in the 10^{-3} range are excellent values for Si-based light emitters, they are still too low for a couple of applications. Generally, there are three different strategies to enhance the EL efficiency [17]: (i) Enhancement of the excitation cross section, (ii) Maximizing the internal quantum efficiency, i.e. the ratio of radiative to non-radiative transitions, and (iii) Maximizing the out-coupling efficiency.

Most studies in literature investigate the first strategy by using sensitizers with large excitation cross section. In case of Er doped SiO_2 it is known that the photoluminescence excitation cross section of Er can be strongly enhanced by using Si nanoclusters (NCs) as sensitizers [29–32]. The main reason is that the direct excitation in case of PL is an optical dipole forbidden transition with the corresponding low excitation cross section. However, in case of EL the use of Si NCs is less advantageous as their incorporation into the SiO_2 network will trigger a couple of competitive processes.

As shown in Fig. 9 the EL mechanism via Si NCs can be divided into the process of Si NC excitation (a), the energy transfer to nearby Er^{3+} ions (b), and the Er^{3+} de-excitation (c). At first, the optical selection rules do not apply in case of electrical excitation, for which reason the difference between the direct electrical excitation cross section and the excitation cross section via Si-NC is smaller. Furthermore, Si NCs introduce additional defects in the SiO_2 network and are scattering centers which lower the average energy of the hot electrons. Thus, hot electrons have a somewhat lower average energy and the fraction of hot electrons having enough energy to generate an exciton (process 1) is lower. In addition, the total number of hot electrons is decreasing as with increasing defect and NC concentration more and more electrons will be transported by Poole-Frenkel or Hopping conduction (process 2). Trapped electrons are not only lost for EL excitation, but in turn they induce further competitive processes lowering the EL efficiency, including the Coulomb screening of Si NCs and Er^{3+} ions (process 3).

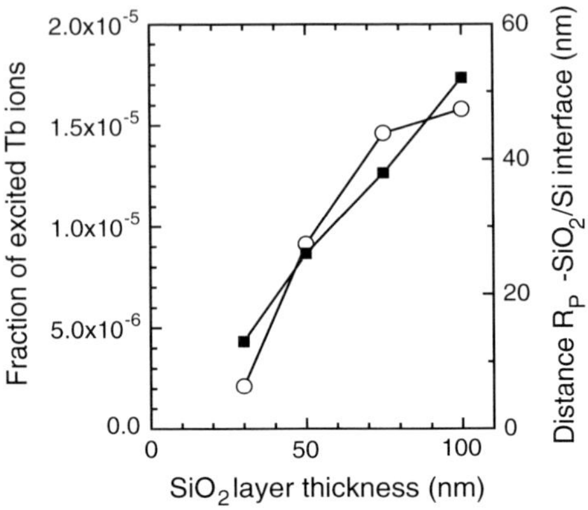

Fig. 8 Fraction of excited Tb ions as a function of the gate oxide layer thickness (*open circles*). For comparison, the distance between the injecting interface and the projected range of Tb implantation (R_p) is given on the right scale (*solid squares*)

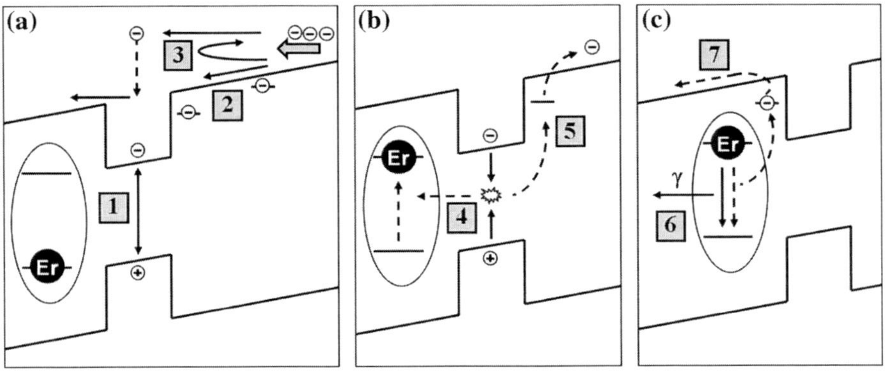

Fig. 9 Schematic view of the Er^{3+} excitation process via a Si NC and competitive processes

If once excited, the NCs can transfer the energy of the exciton recombination to nearby Er^{3+} ions (process 4) but also to trapped electrons in the NC vicinity (process 5). The energy transfer is an additional process compared to the direct excitation of Er and lowers always the total EL efficiency. In the best case the energy transfer is dominating leading only to an insignificant lowering of the total EL efficiency. Finally, the excited Er^{3+} ion must radiatively de-excite to exhibit EL (process 6). However, it was experimentally demonstrated [33, 34] that trapped electrons in the vicinity of Er^{3+} can induce an efficient Auger quenching in which the energy of the excited Er^{3+} ion is used again to release trapped electrons (process 7).

The question whether the larger excitation cross section of Si NCs is outbalanced by the aforementioned disadvantageous processes or not depends on the specific material system. Consequently, most studies found that the incorporation of Si NC into SiO_2 leads to a decrease of the EL efficiency [3, 30], although the EL excitation cross section in the order of 10^{-14} cm^2 is two orders of magnitude higher than in case of PL [33]. Another possibility is the change of the host matrix which is discussed in more detail at the end of the next section.

Besides the use of Si-rich materials, further strategies to increase the EL power efficiency include the pumping by other RE elements, as successfully demonstrated for the case of Gd and Ce in [35], and the co-implantation of fluorine [36]. The latter method takes advantage of the beneficial effect of fluorine on the defect formation, namely the termination of dangling Si bonds and the replacement of Si–H bonds by Si–F bonds [37, 38].

4.3 Improving the Operation Lifetime

Finally, the problem of low operation lifetimes should be discussed. This is mainly a problem of the ambivalent nature of hot electrons: on one side they are needed for hot impact excitation, but on the other hand they produce defects which accumulate and steadily head for the dielectric breakdown. On a pessimistic point of view it can be stated that any lifetime improvement someone will achieve is paid by a corresponding loss in efficiency. More optimistic, the challenge is to find material systems and device layouts where the defect creation by hot electrons is minimized and, if this cannot be avoided, to find materials and designs which are more robust and tolerant to defect creation.

In case of rare earth-implanted MOSLEDs it was shown that the use of LOCOS technology and an additional SiON buffer layer [39] can increase the operation lifetime by more than two orders of magnitude. The presence of a SiON layer cannot prevent the degradation processes in the bulk of the SiO_2 layer, but it decelerates hot electrons after passing the SiO_2 layer. In doing so, it avoids defect creation at the critical gate interface and relocates it to the less critical SiO_2–SiON interface. In addition, the energy a hot electron can release is smaller at the SiO_2–SiON interface that at the SiO_2–ITO interface due to the different conduction band offsets.

Another successful example is the co-implantation of rare earth ions with potassium [40]. As shown in Fig. 10 the EL intensity of Gd-doped MOSLEDs decreases by a factor of 5–6 if co-implanted with potassium. However, this decrease is more than outbalanced by the fact that the usual EL quenching phase is avoided and that the dielectric breakdown is delayed by more than one order of magnitude. In addition, the operation voltage is lowered with increasing K concentration. In the present case it was found that K^+ is accumulated at the Si–SiO_2 interface during annealing which lowers the injection barrier. As K^+ is sufficiently mobile (but not too much) it can migrate under high electric fields and is able to neutralize trapped electrons. Even if this example is not the panacea for long-living devices (potassium is an unloved guest in microelectronics) it demonstrates the possibility to achieve strong lifetime improvements for acceptable or even negligible losses in efficiency.

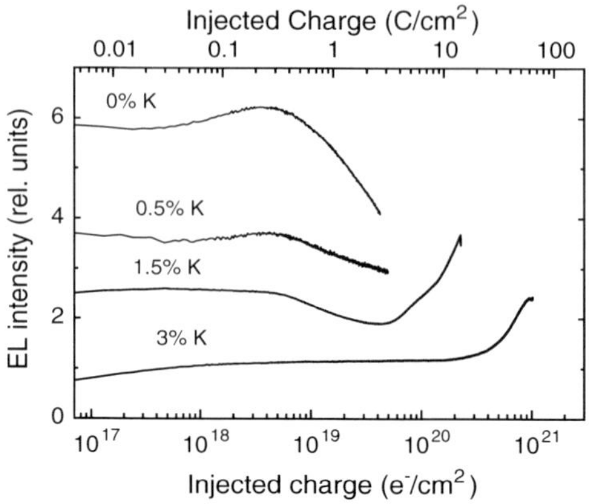

Fig. 10 EL intensity as a function of the injected charge under constant current injection of 0.016 Acm^{-2} for a MOSLED implanted with 2 % Gd and co-implanted with different potassium concentrations (original data from [40])

In this light the use of Si NC, or Si-rich dielectrics in a more general sense, is gaining in importance. According to this, Si NCs do not only increase the excitation cross section of rare earths, but can also increase the operation lifetime and lower the operation voltage. Indeed, there are a couple of newer studies investigating the EL properties of MOSLEDs with Er in Si-rich oxide [41–43], Er in Si-rich nitride [44–46] or an Er silicate layer [47, 48]. In [43] it was found that the external quantum efficiency of the EL of Er-doped Si-rich oxide sensitively depends on layer thickness and Si excess concentration, and that there is often an anti-correlation between EL intensity and conductivity. A comparison of low and high Si excess concentrations in Er-doped Si-rich oxide revealed that the device with a low Si excess concentration has a higher threshold voltage of about 30 V, but also features a power efficiency three orders of magnitude higher than the device with a high Si excess concentration [41]. For Er-doped Si-rich nitride EL excitation cross sections in the 10^{-15} cm^2 range were reported [44]. Furthermore, another study revealed that maximum EL power efficiency in Si-rich nitride is achieved for Si excess concentrations in the order of 10 % [46]. Although the operation lifetime of the aforementioned devices is not known, these materials seem to have the potential to reduce the discussed drawbacks of rare earth-implanted MOSLEDs.

5 Applications

After this discussion the question arises for which applications these types of light emitters are suitable. The list of potential applications covers many fields including that of optical data communication, solid state lighting and optical sensor technology. In the light of the current, enormous efforts to reduce power consumption in microelectronic circuits the requirements for integrated light sources are quite draconic, and the way of tailoring MOSLED devices for this application appears to be very difficult. However, for solid state lighting there are no special or tightened voltage requirements, and the non-toxicity of the used materials is an additional bonus. Nevertheless, high efficiencies and long operation lifetimes still remain a mandatory criterion.

In case of optical sensor technology the situation is less problematic. Similar to solid state lighting there are no special voltage requirements, and the EL efficiency is not any longer the major criterion. Instead of that the absolute EL power a device can emit will presumably determine the limit of detection. The question whether the operation lifetime is an important issue or not depends on the specific measuring task: for long-time monitoring it matters, for disposable measurement chips probably not. Thus, sensor technology seems to have the lowest entrance barriers for the application of Si-based light emitters.

One vision for sensor technology is the fully integrated, Si-based photonic chip for the detection of organic molecules in waterish solutions [17]. In this concept, the light of an integrated MOSLED is coupled into a waveguide on the chip surface, passes through the waveguide, and is recorded by an integrated photodiode. There are several possibilities to prepare the waveguide surface for bio-detection, ranging from thin metal layers for surface plasmon resonance spectroscopy to photonic crystals. If the prepared waveguide surface is in contact with a microfluidic channel where the analyte solution is flowing through, the molecules to be detected can immobilize on the waveguide surface. These selective binding events will change the effective refraction index of the cladding layer of the waveguide, which in turn alters the transmission through the waveguide. This change is a measure for the concentration of the analyte molecules in the solution.

However, there is only a limited number of publications dealing with the full photonic integration including the light emitter at present. Besides the possibility of hybrid integration of III–V materials [49] and organic LEDs [50] one study about an integrated photonic circuit is known where Si and $Si/Si_{1-x}Ge_x/Si$ p-i-n diodes were used as emitter and receiver, respectively [51]. Current investigations in literature focus on Er-doped MOS-based light emitters in a slot waveguide configuration [52–54]. Our own work currently focuses on the optimum coupling between the integrated MOSLED and an integrated waveguide [14].

6 Conclusions

In summary, the advantages and drawbacks of rare earth-implanted MOSLEDs were discussed. Current MOSLEDs implanted with Tb and Er reach power efficiencies above 10^{-3}, but need operation voltages in the order of 100 V and are limited in their operation lifetime. It was shown that the downscaling of the operation voltage is limited due to the need to accelerate electrons to sufficiently high energies. The use of

Si-rich matrices, namely Si-rich oxides and nitrides, can considerably lower the operation voltage and probably increase the operation lifetime, but this is paid by efficiency. However, recent results in literature and own investigations indicate that the loss of efficiency is much smaller than the gain achieved with voltage downscaling and lifetime improvement under certain conditions.

Acknowledgments This chapter is the result of longstanding research activities involving many scientists contributing in many different ways. Special thanks go to F. Bregolin, C. Cherkouk, S. Germer, M. Helm, A. Kanjilal, J. Lehmann, A. Mrotzek, S. Prucnal, W. Skorupa, J.M. Sun, M. Voelskow, R. Wutzler and R.A. Yankov (presently or formerly at the Research Centre Dresden-Rossendorf); A. Nazarov, I.N. Osiyuk, I.P. Tjagulskii, and S. Tjagulskii (Nat. Academy of Science, Kyiv); C. Buchal and S. Mantl (Research Centre Jülich); H. Fröb and K. Leo (Technical University of Dresden); J. Biskupek and U. Kaiser (University of Ulm); Y. Berencén and B. Garrido (University of Barcelona). The authors also acknowledge the Rossendorf and Jülich Implantation Group for performing ion implantation, T. Schumann for flash lamp annealing and H. Hilliges, C. Neisser and G. Schnabel for their careful semiconductor preparation work.

References

1. Wang, S., Eckau, A., Neufeld, E., et al.: Hot electron impact excitation cross-section of Er^{3+} and electroluminescence from erbium-implanted silicon metal-oxide-semiconductor tunnel diodes. Appl. Phys. Lett. **71**, 2824–2826 (1997)
2. Kenyon, A.J.: Erbium in silicon. Semicond. Sci. Technol. **20**, R65–R84 (2005)
3. Jambois, O., Berencen, Y., Hijazi, K., et al.: Current transport and electroluminescence mechanisms in thin SiO_2 films containing Si nanocluster-sensitized erbium ions. J. Appl. Phys. **106**, 063526 (2009)
4. Saleh, B.E., Teich, M.C.: Fundamentals of Photonics. Wiley, Hoboken (2007)
5. Sun, J.M., Skorupa, W., et al.: Efficient ultraviolet electroluminescence from a Gd-implanted silicon metal-oxide-semiconductor device. Appl. Phys. Lett. **85**, 3387–3389 (2004)
6. Skorupa, W., Sun, J.M., Prucnal, S. et al.: Rare earth ion implantation for silicon based light emission. In: Pichaud, B., Claverie, A., Alquier, D., Richter, H., Kittler, M. (eds.) Gettering and Defect Engineering in Semiconductor Technology Xi, pp. 755–760 (2005)
7. Yoshihara, M., Sekiya, A., et al.: Rare-earth-doped SiO_2 films prepared by plasma-enhanced chemical vapour deposition. J. Phys. D-Appl. Phys. **30**, 1908–1912 (1997)
8. Wang, S., Amekura, H., Eckau, A., et al.: Luminescence from Er and Tb implanted into MOS tunnel diodes. Nucl. Instrum. Methods Phys. Res. Sect. B-Beam Interact. Mater. At. **148**, 481–485 (1999)
9. Sun, J.M., Skorupa, W., Dekorsy, T., et al.: Bright green electroluminescence from Tb^{3+} in silicon metal-oxide-semiconductor devices. J. Appl. Phys. **97**, 123513 (2005)
10. Prucnal, S., Sun, J.M., et al.: Switchable two-color electroluminescence based on a Si metal-oxide-semiconductor structure doped with Eu. Appl. Phys. Lett. **90**, 181121 (2007)
11. Rebohle, L., Wutzler, R., Germer, S. et al.: Er- and Nd-implanted MOS light emitting devices and their use for integrated photonic applications. In: Vivien, L., Honkanen, S.K., Pavesi, L., Pelli, S. (eds.) Silicon Photonics and Photonic Integrated Circuits III, pp. 3151–3153 (2012)
12. Dieke, G.H.: Spectra and Energy Levels of Rare Earth Ions in Crystals. Interscience Publ, New York (1968)
13. Liu, G., Jacquier, B.: Spectroscopic Properties of Rare Earths in Optical Materials. Springer, Berlin (2005)

14. Germer, S., Cherkouk, C., Rebohle, L. et al.: Basic structures of integrated photonic circuits for smart biosensor applications. In: Proceedings of the SPIE 8774, Optical Sensors 2013, 87740P (2013)
15. Prucnal, S., Sun, J.M., Muecklich, A., et al.: Flash lamp annealing versus rapid thermal and furnace annealing for optimized metal-oxide-silicon-based light-emitting diodes. Electrochem. Solid State Lett. **10**, H50–H52 (2007)
16. Sun, J.M., Prucnal, S., Skorupa, W., et al.: Electroluminescence properties of the Gd^{3+} ultraviolet luminescent centers in SiO_2 gate oxide layers. J. Appl. Phys. **99**, 103102 (2006)
17. Rebohle, L., Skorupa, W.: Rare-Earth Implanted MOS Devices for Silicon Photonics: Microstructural, Electrical and Optoelectronic Properties. Springer, Berlin (2010)
18. Rebohle, L., Lehmann, J., Prucnal, S., et al.: Physical limitations of the electroluminescence mechanism in terbium-based light emitters: matrix and layer thickness dependence. Appl. Phys. B-Lasers Opt. **98**, 439–442 (2010)
19. Wutzler, R.: Influence of Implantation and Annealing Conditions on the optoelectronic properties of Nd and Er doped MOS layers. Bachelor Thesis, Technical University Bergakademie Freiberg (2012)
20. DiMaria, D.J., Cartier, E., Buchanan, D.A.: Anode hole injection and trapping in silicon dioxide. J. Appl. Phys. **80**, 304–317 (1996)
21. Sah, C.T., Sun, J.Y.C., Tzou, J.J.T.: Study of the atomic models of 3 donorlike defects in silicon metal-oxide-semiconductor structures from their gate material and process dependencies. J. Appl. Phys. **55**, 1525–1545 (1984)
22. Stathis, J.H.: Physical and predictive models of ultrathin oxide reliability in CMOS devices and circuits. IEEE Trans. Device Mater. Reliab. **1**, 43 (2001)
23. Nazarov, A.N., Tiagulskyi, S.I., Tyagulskyy, I.P., et al.: The effect of rare-earth clustering on charge trapping and electroluminescence in rare-earth implanted metal-oxide-semiconductor light-emitting devices. J. Appl. Phys. **107**, 123112 (2010)
24. Tiagulskyi, S., Nazarov, A., Tyagulskii, I. et al.: Shell model for REOx nanoclusters in amorphous SiO_2: charge trapping and electroluminescence quenching. In: Dappe, Y., Jelinek, P., Chab, V. (eds.) Physica Status Solidi C: Current Topics in Solid State Physics, vol 9, No 6. p 1468–1470 (2012)
25. Nazarov, A., Osiyuk, I., Tyagulskii, I., et al.: Charge trapping phenomena in high-efficiency metal-oxide-silicon light-emitting diodes with ion-implanted oxide. J. Lumin. **121**, 213–216 (2006)
26. Prucnal, S., Rebohle, L., Nazarov, A.N., et al.: Reactivation of damaged rare earth luminescence centers in ion-implanted metal-oxide-silicon light emitting devices. Appl. Phys. B-Lasers Opt. **91**, 123–126 (2008)
27. Arnold, D., Cartier, E., Dimaria, D.J.: Theory of high-field electron-transport and impact ionization in silicon dioxide. Phys. Rev. B **49**, 10278–10297 (1994)
28. Dimaria, D.J., Fischetti, M.: Hot electron transport and trapping in silicon dioxide. In: Ferradini, C., Jay-Gerin, J.P. (eds.) Excess Electrons in Dielectric Media. CRC Press, Boca Raton, p. 315 (1991)
29. Kenyon, A.J., Trwoga, P.F., Federighi, M., et al.: Optical properties of PECVD erbium-doped silicon-rich silica—evidence for energy transfer between silicon microclusters and erbium ions. J. Phys. Condens. Matter **6**, L319–L324 (1994)
30. Nazarov, A., Sun, J.M., Skorupa, W., et al.: Light emission and charge trapping in Er-doped silicon dioxide films containing silicon nanocrystals. Appl. Phys. Lett. **86**, 151914 (2005)
31. Franzo, G., Vinciguerra, V., Priolo, F.: Room-temperature luminescence from rare-earth ions implanted into Si nanocrystals. Philos. Mag. B **80**, 719–728 (2000)
32. Franzo, G., Pecora, E., Priolo, F., et al.: Role of the Si excess on the excitation of Er doped SiOx. Appl. Phys. Lett. **90**, 183102 (2007)
33. Iacona, F., Irrera, A., Franzo, G., et al.: Silicon-based light-emitting devices: properties and applications of crystalline, amorphous and Er-doped nanoclusters. IEEE J. Sel. Top. Quantum Electron. **12**, 1596–1606 (2006)

34. Priolo, F., Presti, C.D., Franzo, G., et al.: Carrier-induced quenching processes on the erbium luminescence in silicon nanocluster devices. Phys. Rev. B **73**, 113302 (2006)
35. Sun, J.M., Prucnal, S., Skorupa, W., et al.: Increase of blue electroluminescence from Ce-doped SiO_2 layers through sensitization by Gd^{3+} ions. Appl. Phys. Lett. **89**, 091908 (2006)
36. Prucnal, S., Sun, J.M., Rebohle, L. et al. (2007) Fourfold increase of the ultraviolet (314 nm) electroluminescence from SiO_2 : Gd layers by fluorine coimplantation and flash lamp annealing. Applied Physics Letters 91: 181107
37. Wright, P.J., Kasai, N., Inoue, S. et al.: Improvement in SiO_2 gate dielectrics with fluorine incorporation. In: 1989 Symposium on VLSI Technology: Digest of Technical Papers: 51–52 (1989)
38. McWilliams, K.P., Halle, L.F., Zietlow, T.C.: Improved hot-carrier resistance with fluorinated gate oxides. IEEE Electron. Device Lett. **11**, 3–5 (1990)
39. Sun, J.M., Rebohle, L., Prucnal, S., et al.: Giant stability enhancement of rare-earth implanted SiO_2 light emitting devices by an additional SiON protection layer. Appl. Phys. Lett. **92**, 071103 (2008)
40. Prucnal, S., Sun, J.M., Reuther, H., et al.: Strong improvement of the electroluminescence stability of SiO2: Gd layers by potassium co-implantation. Electrochem. Solid State Lett. **10**, J30–J32 (2007)
41. Muscara, A., Castagna, M.E., Leonardi, S., et al.: Design and electro-optical characterization of Si-based resonant cavity light emitting devices. IEEE J. Quantum Electron. **47**, 1362–1368 (2011)
42. Jambois, O., Ramirez, J.M., Berencen, Y., et al.: Effect of the annealing treatments on the electroluminescence efficiency of SiO_2 layers doped with Si and Er. J. Phys. D-Appl. Phys. **45**, 045103 (2012)
43. Cueff, S., Labbe, C., Jambois, O., et al.: Structural factors impacting carrier transport and electroluminescence from Si nanocluster-sensitized Er ions. Opt. Express **20**, 22490–22502 (2012)
44. Yerci, S., Li, R., Dal Negro, L.: Electroluminescence from Er-doped Si-rich silicon nitride light emitting diodes. Appl. Phys. Lett. **97**, 081109 (2010)
45. Warga, J., Li, R., Basu, S.N., et al.: Erbium-doped silicon nanocrystals in silicon/silicon nitride superlattice structures: light emission and energy transfer. Physica E **41**, 1040–1043 (2009)
46. Yin, Y., Xu, W.J., Ran, G.Z., et al.: Er^{3+} infrared photo- and electroluminescence from Er-doped Si-rich silicon nitride films with varying Si and Er content. Funct. Mater. Lett. **4**, 255–259 (2011)
47. Yin, Y., Xu, W.J., Wei, F., et al.: Room temperature Er^{3+} 1.54 μm electroluminescence from Si-rich erbium silicate deposited by magnetron sputtering. J. Phys. D-Appl. Phys. **43**, 335102 (2010)
48. Irrera, A., Galli, M., Miritello, M., et al.: New approaches for enhancing light emission from Er-based materials and devices. Physica E **41**, 891–898 (2009)
49. Halioua, Y., Bazin, A., Monnier, P., et al.: Hybrid III–V semiconductor/silicon nanolaser. Opt. Express **19**, 9221–9231 (2011)
50. Ramuz, M., Burgi, L., Stanley, R., et al.: Coupling light from an organic light emitting diode (OLED) into a single-mode waveguide: Toward monolithically integrated optical sensors. J. Appl. Phys. **105**, 084508 (2009)
51. Yamada, A., Sakuraba, M., Murota, J.: Photo detection characteristics of Si/Si1-xGex/Si p-i-n diodes integrated with optical waveguides. Thin Solid Films **508**, 399–401 (2006)
52. Miller, G.M., Briggs, R.M., Atwater, H.A.: Achieving optical gain in waveguide-confined nanocluster-sensitized erbium by pulsed excitation. J. Appl. Phys. **108**, 063109 (2010)
53. Ramirez, J.M., Lupi, F.F., Berencen, Y., et al.: Er-doped light emitting slot waveguides monolithically integrated in a silicon photonic chip. Nanotechnology **24**, 115202 (2013)
54. Tengattini, A., Gandolfi, D., Prtljaga, N., et al.: Toward a 1.54 μm electrically driven erbium-doped silicon slot waveguide and optical amplifier. J. Lightwave Technol. **31**, 391–397 (2013)

Part IV
NanoSensors and MEMS/NEMS

Silicon and Germanium Junctionless Nanowire Transistors for Sensing and Digital Electronics Applications

Yordan M. Georgiev, Ran Yu, Nikolay Petkov, Olan Lotty, Adrian M. Nightingale, John C. deMello, Ray Duffy and Justin D. Holmes

Abstract In this chapter, we introduce two specific types of junctionless nanowire transistors (JNTs): (i) silicon-on-insulator (SOI) back-gated JNTs for sensing applications and (ii) germanium-on-insulator (GeOI) top-gated JNTs for digital logic applications. We discuss in detail the suitability of junctionless architecture for these particular applications and present results on device fabrication and characterisation. Back-gated JNTs of 45 different channel geometries (different numbers, lengths, and widths of channel nanowires) have been designed and fabricated with very high precision (down to 10 nm widths of the nanowires) on SOI wafers using a fully CMOS-compatible fabrication process. Electrical characterisation of the fabricated devices has demonstrated their excellent performance as back-gated JNTs. Furthermore, data from pH and streptavidin sensing experiments have proven their good sensing properties. These JNTs are among the smallest top-down fabricated nanowire sensing devices reported to date. Top-gated JNTs with Ge nanowire channels of widths down to 20 nm have been fabricated by a simple CMOS-compatible process on GeOI wafers with a highly *p*-doped ($\sim 1 \times 10^{19}$ cm^{-3}) top germanium layer. The fabricated devices have demonstrated

Yordan M. Georgiev—On leave of absence from the Institute of Electronics at the Bulgarian Academy of Sciences, Sofia, Bulgaria

Y.M. Georgiev (✉) · N. Petkov · O. Lotty · J.D. Holmes
Materials Chemistry and Analysis Group, Department of Chemistry and Tyndall National Institute, University College Cork, Cork, Ireland
e-mail: yordan.georgiev@tyndall.ie

Centre for Research on Adaptive Nanostructures and Nanodevices (CRANN), Trinity College Dublin, Dublin 2, Ireland

R. Yu · R. Duffy
Silicon Research Group, Tyndall National Institute, Lee Maltings, Dyke Parade, Cork, Ireland

A.M. Nightingale · J.C. deMello
Nanostructured Materials and Devices Group, Department of Chemistry, Imperial College London, London, UK

A. Nazarov et al. (eds.), *Functional Nanomaterials and Devices for Electronics, Sensors and Energy Harvesting*, Engineering Materials,
DOI 10.1007/978-3-319-08804-4_17

decent output and transfer characteristics with relatively high I_{on}/I_{off} current ratios of up to 2.0×10^5 and steep subthreshold slopes of 189 mV/dec. To the best of our knowledge, these are the first reported Ge JNTs.

1 Introduction

Traditional scaling of classical complementary metal–oxide–semiconductor (CMOS) devices is facing significant challenges as the microelectronics industry is reaching their scaling limits. The conventional approach of reducing the gate length and gate dielectric thickness, whilst simultaneously increasing the channel doping is no longer able to meet the application requirements for improved performance at sustainable (low) power consumption. This approach has been compromised primarily by the fact that the supply voltage (V_d) has not decreased proportionately to the transistor density increase, leading to unacceptable leakage currents and power consumption. As a result, power density now constrains CMOS chip design. This, together with the rapidly increasing market share of low-power portable electronic systems, motivates the need for conception of novel devices having high performance at reduced power levels.

There are a number of possible routes to meet these requirements by extension of dimensional and functional scaling of CMOS devices (often called "More Moore") as well as by introducing new information processing devices and architectures (usually referred to as "Beyond CMOS" devices). Nevertheless, there is a general agreement within the microelectronic community that all of the possible "parallel paths" require development of new technologies based on one or more of (at least) the following:

1. Optimised stacks of high dielectric constant (high-k) gate dielectrics with metal gate electrodes (HKMG) for threshold voltage (V_{th}) tuning and better channel potential control. Equivalent oxide thickness (EOT) below 0.7 nm (even down to 0.5 nm) and dielectric constants $k > 30$ are envisaged here [1].
2. Novel device structures for improved electrostatic control. Three-dimensional (3D) multigate field effect transistors (MuGFETs) such as the fin-based tri-gate FETs (FinFETs) or the nanowire (NW) MuGFETs are among the most probable candidates to replace the conventional planar device structure.
3. Alternative high mobility (high-μ) channel materials for higher drive currents at low power supply: most prominently germanium (Ge), III-V compound semiconductors, carbon nanotubes (CNTs), and graphene.

In three-dimensional (3D) FETs, the planar channel in the conventional metal–oxide–semiconductor FETs (MOSFETs) is replaced with 3D semiconducting structure(s): fin(s), ribbon(s), or nanowire(s). They offer an appealing approach to extended scaling of CMOS devices because of the much better gate control of the

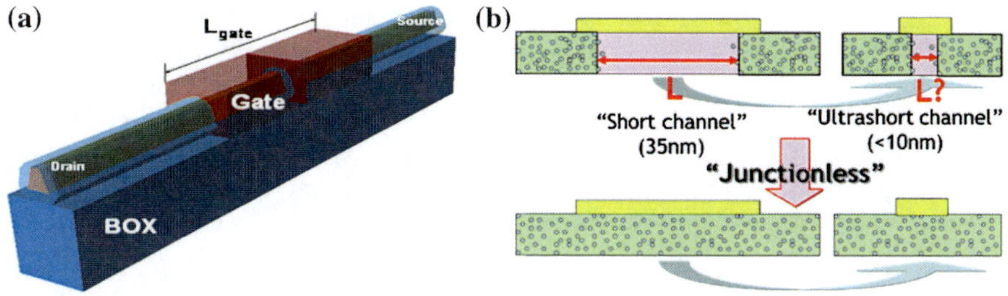

Fig. 1 Schematic presentation of a JNT: **a** 3D and **b** cross-sectional comparison between classical and junctionless transistors

channel and suppressed short channel effects (SCE), leading to reduced off-leakage current.

Intel has recently announced the introduction of FinFETs into their latest technology [2]. It is expected that in the future FinFETs will change to nanowire FETs, because of their yet greater control of SCE and larger channel area for the NW surface per unit area. Moreover, at low diameters, the nanowires exhibit quantum confinement behaviour, i.e. quasi one dimensional (1D) ballistic conduction, that may permit further reduction of short channel effects and other limitations to the scaling of planar MOSFETs.

Another advantage of the NW channel geometry is the possibility to implement multi-layered (vertically stacked) nanowires [3, 4], significantly increasing the number of device channels and, hence, the drive current I_d.

From a production viewpoint, the number of NW FET fabrication processes is expected to be smaller than those for conventional planar CMOS processes. It is assumed that no channel and halo implantations are necessary, because of the good SCE control in the case of nanowire structure, while threshold voltage control can be achieved by work function control of the gate stack.

Recently, junctionless transistor architecture has been proposed and demonstrated for the first time as a disruptive upgrade of the advanced 3D device structure [5–8]. The junctionless nanowire transistor (JNT) is basically a gated semiconductor resistor where the source, channel and drain have the same type of doping without any junctions between them. In fact, this device does not contain any doping concentration gradient at all (see Fig. 1). The gate controls the carrier density and hence the current flow in the resistor, as in a MOSFET. The JNT is the simplest transistor structure possible and probably the most scalable of all FET structures, as demonstrated by both ab-initio simulations [9] and experimental devices [10] with gate lengths as small as 3 nm. Such a device is much easier to fabricate since it does not need separate doping of source and drain regions and the highly challenging formation of abrupt ultra-shallow junctions (see Fig. 1b). In this way, JNTs can be produced at a substantially reduced mask count compared to traditional FET devices, greatly relaxing the thermal budget and also leading to a decrease in the overall power consumption of generating integrated circuits.

However, the commercial implementation of the junctionless architecture in devices requires: (i) cross-section of the channel (its height and width) that is small enough to allow for full depletion of carriers when the device is turned off, i.e. a nanoscale 3D (fin/nanowire/nanoribbon) device structure is needed and (ii) heavy doping (up to 1×10^{19} cm^{-3} and even higher) of the whole device (source, channel and drain, see Fig. 1b) to allow for sufficiently high current flow when the device is turned on.

Both experimental and theoretical data show that beside the significantly simplified fabrication, the JNT possesses a number of other advantages over the classical MOSFETs, i.e. (i) bulk conductance near the centre of the channel (NW) in a JNT, compared to conductance in a thin surface inversion or accumulation layer near the gate in the inversion mode (IM) or accumulation mode (AM) transistors, which leads to higher drive currents; (ii) the absence of a strong electric field attracting majority carriers toward the gate interface when the transistor is on, leading to a charge carriers mobility which is equal to the bulk values or even higher due to the screening effect; (iii) inherent SCE immunity and excellent drain-induced-barrier-lowering (*DIBL*) together with extremely low leakage currents of less than 10^{-15} A at a supply voltage $V_d = 1$ V due to the fact that the effective gate length L_{ge} of JNTs is smaller than the physical gate length L_g in the on-state and larger in the off-state; (iv) near-ideal subthreshold slope of $1.06 \times kT/q$ over a wide range of temperatures (20–200 °C; 64 mV/dec at room temperature); (v) less degradation of charge carrier mobility with gate voltage and temperature [5–8, 11, 12].

The junctionless architecture has also a great potential for incorporating alternative channel materials with enhanced carrier transport properties. Junctionless transistors have already been demonstrated in poly-Si, Ge, gallium arsenide (GaAs), indium gallium arsenide (InGaAs)-on-Insulator, and indium-tin-oxide (ITO) [13–17].

JNTs are criticised for having variability issues. The conventional understanding is that the variability of device parameters with fabrication parameters as well as the random doping fluctuation variability increases with increasing the concentration of channel doping. It has been shown both experimentally [18] and theoretically [19, 20] that the heavily doped JNTs have disadvantages when compared to undoped IM devices. There is, however, an alternative approach that considers many other parameters, beside channel doping concentration, as important for device variability, e.g. proper effective oxide thickness (EOT) ($\leq$1) and gate configurations (tri-gate or gate-all-around), NW diameter, short-channel effects, screening effect, etc. It has been shown that when all these factors are taken into account, JNTs have lower variability than IM devices [21]. It is worth noting that since the JNTs are novel devices, the published experimental data on their variability is still scarce and currently insufficient to make clear conclusions on how they compare to classical IM transistors.

2 Silicon JNTs for Sensing Applications

An attractive application of junctionless nanowire transistors, which has not yet been studied, is as sensing devices. Nanowires receive lately large interest due to their exceptional electrical and mechanical properties and large surface area to volume ratio. As mentioned above, semiconductor NWs, especially silicon NWs (Si NWs), can be used as channels for field effect transistors [22]. NW FETs have promising application not only in digital logic but also as sensing devices [23]. Instead of a physical top gate, however, such devices are gated by the attached electrically charged molecules, e.g. streptavidin, as shown in Fig. 2. The NW surface can be functionalised to make it selectively sensitive to certain molecules. In this way, depending on their functionalisation, NW FETs may be used as gas sensors, chemical sensors, and biosensors. They have a potential for fast, low-cost, low-power, label-free detection, real-time response, high throughput analysis, providing insight into biological processes while not requiring large sampling quantities [23, 24]. Nanometre-scale cross-sections lead to depletion or accumulation of carriers in the bulk of the device channel when a charged molecule binds to the surface, versus surface-only modulation with traditional planar ion sensitive field effect transistor (ISFET) sensors [25–28].

Research into the use of Si NWs as FET-type sensors began by using chemically grown NWs [23] and this is still the main route for fabrication of such devices. Although the bottom-up approach is capable of a massive production of sub-10 nm NWs, their integration into functional devices remains challenging. The main issues here are the proper contacting as well as precise positioning and alignment of grown NWs to other nanostructures. These issues can be naturally addressed by the top-down approach used in semiconductor device fabrication. In addition, this approach is CMOS compatible and improves the control on device parameters and, hence, on reproducibility.

FET-type sensors exhibit the highest possible sensitivity when operated in the subthreshold regime [29, 30]. Usually this can be achieved e.g. by applying an appropriate backgate potential at the expense of increased power consumption. In the case of JNTs, however, their performance depends strongly on the geometry (height, width, and length) as well as on the doping level of the channel. Therefore, these parameters can be used to finely tune the operation point of devices into the subthreshold regime which makes the application of JNTs in low power autonomous sensing systems very promising [31].

2.1 Layout Design and Fabrication of Silicon JNTs

To study the impact of channel geometry on the sensing properties of JNTs, a complicated layout has been designed containing devices with three different NW densities (1, 3, and 20 NWs) and lengths (0.5, 1 and 10 μm) as well as five

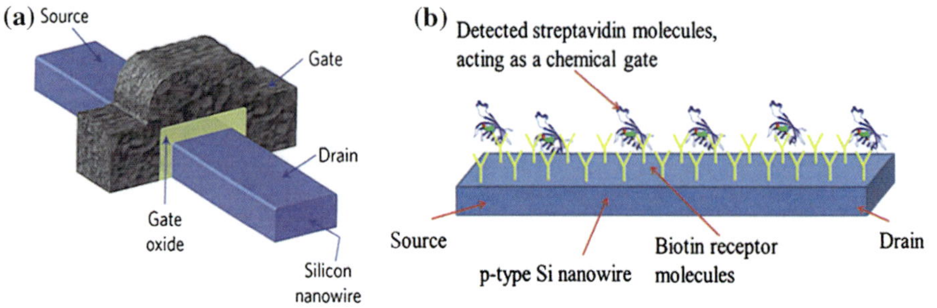

Fig. 2 Schematics of a p-channel nanowire FET (**a**) and the p-type JNT architecture used in this study (**b**). The negatively charged streptavidin molecules will bind to the biotin receptor molecules and act as a chemical gate

Fig. 3 Overview of the design layout

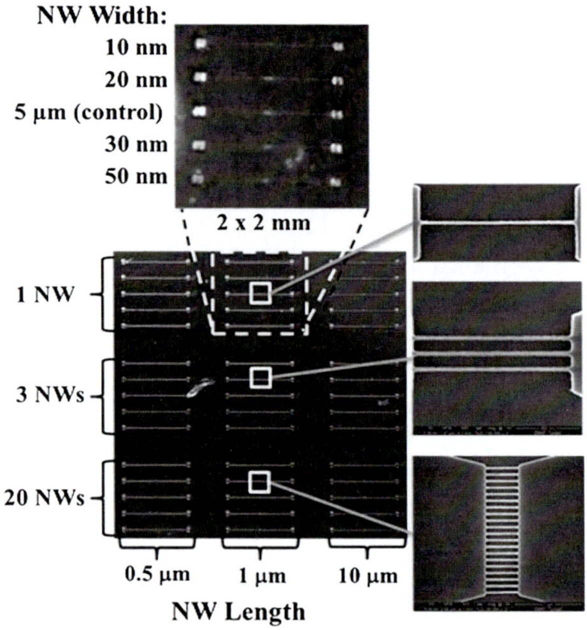

different NW widths: 10, 20, 30 and 50 nm and 5 μm (a large control device). The full combination of all these parameters results in a total of 45 different devices (see Fig. 3).

Devices have been fabricated on SOI wafers with a *p*-doped ($\sim 1 \times 10^{18}$ cm^{-3}) top Si layer of 45 nm, a buried oxide (BOX) SiO_2 layer of 145 nm and a base Si layer of 500 μm (Soitec).

Electron beam lithography (EBL) with a JEOL JBX 6000FS Gaussian beam direct write system operated at 50 kV has been used as a first patterning step. The nanowire devices (nanowires together with the source and drain contact pads) have

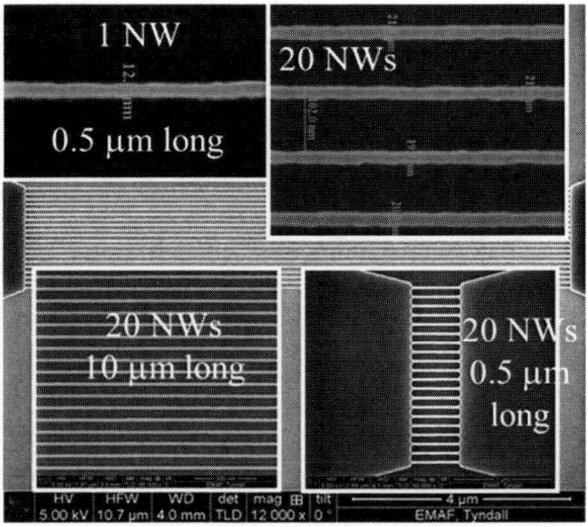

Fig. 4 Collage of *top-view* SEM images of devices having different number and length of NWs and 10 and 20 nm widths

been defined in a 50 nm thick layer of hydrogen silsesquioxane (HSQ) negative tone resist (XR-1541 from Dow Corning Corp.) spun onto a full 4″ SOI wafer. The resist has been developed using an original high-contrast and low roughness development process, based on 25 % tetramethylammonium hydroxide (TMAH) as the main development step [32, 33].

The top Si layer has been etched through the HSQ mask using reactive ion etching (RIE) with chlorine (Cl) chemistry in a Plasmalab System 100 from Oxford Instruments. The precise implementation of the EBL and RIE processes has led to excellent patterning results. The NWs exhibited minimal roughness (between 0.8 and 1.6 nm) and negligible deviation from the prescribed widths as demonstrated in Fig. 4. High quality of the NWs is very important for minimising charge carrier scattering on their sidewalls and hence for high NW conductance.

In the subsequent fabrication steps, an additional 200 nm SiO_2 layer has been deposited on the whole surface, except the device regions, in order to minimise the leakage current through the buried oxide. Then, the metal contacts and interconnection have been created by depositing a 100 nm thick nickel (Ni) layer and thermal annealing for 30 min at 400 °C in forming gas (10 % H_2/90 % N_2). The annealing step not only helps in forming Ni silicide for low resistance Ohmic contacts, but improves the conductance of the NWs as well [34]. The SiO_2 and Ni depositions have been done by electron beam evaporation in Leybold Lab 600 and Temescal FC-2,000 systems, respectively, and have been combined with two steps of photolithography and lift-off. The photolithography exposures have been aligned to the previously exposed EBL pattern using etched alignment marks. Finally, a passivation layer of 2 μm SU8 negative photoresist has been deposited on the whole substrate and patterned by photolithography to open small windows over the device regions and metal pads. A cross-section of fabricated devices is schematically presented in Fig. 5.

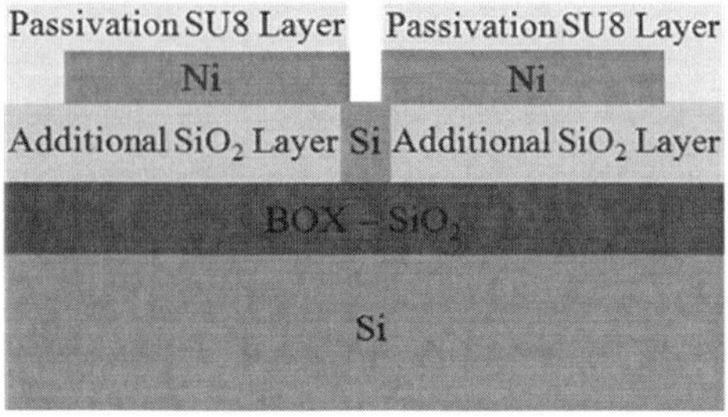

Fig. 5 Cross-sectional schematics of fabricated devices

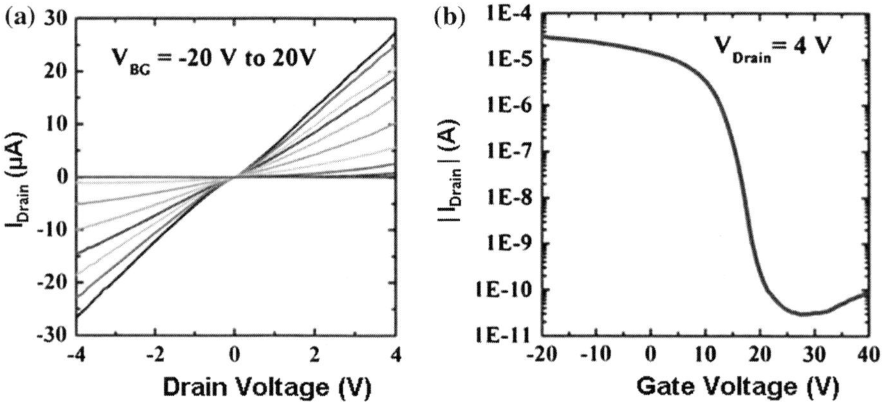

Fig. 6 Output characteristics (**a**) and transfer characteristics (**b**) of devices having 20 nanowires of 10 μm length and 50 nm width

2.2 Electrical Characterisation of Fabricated Silicon JNTs

Electrical measurements of devices have been performed using cascade manual probe station and Agilent semiconductor analyser B1500.

In Fig. 6, examples of output characteristics (drain current I_d as a function of the drain potential V_d, Fig. 6a) and transfer characteristics (drain current I_d as a function of the backgate potential V_{bg}, Fig. 6b) of devices having 20 nanowires of 10 μm length and 50 nm width, are presented. In Fig. 6a, the $I_d - V_d$ lines are plotted for different backgate potentials V_{bg}, ranging from −20 to 20 V. These straight and symmetric lines clearly demonstrate the very good ohmic contacts between the metal and the Si pads and the high on-currents of our devices, which is clear evidence of the excellent overall process quality. The $I_d - V_{bg}$ curve in Fig. 6b is plotted for a $V_d = 4$ V and shows the excellent properties of the fabricated devices as backgated JNTs. The on-currents are high, in the range of

30 μA. The backgate potential allows good control over the current through the multi-nanowire channel, leading to a steep subthreshold slope (*SS*) and high ratios between the on-state and off-state currents (I_{on}/I_{off}) of $\sim 10^6$. These characteristics make the devices promising candidates for sensing applications.

2.3 Nanowire Functionalisation

For a target chemical- or bio-molecule in solution to be sensed with a FET-type Si NW device, it should be able to bind to the NW surface. To enable this, the Si NWs need to be functionalised with a probe molecule that will bind exclusively to the target molecule of interest. In this study, streptavidin-biotin has been chosen as a model target-probe pair, with streptavidin in solution being sensed using biotin covalently tethered to the NW surface.

The functionalisation of the Si JNT sensors to covalently attach biotin involves first covalently attaching (3-Aminopropyl)triethoxysilane (APTES), which is used as the linker molecule between the native oxide of the Si NWs and the biotin molecule. Samples have been immersed for 3 h in a 50 °C solution of 5 v/v % APTES in anhydrous toluene and then rinsed with anhydrous toluene, deionised water and dried under nitrogen. In addition to the role of a linker between the SiO_2 and the biotin, APTES has been shown to be a suitable functionalisation layer for sensing pH [35]. Consequently, it has been decided to test a part of the APTES functionalised sensors for their response to pH.

The other part of the devices have been subjected to the second functionalisation step, where the aminosilanised nanowires (Si-APTES) have been immersed in 2 ml phosphate buffered saline (PBS, pH 7.5) and 100 μl of E,Z link-NHS-LC-Biotin in dimethylformamide (DMF) (1 mg/ml) has been added. The samples have been left to react for 3 h at room temperature. Surfaces have been rinsed with PBS and deionised water and dried under nitrogen. Figure 7 outlines the reaction scheme used.

2.4 Microfluidic Delivery of Analyte

The aim of this study is to deliver miniature self-contained sensing devices. Polydimethylsiloxane (PDMS) stamps with microfluidic channels attached to the device allow for administration of fluid directly to the Si NW structures. Furthermore, only a small quantity (μLs) of analyte is required for analysis. The propulsion of the fluid through the channel can either be achieved by passive delivery via capillary action or using external pumping equipment. While the former option is far preferable for practical application of these devices (e.g. point-of-care health diagnostics) the latter method is more convenient for testing (and is used in the work described herein) as we can keep the device isolated in a probe-station whilst

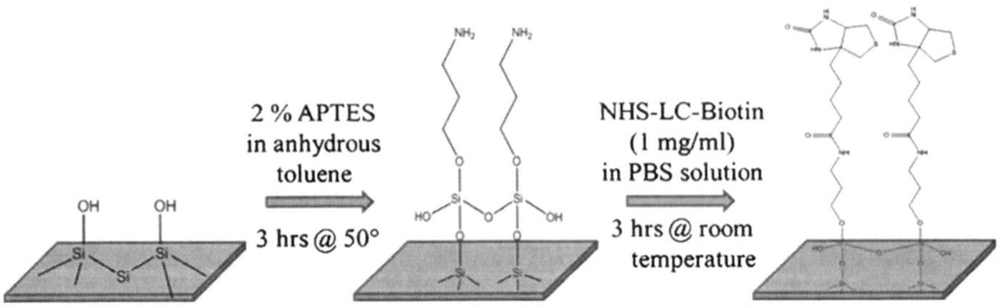

Fig. 7 Schematic of the functionalisation route of the silicon surface of the JNTs from the native oxide termination to the biotin terminated surface

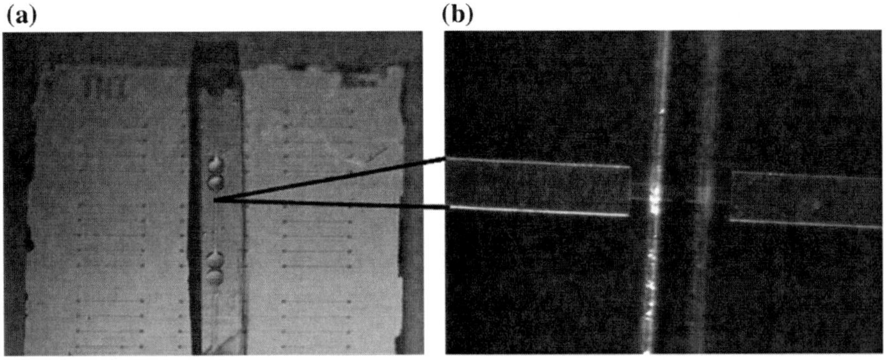

Fig. 8 **a** Image of the integrated device. Three microfluidic channels are shown attached to the chip-based sensor architecture. **b** Microscopic image showing the microfluidic channel positioned above a separate Si NW sensor device

filling/cleaning the channel at will. Such hydrodynamic flow has been deemed to be a more appropriate method of fluid delivery. The inlet/outlets of the microfluidic stamp have been connected to lengths of poly(tetrafluoroethylene) (PTFE) tubing so that fluid could be administered/collected from outside the probe station.

The method of stamps fabrication and subsequent attachment to the sensing device is only briefly described here. Each microfluidic PDMS stamp has been fabricated using standard procedures [36] and comprised a single 200 μm-wide channel with 400 μm access holes drilled in the top to allow interface between the microfluidic channel and external tubing. The stamps have been attached to the devices using the "stamp and stick" technique [37] in which a thin layer of wet PDMS is added to the underside of the stamp before it is positioned on the device. Positioning of the devices has been achieved using a micrometer-controlled positioning rig built in-house. Following curing at 60 °C for 2 h, a strong but non-permanent bond has been formed to yield integrated sensing/fluid delivery devices (Fig. 8).

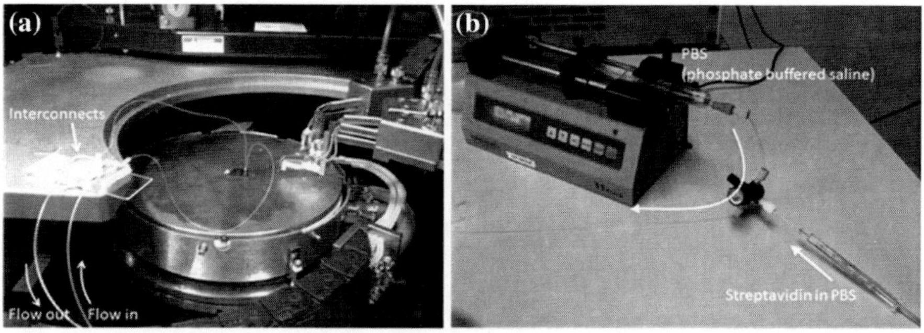

Fig. 9 **a** Image showing an integrated device in the probe station before testing. The device is connected to tubing ready for fluid delivery. **b** Image showing the syringes and syringe pump used to deliver fluid for the streptavidin sensing experimentation

2.5 Experimental Setup for Sensing Experiments

Using the microfluidic PDMS stamps described above, phthalate buffered solutions (Fisher) of different pH levels (pH 4, pH 7 and pH 10) as well as PBS solution have been delivered to the sensor at a rate of 150 or 200 μL/min from a syringe (BD Plastipak, 10 ml) propelled by a syringe pump (Harvard, Pump 11+). From the syringe, the flow has passed through polyethylene (PE) tubing (ID 0.4 mm, OD 1.0 mm). Approximately 10 cm from exiting the syringe, the flow has passed through a T-junction where a small separate solution of streptavidin in PBS could be added into the main PBS flow as required (see Fig. 9b). From the T-junction the flow has continued along the PE tubing (~ 1 m) into the probe station. Standard fittings from Upchurch Scientific have been used for all connections.

Inside the probe station (Fig. 9a), the tubing has been downsized to OD 0.4 mm ID 0.1 mm PTFE tubing using interconnect junctions fabricated in-house from PDMS. Commercially available interconnects could also have been used (e.g. from Upchurch Scientific). The smaller tubing has been required to interface to the PDMS stamp but could not be used for the entire length of the fluid supply lines as excessive backpressure would be generated.

2.6 Results from Sensing Experiments

To test the sensing properties of the fabricated JNTs, two different sensing experiments have been performed: (i) pH sensing with APTES-functionalised devices and (ii) streptavidin sensing with biotinated devices.

In Fig. 10, typical examples of the sensing results are presented. Figure 10a shows the measurement of the drain current I_d as a function of time t ($I_d - t$) for an APTES-functionalised device with 20 NWs of 0.5 μm length and 50 nm width,

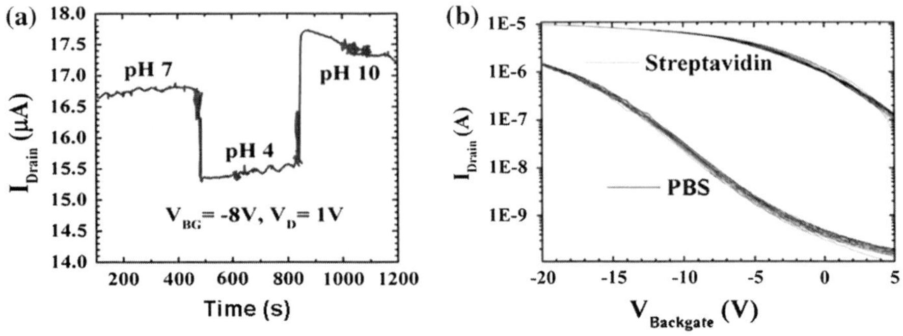

Fig. 10 **a** $I_d - t$ measurements for an APTES-functionalised device with 20 NWs of 0.5 μm length and 50 nm width, operated at a backgate voltage $V_{bg} = -8$ V and a drain voltage $V_d = 1$ V, during consequent administration of pH 7, pH 4 and pH 10 buffers. **b** $I_d - V_{bg}$ curves of a biotinated device with 20 NWs of 0.5 μm length and 50 nm width at a drain voltage $V_d = 1$ V, following PBS and streptavidin injection

operated at a backgate voltage $V_{bg} = -8$ V and a drain voltage $V_d = 1$ V. A steady stream (150 μL/min) of a pH 7 buffer has first been injected into the dry microfluidic channel. This yielded a signal of ~16.7 μA magnitude. Around 480 s from the beginning of the measurement, a pH 4 buffer has been injected into the microfluidic channel, leading to a significant drop of I_d down to ~15.3 μA (~9 % lower drain current). Then, around 360 s later a pH 10 buffer has been administrated into the microfluidic channel and as a result the drain current jumped up to ~17.7 μA (~14 % higher value). These data reflect the protonation (at pH 4) and deprotonation (at pH 10) of the APTES molecule attached to the device channel (NW) surface [23], causing respectively depletion and accumulation of majority carriers in the p-type channel and, hence, the decrease and increase of the drain current observed.

Figure 10b presents typical transfer characteristics (drain current I_d as a function of the backgate potential V_{bg}, $I_d - V_{bg}$ curves) of a biotinated device with 20 NWs of 0.5 μm length and 50 nm width at a drain voltage $V_d = 1$ V. First, a steady stream (150 μL/min) of PBS has been injected into the dry microfluidic channel and a series of ten consequent $I_d - V_{bg}$ measurements has been performed, yielding on-currents I_{on} of ~1.5 μA and off-currents I_{off} of ~0.15 nA (I_{on}/I_{off} ratio of ~10^4). Then, 0.1 ml of a 0.02 mg/ml (0.4 μM) solution of streptavidin in PBS has been administered and again a series of ten consequent $I_d - V_{bg}$ measurements has been done. As a result, $I_d - V_{bg}$ curves with much smoother subthreshold slope (*SS*) and significantly higher currents, I_{on} ~ 10 μA and I_{off} ~ 0.1 μA (I_{on}/I_{off} ratio of ~10^2), have been obtained. These data reflect the successful binding of streptavidin molecules to the biotinated NW surface. Since the streptavidin molecules have a negative electrical charge, they are causing accumulation of majority carriers (*holes*) in the p-type channel end, hence, a significant increase of the drain current (up to a factor of 10 for the I_{on} and a factor of 10^3 for the I_{off}).

Figure 10 is a clear demonstration of the excellent sensing properties of the fabricated JNTs. To the best of our knowledge, they are among the smallest top-down fabricated Si NW sensing devices reported to date.

3 Germanium JNTs

A prudent way to further boost the assets of the junctionless transistor architecture is by introducing germanium (Ge) as a high-mobility channel material. Ge is among the most attractive alternative materials because it has a higher mobility of charge carriers, especially of holes, and a smaller band gap than silicon (Si) (see Table 1). In addition, Ge is the most Si-compatible material as compared to the other high-mobility materials.

Theoretically, germanium outperforms Si in nearly every way. In Ge, the lower transport mass (m^*) of the electron and hole is responsible for the higher electron and hole mobilities. Electrons have a *bulk* mobility that is 4 and 10 times higher than in silicon for doping concentrations of 10^{16} and 10^{19} cm^{-3}, respectively. In addition, germanium has the highest hole mobility of all semiconductors. Nevertheless, due to carrier velocity saturation at high electric fields [38], the advantage of Ge mobility is only valid below certain values of electric field at room temperature, which limits the performance of Ge short channel devices. As the transistors enter the sub-22 nm nodes, however, Ge can be used in the ultra-short channel MOSFETs exhibiting (quasi-) ballistic transport [39]. In this case the drive current is determined by the carrier injection probability, which is related to the mobility at low electric field. Thus, Ge again becomes a very promising channel material and has attracted much attention recently.

High-performance p-channel germanium MOSFETs have been realised and reported by several research groups. The maturity of these devices has been recognised by the latest edition of ITRS [1] where the materials for them and the n-III-V devices have been transitioned from the Emerging Research Materials (ERM) chapter to the Process Integration, Devices, and Structures (PIDS) and Front End Processes (FEP) chapters. They are forecasted to be in production in 2018.

The situation with the n-channel Ge MOSFETs is, however, completely different. They face significant challenge due to three main reasons: (i) the presence of high density of interface traps near the conduction band, resulting from germanium oxide instability, causing significant degradation of electron mobility in

Table 1 Mobility and band gap values for Si and some alternative channel materials

	Si	Ge	GaAs	InAs	Graphene	InGaAs	InSb
Mobility, electron	1,600	3,900	9,200	40,000	~ 10,000	11,200	77,000
Mobility, hole	430	1,900	400	500	~ 10,000	300	850
Band gap	1.12	0.66	1.42	1.42	0		0.17

surface inversion channels; (ii) poor results of incorporation and activation of *n-type dopants in the source and drain regions* that do not reach the required values of $>10^{20}$ cm^{-3}; (iii) *junction leakage currents* are very high because of the *small bandgap of Ge*.

While significant efforts are currently being made to tackle these problems by developing, e.g. ozone (O_3) oxidation, alternative passivation approaches and high-*k* gate dielectrics, metal-induced dopant activation, etc. the junctionless transistor architecture seems to naturally solve them:

- Carrier transport is in the *bulk* of the device, not in a surface channel, thereby reducing the impact of imperfect semiconductor-insulator interfaces on electrical characteristics.
- There is *no need to separately implant source and drain* regions and the optimum active doping concentrations for JNTs are much lower: in the upper 10^{18} to lower 10^{19} cm^{-3}.
- *Leakage current* in JNTs is determined by the electrical squeezing of the channel by the gate potential and not by a reverse-biased junction as in classical MOSFETs. Therefore, tri-gate or even gate-all-around architecture is essential here to achieve best electrostatic control of the channel.

The successful elaboration of both p- and n-type Ge MOSFETs will allow the fabrication of Ge CMOS devices. This is highly appealing since it will significantly ease the severe integration challenges in case of co-integration of III-V compounds and Ge with CMOS devices. In addition, Ge has the smallest difference between the mobilities of electrons and holes among all semiconductor materials and, hence, the mismatch between the Ge p-MOSFET and n-MOSFET should also be low. An “all Ge” solution offers a distinct advantage over the Si option where hybrid Ge (pMOST) and III-V (nMOST) is proposed with commensurately far more challenging processing.

Ge processing is, however, quite different from Si processing and developing a JNT process on Ge is by no means trivial. Nevertheless, it is worth noting that the junctionless device architecture relaxes significantly the requirements to some of the most critical process steps. Classical inversion-mode (IM) MOSFETs are very sensitive to the quality of the gate dielectric. Moreover, both industry and academic data show rapid degradation in mobility with decreasing electrical oxide thickness in Ge MOSFETs and this issue exists even if a high-k dielectric is used [40]. The JNT, on the other hand, is expected to be less sensitive to the interface imperfections due to its bulk conduction mechanism and reduced vertical electric field [41]. This will result in relaxed requirements to Ge surface passivation, EOT scaling, and overall quality of the gate stack. In addition, junctionless architecture avoids the need for separate doping of source and drain regions to high impurity concentrations ($>1 \times 10^{20}$ cm^{-3}) which is extremely challenging especially in the case of Ge n-MOSFETs. This fact significantly relaxes the requirements to doping and annealing processes too, since the doping concentrations relevant to JNTs are in the range of 1×10^{19}–5×10^{19} cm^{-3}.

Heterogeneous integration of alternative high-mobility channel materials, including crystalline Ge layers, on Si substrates is a challenging task. A possible elegant solution in the case of Ge might be the novel germanium-on-insulator (GeOI) platform. Like the silicon-on-insulator (SOI) wafers are the substrate material of choice in the case of Si JNTs, in the same way the GeOI wafers offer an excellent material platform for the fabrication of Ge JNTs. Significant research efforts were made to fabricate GeOI wafers. Nakaharai et al. reported 7-nm-thick strained GeOI made using a Ge-condensation technique [42]. Another approach applies the Smart-Cut method used for making SOI substrates to fabricate GeOI wafers [43, 44]. However, fabrication of ultra-thin Ge layers is difficult, which motivates the need of further process development.

Only one group has reported on Ge junctionless transistors to date [14]. These are, however, large planar p-type devices having channel length, width, and thickness of 160, 130 μm and 11 nm, respectively. Herein we present our initial results on fabrication and electrical characterisation of the first p-type Ge junctionless nanowire transistors, which significantly outperform the ones reported in [14].

3.1 Fabrication of Ge JNTs

GeOI wafer with highly p-doped ($\sim 1 \times 10^{19}$ cm^{-3}) top Ge layer of 39 nm and buried oxide (BOX) SiO_2 layer of 145 nm (IQE) has been used in this work. The process flow has been kept as simple as possible to ensure the robustness and reliability of device fabrication. It contains the following main steps:

1. Electron Beam Lithography.
2. Reactive Ion Etching.
3. GeO_2 passivation layer formation by rapid thermal oxidation.
4. Atomic layer deposition (ALD) of gate oxide (Al_2O_3) layer.
5. Source/drain electrodes and metal gate formation.

The challenges with Ge processing appear already at the first patterning steps. In this work, HSQ has been used as the primer resist for EBL since it has very high etch resistivity and the highest resolution among all commercially available EBL resists [32, 33, 45]. The definition of sub-20 nm features by EBL on Ge using HSQ is, however, a challenging task. The complex native oxide GeO_x is soluble in the HSQ aqueous developers. As a result, lift-off of any sub-20 nm features written by EBL occurs during development [46] (see the inset in Fig. 11). This issue can be solved in two ways: (i) by removal of GeO_x and chlorine (Cl) termination of Ge surface prior to HSQ deposition [45] or (ii) application of a buffer layer between GeO_x and HSQ (SiO_2, Si_3N_4, Al_2O_3, etc.) which can also be used as a hard mask during subsequent dry etching. We have sufficiently elaborated both approaches and arrays of sub-20 nm HSQ lines have successfully been fabricated on Ge by surface treatment, Fig. 11, as well as by buffer layer deposition, Fig. 12a for a buffer layer of Al_2O_3 and Fig. 12b and c for a buffer layer of Si_3N_4. These are

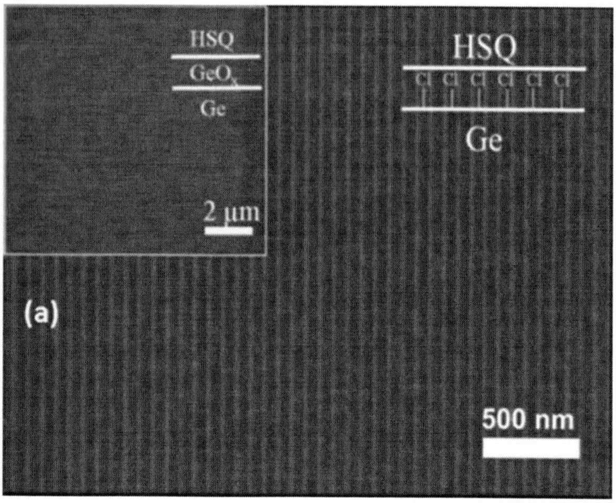

Fig. 11 HSQ lines, 20 nm wide, at a pitch of 70 nm, written on a Ge < 100 > substrate. (Inset) HSQ lines, 20 nm wide, at a pitch of 70 nm, written on a Ge < 100 > (GeOx surface) substrate at the same area dose [46]

among the smallest structures created by direct patterning on Ge and can be used as a hard mask for top-down fabrication of very small Ge NWs. A patent has been filed on the technology for GeO_x removal and chlorine termination of Ge surfaces [47].

In this work, the process with the Si_3N_4 buffer layer has been used. The nanowire structures of widths down to 20 nm have been defined again by a JEOL JBX 6000FS EBL system in the HSQ resist. The lithographic HSQ pattern has then been transferred into the Si_3N_4 layer by an RIE process with CH_4:CHF_3 gas in an STS inductively coupled plasma (ICP) etcher. Finally, the Si_3N_4 mask has been transferred into the top Ge layer of the GeOI substrate using the same Cl-based RIE process that has been used for etching the Si JNTs. This patterning process is discussed in detail elsewhere [48].

As mentioned above, the high quality gate dielectric for surface passivation is one of the major challenges in realising Ge MOSFETs. In this work, GeO_2 grown by a rapid thermal oxidation (RTO) has been used as an interfacial passivation layer for the high-*k*/Ge gate stack. The oxidation behaviour of Ge by RTO has been investigated for various temperatures, times, and gas flows. The results have shown that oxidation temperatures higher than 550 °C lead to rapidly increasing GeO_2 surface roughness, reaching values as high as 14.5 nm at 600 °C, which can be related to pronounced GeO desorption. In contrast, the interfacial GeO_2 layers formed at higher temperatures exhibit systematically decreased interface states density (D_{it}) values [49]. Post-metallisation forming gas annealing (FGA) of MOS devices has also been found to reduce the D_{it} values, which indicates that FGA can result in significant reduction of the interface states and compensate the charges inside the oxide [50]. Eventually, oxidation temperature of 550 °C for 1 min has been chosen and the forming gas anneal has been performed at 400 °C for 30 min.

After the GeO_2 passivation layer formation by RTO, a gate oxide (Al_2O_3) of ~8 nm thickness has been deposited by ALD. Contact windows for the source

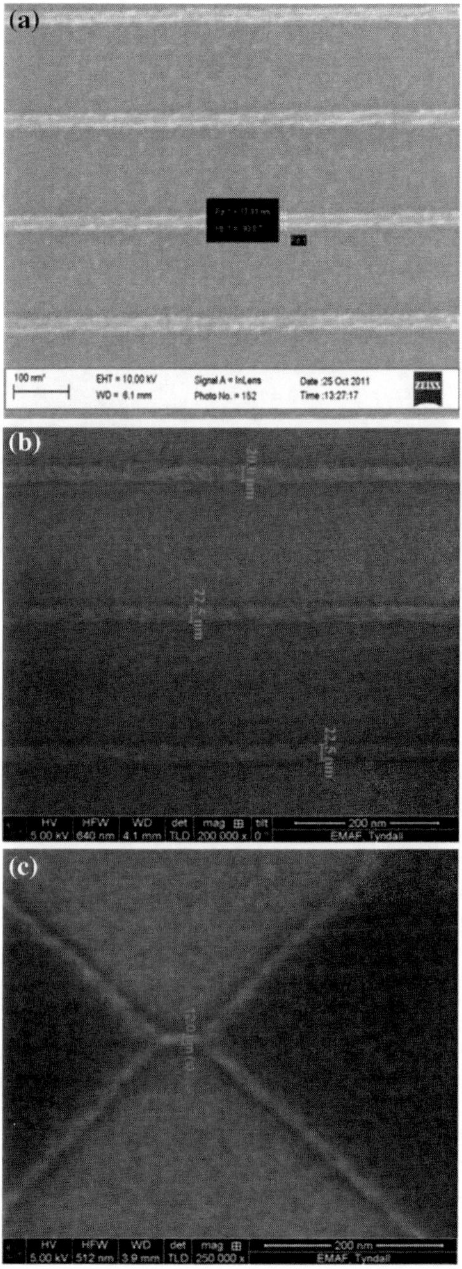

Fig. 12 **a** HSQ lines, 20 nm wide, written on a GeOI substrate with a buffer layer of Al_2O_3. **b** HSQ lines, 20 nm wide, written on a GeOI substrate with a buffer layer of Si_3N_4. **c** Short HSQ line, 12 nm wide, written on a GeOI substrate with a buffer layer of Si_3N_4

and drain regions have been opened by a wet etch in a buffered oxide etch (BOE) diluted with deionised water (DI) to the ratio of 5:1 DI:BOE. The source/drain electrodes and the metal gate have been defined simultaneously by the same

Fig. 13 SEM *top view* of one single nanowire Ge JNT with GeO_2/Al_2O_3 as a gate stack and Ni as a metal gate

lithography mask. Thus, both the source/drain electrodes and the gate have been formed using one metallisation step, namely electron-beam evaporation of a 100 nm thick nickel (Ni) layer. As mentioned above, FGA has been performed at 400 °C for 30 min. In Fig. 13, a top view of a fabricated Ge JNT with a Ni metal gate on the top of the Ge nanowire is shown.

3.2 Electrical Characterization of Ge JNTs

Electrical measurements of the fabricated devices have been carried out at room temperature by a cascade probe station and an Agilent semiconductor analyser B1500.

Figure 14a presents sample output characteristics $I_d - V_d$ of of a 2.5 μm gate length Ge JNT with fin height of 39 nm and fin width of 20 nm. The curves clearly exhibit the linear and the saturation regions. The increase of current is relatively proportional to $(V_g - V_{th})^2$, which indicates that the carriers in the channel have not yet reached the velocity saturation.

In Fig. 14b, transfer characteristics $I_d - V_g$ of the same transistor are shown. It can be seen that when the V_d decreases from -0.7 to -0.4 V, the leakage current is dramatically reduced. A possible explanation of this effect is that when the V_d is larger than the Ge bandgap, band to band tunnelling current might occur even in junctionless structures. Therefore, leakage currents appear at some level for $V_d = -0.7$ and -1 V. This Ge JNT exhibits a subthreshold slope *SS* of 216 mV/dec with an I_{on}/I_{off} current ratio of 1.2×10^3 at $V_d = -1$ V and 189 mV/dec with an I_{on}/I_{off} current ratio of 2.0×10^5 at $V_d = -0.1$ V. The *DIBL* is calculated to be 87 mV from $V_d = -0.1$ V and -1 V.

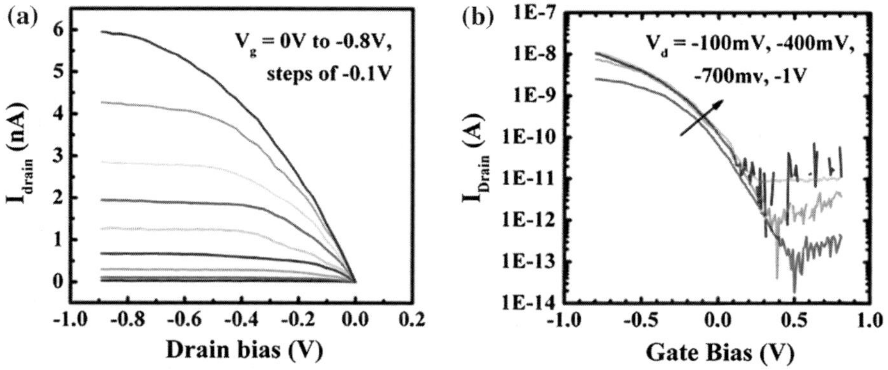

Fig. 14 **a** Output characteristics $I_d - V_d$ at $V_g = 0$ to -0.8 V of a Ge JNT with 1 Ge NW of 5 μm length, $T_{Ge} = 39$ nm, $W_{Ge} = 20$ nm, $L_g = 2.5$ μm and **b** transfer characteristics $I_d - V_g$ at $V_d = -0.1$ to -0.8 V of the same transistor

Table 2 Comparison of Ge junctionless transistor performance in Ref. [14] and present work

		Ref. [14]	Present work
W (channel width)	μm	130	0.04
L (channel length)	μm	160	3
W/L		0.8125	0.013
I_{off} @$V_D = -1$ V	A	7.00E-08	1.00E-11
I_{on} @$V_D = -1$ V and $V_G(I_{off}) - 1.5$ V	A	8.00E-06	1.20E-08
I_{on}/I_{off} @ $V_D = -1$ V and $V_G(I_{off}) - 1.5$ V		114	1,200
I_{on}/I_{off} @ low V_D and $V_G(I_{off}) - 1.5$ V		3,000	200,000
I_{off} @$V_D = -1$ V normalized to *W/L*	A	8.62E-08	7.70E-10
I_{on} @$V_D = -1$ V and $V_G(I_{off}) - 1.5$ V normalised to *W/L*	A	9.85E-06	0.92E-06
Subthreshold slope @ $V_D = -1$ V	mv/dec	700	216
Subthreshold slope @ low V_D	mV/dec	1,000	189

To the best of our knowledge, these are the first reported Ge JNTs.

Table 2 compares the electrical characteristics of our JNTs with those of the Ge junctionless devices reported in Ref. [14].

4 Conclusions

Junctionless nanowire transistors are novel devices having the simplest possible transistor architecture. Beside the significantly simplified fabrication, they possess a number of other advantages over the classical MOSFETs, which are briefly discussed in this chapter.

An attractive application of JNTs is as sensing devices. Their performance depends strongly on the geometry (height, width, and length) as well as on the doping level of the nanowire channel. These parameters can be used to finely tune the operation point of devices into the most preferable subthreshold regime which makes the application of JNTs in low power autonomous sensing systems very promising. In this chapter we present back-gated Si JNTs of 45 different channel geometries having NW widths down to 10 nm. They are fabricated by a fully CMOS-compatible process and exhibit excellent electrical characteristics. Moreover, their good sensing properties are clearly demonstrated by the data from pH and streptavidin sensing experiments. These JNTs are among the smallest top-down fabricated Si nanowire sensing devices reported to date. Further sensing experiments with them are planned in the near future.

Junctionless architecture has also a great potential for incorporating alternative channel materials with enhanced carrier transport properties, particularly germanium. We present here the first reported Ge JNTs, which are fabricated by a simple CMOS-compatible process on GeOI wafers with a highly p-doped ($\sim 1 \times 10^{19}$ cm^{-3}) top Ge layer. They have channel widths down to 20 nm and demonstrate decent output and transfer characteristics with I_{on}/I_{off} ratios of up to 2.0×10^5 and steep subthreshold slopes of 189 mV/dec. Fabrication of Ge JNTs with improved design is currently underway.

Acknowledgments This work has been funded by the EU 7th Framework Programme under the SiNAPS project (no. 257856) and Science Foundation Ireland (SFI) under the grants no. 09/IN.1/I2602, 09/SIRG/I1623 and 10/IN.1/I2992.

References

1. International Technology Roadmap for Semiconductors (ITRS). http://www.itrs.net/ (2011)
2. http://newsroom.intel.com/docs/DOC-2032
3. Ernst, T., et al.: Novel 3D integration process for highly scalable nano-beam stacked-channels GAA (NBG) Fin FETs with HfO_2/TiN gate stack. In: IEEE International Electron Device Meeting (IEDM) Technical Digest, pp. 997–1001 (2006)
4. Buitrago, E., Fernandez-Bolaños, M., Georgiev, Y.M., Yu, R., Lotty, O., Holmes, J.D., Nightingale, A.M., Ionescu, A. M (2013) 3D silicon nanostructures for biosensing applications: functionalized 3D 7x20-array, vertically stacked SiNW FET for streptavidin sensing. In: 71st Annual Device Research Conference, DRC, Notre Dame, IN, USA (2013)
5. Colinge, J.P., et al.: SOI Gated Resistor: CMOS without Junctions. In: Proceedings of the IEEE International SOI Conference, paper 11.1 (2009)
6. Colinge, J.P., et al.: Nanowire transistors without junctions. Nat. Nanotechnol. **5–3**, 225–229 (2010)
7. Ferain, I., Colinge, C.A., Colinge, J.P.: Multigate transistors as the future of classical metal–oxide–semiconductor field-effect transistors. Nature **479**, 310–316 (2011)
8. Colinge, J.P., et al.: Junctionless nanowire transistor (JNT): Properties and design guidelines. Solid-State Electron. **65–66**, 33–37 (2011)
9. Ansari, L., et al.: Simulation of junctionless Si nanowire transistors with 3 nm gate length. Appl. Phys. Lett. **97**, 062105 (2010)

10. Migita, S., et al.: Electrical performances of junctionless-FETs at the scaling limit (LCH = 3 nm). In: IEEE International Electron Device Meeting (IEDM) Technical Digest. pp. 191–194 (2012)
11. Park, C. –H., et al.: Comparative study of fabricated junctionless and inversion-mode nanowire FETs. In: Proceedings of the 69th Annual Device Research Conference, pp. 179–180. Santa Barbara, CA, USA (2011)
12. Gnani, E., et al.: Theory of the junctionless nanowire FET. IEEE Trans. Electron. Devices **58**, 2903–2910 (2011)
13. Su, C.-J., et al.: Gate-all-around junctionless transistors with heavily doped polysilicon nanowire channels. IEEE Electron Device Lett. **32**, 521–523 (2011)
14. Zhao, D.D., et al.: Junctionless Ge p-Channel metal–oxide–semiconductor field-effect transistors fabricated on ultrathin Ge-on-insulator substrate. Appl. Phys. Express **4**, 031302 (2011)
15. Cho, S., et al.: Silicon-compatible bulk-type compound junctionless field-effect transistor. In: Proceedings of ISDRS 2011, College Park, MD, USA, 7–9 Dec 2011
16. Yokoyama, M., et al.: Sub-10-nm extremely thin body InGaAs-on-Insulator MOSFETs on Si wafers with ultrathin Al_2O_3 buried oxide layers. Electron Device Lett. **32**, 1218–1220 (2011)
17. Jiang, J., et al.: Junctionless in-plane-gate transparent thin-film transistors. Appl. Phys. Lett. **99**, 193502 (2011)
18. Choi, S.-J., et al.: Sensitivity of threshold voltage to nanowire width variation in junctionless transistors. IEEE Electron Device Lett. **32**, 125–127 (2011)
19. Aldegunde, M., et al.: Study of discrete doping—induced variability in junctionless nanowire MOSFETs using dissipative quantum transport simulations. IEEE Electron Device Lett. **33**, 194–196 (2012)
20. Taur, Y., et al.: On–off charge–voltage characteristics and dopant number fluctuation effects in junctionless double-gate MOSFETs. IEEE Trans. Electron Devices **59**, 863–866 (2012)
21. Dehdashti, A.N., et al.: Random dopant variation in junctionless nanowire transistors. In: Proceedings of the IEEE International SOI Conference, pp. 55–56 (2011)
22. Cui, Y., Lieber, C.M.: Functional nanoscale electronic devices assembled using silicon nanowire building blocks. Science **291**(5505), 851–853 (2001)
23. Cui, Y., Wei, Q., Park, H., Lieber, C.M.: Nanowire nanosensors for highly sensitive and selective detection of biological and chemical species. Science **293**(5533), 1289–1292 (2001)
24. Chen, K.I., Li, B.R., Chen, Y.T.: Silicon nanowire field-effect transistor-based biosensors for biomedical diagnosis and cellular recording investigation. Nano Today **6**(2), 131–154 (2011)
25. Elfström, N., Juhasz, R., Sychugov, I., Engfeldt, T., Karlström, A.E., Linnros, J.: Surface charge sensitivity of silicon nanowires: size dependence. Nano Lett. **7**(9), 2608–2612 (2007)
26. Stern, E., et al.: Label-free immunodetection with CMOS-compatible semiconducting nanowires. Nature **445**(7127), 519–522 (2007)
27. Heitzinger, C., Klimeck, G.: Computational aspects of the three-dimensional feature-scale simulation of silicon-nanowire field-effect sensors for DNA detection. J. Comput. Electron. **6**(1), 387–390 (2007)
28. Kim, A., et al.: Ultrasensitive, label-free, and real-time immunodetection using silicon field-effect transistors. Appl. Phys. Lett. **91**(10), 103901–103903 (2007)
29. Gao, P.A., Zheng, G., Lieber, C.M.: Subthreshold regime has the optimal sensitivity for nanowire FET biosensors. Nano Lett. **10**, 547–552 (2009)
30. Knopfmacher, O., Tarasov, A., Fu, W.Y., Wipf, M., Niesen, B., Calame, M., Schonenberger, C.: Nernst limit in dual-gated Si-nanowire FET sensors. Nano Lett. **10**, 2268–2274 (2010)
31. Buitrago, E., Fagas, G., Fernández-Bolanos, M.B., Georgiev, Y.M., Berthomé, M., Ionescu, A.M.: Junctionless silicon nanowire transistors for the tunable operation of a highly sensitive, low power sensor. Sens. Actuators, B **183**, 1–10 (2013)
32. Henschel, W., Georgiev, Y.M., Kurz, H.: Study of a high contrast process for hydrogen Silsesquioxane as a negative tone electron beam resist. J. Vac. Sci. Technol., B **21**(5), 2018–2025 (2003)

33. Georgiev, Y.M., Henschel, W., Fuchs, A., Kurz, H.: Surface roughness of hydrogen silsesquioxane as a negative tone electron beam resist. Vacuum **77**(2), 117–123 (2005)
34. Park, I., Li, Z., Li, X., Pisano, A.P., Williams, R.S.: Towards the silicon nanowire-based sensor for intracellular biochemical detection. Biosens. Bioelectron. **22**, 2065–2070 (2007)
35. Chen, Y., Wang, X., Erramilli, S., Mohanty, P., Kalinowski, A.: Silicon-based nanoelectronic field-effect pH sensor with local gate control. Appl. Phys. Lett. **89**(22), 223512 (2006)
36. Duffy, D.C., McDonald, J.C., Schueller, O.J.A., Whitesides, G.M.: Rapid prototyping of microfluidic systems in Poly(dimethylsiloxane). Anal. Chem. **70**(23), 4974–4984 (1998)
37. Satyanarayana, S., Karnik, R.N., Majumdar, A.: Stamp-and-stick room-temperature bonding technique for microdevices. J. Microelectromech. Syst. **14**(2), 392–399 (2005)
38. Pantisano, L., Trojman, L., Mitard, J., DeJaeger, B., Severi, S., Eneman, G., Crupi, G., Hoffmann, T., Ferain, I., Meuris, M., Heyns, M.: Fundamentals and extraction of velocity saturation in sub-100 nm (110)-Si and (100)-Ge. In: IEEE Symposium on VLSI Technology Digest of Technical Papers, pp. 52-53 (2008)
39. Natori, K.: Ballistic metal-oxide-semiconductor field effect transistor. J. Appl. Phys. **76**, 4879 (1994)
40. Kuhn, K.J., et al.: Past, present and future: SiGe and CMOS transistor scaling. ECS Trans. **33**(6), 3–17 (2010)
41. Colinge, J.-P., et al.: Reduced electric field in junctionless transistors. Appl. Phys. Lett. **96**(7), 073510 (2010)
42. Nakaharai, S., et al.: Characterization of 7-nm-thick strained Ge-on-insulator layer fabricated by Ge-condensation technique. Appl. Phys. Lett. **83**(17), 3516–3518 (2003)
43. Deguet, C., et al.: Fabrication and characterisation of 200 mm germanium-on-insulator (GeOI) substrates made from bulk germanium. Electron. Lett. **42**(7), 415–417 (2006)
44. Yu, R., et al.: Fabrication of germanium-on-insulator by low temperature direct wafer bonding. In: 10th IEEE International Conference on Solid-State and Integrated Circuit Technol (ICSICT): 953 (2010)
45. Namatsu, H., et al.: Three-dimensional siloxane resist for the formation of nanopatterns with minimum linewidth fluctuations. J. Vac. Sci. Technol., B **16**, 69 (1998)
46. Hobbs, R.G., Schmidt, M., Bolger, C.T., Georgiev Y.M., Xiu, F., Wang, K.L., Fleming, P., Morris, M.A., Djara, V., Yu, R., Colinge, J.-P., Petkov, N., Holmes, J.D.: Resist-substrate interface tailoring for generating high density arrays of Ge and Bi_2Se_3 nanowires by electron beam lithography. J. Vac. Sci. Technol. B **30**(4), 041602(1)–041602(7) (2012)
47. Hobbs, R. G., Petkov, N., Holmes, J. D.: Methods and materials for lithography of a high resolution HSQ resist, European Patent Number: 11163598.3 (2012)
48. Yu, R., Das, S., Hobbs, R., Georgiev, Y., Ferain, I., Razavi, P., et al.: Top-down process of germanium nanowires using EBL exposure of hydrogen Silsesquioxane resist. In: IEEE 13th International Conference on Ultimate Integration on Silicon (ULIS). pp. 145–148 (2012)
49. Matsubara, H., Sasada, T., Takenaka, M., Takagi, S.: Evidence of low interface trap density in GeO_2/Ge metal-oxide-semiconductor structures fabricated by thermal oxidation. Appl. Phys. Lett. **93**, 032104 (2008)
50. Xie, R., He, W., Yu, M., Zhu, C.: Effects of fluorine incorporation and forming gas annealing on high-k gated germanium metal-oxide-semiconductor with GeO_2 surface passivation. Appl. Phys. Lett. **93**, 073503–073504 (2008)

SOI-Based Microsensors

Daniel Tomaszewski, Michał Zaborowski, Krzysztof Kucharski, Jacek Marczewski, Krzysztof Domański, Magdalena Ekwińska, Paweł Janus, Tomasz Bieniek, Grzegorz Głuszko, Bohdan Jaroszewicz and Piotr Grabiec

Abstract A silicon-on-insulator (SOI) technology has become a key tool for manufacturing of microsensors. A number of the SOI-based microsensors have been presented in the chapter. The developments are related mostly to non-electrical sensors. Most of them have been developed at Instytut Technologii Elektronowej (ITE) Warsaw, in collaboration with numerous partners. The chapter is divided into two main sections describing the technology and device issues. First, ITE expertise in the area of SOI technologies is described. The SOI CMOS processes based on wafers with a thick and thin device layers are presented. The CMOS technology has become an origin for a PaDEOx technique developed for narrow (order of 100 nm width) line fabrication on the silicon wafers, which allows for development of nanowire-based devices (e.g. multi-gate MOSFETs). The second branch of the SOI technologies, being under development in ITE, includes micromachining techniques. Based on the SOI technologies, a number of devices have been developed with the ITE participation. The following solutions have been presented: monolithic pixel detectors of ionizing radiation, sub-THz radiation detectors based on a concept of radiation-induced plasmon oscillations in the MOSFET channels, smart antennas with a reconfigurable aperture. Next, devices fabricated based on micromachining techniques have been mentioned, e.g. microactuators and probes for scanning thermal microscopy. Finally, development of SOI-based biochemical sensors for small volume sample testing is more widely described. These devices have been manufactured based on the PaDEOx technique and have been applied for analysis of hydrogen an metal ions in water solutions.

D. Tomaszewski (✉) · M. Zaborowski · K. Kucharski · J. Marczewski · K. Domański · M. Ekwińska · P. Janus · T. Bieniek · G. Głuszko · B. Jaroszewicz · P. Grabiec
Division of Silicon Microsystem and Nanostructure Technology, Instytut Technologii Elektronowej (ITE), ul. Puławska 34, 05-500 Piaseczno, Poland
e-mail: dtomasz@ite.waw.pl

A. Nazarov et al. (eds.), *Functional Nanomaterials and Devices for Electronics, Sensors and Energy Harvesting*, Engineering Materials,
DOI 10.1007/978-3-319-08804-4_18

1 Introduction

A Silicon-On-Insulator (SOI) technology has become one of the main streams of the state-of-the art micro- and nanoelectronics [1]. It offers a number of features helpful for design and fabrication of most of the devices based on silicon. As compared to the CMOS devices and integrated circuits (ICs) fabricated in the bulk technology their SOI counterparts take advantage of (i) better gate control over channel conduction, (ii) much lower areas of source and drain junctions, and (iii) ideal electrical insulation from the substrate. Thanks to the feature (i) the MOS transistors exhibit steeper subthreshold slope, which makes them predestined for low voltage blocks and reduces static and dynamic power dissipated in the SOI circuits. Thanks to the properties (i), (ii) the MOS devices demonstrate lower leakage current, which makes the SOI MOSFETs suitable for low power applications, reduces a static power dissipated by the SOI circuits and allows SOI CMOS ICs to operate at higher temperatures. Thanks to the features (ii), (iii) the parasitic capacitances are lower in the SOI MOSFETs, which allows for higher operation frequency of the SOI ICs and reduces the dissipation of a dynamic power. Also, thanks to the properties (ii), (iii) the SOI MOSFETs exhibit higher resistance against ionizing radiation. Furthermore the SOI technology offers a flexibility in terms of device operation principle and IC design. E.g. an accumulation-mode p-channel MOSFET operation is not possible on bulk wafers. Also a wide family of so-called multi-gate MOSFETs has been developed based on the SOI technology [2]. In terms of the IC design one of the most important advantages is a full compatibility of the SOI and bulk processes, which allows for a direct adaptation of the bulk IC design methods, decreasing at the same time a number of photolithography steps (e.g. wells become useless, or even harmful, so they are eliminated). The SOI technology offers also many advantages for micro-electro-mechanical system (MEMS) development. The most important one is a natural etch-stop introduced by a buried oxide (BOX) layer. This etch stop significantly facilitates procedures for membrane, cantilever, and other micromechanical element fabrication and for movable element release allowing for fabrication of a very thin, free standing silicon structures. Finally, SOI technology allows for much more effective integration of CMOS blocks and MEMS structures into complex heterogenous systems. SOI-based microsensors are one of the categories of such systems. In the presented chapter a number of concepts related to the SOI-based microsensor development have been reported.

2 SOI Technology for Heterogenous Applications

The SOI wafers and processes have been used successfully for development of different device categories, i.e. detectors of ionizing radiation, detectors of mm-wavelength electromagnetic radiation, temperature sensors, and chemical sensors.

Two other SOI-based technologies for manufacturing of CMOS circuits, and smart antennas have been also mentioned. As far as possible the process/device descriptions (in particular for chemical sensors) are accompanied by experimental data: measurement setup specifications, electrical measurements, etc. Realization of these projects has led and still leads towards growth of expertise in the SOI technology and to the SOI-based microsensor development capabilities.

2.1 Monolithic Pixel Detectors of Ionizing Radiation in SOI Technology

Detection of elementary particles is an important tool for observation of experimental results in the field of physical studies of the matter. Fully depleted pixel detectors of ionizing radiation have been widely used for several experiments including the most recent ones carried out at CERN on the Large Hadron Collider. The main trend noticeable in sensor technology consists in integration of detectors with read-out electronics. The integrated sensors have usually not only better performance but are smaller, more reliable and also cheaper than standard sensor solutions based on separately designed detectors and data acquisition systems. Two types of the integrated detectors are in a wide use. Hybrid active pixel sensors (HAPS) consist of sensor and read-out circuitry chips manufactured separately and connected together using a flip-chip technology [3]. In the case of monolithic active pixel sensors (MAPS) both detector and read-out circuit parts are manufactured in the same semiconductor substrate [4]. It may be stated that the hybrid sensors introduce more material (mostly silicon) that may interact with impinging particles and affect precision of tracking. On the other hand the monolithic sensor concept inherited explicitly from pixel detectors used in visible light cameras suffers from a small volume of the active area (this implicates small output signal) and its large distance from the irradiated detector surface (a limiting factor for very low energy detection).

These serious limitations of MAPS have been overcome by integration of the read-out circuitries and the pixel detectors in the SOI wafers. The fully depleted sensing diodes are manufactured under buried oxide (BOX) layer, while the read-out circuitry occupies upper silicon layer (device layer). For the first time this idea was reported in [5], however its implementation appeared unsuccessful. An improved concept was proposed next within the EC FP5 project SUCIMA [6] by the Instytut Technologii Elektronowej—ITE (Warsaw, Poland) team in collaboration with an international team including University of Mining and Metalurgy (Krakow, Poland), and University of Insubria (Como, Italy). A structure of the proposed device is shown schematically in Fig. 1.

The sensors have been manufactured using wafer-bonded SOI substrates. This allows for taking advantage of combination in one SOI wafer of two silicon layers, each optimized for another circuitry. The matrices of radiation sensitive junctions

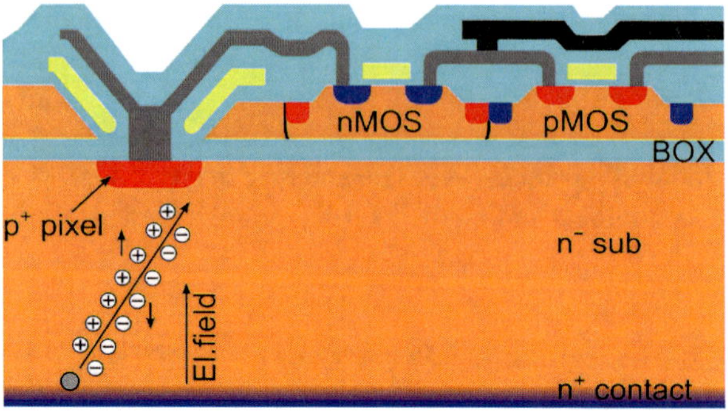

Fig. 1 Cross-section and principle of operation of a monolithic active pixel detector fabricated in SOI technology

(biased with tens of volts) have been formed in a high resistivity ($\rho > 4$ kΩ cm) 400 μm thick n-type substrate of the SOI wafer. The CMOS read-out and peripheral blocks operating at 5 V supply bias have been built within the 2 μm thick device layer. Both silicon areas are separated by the 1 μm thick BOX layer.

With such a large thickness of the device layer the bulk CMOS technology rules have been used. More detailed description of the technology may be found elsewhere [7, 8]. It is worthwhile to mention that in this very first approach p-type wells containing n-channel MOSFETs have touched the BOX layer. This solution has led to pockets of potential in the support layer beneath the wells, limiting the effective efficiency of charge collection by the pixel p^+-n junctions. The problem has been solved by increase of the device layer thickness to 5 μm that enables keeping the bottom part of the p-well far away from the BOX [9]. As a result, for the new devices, the effective charge collection efficiency has been significantly increased, exceeding 99 %. It is also worthwhile to mention that for both considered designs the device layer is thick enough to protect the MOSFET threshold voltages from being affected by the high voltage supplied at the bottom contact in order to fully deplete the substrate area.

Four detector matrices shown in Fig. 2 (24×24 cells each) monolithically integrated with read-out circuitry making one of the first in the world real 3-D integrated devices were used in experiments in which they were exposed to beta radiation. A response of a pixel cluster to the beam is shown in Fig. 3.

The main disadvantage of the devices described above is a fairly thick device layer not routinely used for submicron devices, which are envisaged for detectors due to their radiation hardness. Therefore, the next approaches proposed by several groups have been based on submicron technologies, which bring opportunity for a smaller pitch of the pixel detectors, thus for improvement of the imaging resolution. However, a number of problems arise from device miniaturization. In the case of a smaller pixel pitch a cross-talk between pixels becomes a serious challenge. However, the main challenge in the case of the submicron CMOS is the back gate effect, mentioned above. Its full elimination is a difficult task. The CMOS foundries have only limited sets of well-developed and characterized processes on standard wafers. Deviations from these standards are usually not allowed. Hence it is hardly possible to charge the foundry with a task of processing

Fig. 2 SOI detector chip fabricated in the ITE (four quadrants of 24 × 24 pixels) with independent sets of supply, control and analog output *lines*, clock frequency 8 MHz), and assembled at a dedicated PCB

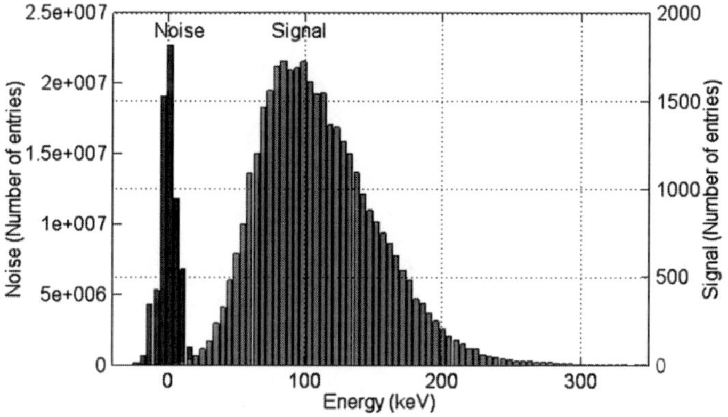

Fig. 3 Distribution of the noise after a pedestal subtraction and histogram of the cluster pulse height exposed to the beta radiation from the ^{90}Sr source

of the non-standard SOI wafers (with a highly resistive handle wafer). One exception is the most advanced group working on the subject formed within KEK—the High Energy Accelerator Research Organization. KEK has been running since several years a project [10] based on submicron CMOS FDSOI technology provided by OKI. Selected results of this undertaking may be found e.g. in [11]. The main effort has been done toward solving the back gate effect. A buried p-well (BPW) below CMOS circuitry appears to be a fruitful solution. The BPWs shield the thin device layer from the back gate effect induced by the substrate high voltage. Moreover they also relax the electric field at the edges of guard-rings resulting in the higher breakdown voltage of the entire device. Furthermore they allow for minimizing an effect of charge trapping in the SOI detector BOX area during irradiation. This has been achieved in two ways. Firstly, the electric field in the BOX layer is reduced. Secondly, the possible trapped charge effect onto CMOS circuits operation may be minimized via a dedicated BPW bias.

There are also other concepts to provide effective shielding of the read-out electronics from the fully depleted sensor. In a nested well structure (e.g. [5]) an n-well fabricated inside a p-well hole collector serves as a shield. In a solution based on double SOI substrates an intermediate silicon layer between two buried oxide films forms a shield (e.g. [11]), which may be efficiently used to compensate electric field generated by the holes trapped in the BOX during irradiation. Both solutions mentioned above have been recently tested by the KEK group. There is also a method for effective shielding by placing the read-out circuitry above the corresponding pixel with its potential pinned-up by an input of a charge amplifier [12].

An alternative solution for the active pixel sensor manufacturing could be provided by the adoption of the 3-D integration approach. In such a technology the CMOS circuits are formed in a separate SOI wafer using a standard processing sequence, whereas the imaging devices are fabricated in another wafer using adequate processing. Then, the device layer with CMOS circuitry is released and transferred on top of the wafer with the fully depleted sensor. Next, deep vias have to be formed by etching to provide access to the imager as well as to the CMOS circuit metal pads. Finally, the metal layer connecting elements of the system has to be defined by a metal deposition and a lithography step. The beauty of this approach is that it offers possibility of stacking together several device layers playing different roles in the final device (sparsing, computing, communication etc.). The 3-D SOI integration technology is extensively tested by FermiLab [13] and KEK.

2.2 Smart Antennas in SOI Technology

Smart antennas are well known since many years and often designed for microwave bands applications [14]. They can be used for radars (replacing function of mechanical scanners), wireless communication (spatially selective transmission), different adaptive systems, etc. Their use is compatible with almost any modulation method, bandwidth, or frequency band. The functionality depends on several parameters. The main feature of these antennas is an ability to influence their directional characteristics. A conventional method of the aperture reconfiguration can be achieved by the opening or closing different connections between conductive areas of the aperture. That may be done without any conventional switches or relays with use of highly conducting plasma induced in surface PIN (S-PIN) devices. The S-PIN is a lateral p-i-n diode. If the diode is forward biased carriers are injected into the intrinsic region making it highly conductive [15].

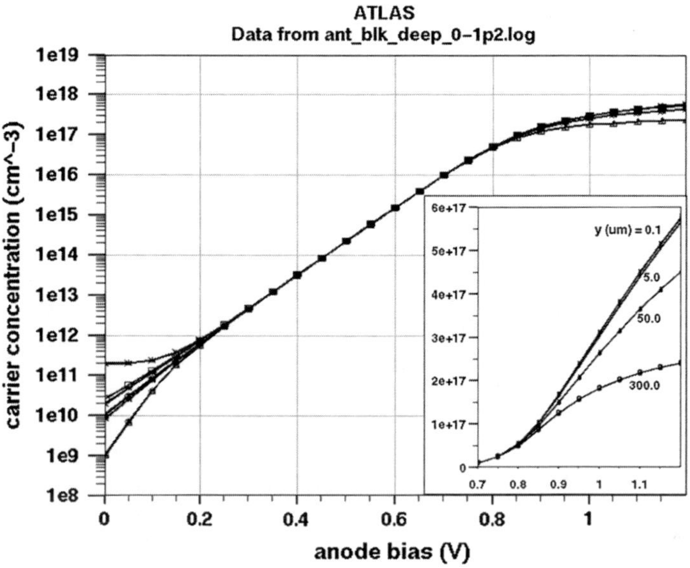

Fig. 4 Distribution of injected carriers in several points of a bulk substrate (x = 0, at depths y = 0.1, 1.5, 5, 50, 100, 200, and 300 μm); the S-PIN diode length is 10 μm

In order to get the low resistance plasma switch one has to use a high level of injection in the whole volume of the switch. However the silicon wafers (standard material for S-PIN diodes fabrication) are usually 300–700 μm thick. Hence even if the high resistivity material obtained with a Floating Zone method is used, which exhibits a very long lifetime of the charge carriers, the highly conductive plasma created in the intrinsic region of a S-PIN surface junction is not able to fill the entire volume of the wafer in a uniform way. It is clearly seen in the simulation results shown in Fig. 4 that at the high forward bias of 1.2 V a difference between the surface and the bulk carrier density appears (inset in Fig. 4). The carriers are spread over nearly the whole thickness of the wafer. Their density at the surface is hardly sufficient to create an ideal switch. But, what is even worse, the carrier density at the bottom of the wafer is high enough to make the silicon there lightly conductive, thus dissipative. It may be stated, that the optimum S-PIN should have well conducting plasma localized in the surface area, leaving the rest of the volume empty of carriers (dielectric). Silicon-On-Insulator (SOI) wafers with S-PIN diodes fabricated in the device layer are the remedy. The high-resistive handle wafer separated by the BOX layer from device film containing SPINs is always non-dissipative for electromagnetic radiation. It should be pointed out that the S-PIN in any case does not represent a short-circuit when is ON, therefore the device layer should be thick enough (a few skin-depths) for the required frequency range.

The simulations have been done for a smart antenna designed and then fabricated with use of a SOI high resistive wafer (both sides). The thickness of the device layer (81 μm) is determined by a depth of microwave penetration. The

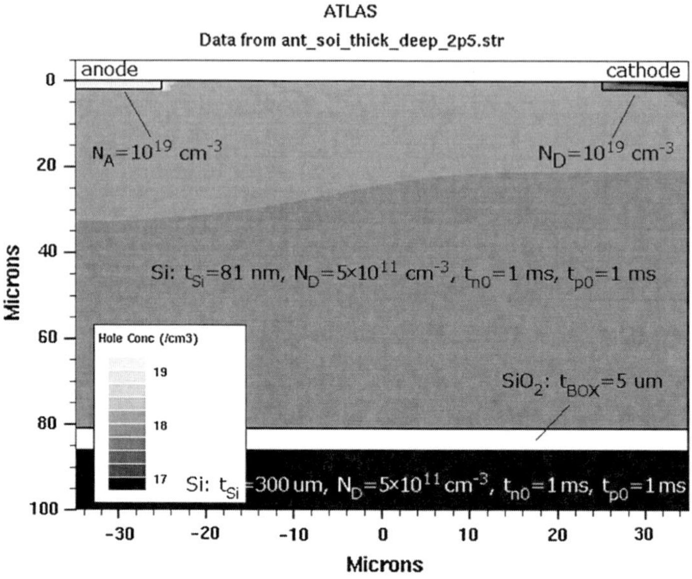

Fig. 5 The concentration of holes of a S-PIN diode fabricated in the *top* silicon layer; the diode is strongly forward biased

results (Figs. 5 and 6) demonstrate that high density of carriers is easily achieved within entire device layer. Also blocking behavior in the SOI-based S-PIN appears to be better that in its bulk counterpart.

Several examples of smart antennas designed at the Warsaw University of Technology, and then manufactured at ITE, Warsaw have been described in [16]. The developments have been finally focused on smart antennas built using the SOI technology. Their operation is based on reconfigurable slot apertures, which were placed on one wall of a metallic rectangular waveguide. The experimental results have confirmed the theoretical predictions. Proper gain and beam characteristics have been obtained, proving the SOI technology advantages i.e. higher plasma density at the surface, lower forward current required to create sufficient carrier concentration, and last but not least much smaller microwave losses in entire device.

2.3 FD SOI CMOS Technology

The CMOS technology is a key process not only widely used for fabrication of integrated circuits, but it is recognized as the most complete tool for evaluation of the clean-room capability and quality of the implemented processes. The fully-depleted (FD) SOI CMOS technology has been developed in 1990s but recently it has drawn a lot of attention and it has been recognized as a potential competitor for

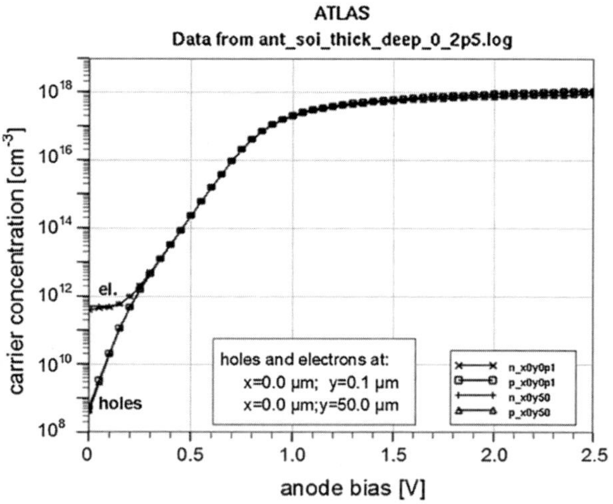

Fig. 6 Concentrations of carriers versus forward bias voltage beneath the surface (0.0 µm; 0.1 µm) and in the volume (0.0 µm; 50.0 µm) of the Si *top* layer

FinFET process even down to 11 nm node [17]. One of the early versions of the CMOS FD SOI technology developed in a laboratory of Université Catholique de Louvain—UCL [18] has been implemented in ITE within the EC FP7 project TRIADE [19] aimed at development of technology building blocks for structural health monitoring sensing devices in aeronautics [20].

There are several issues particularly important for the FD SOI technology development. First, a good quality of the SOI substrate wafers is essential, thus SOITEC wafers have been used. Next, an accurate setting of the Si thin layer initial thickness equal to 108 nm with a very good uniformity is required. This has been originally achieved by silicon film thinning in a dilute $NH_4OH + H_2O_2$ bath. However a method of thinning by thermal oxidation has been also developed and tested. Finally, the channel implantations setting the appropriate values of the threshold voltages are very important steps. The energy levels and doses must accurately fit the obtained earlier thickness of the active region below the transistor gates. While the technology utilizes inversion-mode nMOSFETs and accumulation-mode pMOSFETs different implantation doses for both device types are required. In the case of the UCL process two doses only are used. These, along with the not implanted and combined implantation variants, offer four threshold voltage variants by direct mask design at no additional cost. The overall process requires 10 photolithography steps. A cross-section of the FD SOI MOSFET pair is illustrated in Fig. 7, whereas a set of input I_D-V_{GS} characteristics of the n- and p-channel MOSFETs of 1.5 µm channel length is shown in Fig. 8.

The n-channel devices have the threshold voltage of 0.3 V and a subthreshold slope of 71 mV/dec, whereas the p-channel MOSFET threshold voltage is −0.45 V and the subthreshold slope is 84 mV/dec. Thus it may be stated that the device performance is satisfactory, though the threshold voltage symmetry should be improved.

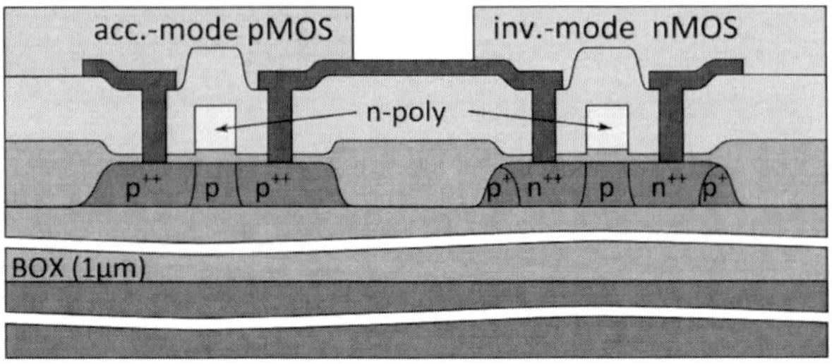

Fig. 7 A schematic view of a n- and p-channel MOSFET pair manufactured based on the FD SOI CMOS process

Fig. 8 I_D-V_{GS} characteristics of n- and p-channel MOSFETs; channel width = 20 μm, nominal channel length = 1.5 μm

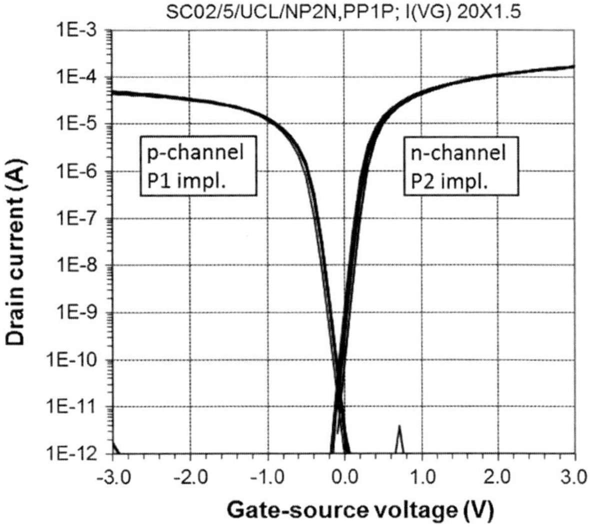

The FD SOI CMOS process transferred from UCL to ITE, allows manufacturing digital and analog-digital blocks. The experience gained by the technology implementation has been also helpful for a development of other technologies dedicated for fabrication of the SOI-based microsensors. They are described in the next sections.

2.4 THz Detectors Based on N-MOS Transistors Fabricated in SOI Technology

A THz and sub-THz frequency range of electromagnetic spectrum has been recently extensively investigated, because devices operating in this frequency range are a promising solution as a new class of detectors for imaging and safety applications. The THz source and the THz detector are basic elements of such systems. They are based on a photovoltaic effect within a two-dimensional electron gas (2DEG) in an inversion channel of a nMOS transistor or a HEMT [21]. It should be pointed out that the effect has been observed for frequencies (300 GHz) much higher than a maximum f_{max} and cut-off f_T frequencies (order of 10 GHz) of the test devices (see below). Essentially, any device containing the 2DEG may serve as the THz radiation detector. According to the THz detection theory if the length of the channel with the 2DEG is in the submicron range, a photovoltaic effect results from a channel plasmon resonance behavior. Otherwise non-resonant detection takes place. A detailed model for non-resonant detection in the field effect transistors may be found in [22].

With the standard approach, the radiation energy is delivered to the FET through a port between the FET source and gate terminals, while the rectified DC output signal appears between the source and drain terminals. Typically only the gate is biased with the gate-source DC voltage close to the FET threshold voltage. A detector responsivity strongly depends on a way a THz signal is coupled with the detector input. E.g. properly designed antennas allow to achieve a noise equivalent power (NEP) as low as in the case of Schottky diode detectors. Another important issue strongly related to the input stage design is radiation losses. A fraction of energy misses the antenna. Moreover it may create unwanted resonant modes in the die or simply may be dissipated.

The THz detectors have been fabricated in ITE, Warsaw according to designs prepared at the Warsaw University of Technology [23]. Two detector types have been developed. It is expected that the radiation propagation modes in the silicon wafer are eliminated, and the radiation energy losses are minimized by utilizing dedicated structures. In the first approach the detector (transistor and antenna) has been placed on a 40 μm thick silicon membrane obtained by means of chemical etching of the back-side of a silicon substrate as shown in Fig. 9. The device layout details are presented in Fig. 9a, b and its cross-section is shown schematically in Fig. 9c. Validity of this concept has been confirmed by extensive numerical simulations [24]. In the second approach the antennas have been placed above the transistor electrode level. A thick layer of SU-8 photoresist has been used as a dielectric separating the transistors and antennas, which are coupled by means of coplanar lines and capacitive couplers as shown in Fig. 10a, b. In Fig. 10c a cross-section of the complete device is schematically presented, whereas in Fig. 10d an image of the packaged and bonded device is shown. In this way the THz radiation energy is confined mostly to the SU-8 layer. For both designs the SOI CMOS process has been used. The SOI substrates with high resistive handle wafer have been chosen to reduce additionally possible energy losses in the volume of the device. Moreover, if the THz detector built on the SOI wafer with a lossless high-resistivity substrate were illuminated from the backside, then it would be possible to additionally increase the detector efficiency by using microlenses at the bottom of the wafer. Such microlenses focus the radiation onto the devices and an excitation of the radiation propagation modes in the substrate is eliminated.

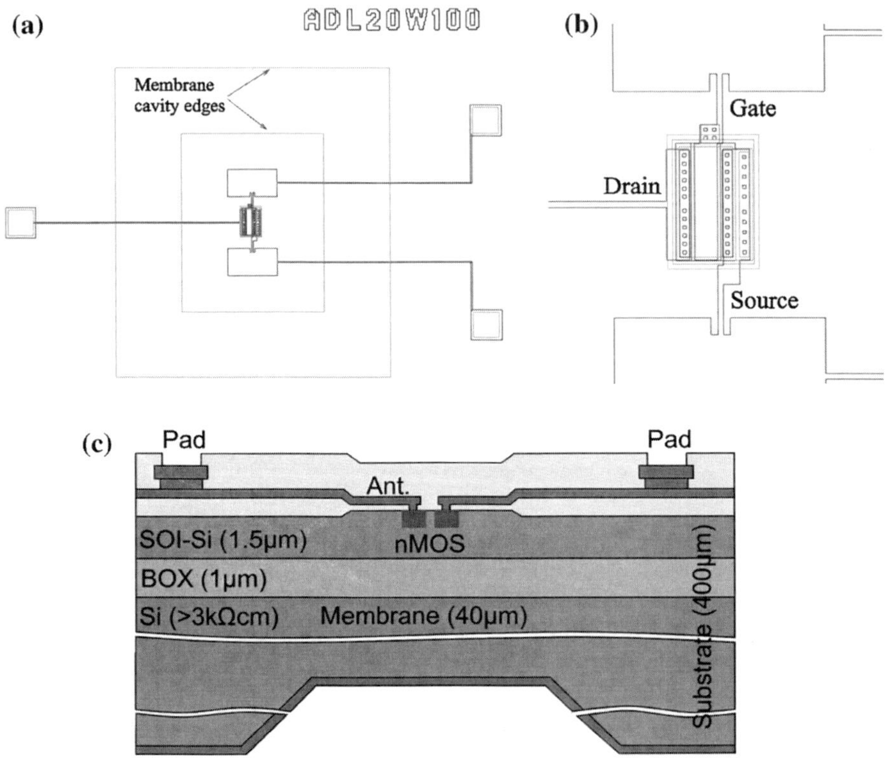

Fig. 9 The idea of placing the detector on the silicon membrane to suppress unwanted propagation modes in the substrate; **a** topography of the complete device with antennas and pads, **b** details of the MOSFFET structure, W = 100 μm, L = 20 μm, **c** a schematic device cross-section

The DC voltages generated at the detector outputs have been measured for a number of devices. The frequency characteristics of both detector types described above are shown in Fig. 11. The E4AL1L5W10 symbol denotes a detector on the membrane with antenna coupled to n-MOS with W/L = 5/10 μm. The AAL5W20 symbol denotes a detectors with the antennas coupled capacitively through 20 or 50 μm SU-8 photoresist. It may be stated that the developed detectors exhibit a satisfactory selectivity. The photoresponse versus gate bias of the membrane detector is shown in Fig. 12. The plot illustrates clearly that the MOSFET-based THz detector sensitivity may be efficiently tuned with the gate DC bias.

2.5 SOI-Based Micromachining for Sensor Applications

As already mentioned, the SOI technology has significant advantages for MEMS fabrication. It provides a natural etch-stop for silicon etching which is the BOX layer. As a consequence, very thin micromechanical structures (beams, comb-drive

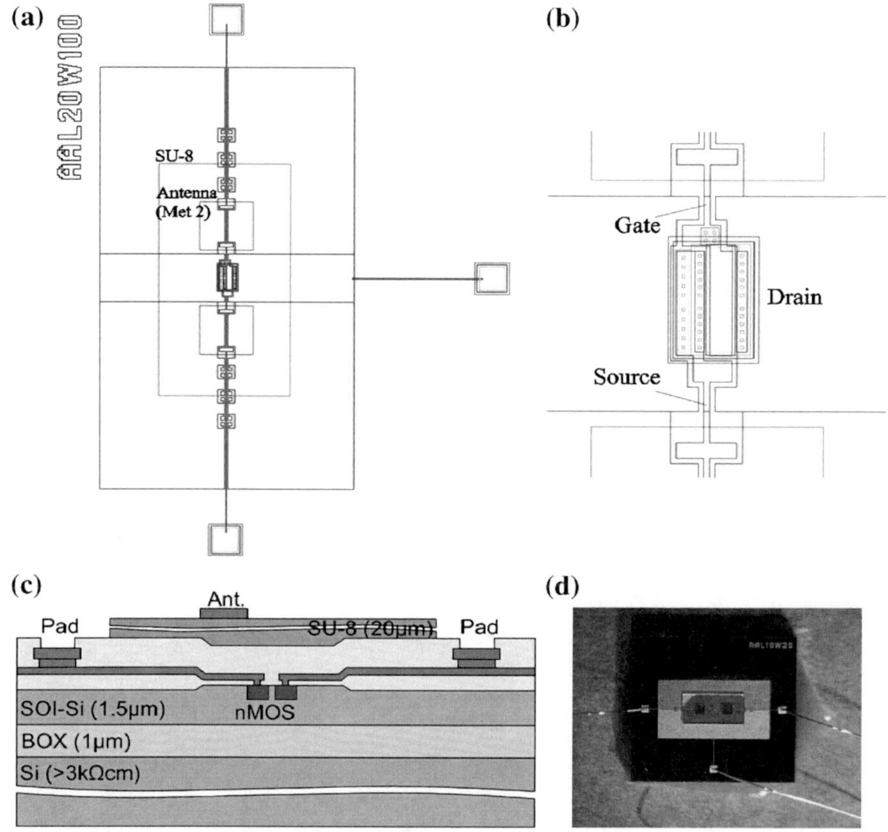

Fig. 10 The detector with the antenna coupled capacitively through SU-8 layer; **a** topography of the complete device with antennas and pads, **b** details of the MOSFFET structure, W = 100 µm, L = 20 µm, **c** a schematic device cross-section, **d** a *top* view at the detector with SU-8 layer and antennas, after packaging and wire bonding

Fig. 11 Photoresponse of the sub-THz radiation detectors versus frequency

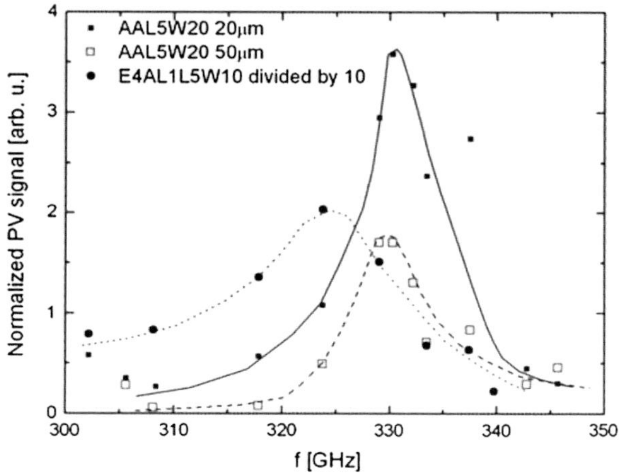

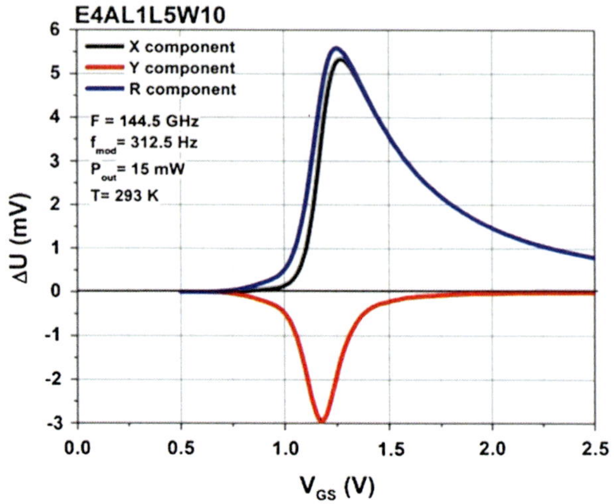

Fig. 12 Photoresponse of the detector manufactured on the silicon membrane (n-MOS W/L = 5/10 μm) versus gate voltage as illuminated with 15 mW source and measured by lock-in amplifier

electrodes etc.) with precisely defined thickness may be manufactured out of the device layer. Two types of SOI based technologies are common in MEMS manufacturing. A dry release plasma technology allows for fabrication of various types of devices based on comb-drive electrodes (micro-grippers, gyroscopes, accelerometers). An example of accelerometer structure manufactured based on SOI process and results of admittance spectroscopy measurements are shown in Fig. 13.

The SOI-based technology can be also applied for manufacturing of various types of microcantilever-beams for AFM (Atomic Force Microscopy) sensors. Quantitative measurements of the interaction between the tip and surface under test require well defined mechanical properties of the microbeam. One of the crucial beam parameter that influences the mechanical stiffness is the beam thickness. In case of application of SOI-based technology this parameter can be well controlled. Figure 14 shows some examples of micromachined AFM-related cantilevers manufactured for thermal and biochemical analyses.

2.6 PaDEOx Process for Fabrication of Biochemical Sensors

Fin-type FETs are expected to be the best candidates for operation in sensor applications. An effective multigate control of channel conductance is their advantage. In ITE a smart technique for manufacturing of the FinFET-like devices with a sub-micrometer width without advanced lithography tools has been developed. The proposed process is briefly illustrated in Fig. 15. The bulk silicon or SOI wafers are thermally oxidized and covered with a nitride layer. A standard photolithography is used for definition of the nitride pattern (Fig. 15a). Next, the

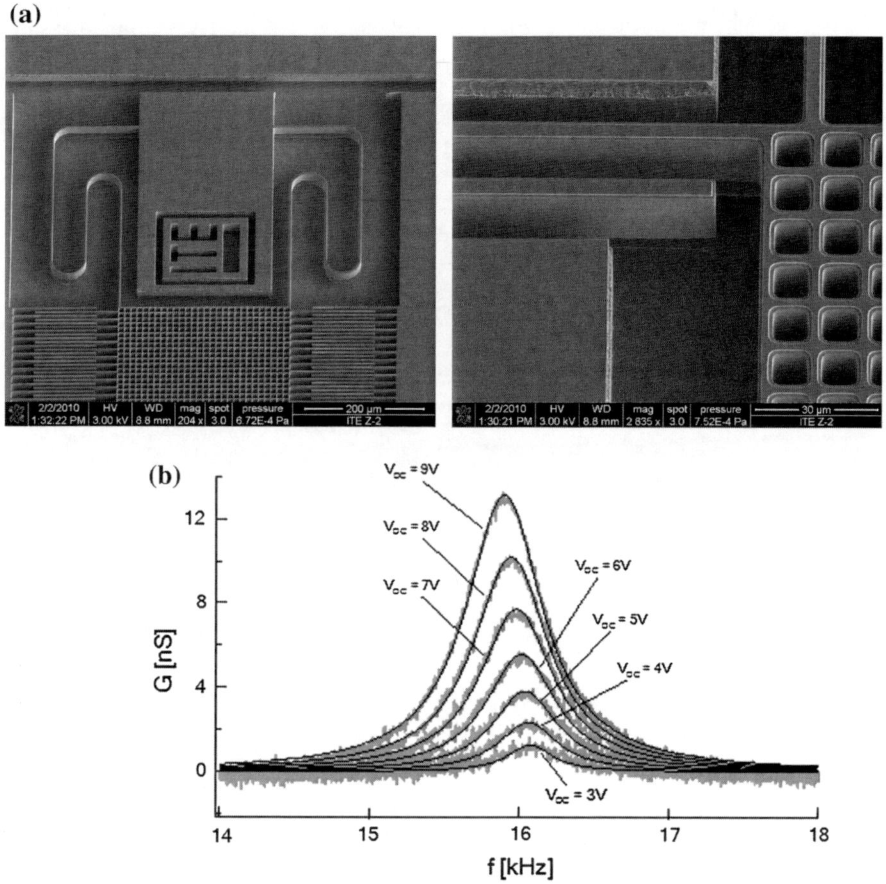

Fig. 13 A picture of the device fabricated based on the SOI process (**a**) and results of mechanical properties measurements of comb-drive based accelerometer (**b**)

technique employs the nitride pattern as a mask against oxidation around the FET active area. At the boundary of the nitride a so-called "bird's beak" appears (Fig. 15b). Locally the SiO_2 layer below the nitride pattern edge is thicker than the neighboring plane field oxide area. This subtle effect has been successfully used in a Pattern Definition by Edge Oxidation method (PaDEOx) [26, 27] for fabrication of narrow Si channels. According to this technique, after the oxidation, the entire nitride layer and the most part of the oxide layer are carefully removed to reveal an oxide path along the "bird's beak" region (Fig. 15c). The remaining oxide path serves as a mask in successive anisotropic etching of a fin-shaped Si narrow path—(Fig. 15d). Next, the oxide is completely etched off and a high quality gate oxide is thermally grown (Fig. 15e). Finally the silicon nitride layer is deposited (Fig. 15f). It protects the oxidized silicon nanowires remaining in contact with aqueous solutions from a water penetration. Additionally, it creates a surface which is

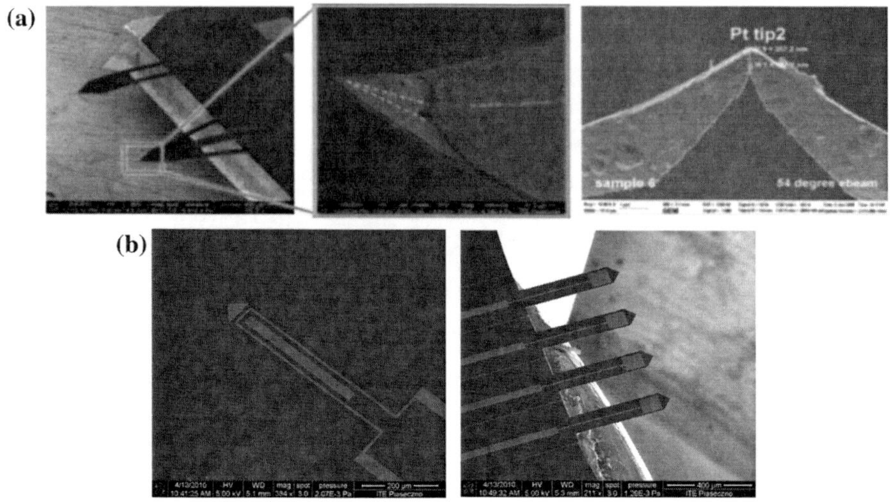

Fig. 14 **a** Scanning thermal microscopy cantilevers with active nanotip manufactured at ITE [25]. **b** Silicon microcantilevers for biochemical applications

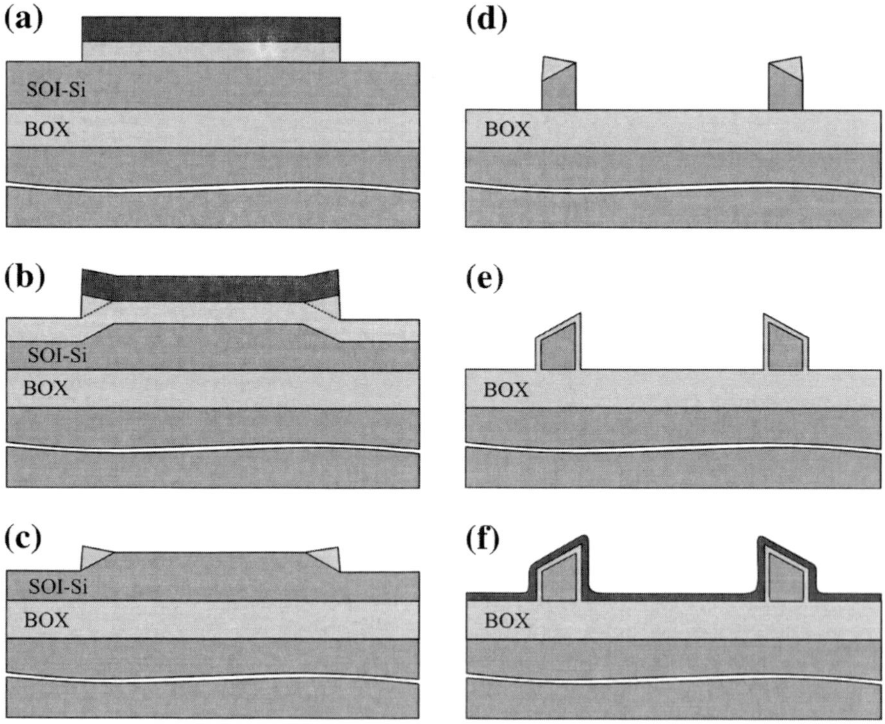

Fig. 15 Basic steps of the PaDEOx process for manufacturing of SOI-based FinFET-type MOSFETs as biochemical sensors (description in the text)

sensitive to hydrogen ion concentration due to charge exchange with the solution. The process is completed with the necessary and standard contact formation, metallization and passivation steps which are not shown in Fig. 15.

It is worthwhile to mention, that the devices have an even number of fins (nanowires—NWs). This is a feature of all the FinFET fabrication methods in which silicon masking shapes originate from materials/phenomena present at the edges of other, usually sacrificial areas (e.g. [28]). The technique described above has been successfully used for fabrication of devices for hydrogen ion measurement and for biochemical investigation. The research done in this field at ITE in collaboration with other groups is described in the next section.

3 Non-Standard FinFET Devices for Small Volume Sample Sensors

A concept of the Ion Sensitive Field Effect Transistors (ISFET) was introduced by P. Bergveld in 1970 in [29]. Since then the ISFETs have become analytical tools widely used e.g. in chemistry, biology, medicine, and for environment monitoring [30, 31]. In all these applications they are applied primarily for measurement of hydrogen ion concentration in the water solution (pH), which is in many cases also an indirect measure of concentration of other ions/species. As compared with conventional Ion Selective Electrodes (ISE) the following features make them attractive for the researchers and engineers: small size, short response time, low output impedance. Moreover, they are suitable for mass fabrication. After extensive investigations it has been found, that by modification of the gate material an adjustment of the ISFET selectivity is possible. A dedicated gate stack composition may increase or change the device sensitivity to ions, thus may improve or change the device functionality. By deposition of ion selective membranes Chemically sensitive ISFETs (ChemFETs) are developed. Polyvinyl chloride, polysiloxanes, polyurethanes and other polymers are suitable as the membrane materials. Additionally biorecognition particles (e.g. enzymes) may be deposited at the gate surfaces covered with the membranes. This allows for detection of certain molecular species. The enzymatic reaction is responsible for high selectivity. On the other hand, the device sensitivity may be also decreased, thus leading to the ion-insensitive devices, i.e. reference field-effect transistors (REFETs) which are usually used together with the ISFETs in differential measurements, where the requirements for the reference electrode stability are weaker. Hence it may be stated, that the ISFETs have become a root for a wide family of biochemical sensors.

A trend for scaling down of all electronic devices, including the ISFETs has caused a quest for tools for detection of small samples of the analyte. One of the concepts consists in placing the ISFET inside a silicon nanowire (NW). This solution takes advantages of great progress in the SOI device technology. Advantages of the NW channel FETs for sensor and detector applications have

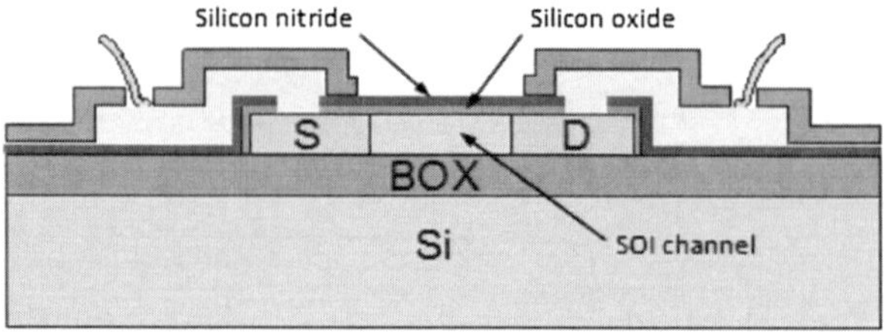

Fig. 16 Longitudinal cross-section of SOI-ISFET devices

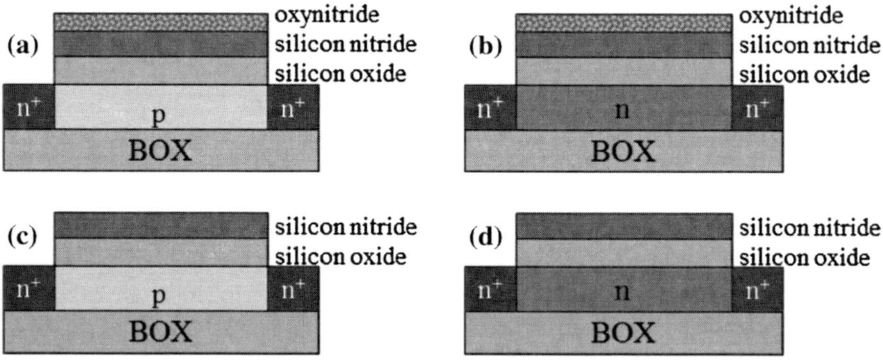

Fig. 17 Schematic view of all the fabricated devices: standard n-type FET (**a**, **c**) and junctionless n^+-n-n^+ FETs (**b**, **d**); sensors a and b are covered with oxynitride dielectric layer, which has been etched off in versions **c** and **d**

been recognized (e.g. [32]). The PaDEOx process described in the Sect. 2 has been used in ITE for fabrication of small sensors. Identically doped 8 NW-channel FETs and 4 μm-wide channel FETs have been fabricated in the same chip. Schematic cross-section, valid for both mentioned devices, is presented in Fig. 16.

Regardless of the variation in topography there are also two groups of devices that differ in the channel doping. In the first group the transistor channels have been formed based directly on the SOI p-type device layer (boron concentration of 8×10^{15} cm^{-3}). In the second device group the channels have been implanted with phosphorus to a concentration level of 2×10^{16} cm^{-3}. The source and drain regions in both groups have been doped with phosphorus (concentration of 1×10^{19} cm^{-3}). The channels have been covered with 14 nm thick thermal oxide layer and then with LPCVD nitride layer of the same thickness. All types of the fabricated devices are schematically shown in Fig. 17.

Next, a Ti:W/Al metallization for source and drain contacts has been grown by sputtering and patterned. The wafer surface has been covered with CVD oxide protection layer. The layer has been removed from gate active areas of the FETs and metal bonding pads by means of photolithography and wet etching. Then a photoresist mask has been ashed out in an oxygen plasma etcher. As a result of this process a thin oxynitride layer has been grown on the gate dielectric surface

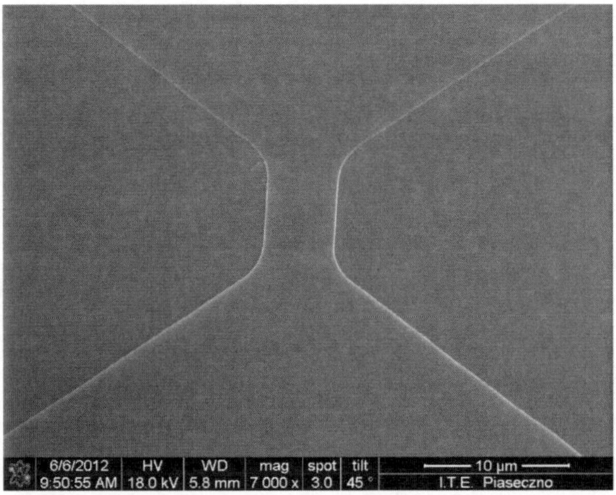

Fig. 18 Photolithographic process SOI FET. Channel width = 4 μm

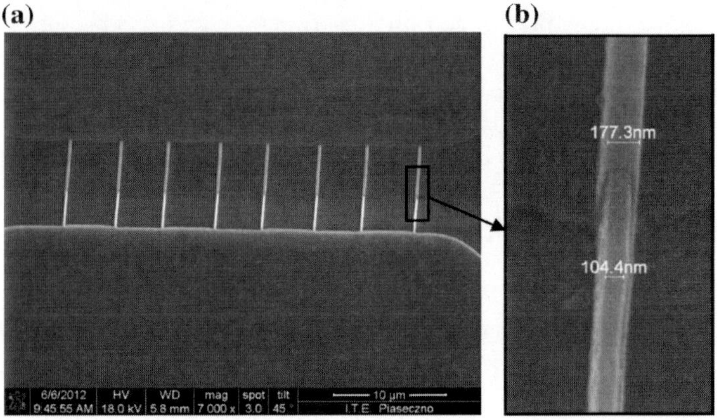

Fig. 19 Eight 100 nm-wide channel nano-wire FET (**a**); the Si wire width and a gate dielectric profile after partial etching of the dielectric layers (**b**)

(Fig. 17a, b). This layer has been intentionally removed in a part of the manufactured devices using wet etching step (Fig. 17c, d).

The active area of the fabricated sensor devices with the channel width defined by the photolithography step is displayed in Fig. 18. A SEM image of the eight-NW FET is shown in Fig. 19a. The details of a single NW are shown in Fig. 19b where the dielectric covering the nanowire has been partially removed in order to reveal the bare Si wire. The NW width determined by the SEM measurements is equal approximately to 100 nm. A mean height of the silicon path after processing is equal to 160 nm in all types of the devices.

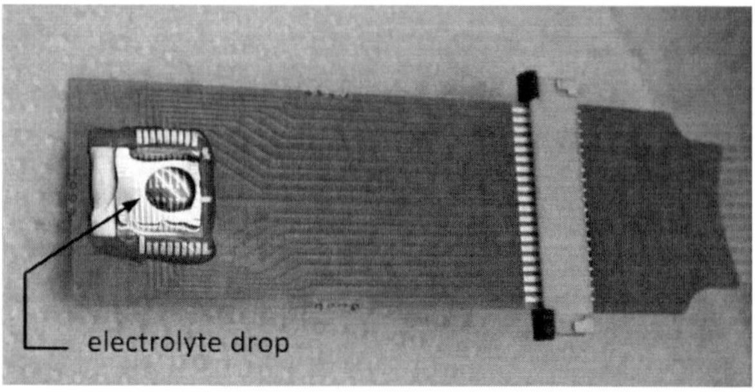

Fig. 20 The sensor chip, assembled on a PCB foil and connected to a standard ZIF socket; a droplet of an electrolyte is visible in the *middle* of the sensor

All the devices mentioned above have been designed together within a single chip. The wafers have been diced into chips and assembled on a PCB foil (Fig. 20). Edges of the chips have been encapsulated with an epoxy resin.

The sensors have been characterized in water solutions using a AgCl reference electrode for the solution DC biasing. Experiments have been carried out using a small volume droplet of a liquid analyte (roughly 20 μl) adhered between the reference electrode and the sensor. The drain current I_{ds} versus drain voltage V_{ds} characteristics have been measured in a pH = 7 buffer solution with the reference electrode voltage V_{ref} and a silicon substrate voltage V_{bs} set as parameters. Results for standard n-type ("enhancement mode" n^+-p-n^+) 4 μm-wide FET with oxidized nitride gate dielectric are presented in Fig. 21. It may be stated, that the wide, "photolithographic" transistor is in the "ON" state if its substrate bias is 0 V. Then even for negative V_{ref} voltage the drain current I_{ds} does not drop below several μA. It appears, that for substrate bias $V_{bs} = 0$ V the parasitic current along the bottom Si-SiO_2 interface may not be switched by the negative V_{ref} bias. However, if the substrate bias V_{bs} is -10 V or less then the back channel conduction is suppressed.

In the case of the nanowire sensors a dependence of the drain current on the reference electrode potential is different than for the wide channel devices. In Fig. 22 the $I_{ds}(V_{ref})$ characteristics of the eight NW n-channel junctionless sensor (Fig. 17b) together with analogous characteristics of the 4 μm wide n-type FET are shown. The NW sensor demonstrates current in the $10^{-12} \div 10^{-11}$ A range in the "OFF" state. Namely, at $V_{ref} = 0.5$ V the drain current is 5×10^{-12} A, whereas for the same bias conditions the drain current of the identically doped wide n-type FET is 5 μA. As has been mentioned in the previous paragraph the substrate bias as low as $-10 \div -20$ V is necessary to sufficiently decrease the drain current in the "OFF" state.

At the next stage of the SOI-based sensor characterization the measurements of concentration of hydrogen ions in aqueous solutions have been carried out, which is their primary application. The setup presented in Fig. 20 has been used. Commercially available buffer solutions (of pH 5, 7 and 9) have been dosed directly to

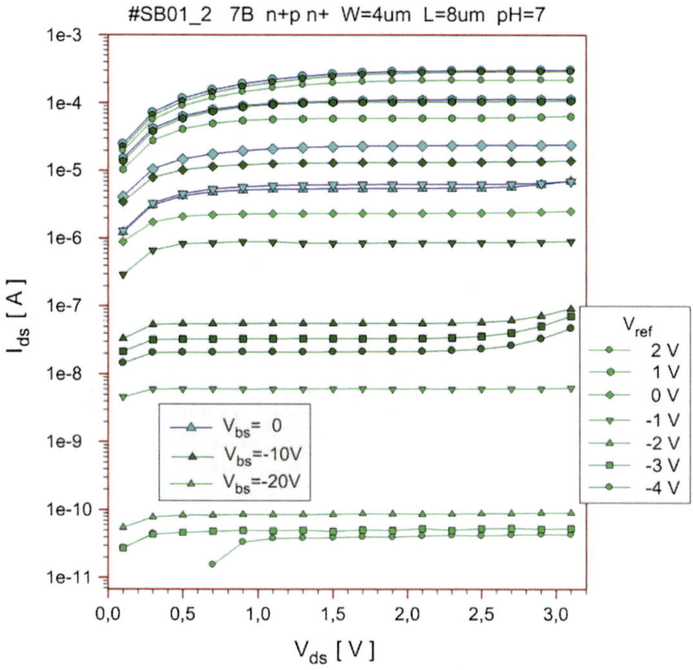

Fig. 21 Output I_{ds}-V_{ds} characteristics of the standard n-type FET sensor with oxidized nitride gate dielectric measured in a droplet of pH = 7 buffer adhered between the reference AgCl electrode and the sensor surface; a reference electrode potential and a bulk substrate potential are parameters

Fig. 22 The drain current I_{ds} versus reference electrode potential V_{ref} characteristics of two n^+-n-n^+ FET sensors in a pH = 7 buffer solution: 4 μm-wide device defined by a standard photolithography, and eight NW (each 100 nm-wide) device; the substrate V_{bs} voltage is a parameter

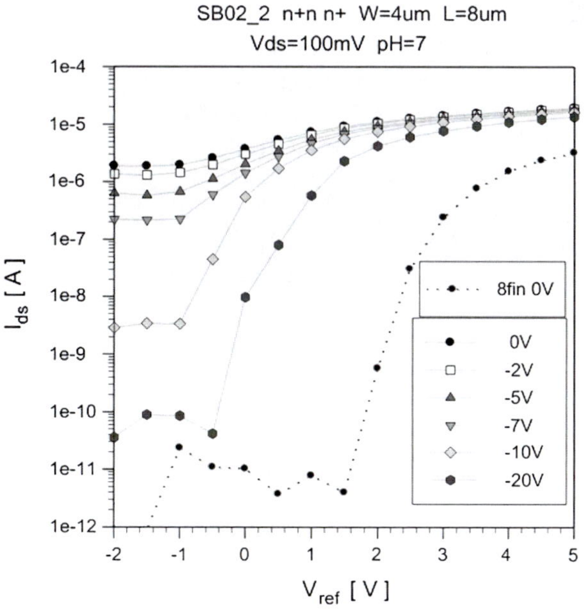

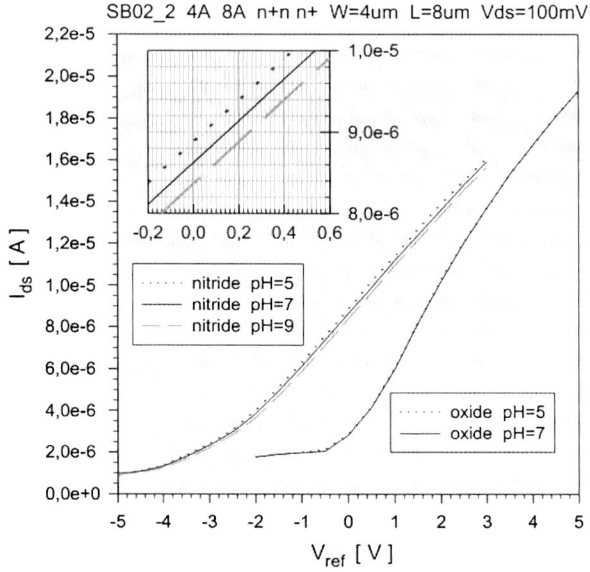

Fig. 23 The I_{ds}-V_{ref} characteristics of two 4 μm wide, 8 μm long junctionless FET sensors, types "oxide" (Fig. 17b) and "nitride" (Fig. 17d), measured in a droplet of buffer solution; in the inset—magnification of "nitride" device characteristics; pH sensitivity = 50 mV/pH

the sensor active areas. A miniature AgCl reference electrode has been used for biasing the liquid. Two 4 μm-wide FET devices placed in the same chip have been compared. In the first type of the ISFETs there is $SiO_2/Si_3N_4/SiO_xN_y$ gate dielectric resulting from the oxygen plasma treatment of the Si_3N_4 surface. In the second device type there is only SiO_2/Si_3N_4 gate dielectric originating from the first one by wet etching of the silicon oxynitride layer. The input characteristics I_{ds} versus V_{ref} of the first and the second junctionless sensors (denoted by "oxide" and "nitride" respectively) are shown in Fig. 23 for various pH of the liquid gate. We may notice, that the oxidized sensor characteristics are not dependent on the pH value. On the contrary, the pure nitride surface sensor reveals a typical ISFET sensitivity towards H^+ ions of 50 mV/pH. A behavior that is similar to the standard FETs, has been demonstrated by measurements of the eight-NW 100 nm wide FET sensors (Fig. 24). In this case for the "nitride"-type ISFETs, the pH sensitivity of 40 mV/pH has been found. A transconductance of the NW sensor has been calculated as 1.8×10^{-6} S, that is 30 % less than for the 4 μm wide sensor. However the NW sensor can be used with a higher current density, which is desirable in application circuits. Moreover, due to the positive threshold voltage the n-type NW ISFET needs a small positive DC bias of the reference electrode during measurement in order to increase its conductance.

The NW ISFET sensors have been used as base devices for modifications aimed at preparation of chemical detectors or sensors, analogously to the widely used CHEMFETs. A so-called gate dielectric functionalization via deposition of a dedicated ion/molecule sensitive membrane is a key issue. During realization of the presented work an ethylenetriamine (ETA) dissolved in ethyl alcohol has proved to be an interesting substance for functionalization purpose [33].

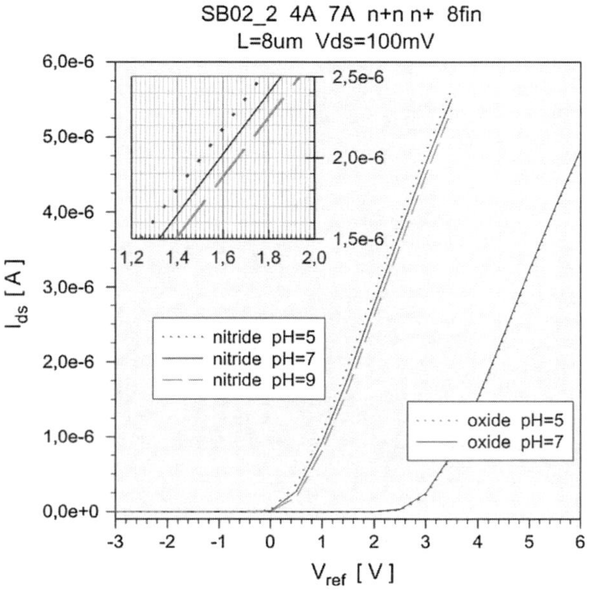

Fig. 24 The I_{ds}-V_{ref} characteristics of two eight-NW 8 μm long junctionless FET sensors, types "oxide" (Fig. 17b) and "nitride" (Fig. 17d), measured in a droplet of buffer solution; in the inset—magnification of "nitride" device characteristics; pH sensitivity = 40 mV/pH

The $I_{ds}(V_{ds})$ characteristics of the four-NW 20 μm long FET functionalized with ETA have been measured in 10 mmol solution of KCl. It has been found that the output characteristics have been noticeably shifted because of the functionalization. Subsequently, the ETA-covered sensors have been used in experiments of metal ion detection. The electrical response of the sensor towards copper ions has been tested. In the measurement setup the device operating point has been set to $V_{ds} = 2$ V and $V_{ref} = 1$ V. Small amounts of Cu^{2+} ions have been successively added, and the device conductance has been measured. A most noticeable response of the sensor has been observed between 10^{-5} and 10^{-4} mol concentrations (Fig. 25). This value may be determined as a threshold for detection of Cu^{2+} ions with use of the sensor. A saturation of the Cu^{2+} influence has been observed for further increase of the ion concentration.

Furthermore, an experiment aiming at a detection of lead ions has been also carried out. In this case the eight-NW, 20 μm-long FET has been used. Its gate dielectric has been functionalized with silane derivative of EDTA (ethylenediaminetetraacetate). The sensor and the reference electrode have been put into a 10 mmol TRIS buffer solution. A DC bias of +2 V has been applied both to the sensor drain and to the reference electrode. It has been found that adding a water solution of lead nitrate noticeably increases the sensor conductance. It has also appeared that the effect is reversible. Finally, the sensor has been scaled towards the Pb^{2+} ions in water solutions in a concentration range from 1×10^{-7} to 1×10^{-3} mol. Sensor conductance, as measured after 50 s delay for the sensor output signal stabilization at each solution, is presented in Fig. 26. Conductance of the sensor increases nearly 1 % per decade of Pb^{2+} ion concentration. Moreover, it

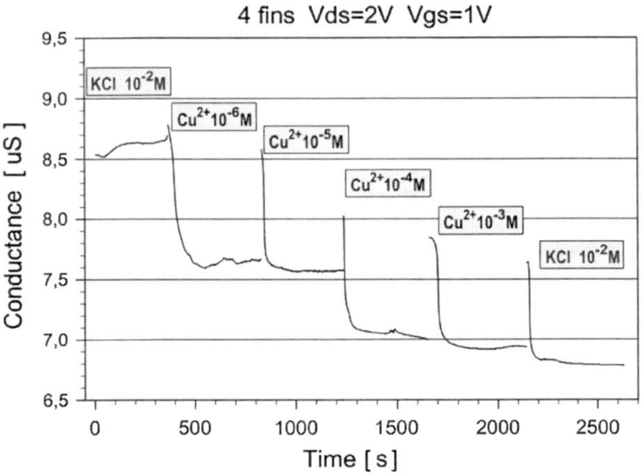

Fig. 25 ETA-functionalized sensor conductance versus time in experiment of Cu^{2+} ion detection; the functionalized device: four-NW, channel length = 20 μm

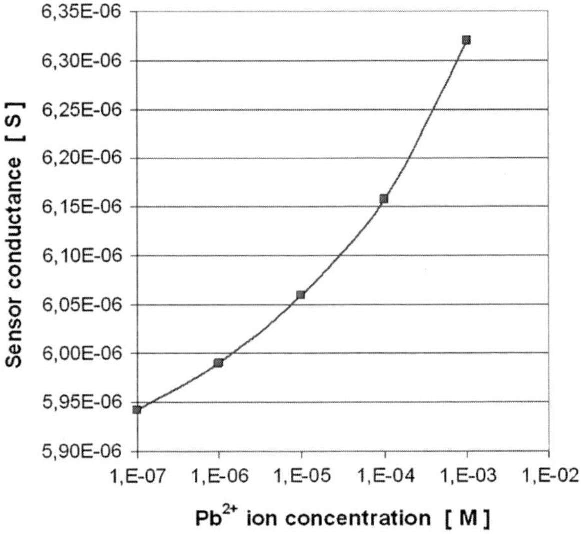

Fig. 26 An influence of Pb^{2+} ion concentration on sensor conductance, obtained for 50 s stabilization time at each ion concentration

is clearly visible in Fig. 26 that the device sensitivity slightly increases along with the lead ion concentration.

Other metal ions have been also detected using the EDTA-functionalized NW sensors. The sensor has been sensitive to Cu^{2+} ions dissolved in TRIS pH = 7 buffer with a maximum sensitivity at 5×10^{-4} mol ion concentration. A similar experiment carried out with Cd^{2+} ions has proved that a detection threshold is not higher than 1×10^{-5} mol. Also mercury ions may be detected using the same setup. EDTA functionalized eight-NW sensor exhibits the highest sensitivity at 1×10^{-5} mol concentration of Hg^{2+} ions.

A number of organic compounds can influence conductance of the sensor too. An ascorbic acid and lysine may be detected at a level of 1×10^{-6} and

2×10^{-6} mol respectively. It should be underlined that in the experiments mentioned above the sensors have been operated rather as detectors than as meters. The ions or organic molecules haven't been detected selectively, so the same signal could result from different elements in an analyte.

4 Summary

In the presented chapter a number of research projects utilizing the SOI technology have been described with special emphasis put on the microsensor applications. The SOI material itself as well as the SOI based processes and specific design methods have many advantages. They have been partly mentioned in the introduction, and illustrated in the following sections. Most of the presented results have been achieved at ITE within a strong collaboration with other groups.

It has to be mentioned, that the SOI technology is highly demanding. Meeting the requirements for the high quality silicon layer/BOX interface and for a very good uniformity of the thin Si layer thickness is very important. The wafer structure and quality are reflected by the SOI material cost, which is significantly higher than of the bulk wafers. On the other hand, a number of photolithography levels is lower in the case of SOI processes than in the case of the corresponding bulk ones. So, a number of the SOI process steps is also lower. Thus, taking into account the overall processing cost, usually the higher SOI wafer price is easily compensated by the simpler SOI processing and its lower cost. Of course, the selected SOI process steps (particularly carried out on thin body wafers) related e.g. to growing/removing of the layers require a strict control, but this is also the case in the state-of the-art bulk technologies, which in many cases do not allow for fabrication of the complex hetrogenous systems.

To summarize, in spite of the challenges mentioned above, there is a confidence, that the SOI technology creates a lot of opportunities and is recommended for further microsensor and heterogenous system developments.

Acknowledgments Authors would like to acknowledge a collaboration with Prof. Teodor Gotszalk, and Dr. Piotr Pałetko, Wroclaw University of Technology in terms of measurements and helpful discussions. Also a financial support by The National Centre for Research and Development, Poland under grant "Multipixel detector of THz radiation based on selective MOS transistors and its application in biology, medicine and safety installations", PBS1/A9/11/2012 is acknowledged.

References

1. Colinge, J.-P.: Silicon-on-insulator technology: materials to VLSI, 3rd edn. Kluwer Academic Publishers, Boston (2004)
2. Colinge, J.-P.: FinFETs and Other Multi-Gate Transistors. Springer, New York (2008)

3. Wermes, N.: Pixel detectors for particle physics and imaging applications. Nucl. Instr. Meth. Phys. Res. A **512**, 277–288 (2003)
4. Deptuch, G., Berst, J.-D., Claus, G., et al.: Design and testing of monolithic active pixel sensors for charged particle tracking. IEEE Trans. Nucl. Sci. **49**(2), 601–610 (2002)
5. Pengg, F.X.: Monolithic silicon pixel detector in SOI technology. Ph.D. thesis, University of Linz (1996)
6. Caccia, M., Alemi, M., Badano, L., et al.: SUCIMA—silicon ultra fast cameras for electron and gamma sources in medical applications. In: Nuclear Science Symposium on Conference Record, vol. 3, pp. 1574–1578. 19–25 October 2003
7. Niemiec, H., Jastrzab, M., Kucewicz, W., et al.: Full-size monolithic active pixel sensors in SOI technology—design considerations, simulations, and measurements results. Nucl. Instr. Meth. Phys. Res. A **568**, 153–158 (2006)
8. Marczewski, J., Domanski, K., Grabiec, P., et al.: Technology development for soi monolithic pixel detectors. Nucl. Instr. Meth. Phys. Res. A **560**, 26–30 (2006)
9. Kucewicz, W., Armstrong, B.M., Gamble, H.S., et al.: development of monolithic active pixel sensor in soi technology fabricated on the wafer with thick device layer. In: Nuclear Science Symposium on Conference Record, pp. 1123–1125. 19–25 October 2008
10. High Energy Accelerator Research Organisation website http://rd.kek.jp/project/soi/research.html
11. Arai, Y., Miyoshi, T., Unno, Y., et al.: Development of SOI pixel process technology. Nucl. Instr. Meth. Phys. Res. A **636**(1), S31–S36 (2011)
12. Obrębski, D., Szymański, A., Tomaszewski, D., et al.: Development of ioniz-ing radiation detectors integrated with readout electronics. In: Proceedings of the 20th International Conferences on Mixed Design of Integrated Circuits and Systems MIXDES 2013, Gdynia, Poland, pp. 229–234. 20–22 June 2013
13. Deptuch, G., Demarteau, M., Hoff, J.R., et al.: Vertically integrated circuits at fermilab. IEEE Trans. Nucl. Sci. **57**(4), 2178–2186 (2010)
14. Bernard, J.T.: Reconfigurable Antennas. Morgan & Claypool Publishers, San Rafael (2007)
15. Taylor, G.C., Rosen, A., Fathy, E., Swain, P.K., Perlow, S.M.: Surface PIN device. US Patent 6,617,670 B2, 9 Sept 2003
16. Yashchyshyn, Y., Marczewski, J., Tomaszewski, D.: Investigation of the S-PIN diodes for silicon monolithic antennas with reconfigurable aperture. IEEE Trans. Microw. Theory Tech. **58**(5), 1100–1106 (2010)
17. SOI Industry Consortium website http://www.soiconsortium.org
18. Flandre, D., Adriaensen, S., Akheyar, A., et al.: Fully depleted SOI CMOS technology for heterogenous micropower, high-temperature or RF microsystems. Solid State Electron. **45**, 541–549 (2001)
19. EC FP7 Project, Development of technology building blocks for structural health monitoring sensing devices in aeronautics ("TRIADE"), ACP7 GA 2008 212859
20. Rouet, V.: Aeronautics structural health monitoring with a smart integrated system the size of a credit card. In: Smart Systems Integration 2012, Zurich, 21–23 March 2012
21. Knap, W., Dyakonov, Y., Coquillat, D., et al.: Field Effect transistors for terahertz detection: physics and first imaging applications. J. Infrared Millimeter Terahertz Waves **30**, 1319–1337 (2009)
22. Preu, S., Kim, S., Verma, R., et al.: An improved model for non-resonant terahertz detection in field-effect transistors. J. Appl. Phys. **111**, 024502 (2012)
23. Kopyt, P., Marczewski, J., Kucharski, K., et al.: Planar antennas for THz radiation detector based on a MOSFET. In: 36th International Conferences on Infrared, Millimeter, and Terahertz Waves (IRMMW-THz), Houston (USA), 2–7 October 2011
24. Kopyt, P.: Modeling of a THz Radiation Detector Built of Planar Antenna Integrated with MOSFET. In: 20th International Conferences Mixed Design of Integrated Circuits and Systems MIXDES 2013, pp. 69–74. Gdynia, Poland, 20–22 June 2013

25. Wielgoszewski, G., Sulecki, P., Gotszalk, T., Janus, P., Szmigiel, D., Grabiec, P., Zschech, E.: Microfabricated resistive high-sensitivity nanoprobe for scanning thermal microscopy. J. Vac. Sci. Technol. B **28**, C6N7 (2010). doi: 10.1116/1.3502614
26. Zaborowski, M., Grabiec, P., Dobrowolski, R., et al.: Nanoscale pattern definition by edge oxidation of silicon under the Si3N4 mask—PaDEOx. Acta Phys. Pol. A **116**, 139–141 (2009)
27. Zaborowski, M., Tomaszewski, D., Dumania, P., Grabiec, P.: Development and characterisation of nanowire-based FETs. Microelectron. Reliab. **51**, 1162–1165 (2011)
28. Choi, Y.-K., King, T.-J., Hu, C.: Spacer FinFET: nanoscale double-gate CMOS technology for the terabit era. Solid State Electron. **46**, 1595–1601 (2002)
29. Bergveld, P.: Development of an ion-sensitive solid-state device for neu-rophysiological measurements. IEEE Trans. Biomed. Eng. **17**(1), 70–71 (1970)
30. Bergveld, P.: Thirty years of ISFETOLOGY. Sens. Actuators B **88**, 1–20 (2003)
31. Jimenez-Jorquera, C., Orozco, J., Baldi, A.: ISFET based microsensors for environmental monitoring. Sensors **10**(1), 61–83 (2010)
32. Nair, P.R., Alam, M.A.: Design considerations of silicon nanowire biosen-sors. IEEE Trans. Electron Dev. **54**(12), 3400–3408 (2007)
33. Zaborowski, M., Dumania, P., Tomaszewski, D., et al.: Development of Si nanowire chemical sensors. Procedia Eng. **47**, 1053–1056 (2012)

Photoexcitation and Recombination of Charge Carriers in Si/Ge Nanoheterostructures

Vladimir S. Lysenko, Sergey V. Kondratenko and Yuriy N. Kozyrev

Abstract In this chapter, we present the study of the photogeneration and recombination of nonequilibrium charge carriers, and the determination of photoconductivity mechanisms in semiconductor SiGe/Si and SiGe/SiO_2/*p*-Si heterostructures with nanoislands. The work generalizes the results of studies of morphological, structural, optical and electrical properties of semiconductor nanoheterostructures. It is shown that the photoconductivity of nanoheterostructures SiGe/Si in the infrared range depending on the component composition, size and magnitude of the mechanical stresses in nanoislands $Si_{1-x}Ge_x$ is determined by interband and intraband transitions involving localized states of the valence band of the nanoislands. The effects of residual conductivity and optical quenching of conductivity in SiGe/SiO_2/*p*-Si heterostructures with SiGe nanoclusters was found to be caused by variations of the electrostatic potential in the near-surafce region of *p*-Si substrate and optically-induced spatial redistribution of trapped positive charges between SiO_2/Si interface levels and localized states of Ge nanoislads.

1 Introduction

Semiconductor heterostructures and especially semiconductor heterostructures with low-dimensional objects, including quantum wells, quantum wires and quantum dots, currently comprise the object of intensive study [1, 2]. Due to spatial

V.S. Lysenko
Institute of Semiconductor Physics, 41 Prospect Nauki, Kyiv 03028, Ukraine

S.V. Kondratenko (✉)
Taras Shevchenko National University of Kyiv, 64/13 Volodymyrs'ka St., Kyiv 01601, Ukraine
e-mail: kondr@univ.kiev.ua

Y.N. Kozyrev
O.O. Chuiko Institute of Surface Chemistry, 17 Generala Naumova St., Kyiv 03164, Ukraine

A. Nazarov et al. (eds.), *Functional Nanomaterials and Devices for Electronics, Sensors and Energy Harvesting*, Engineering Materials,
DOI 10.1007/978-3-319-08804-4_19

confinement of charge carrier's motion in one, two or three directions, respectively, such nanostructures have unique fundamental properties and technological applications. Particular interest is attracted by nanoelectronic devices and systems grown using epitaxy methods—vapor-phase, molecular-beam and liquid-phase—in which the formation and spatial arrangement of nanoscale elements is carried out using the effects of self-organization. Knowledge of the electronic spectrum, transport, recombination, and photogeneration in self-organized nanostructures is essential for creation of novel electronic and photonic devices.

Low-dimensional Ge/Si heterostructures have attracted considerable research interest in recent years due to their significant potential to impact new electronic devices which are compatible with the available silicon technology. Optoelectronic devices based on SiGe dots grown on a Si substrate have been already proposed [3, 4]. In particular, the use of interband or intraband transitions involving localized states in valence band of SiGe can increase the photoconversion efficiency in the near infrared range. The low-dimensional silicon-germanium alloys have a wide range of applications, including quantum dot IR photodetectors, memory cells and spintronic devices. Widespread application of such system is the arrangement of SiGe quantum dots in the space-charge region of heterojunctions, Schottky diodes, p–n junctions or metal-oxide-semiconductor structures.

Morphology, optical and electrical properties of self-assembled quantum dots in semiconductor heterostructures studied intensively over the past decade. It has been shown that the optical and electrical properties of these nanoheterostructures are defined by quantum dots (QDs) morphology, composition and spatial distribution of strain near nano-objects. The results of quantum-size heterostructures showed prospects for their use as active elements of the new solar cells, photodetectors and transistors. In particular, significant increase of photoconversion efficiency was achieved through the use of quantum size effects in solar cells with quantum wells [5], which used optical transitions involving quantum-sized states [5]. According to theoretical calculations, the maximum efficiency of an ideal solar cell based heterostructures, which involves appropriate energy levels nano-objects, is 63.2 %. This exceeds the efficiency of 40.7 %, which corresponds to the theoretical limit of solar cells based on a semiconductor [5, 6].

Strained Ge/Si heterostructures with Ge QDs, in which holes are confined in Ge QDs leading to essential offset of the valence band, are considered as the second type heterojunctions. It is supposed that quantum size effects and considerable elastic strains in Ge nanoislands produce additional energy levels in the valence band. Hole transitions from the localized states in the QDs into delocalized states of the valence band may be used for development of infrared photodetectors similar to already fabricated on the basis of InAs/GaAs heterostructures. Doing this, a special attention should be paid to elaboration of lateral photodetectors that offer large practical application due to possibilities of a simple realization of infrared devices operating at the normal incident radiation with a sensitive surface large area. More detailed investigation of optical and photoelectrical properties of Ge QDs and adjacent Si layers is required to improve the technical parameters of such systems. A large number of experimental works were devoted to investigation of the energy-band structure in a quantum dot. It is supposed that the positions of energy levels in a quantum dot are determined by its size, shape and composition, as well as by the value of inhomogeneous strain inside it caused by the lattice mismatch.

A very important point affecting essentially the photoresponse value of Si–Ge heterostructures is the in-plane inhomogeneity of their physical properties. The lack of homogeneity is caused by nonuniform strain, variations of quantum dot content, variations of impurity concentrations, charge accumulation by nanoislands themselves, etc. This leads to spatial modulation of the potential in a near-surface region. A scale of fluctuations is determined by the density and lateral parameters of the quantum dots. In addition, the surface potential fluctuations may cause the appearance of inner electric fields affecting the transport and recombination of nonequilibrium charge carriers.

In this review we present the study of the photogeneration and recombination of nonequilibrium charge carriers, and the determination of photoconductivity mechanisms in semiconductor SiGe/Si and Ge/SiO_2/Si heterostructures with nanoislands. The work generalizes the results of studies of morphological and photoelectrical properties.

2 Sample Preparation and Measurement Technique

The molecular beam epitaxy (MBE) technique ("Katun'-B" set-up, produced in Novosibirsk, Russia) was used to prepare multilayer Ge–Si(100) nanocluster arrays with the islands of various sizes and surface density. The (100) oriented wafers of *n*-Si with 7.5 and 20 Ω cm resistivity and diameter of 76 mm were used as substrates. In order to prepare multilayer quantum dot systems with regular nanoisland distribution over the substrate surface, we have proposed to use a system of $Si_{1-x}Ge_x$ intermediate layers with a sub-critical thickness [7]. The Ge mole fraction x was gradually increased from layer to layer grown at gradually decreasing substrate temperature started from $T_s = 500$ °C. The growth process, in particular the moment of the $2D \rightarrow 3D$ transition in the Stranski-Krastanov growth regime, was controlled via RHEED (reflection high energy electron diffraction). To study the surface morphology, atomic force microscopy (AFM) measurements were carried out using an Ntegra AFM from NT-MDT with a closed loop scanner. Standard Si cantilevers with tips having a half opening angle of 10° were employed as probes. The growth of each Si intermediate layer was continued until a high-contrast Si(100)2 × 1 RHEED pattern was produced typical of clean Si. Thus, the multilayer Ge–Si(100) nanocluster arrays were grown at the temperature $T_s = 500$ °C.

The Stranski-Krastanow growth of Ge nanoislands on Si(001) surface is an intermediary process characterized by both 2D wetting layer (WL) and 3D island formation. Transition from the layer-by-layer epitaxy to nanoisland structure growth occurs at a critical layer thickness which is highly dependent on surface

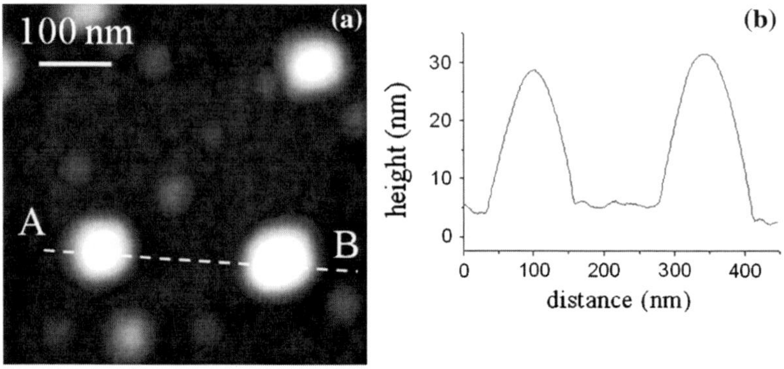

Fig. 1 AFM images of different SiGe nanoclusters grown on Si (001) (**a**); QDs height distribution along AB direction (**b**)

energies and lattice parameters. Figure 1 shows AFM images of different SiGe nanoclusters grown on Si(001).

Germanium nanoclusters grown on/in silicon or silicon dioxide have been successfully applied in new nanoelectronic, optoelectronic and memory devices due to quantum confinement effect and possibility of integration within Si-based technology [1, 8]. Heterostructures with epitaxial Ge nanoclusters isolated from Si sub-strate by ultrathin silicon oxide layer, would be practically promising due to their nanoscale size, tunability and high density. The interest in optoelectronic and solar cells application stems from observation of infrared photoluminescence and photoconductivity caused by optical transitions through confined states of Ge nanoislands (NI). Other important applications of Ge on SiO_2 structures include CMOS transistors and nanocrystal nonvolatile memory [9, 10].

The technique of Ge nanoisland growth on Si(100) covered with ultra-thin SiO_2 layers is widely accepted now [11]. This technique enables increasing the nano-island density up to 10^{12}–10^{13} cm^{-2} due to thermal decomposition of the ultrathin oxide layer and formation of "defects" at the surface, that are nucleation centers for both epitaxial and non-epitaxial Ge nanoislands with high aspect ratio varying between 0.2 and 0.6 [12, 13]. Moreover, structures with Ge nanoislands grown by this technique do not contain underlying germanium wetting layer as for Stranski–Krastanow mode.

Non-epitaxial Ge nanoislands which are separated from the substrate attract special interest due to spatial separation of electron-hole pairs leading to reduction of recombination rate [14]. NI's growth at the silicon surface covered with ultrathin silicon oxide layer is mainly determined by the dynamics of changes of the SiO_x film structure and physical properties during Ge deposition and is principally possible at temperatures below $\sim$400 °C, when the formation of voids in ultrathin SiO_2 films is suppressed [15]. Epitaxy at such low temperatures puts some limitations on the crystallinity and structural perfection of the obtained nanoclusters. Increasing of growth temperature up 430 °C allows for growing epitaxial crystalline NI's on silicon, while silicon oxide is destroyed due to thermal

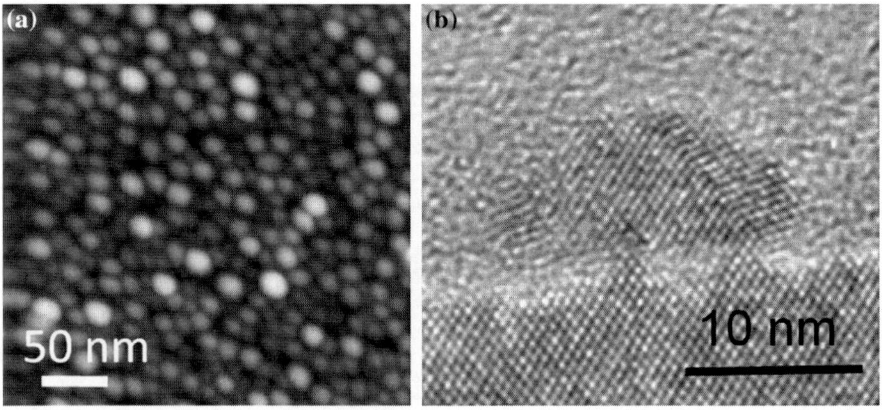

Fig. 2 **a** AFM image of Ge NCs and sample view for in-plane PC measurements. **b** HR-TEM images of Ge NCs grown on silicon oxide

decomposition effect [12]. Possibility for high temperature growth of crystalline Ge NI's on top of silicon oxide has not been studied in detail. Question under study is what type of structure of Ge nanoclusters is formed after high temperature growth in the presence of tetragonal silicon oxide film providing a poor epitaxial relationship to the underlying Si(100) surface.

The Ge nanocluster structures were grown using a molecular beam epitaxy (MBE) technique on boron-doped ($N_a \sim 10^{17}$ cm^{-3}) *p*-Si(100) substrates with the resistivity of 7.5 Ω cm. A pre-epitaxial chemical oxidation of silicon resulted in formation of a 2 nm thick SiO_2 layer on the substrate. The surface modifications were monitored *in situ* using reflection high-energy electron diffraction (RHEED) technique. The Debay rings in electron diffraction image showed the presence of amorphous silicon oxide layer. Prior to Ge deposition the oxidized silicon surface was annealed in vacuum ($\sim 10^{-10}$ Pa) at the temperature of about 800 °C for an hour. The diffraction pattern has changed. Appearance of clear and bright Kikuchi lines allows us to conclude, taking into account the film thickness and the annealing temperature, that phase separation and crystallization in oxide layer have occurred [16]. Therefore, the film can be considered as a silicon suboxide SiO_x ($0 < x < 2$) with silicon rich regions. Then the substrate temperature was lowered to 700–730 °C, and the deposition of germanium was performed, leading to the formation of Ge nanoclusters on top of SiO_x layer.

AFM image in Fig. 2a shows that deposition of Ge onto the SiO_2 film at 700 °C results in formation of NCs with the hemispherical top surface and the surface density of $\sim 3 \times 10^{11}$/cm^2. The base diameter distribution of Ge NCs is approximated by Gaussian function with a maximum at 16 nm and a full width at half maximum of 6 nm. The mean height was about 10 nm. The HR-TEM image shows the presence of crystalline Ge NCs, which are separated from Si(100) by non-uniform silicon oxide layer containing crystalline voids with diameter about 1–2 nm (Fig. 2b). The oxide thickness $\delta(x,y)$ varies from 0 to 1 nm. The Ge nanoclusters have a crystalline structure, but its quality is strongly affected by twin boundaries, which are observed close to crystalline voids in oxide film. We can see that 0.6 (height/base) aspect ratio for Ge islands on silicon oxide is much larger than typical values 0.1–0.2 for Stranski-Krastanow islands [1].

The use of chemically oxidized Si surface with the initial 2-nm-thick SiO_2 film allows formation of crystalline Ge nanoclusters, which are separated from the substrate by thin oxide layer, at high ($\sim$700 °C) temperature. One of key differences of the proposed technique for the formation of Ge nanoclusters is preliminary modification of the chemically oxidized Si(100) surface. During the high-vacuum annealing of SiO_2 film at 800 °C for an hour, the thermal decomposition and desorption of the silicon dioxide film take place according to the reaction

$$SiO_2 + Si \rightarrow 2SiO\ (gas) \tag{1}$$

This process takes place at SiO_2/Si interface leading to formation of voids in SiO_2 with further lateral widening [17]. This process (Eq. 1) can be successfully used for Si surface cleaning prior to epitaxial growth. In our case, we stop desorption and decomposition of the oxide film at the initial stage, when the voids have appeared on top surface only. As a result, the non-stoichiometric SiO_x ($x \leq 2$) layer is formed containing nonbridging oxygen hole centers and first oxygen-deficiency centers (E' centers). Two neighboring E' centers can be transformed into the neutral oxygen vacancy, in which two incomplete tetrahedra are linked by Si–Si bond [18]. The Si–Si bonds on top of oxide film are the nucleation centres for Ge nanocluster formation. During deposition of the first germanium monolayer onto the silicon oxide, the absorption layer is formed, which, starting just from the second monolayer, is transformed into nanoclusters that nucleate randomly on top of chemical SiO_2. Such a type of surface reconstruction is energetically favorable in the case if the direct contact with the Si(100) substrate is poor, i.e. when the oxide layer is still present [19].

3 Electronic Spectra of Ge/Si and Ge/SiO_2/Si Heterostructures with Nanoislands

The Ge/Si generally referred to the second type heterostructures, in which holes are localized in the quantum dots, while electrons are supposed to be free in the conduction band. A large number of experimental works was devoted to investigation of the energy-band structure in a quantum dot. It is supposed that the positions of energy levels in a quantum dot are determined by its size, shape and composition, as well as the value of inhomogeneous strain inside it caused by the lattice mismatch. Morphology studies may make possible the optimization of epitaxial conditions and growth mode of the heterostructure formation. Another important parameter that essentially influences optical and photoelectric properties of Ge/Si heterostructures is the value of the valence band offset. The average valence band offset of $Si/Si_{1-x}Ge_x$ heterojunction was estimated to be 0.54 x (eV), where x is Ge concentration. This value describes rather accurately the properties of heterojunctions of different compositions. The significant valence band offset between silicon and $Si_{1-x}Ge_x$ alloys determines a wide range of spectral sensitivity variation of photodetectors with the SiGe quantum dots.

The Ge/Si system with nanoclusters (NC) on/in silicon oxide has attracted considerable attention for their practical application in optoelectronic and memory devices [1]. Idealizing Ge–NC grown on oxidized silicon surface as well as Ge/Si heterostructures can be considered as a giant trap for holes due to deep potential well in the valence band of Ge [20, 21]. However, real Ge/SiO_2 and SiO_2/Si interfaces contain localized electronic states, which may capture electrons and holes from Ge–NC and Si. Charge of trapped carriers may change the band bending in underlying substrate, affecting electronic and transport properties of $Ge–SiO_2$ structures [21]. The presence of silicon oxide between Ge and Si substrate favours spatial separation of electron-hole pairs, prevents their recombination, and increases lifetime. Very little is known about possible effect of charge trapping by localized states of Ge–NC and interface levels on surface conductivity, including optically-induced recharging phenomena.

3.1 $Si_{1-X}Ge_X/Si$ Nanoheterostructures

The $Si_{1-x}Ge_x/Si$ nanoheterostructures refer to the second type, in which the potential well for holes is in the valence band of $Si_{1-x}Ge_x$. Energy diagram of the heterojunction is primarily determined by the values of the band gap and electron affinity of the contacting materials. In unstrained $Si_{1-x}Ge_x$ alloys the bandgap decreases monotonically with increasing of Ge content. The dispersion law E(k) depends on the Ge content also. Thus, at $x > 0.85$ the band structure of $Si_{1-x}Ge_x$ alloys is similar to Ge minimum of the conduction band at L-point. Alloys with lower Ge content with $x < 0.85$ have a band structure similar to that of the Si conduction band minimum at Δ-point [22].

Important parameters in terms of construction of an energy diagram is valence band offset ΔE_v and conduction band offset ΔE_c, which depend on $Si_{1-x}Ge_x$ alloy composition and values of strain near $Si_{1-x}Ge_x/Si$ heterojunction. According to theoretical calculations [23] average (subband for heavy holes, light holes, and split-off subband) valence band gap discontinuity for heterojunction between unstrained Si and Ge is $\Delta E_{v,av0} = 0.58$ eV. The best agreement between theory and experimental results for $Si_{1-x}Ge_x/Si$ heterojunction has been achieved if we assume a linear dependence of the average offset for valence band from germanium content x [22]:

$$\Delta E_{v,av} = 0.58x \tag{2}$$

To evaluate the conduction band discontinuity, the bandgap value of materials that come into contact should be taken into account:

$$\Delta E_c = \varepsilon_{g,Si} - \varepsilon_{g,SiGe} - \Delta E_{v,av} \tag{3}$$

Taking into account band gap values for Ge ($\varepsilon_{g,Ge} = 0.66$ eV) and Si ($\varepsilon_{g,Si} = 1.1$ eV) the conduction band offset for Si/Ge heterojunction was found to be $\Delta E_c = -0.14$ eV at room temperature.

It should be noted that the above theoretical estimations are simplified. Real heterojunction is characterized by transition regions, the space charge layer and band bending near the interface. The barrier profile can be significantly distorted by electric charges that are trapped by interface states. In addition, mechanical stresses exist near the interface between the $Si_{1-x}Ge_x$ nanoislands and silicon surrounding thus affecting the band structure of semiconductors and the shape of the potential barrier.

The valence band of Si and Ge consists of three subbands: $E_{v,hh}$ heavy holes, $E_{v,lh}$ light hole, and $E_{v,so}$ split-off subband. In the absence of spin-orbit interaction and strain the valence band of these materials is triply degenerate. In unstrained materials spin-orbit coupling removes the degeneracy and split-off one of the zones on value of Δ_0. Namely, two subbands (heavy and light holes) are shifted upward on $\Delta_0/3$, while third subband (split-off) is shifted down on $2\Delta_0/3$. The value of the spin-orbit splitting for Si is rather small $\Delta_0 = 0.04$ eV. In the case of Ge, this value is much higher $\Delta_0 = 0.30$ eV.

It is known that in the system of Ge islands grown on Si substrate the nanoislands are subject to compressive strain in the growth plane due to a 4 % mismatch of Si and Ge lattice constants. The strains are maximal near the nanoisland base and partially relax from the base to the top of nanoisland. Covering such a nanoisland structure with a silicon layer causes additional strains both in the growth plane and in the growth direction which leads to an appearance of a hydrostatic strain component. The silicon matrix near the QDs is subject to tensile deformations with the maximum value on the interface of the heterojunction between the QDs and Si. Hydrostatic strain does not change the semiconductor's symmetry and degenerancy degree of electron and hole states. Hydrostatic pressure changes the average energy. As a result, the average valence band offset is changed on $a_v \cdot \Delta V/V$ value [22]:

$$\Delta E_v = \Delta E^0_{v,av} + a_v \frac{\Delta V}{V} = \Delta E^0_{v,av} + a_v(2\varepsilon_{xx} + \varepsilon_{zz}) \tag{4}$$

where a_v—hydrostatic deformation potential, ε_{xx}, ε_{zz}—strain values.

Uniaxial strain leads to additional splitting of subbands of light and heavy holes in the point Γ_8, which is proportional to the strain values and can be described using the deformation potentials. The energy of spin-orbit splitting changes also.

The shift (relative to average value) of subbands of light and heavy holes as well as split-off subband for uniaxial strain along the [001] are defined as follows [24, 25]:

$$\Delta E_{hh} = -b(\varepsilon_{xx} - \varepsilon_{zz}) \tag{5}$$

$$\Delta E_{lh} = -\Delta E_{hh} - \frac{1}{2}\left[(\Delta_0 + \Delta E_{hh}) - \sqrt{(\Delta_0 + \Delta E_{hh})^2 + 8(\Delta E_{hh})^2}\right] \tag{6}$$

$$\Delta E_{SO} = \frac{1}{2}\left[(\Delta_0 + \Delta E_{hh}) - \sqrt{(\Delta_0 + \Delta E_{hh})^2 + 8(\Delta E_{hh})^2}\right] \tag{7}$$

where b is deformation potential for uniaxial strain along [001]. The values of spin-orbit splitting of the valence band Δ_0, deformation potential and elastic constants for $Si_{1-x}Ge_x$ alloys are determined by linear interpolation of corresponding values for Si and Ge [23, 26]. Effect of strain on the conduction band of Si and Ge is more complex. Thus, the hydrostatic pressure shifts the energy position of corresponding subbands, therefore uniaxial or biaxial strain break the degeneracy. In the presence of hydrostatic strain, a shift of the conduction band valley of SiGe structures occurs is determined by the following expression [27]:

$$\Delta E_{c,av} = \left(\Xi_d + \frac{1}{3}\Xi_u\right)(\varepsilon_{zz} + 2\varepsilon_{xx}) \tag{8}$$

where $\Xi_d + \frac{1}{3}\Xi_u$ is the conduction band hydrostatic deformation potential.

There are 8 and 6 equivalent valleys near the points L and Δ in the conduction band of Ge and Si, respectively. Due to the presence of compressive strain in the growth plane of the $Si_{1-x}Ge_x$, the sixfold degenerated conduction band Δ valley splits. As a result, a fourfold degenerated Δ_4 valley shifts towards lower energies and a twofold degenerated Δ_2—towards higher energies. The shifts of the Δ valleys with respect to the average $\Delta E_{c,av}$ for the [001] direction is given by:

$$\Delta E_C(\Delta_2) = \frac{2}{3}\Xi_u^{\Delta}(\varepsilon_{zz} - \varepsilon_{xx}) \tag{9}$$

$$\Delta E_C(\Delta_4) = -\frac{1}{3}\Xi_u^{\Delta}(\varepsilon_{zz} - \varepsilon_{xx}) \tag{10}$$

where Ξ_u^{Δ} is the deformation potential of the conduction band Δ valley.

For $Si_{1-x}Ge_x$, at germanium content $x < 0.85$ the value of the conduction band Δ valley deformation potential can be obtained by linear interpolation of the corresponding potentials for pure Si and Ge [22].

In conclusion, the band offset values for valence band and conduction band of $Si_{1-x}Ge_x$/Si heterojunction depend on the Ge content and the values of uniaxial or biaxial strain. The strain values for heterojunction between strained $Si_{1-x}Ge_x$ and unstrained Si(001) substrate can be estimated by:

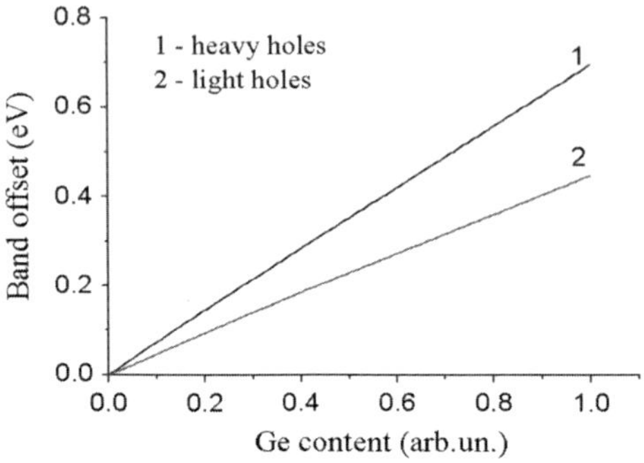

Fig. 3 The calculated band offsets for heavy holes (*curve 1*) and light holes (*curve 2*), depending on Ge content for heterojunction between strained $Si_{1-x}Ge_x$ alloy and Si surrounding

$$\varepsilon_{xx} = \varepsilon_{yy} = (a_{Si} - a_{SiGe})/a_{SiGe} \approx -0.04x \tag{11}$$

$$\varepsilon_{zz} = -2\varepsilon_{xx}C_{12}/C_{11} = -2\nu/(1-\nu)\varepsilon_{xx} = -0.77\varepsilon_{xx} \approx 0.03x \tag{12}$$

where a_{Si} and a_{SiGe} is lattice constants for Si та $Si_{1-x}Ge_x$, correspondingly, ν—Poisson's ratio, C_{11} and C_{12}—elastic constants.

Figure 3 shows calculated band offsets for heavy holes (curve 1) and light holes (curve 2), depending on Ge content for heterojunction between strained $Si_{1-x}Ge_x$ alloy and Si surrounding.

In structures with Ge nanoislands, as opposed to two-dimensional $Si_{1-x}Ge_x$ films on Si, there is relaxation of elastic strains. Such relaxation is the driving force of nanoislands formation, accompanied by the appearance of nonuniform stress fields in both nanoislands and Si surrounding. A numerical calculation of mechanical stress field distribution was carried out in a system of four $Ge_{0.7}Si_{0.3}$ QD layers with the height of 1.5 nm and lateral size of 23 nm separated by Si spacers [28]. It was shown that as a result of such deformations the Δ_2 valley of the silicon matrix valence band in the region of $Ge_{0.7}Si_{0.3}$/Si heterojunction shifts with respect to the valence band valley of unstrained silicon by about 100 meV. Strain fields near nanoislands are nonuniform and reach maximum values near base and top of the nanoislands. As a result, the strain field creates potential well for electron in the Si near QDs [29] (Fig. 4).

Thus, the energy diagram of $Si_{1-x}Ge_x$-QD/Si heterojunction depends on Ge content, inhomogeneous stress fields, and QDs shape. Effect of strain on band offsets, dispersion law, and effective mass of Si, Ge, and $Si_{1-x}Ge_x$ should be taken into account at solving of the quantum-mechanical problem on finding the eigen values of energy holes in $Si_{1-x}Ge_x$ nanoislands in Si matrix.

To evaluate the effect of quantum confinement on energy spectrum of holes in strained $Si_{1-x}Ge_x$/Si heterostructures with QDs we should consider the motion of quasi-free charge carriers in three-dimensional rectangular potential well, the depth of which is given by $V_0 = \Delta E_v$.

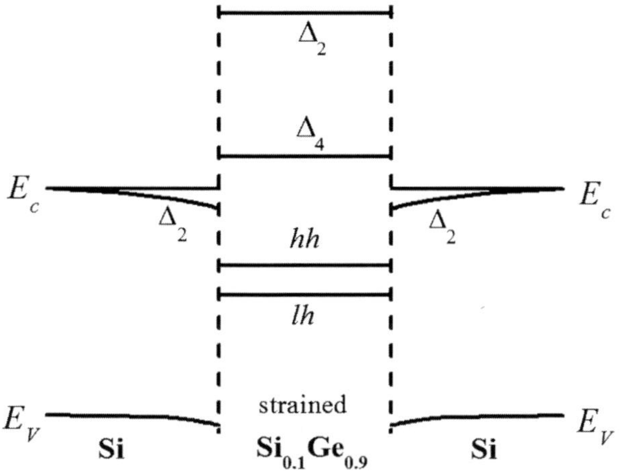

Fig. 4 Band diagram of strained SiGe–QD/Si heterostructure [30]

Typically, the lateral size of $Si_{1-x}Ge_x$ QDs grown by MBE technique is greater by an order than their height. In most cases, the actual quantization of motion of holes in the lateral direction is negligible compared to confinement in growth direction. In this case, the analysis of the energy spectrum of holes in the valence band of $Si_{1-x}Ge_x$ QDs can be done in the model of one-dimensional potential well of finite depth. For symmetric one-dimensional potential well problem is reduced to the solution of stationary Schrödinger equation using the effective mass approach:

$$-\frac{\hbar^2}{2m^*}\Psi'' + (U(z) - E)\Psi = 0 \tag{13}$$

$$U(x) = \begin{cases} V_0, & x<0.5d, \\ 0, & |x|<0.5d, \\ V_{0,} & x>0.5d, \end{cases} \tag{14}$$

where d is the width of potential well, m^* is the effective mass of holes in strained $Si_{1-x}Ge_x$ alloy.

It should be noted that the above approach allows us to accurately determine the energy of the ground state. As shown by theoretical calculations of the excited states of SiGe nanoislands with different shapes and sizes [31, 32], an account of their real shape slightly changes the positions of the energy levels. However, this simplified consideration allows explaining the optical absorption and photoconductivity of quantum-sized Si–Ge heterostructures.

Figure 5 shows the results of numerical calculations of the energy spectra of holes in $Si_{1-x}Ge_x$ quantum wells with width of 2 nm for different Ge contents. The analysis shows that the energy position of localized states with respect to top of Si valence band increases nonlinearly with x due to the dependence of the effective mass of holes from the strain values in this system.

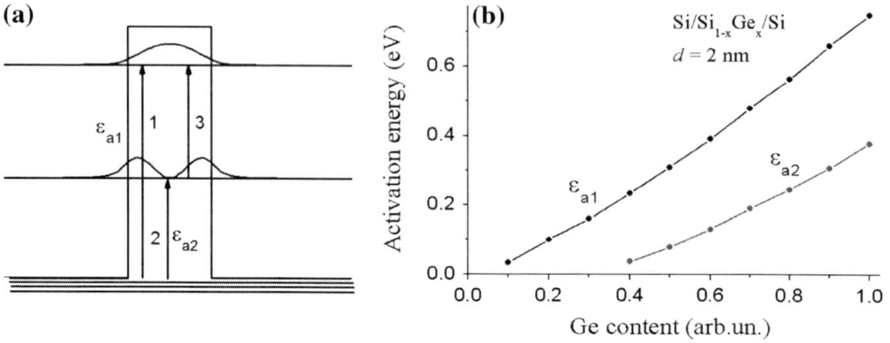

Fig. 5 Valence band potential profile and intraband transitions (**a**). The activation energies for localized holes of $Si_{1-x}Ge_x$ quantum wells with width of 2 nm and different content of Ge (**b**)

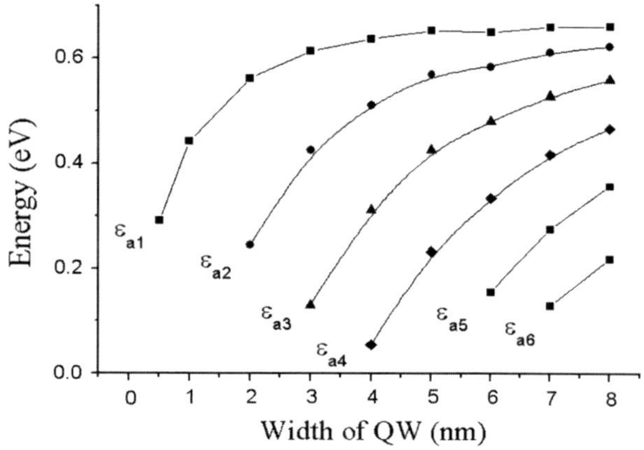

Fig. 6 Estimated depth for heavy hole states for different width of $Si_{0.2}Ge_{0.8}$ QWs relative to top the valence band of c-Si

Assuming the existence of filled by holes the quantum-sized states in heterostructures $Si_{1-x}Ge_x$/Si, two types of intraband transitions are possible in the valence band of $Si_{1-x}Ge_x$/Si: 1) bound-to- continuum transition (transitions 1 and 2 in Fig. 5a) and the transition between localized states (transition 3 in Fig. 5a) [33–35]. These intraband transitions are observed when the electric vector of electromagnetic wave has nonzero z-component (oscillation along confinement direction.

The results of calculation of energy levels position (relative to the top of the valence band of *c*-Si) for heavy holes in $Si_{0.2}Ge_{0.8}$ QWs is presented in Fig. 6. As follows from calculations, with increasing QW width the number and position of localized states in the valence band increases.

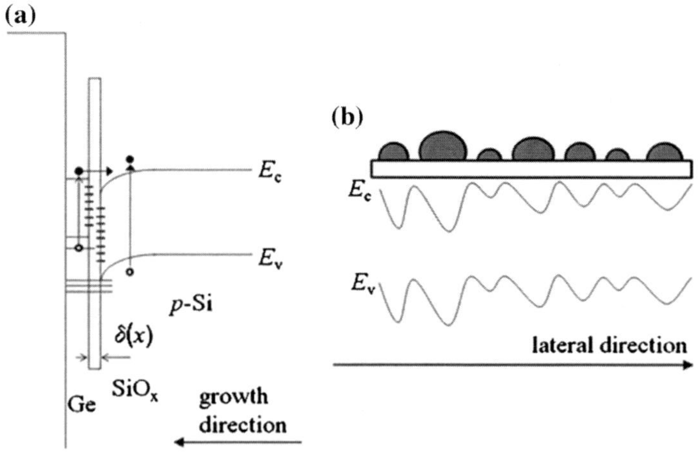

Fig. 7 Band diagram of Ge–NC/SiO_2/Si structure. **a** The arrays show interband electronic transitions in *c*-Si and Ge NC. **b** Electrostatic in-plane fluctuations in the space charge region of underlying *p*-Si substrate

3.2 Ge/SiO_2/Si Nanoheterostructures

The energy band diagram for Ge–NC/SiO_2/Si structure (Fig. 7a) has more complex electronic spectrum as compared to well-known SiO_2/Si or Ge/Si systems. Presented diagram shows bands along [001] direction, through the base of Ge–NC in the place, where oxide thickness equals to $\delta(x,y)$. It should be noted that the value of $\delta(x,y)$ varies from place to place in the range of 0–1 nm. The studied structures as well as Ge/Si heterojunctions have a large ~0.7 eV valence band offset (type II) leading to effective confinement of holes in Ge regions, while electrons are localized in silicon surrounding [36]. Thereby, Ge NCs can be considered as giant traps for holes [1, 35]. An intermediate ultrathin SiO_2 layer favors to electron-hole separation with respect to Ge nanoislands grown directly on Si. Additionally, chemically oxidized *p*-Si accumulates positive charge at SiO_2/*p*-Si interface [37], which induces downward bend bending up to 0.5 eV in the underlying Si with acceptor concentration $N_a = 10^{15}$ cm^{-3}. Surface photovoltage measurements give downward band bending value in the range from 300 to 500 meV. Besides localized states in the valence band of Ge, a high density of traps was reported for Ge–NC/SiO_2 interface [21]. Thereby, strong effect on band bending and surface conductivity in the underlying *p*-Si should have charge carriers captured by Ge/SiO_2 and SiO_2/Si interface states. In addition, NCs with areal density of $\sim 3 \times 10^{11}$/cm^2 populated by holes may induce additional fluctuations of surface potential up to few hundreds meV (Fig. 7b). Optical recharging of these states will have effect on band bending, electrostatic potential fluctuations, and surface conductivity.

4 Photocurrent Spectroscopy of Ge/Si and Ge/SiO_2/Si Nanoheterostructures with Nanoislands

Analyzing the energy diagrams of $Si_{1-x}Ge_x$/Si heterojunction we can conclude that the photosensitivity range of these structures is determined by the position of the Fermi level in the heterostructure, i.e. the dopant concentration in Si substrates and epitaxial films (Fig. 8a). Interband optical transitions are realized in the presence of electrons in quantum-sized states of the valence band nanoislands. For intraband transitions in the valence band, the Fermi level must be below at least the ground state of nanoislands. Development of efficient optoelectronic devices requires information on energy, oscillator strengths, and selection rules for interband and intrabend transitions. A large number of publications has been devoted to the study of optical properties of $Si_{1-x}Ge_x$ nanoislands, including absorption [38] and photoluminescence spectroscopy [39]. Fluorescent measurements do not reflect all transitions possible in heterogeneous in size and composition of deformations heterostructures. Opportunities of absorption spectroscopy are severely limited by the fact that the radiation passing through nanoscale quantum dot layer is absorbed only by its small part ($\sim 10^{-4}$–10^{-5}). As a result, the direct measurement of the absorption spectra of quantum dots is rather difficult task which requires a very sensitive technique and long-time measurements. One of methods which makes possible to study the absorption spectra in nanoscale semiconductor structures is an in-plane photocurrent spectroscopy [40, 41]. The value of photoconductivity is proportional to the number of photogenerated charge carriers, and thus the absorption coefficient. Photocurrent spectroscopy is a direct, sensitive and relatively simple method of studying the shape of optical absorption spectra and energy and interband transitions possible in heterostructures with nanoscale objects.

For in-plane photoconductivity (PC) studies, Ohmic contacts separated by distance of 10 mm were formed by annealing of Au at 370 °C. As a result, AuSi eutectic mixture gives electric contact to *p*-Si substrate (Fig. 8b). Ohmic current-voltage characteristics were linear in a wide temperature range of 50–290 K for applied biases from −1.0 to +1.0 V.

4.1 Intravalence Band Transitions in Ge/Si Nanoheterostructures

Spectral dependences of lateral photocurrent of Ge/Si heterostructure with quantum dots were measured at normal incidence and side illumination at 77 K (Fig. 9a). The photocurrent is caused by intraband transition of holes between localized states in the valence band of nanoislands (Fig. 9b).

For in-plane photoconductivity (PC) studies, Ohmic contacts separated by distance of 6 mm were formed by annealing of Au at 370 °C. As a result, AuSi

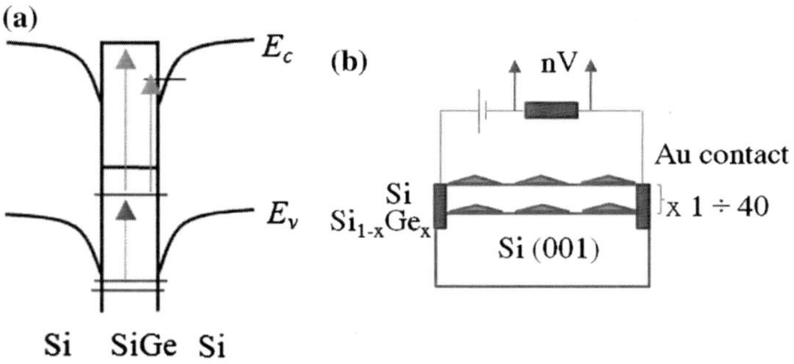

Fig. 8 **a** Different types of optical transitions in Si/Ge heterostructures, which causes photoconductivity: interband transition through localized states of Ge QD (*red arrows*), intraband transition (*blue arrow*). **b** Set-up for in-plane photoconductivity measurement

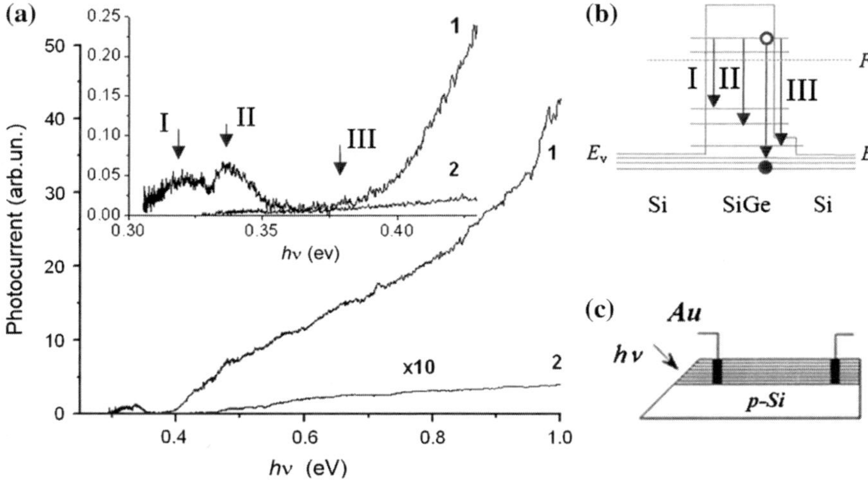

Fig. 9 **a** Spectral dependence of lateral photoconductivity of Ge/Si heterostructure with Ge quantum dots measured at side excitation and normal incidence geometry at 77 K. The inset shows in details the low-energy part of the spectra. **b** Energy level scheme for valence band of SiGe QDs. **c** Measurement geometry for in-plane PC measurements

eutectic mixture gives electric contact to *p*-Si substrate (see Fig. 9c). Ohmic current-voltage characteristics were linear for wide temperature range of 50–290 K for applied biases from −1.0 V to +1.0 V. PC spectral dependences were measured using an infrared spectrometer under illumination of a 250 W halogen lamp. Spectral dependences were normalized to the constant number of exciting quanta using a nonselective pyroelectric detector. The dark current and the photocurrent were measured in temperature range of 50–290 K using current amplifier and standard detection technique of the direct current.

In the case of side excitation, when nonpolarized light spreads along the basis of Ge nanoislands, there exists a component of the vector **E** oscillating along the growth direction (z-component), in which the confinement is the most pronounced. As it is known, intraband transitions in potential wells may be caused only by z-component of the vector **E**. The optical absorption coefficient for the transition from level i to f in a QDs is proportional to the matrix element:

$$\alpha \propto \left|\langle \Psi_i | \mathbf{p} | \Psi_f \rangle\right|^2 \tag{15}$$

where **p** is the momentum operator for corresponding transition. The integral defining the transitions between the levels has the form:

$$P_{mn} = -i\hbar \int\int\int \Psi_i^* \frac{\partial}{\partial x_k} \Psi_f dxdydz \tag{16}$$

where $x_k = x,y,z$. Polarization selection rules will be specified by nonzero terms of this integral. Only transitions with $\Delta n = 1$, i. e. between the first and second levels are dominant. The transitions $E_{111} \rightarrow E_{112}$, E_{121},$E_{122} \rightarrow E_{122}$,$E_{212}$ and E_{221},E_{222} are allowed with polarization along the growth axis.

Selection rules can be changed if a potential well has a finite depth and effective masses of carriers in the well and barrier layer are different. As a consequence, electron transitions under the irradiation of light polarized in the lateral structure direction become possible (under the action of x- and y-components of the vector **E**). For GaAs/AlGaAs quantum wells, the absorption coefficient for the light polarized along quantum dimensional layers is several orders smaller than for the light with z-polarization. However, selection rules are modified in quantum dots making it possible to observe intraband transitions at one band also at normal incidence of exciting irradiation. As a result, the photocurrent value in our photoconductivity experiments depended essentially on the way of irradiation of the heterostructure. The photocurrent in the range from 0.3 to 1.0 eV at normal incidence of exciting irradiation was much smaller (Fig. 9, curves 2).

In the case of side excitation in the spectral range from 0.3 to 0.37 eV, two peaks of current were observed at 0.32 and 0.34 eV and the photocurrent increased monotonically as the quantum energy increased in the range $h\nu > 0.38$ eV. The peaks of current observed in the spectral range from 0.3 to 0.37 eV can be attributed to hole transitions between the levels in the valence band of nanoislands. It can be seen that these values, by order of magnitude, lie within the range of excitation energies which follow from the level positions estimated above for SiGe quantum dots. The photoresponse in the range $h\nu > 0.38$ eV may be caused by hole transitions from the ground state of the valence band of SiGe nanoislands to two-dimensional continuum states of the valence band of a wetting layer or intermediate layers of Si surrounding [42–44]. Nonequilibrium carriers excited by such transitions may contribute to the observed lateral photocurrent, as far as no potential barriers exist for electron transport in the lateral direction. However, it is

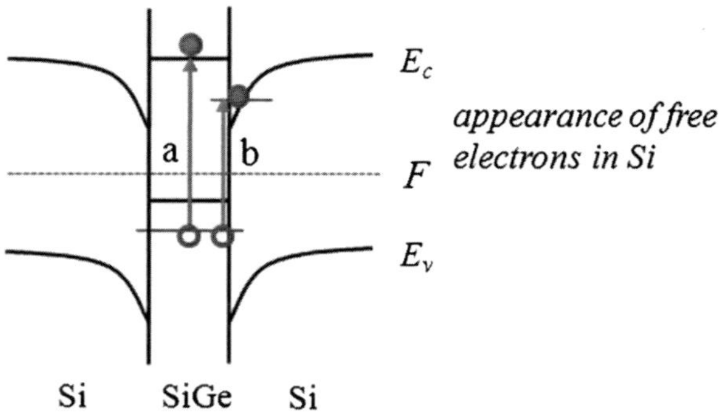

Fig. 10 Two types of interband transitions in SiGe/Si heterostructures with QDs

impossible to distinguish a contribution of bound-to-continuum transitions to the states existing in a wetting layer or Si intermediate layers in the conditions of given experiment.

4.2 Interband Transitions in Ge/Si Nanoheterostructures

As shown above, interband transitions between localized states of the valence band of SiGe QDs and delocalized states of the conduction band of silicon surrounding can be observed in low-dimentional Si–Ge heterostructures. Such transitions are possible if ground states are partially filled by electrons (Fig. 10). These transitions cause the appearance of nonequilibrium electrons in the Si spacer layers and WLs, which are transport channel. Observed photoconductivity is monopolar. The spectral range of interband transitions is determined by Ge content of QDs, strain value and confinement energy for holes in the valence band.

Figure 11a shows the spectral dependence of the lateral photoconductivity of the Ge/Si nanoisland structure measured at room temperature. A photoresponse was observed under the illumination with energies $h\nu > 0.81$ eV. Photoconductivity spectra were also measured for the sample *c*-Si without Ge QDs in order to compare the results. The main distinction of the measured spectral photocurrent dependencies was found in the range from 0.81 to 1.02 eV. The samples contain 5 periods of layers with QDs separated by Si layers with thickness of about 50 nm. TEM image of the first layer of Ge QDs grown on the system of $Si_{1-x}Ge_x$ layers and size distribution of QDs are given in Fig. 11b. AFM image is given in Fig. 11c. As one can see, the nanoislands are shaped as a tetrahedral pyramid with the base of about 50 nm and height of about 4 nm. The average distribution density is about 10^9 cm^{-2}. To explain these results, the shape of experimental spectra should be analyzed in detail.

To determine a shape of a photocurrent spectrum theoretically, one should solve the continuity equation for photoexcited carries with corresponding boundary conditions [46]. An approximate expression for the photocurrent in the case of uniform absorption $\alpha d <<1$ can be given as follows:

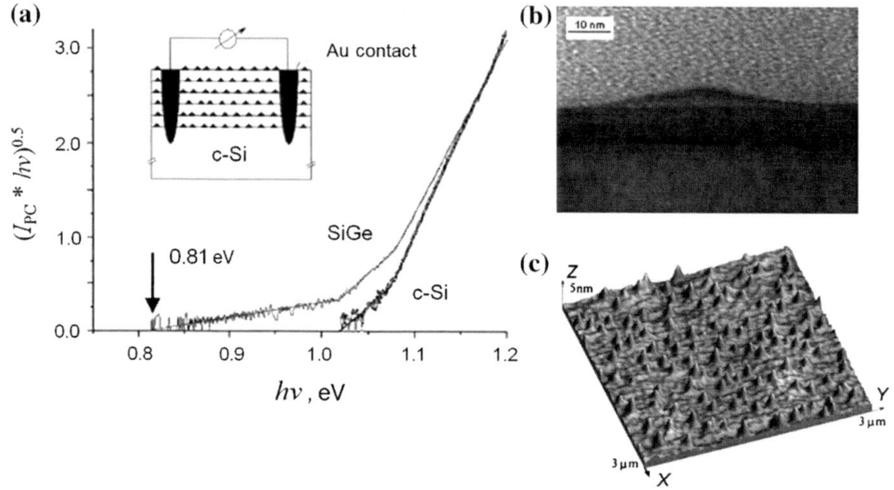

Fig. 11 **a** Spectral dependences of the lateral photocurrent of the Ge/Si heterostructure with Ge QDs (5 layers, growth temperature 450 °C, base 50 nm, height 4 nm and average density 10^9 cm^{-2}, Ge fraction 0.75–0.85) and *c*-Si substrate measured at 290 K. **b** AFM image of the surface of Ge/Si heterostructure with Ge QDs; **c** TEM image of the first layer of Ge QDs grown on the system of $Si_{1-x}Ge_x$ intermediate layers [45]

$$I_{PC} \approx \frac{I_0(1-R)d}{h\nu}\alpha \tag{17}$$

where I_0 is the intensity of exciting radiation, α is optical absorption coefficient, R is reflection coefficient and d is the sample thickness. It is supposed that the absorption is uniform for the investigated samples in the range from 0.81 eV to 1.2 eV. In this case the shape of photocurrent spectrum is determined by the component $\alpha(h\nu)/h\nu$. The distinction of the spectra for GeSi nanoisland samples and *c*-Si sample without nanoislands is conditioned by changes in the optical absorption.

Silicon and germanium, as well as solid alloys $Si_{1-x}Ge_x$ are known as semiconductors with indirect bands. The optical absorption coefficient for indirect allowed transitions with phonon energy $h\nu_k$ is determined as follows:

$$\alpha(h\nu) = \frac{A}{h\nu}\left[1+N_q\right](h\nu-\varepsilon_G-h\nu_k)^2+\frac{A}{h\nu}N_q(h\nu-\varepsilon_G+h\nu_k)^2 \tag{18}$$

where ε_G is the band gap energy, $N_q = \left[\exp\left(\frac{h\nu_k}{kT}\right)-1\right]^{-1}$ is a number of phonons with energy $h\nu_k$ [46]. The first component in the Eq. (18) corresponds to photon absorption with phonon emission; the second component corresponds to photon absorption with phonon absorption. The spectral dependence of the optical absorption can be well approximated by two straight lines in coordinates $h\nu$, $\sqrt{h\nu \cdot \alpha}$. The straight line for small values of the absorption coefficient intersects the $h\nu$ axis at value $h\nu = \varepsilon_G - h\nu_k$. The straight line for larger values of the absorption coefficient has a different slope and begins contributing at $h\nu = \varepsilon_G + h\nu_k$, allowing to determine the values of ε_G and $h\nu_k$.

Thus, the photocurrent spectral dependencies given in coordinates $h\nu$, $h\nu\sqrt{I_{PC}}$ have linear regions indicating indirect phonon transitions. The photocurrent spectral dependence for *c*-Si sample in such coordinates can be approximated by two straight lines corresponding to contribution of indirect transitions with absorption and emission of phonon, correspondingly. The same straight lines were observed for the QD structures in the range from 1.02 to 1.2 eV. Such an effect can be conditioned by the emission being absorbed by *c*-Si substrate and intermediate layers. It was revealed that the nanoisland GeSi structure was photosensitive in the range from 0.81 to 1.02 eV, where the sample *c*-Si without the nanoislands is transparent.

The shape of spectral dependence in coordinates $h\nu$, $h\nu\cdot\sqrt{I_{PC}}$ was linear in this range. The adsorption spectra are approximated by equation:

$$\alpha(h\nu) = \frac{A}{h\nu} N_q (h\nu - \varepsilon_G)^2, \qquad (19)$$

with $\varepsilon_G = 0.81$ eV.

This type of spectral dependence is typical for nonuniform or quantum-sized semiconductors. The photocurrent of Ge/Si heterostructures with Ge nanoislands in the range from 0.81 to 1.02 eV is caused by interband transitions in the SiGe nanoislands. One should pay attention to the fact that the determined limited energy value 0.81 eV exceeds Ge band gap energy 0.66 eV at 290 K. Such a shift can be explained by quantum size confinement of the carriers in the nanoislands and by the increase of band gap of SiGe alloys due to Si–Ge intermixing.

4.3 Photocurrent Spectroscopy of Ge–NC/SiO_2/Si Structures

Measurements of infrared photoconductivity in Ge–NC/SiO_2/Si structures made it possible to evaluate their electronic spectrum. The PC spectra measured at temperatures 50, 80, and 120 K (Fig. 12) give information about energies of electronic transitions in Ge–NC/SiO_2/*p*-Si structure. The in-plane photocurrent in the range $h\nu > \varepsilon_{G,Si}$ is mainly originated from band-to-band transitions in *c*-Si. For light excitations with photon energy below band gap of Si $h\nu < \varepsilon_{G,Si}$ ($\varepsilon_{G,Si} = 1.17$ eV at 77 K [47]), the electronic transitions from valence band to conduction band of NCs give main contribution to PC. However, generation of photocurrent in the range $0.8 < h\nu < \varepsilon_{G,Si}$ for Ge–NC/SiO_2/Si is also possible due to transitions between tails of the density of states in the near-surface *c*-Si [48], the optical absorption spectra of which are described by Urbach law [49]. The electron transitions through the states of Ge–NC/SiO_2 and Si/SiO_2 interfaces may also be observed. However their contribution to PC is expected to be small due to high probability of recombination through interface state.

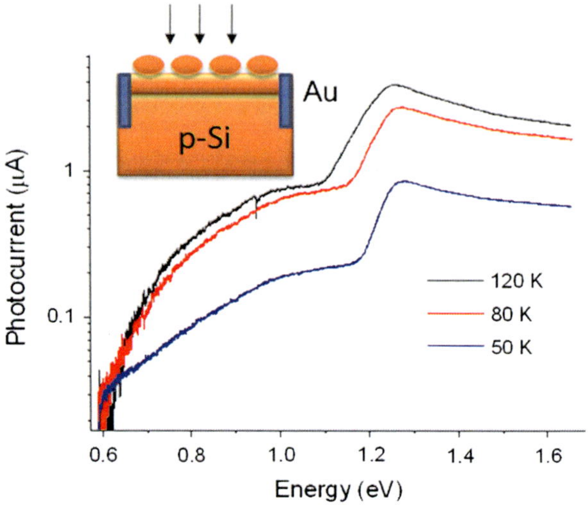

Fig. 12 In-plane PC spectra measured at 50, 80, and 120 K

The contribution of electron-hole pairs photoexcited in Si is observed, when the quanta energy exceed the band gap value. In the spectral range $h\nu < 1.1$ eV, in which *c*-Si is transparent, interband indirect transitions take place via the states in the valence and conduction bands of nanoclusters. Non-equilibrium carriers photoexcited in nanoclusters do not contribute into carrier transport directly. In order to contribute into the lateral current, the non-equilibrium electrons and holes should be spatially separated. As for Ge/Si heterojunctions, studied systems referred to type II, where strong confinement for holes in the region of Ge nanoclusters occurs. In the studied heterostructures, electrons can tunnel through the oxide SiO_x film into the near-surface silicon region and make contribution into conductivity. At the same time, non-equilibrium holes are localized in the valence band of Ge nanoclusters, however, they can affect the potential relief in the near-surface region of Si substrate, and hence, make an indirect effect on the system conductivity.

Thus, photoconductivity of the structures in the range of Si transparency is unipolar—intrinsic absorption of light in nanoclusters leads to an increase of the electron concentration in the Si potential well near the SiO_x-Si interface and to an increase of the surface conductance. In this case, the shape of lateral photoconductivity spectra reflects main features of intrinsic absorption of light in nanoclusters. The edge of PC spectrum of the investigated structures at $h\nu > \varepsilon_0$ is described by the dependence typical for the indirect band semiconductors:

$$\alpha(h\nu) = \frac{C}{h\nu}(h\nu - \varepsilon_0)^2 \tag{20}$$

where C is a constant, ε_0 is the width of the optical band gap. At excitement with quanta $h\nu < \varepsilon_0$ the Urbach tail is observed due to the crystal structure disorder [9].

Photocurrent spectroscopy and X-ray diffraction demonstrate that the nanoclusters have the local structure of body-centered-tetragonal Ge, which exhibits an optical adsorption edge at $\varepsilon_0 = 0.48$ eV. Taking into account quantum-size effect, this is in a good agreement with the theoretical calculations of electronic and optical properties of bulk body-centered-tetragonal Ge and Si, according to which the band gap for the mentioned polytypes is 0.38 and 0.86 eV, respectively [50].

5 Photoinduced Conductivity Changes in Ge–NC/SiO_2/Si Structures: Persistent and Quenching Effects

Selective excitation of Ge–NC/SiO_2/Si system by photons with different energy shows possibility to induce two non-equilibrium modes with excess and shortage of conductivity value as compared to the equilibrium one I_0 at temperatures below 150 K. PC relaxation process depends on photon energy and previous excitation history. Illumination by photons with energy $h\nu > \varepsilon_{G,Si}$ causes conductivity growth. Residual conductivity with level I_1 higher than equilibrium value I_0 persists for long time after the optical excitation is removed. In Fig. 13 we show persistent photoconductivity (PPC) decay curves measured at 50 K for $h\nu = 1.5$ eV excitation, which can be expressed as:

$$I_{PPC}(t) = I_{PPC}(0)\exp\left(-\left(\frac{t}{\tau}\right)^{\beta}\right) \tag{21}$$

where $\beta = 0.83$ is the decay exponent, t is time after end of photoexcitation, τ is the characteristic decay time constant. The stretched-exponential decay usually describes the relaxation of a wide class of disordered systems toward equilibrium. The new steady-state value exceeds initial dark current I_0 by 8 %. The main peculiarities for PPC effect: long-term conductivity, thermal quenching, and independence from bias voltage breaking were confirmed.

Excitations by photons with energy below *c*-Si band gap results in current spike and subsequent slow decay of PC under stationary illumination. In the case when photons with $h\nu \leq 0.7$ eV excite Ge–NC/SiO_2/Si structure, the total conductivity value becomes lower than the initial equilibrium level, which was established after cooling in the dark from 290 to 50 K. Removal of 0.7 eV photoexcitation leads to further decay of current, and in consequence the steady-state conductivity I_2 remains lower than initial equilibrium value I_0. Subsequent illumination by 1.5 eV photons switches system to the high-conductivity persistent mode. Figure 13 shows transient PC upon 0.7 eV and 0.9 eV illumination performed after transfer of system to persistent conductivity state I_1 by 1.5 eV photons. PC transients upon excitation are characterized by a relatively fast current growth with following slow decay to a quasi-stationary value lower than I_1. Termination of 0.7 eV or 0.9 eV illuminations gives the lowest value of in-plane conductivity (I_2). The optical quenching values $(I_1 - I_2)/I_1$ were found to be 6.1 and 3.2 % for 0.7 and 0.9 eV, respectively.

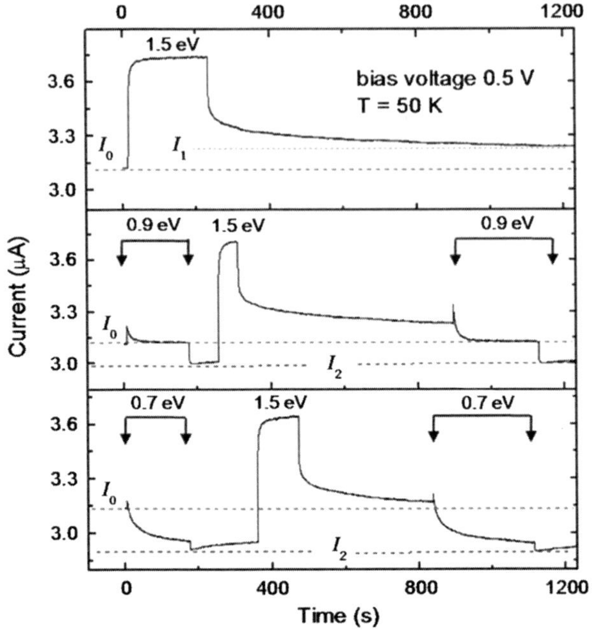

Fig. 13 Transient in-plane photocurrent upon exposure of structure with Ge NCs by photons with energy 1.5; 0.9 and 0.7 eV at temperature of 50 K. Bias voltages was 500 mV

The curves of PC transient depend on previous excitation history. For example, if after optical quenching by 0.7 eV or 0.9 eV photons free electron-hole pairs in the depletion layer of underlying Si are generated by photons with energy $h\nu > \varepsilon_{G,Si}$, the system goes back to the nonequilibrium mode with persistent conductivity ($I_1 > I_0$). Photoexcitation by photons with energy $h\nu < \varepsilon_{G,Si}$ can be used to induce the low-conductivity steady-state again, i.e. to transfer system from persistent state to low-conductivity steady state ($I_1 \rightarrow I_2$).

Several models have been proposed to explain the origin of persistent conductivity. In the microscopic local-potential model, the local fields of inhomogeneity separate photoexcited electrons and holes and thus delay recombination [51]. The other dominant mechanism of PPC involves macroscopic potential barriers, which prevent recombination in the other way [52, 53]. One type of carriers should be localized by traps, while the other ones are free and separated spatially. As reported in [51, 54], both long decay and high conductivity values can not be explained by usual models of Shockley-Read recombination centers. After the optical illumination is removed, the long-term nonequilibrium modes with reduced and increased conductivity as compared to the equilibrium one were observed due to recombination-preventing potential barrier, which separates spatially the regions with trapped carriers and the conductivity channel in *p*-Si substrate. Described macroscopic barrier is originated from the space-charge region of near-surface Si. In addition, the presence of Ge–NCs induces in-plane fluctuations of electrostatic potential in conductivity channel, which favor spatial separation of photoexited electron-hole pairs in lateral direction.

We believe that large-scale electrostatic fluctuations could exist in the near-surface region of Si substrate, which are the result of spatial distribution of trapped electrons and holes, inhomogeneity of the oxide film composition, interface imperfections, and non-uniform strain in Ge–NCs and Si substrate. Moreover, strain-modified confinement potential for electrons in the underlying silicon may enhance in-plane fluctuations in Ge–NCs/SiO_2/Si structures [55]. Persistent photoconductivity in systems with fluctuating potential can arise in the case, when the thermalization of electrons and holes is faster than the recombination, and activation energy of recombination exceeds the activation energy of conduction, i.e. the energy of charge carrier's emission to percolation level [56].

Let us analyze the peculiarities of PPC originated from electron-hole pairs photoexcitation in the near-surface region of Si by 1.5 eV photons. The electric field of space charge region of *p*-Si shifts photoexcited holes to bulk, while the nonequilibrium electrons drift to surface filling the minima of the potential relief in underlying *p*-Si. In turn, recombination rate of these electrons and holes trapped by SiO_2/Si interface states would be enhanced. Moreover, excess of electrons near SiO_2/Si increases a probability for electron penetration into NC's region and their subsequent recombination with trapped holes. As a consequence, there are reductions of positive charge trapped by NCs and SiO_2/Si interface states. Removal of excitation does not return the initial concentrations of trapped holes; therefore downward band bending becomes smaller than initial one. The reason is the presence of capture-preventing barrier near SiO_2/Si interface centers for holes in *p*-Si substrate. In other words, the electric field of *p*-Si depletion region limits the flow of holes to SiO_2/Si interface from substrate and restricts their subsequent capture by NC's or interface states (see Fig. 7a). As a consequence, the increased conductivity persists due to the lowest value of trapped positive charge and the presence of a certain number of non-equilibrium electrons in the near-surface depletion region of *p*-Si.

It should be noted that described PPC effect originates from spatial separation of carriers photoexcited in Si substrate mainly. Fundamentally different effect of photoconductivity quenching was observed under excitation by photons, which can not photogenerate free electron-hole pairs in Si. The conductivity of Ge–NC/SiO_2/Si system can be quenched to level $I_2 < I_0$, which is lower than equilibrium value (see Fig. 13) by photons with energy sufficient for electronic transition from the valence band to the conduction band of Ge–NCs only ($\varepsilon_{G,NC} < h\nu < \varepsilon_{G,Si}$). Observed photoconductivity is monopolar: electrons photoexcited in Ge–NC have high probability to transfer through interface states into underlying *p*-Si, giving contribution to lateral PC (see Fig. 7a). As a consequence, current spike is observed at initial stage of illumination by 0.7 eV photons. Photoexcited holes, which are localized in the deep potential well of Ge–NCs, can not contribute into carrier transport. Appearance of excess electrons in near-surface region of *p*-Si enhances recombination rate in the space charge region of Si and decreases positive charge trapped by SiO_2/Si interface states. Taking into account bipolar photoexcitation and neutrality principle we can conclude that increasing of positive charge trapped by Ge–NCs must be accompanied by decreasing of holes in

SiO_2/Si interface traps. In other words, an optically induced holes redistribution between SiO_2/Si interface traps and localized states of Ge–NCs occurs. Accumulation of positive charges by Ge–NCs increases gradually the fluctuations of downward band bending in the underlying *p*-Si, thereby reducing the recombination level and decreasing the surface conductivity upon excitation.

Removal of illumination stops the photoexcitation of electrons and holes in Ge–NCs, therefore photoconductivity drops (see Fig. 13). However, thermal emission of excess holes localized by deep potential well of Ge–NCs to underlying silicon restores slowly the equilibrium value of positive charge trapped by SiO_2/Si interface states tending to set equilibrium level of conductivity. In addition, some part of excess electrons with energy greater than the recombination level has possibility to recombine through interface states or penetrate to NCs quantum region, where they recombine with non-equilibrium holes. However, photoexcited electrons and holes are separated spatially by an oxide layer and local electric fields, which can greatly limit probability of their recombination. As a consequence, excess holes remain trapped by Ge–NCs and SiO_2/Si interface states a long time at low temperatures, while electrostatic fluctuations in the space-charge region of underlying *p*-Si provide a sufficient number of local traps serving as long-lived charged states. Excess of trapped holes induces residual downward band bending, the magnitude of which is greater than the equilibrium one. It causes setting of non-equilibrium mode $I_2 < I_0$ with lowest surface conductivity value in *p*-Si space charge region due to local field effect on surface conductivity.

Let us analyze some difference of optical quenching under photoexcitation by 0.7 and 0.9 eV photons. As mentioned above, such values of $h\nu$ are enough for interband transitions in Ge–NCs, while *p*-Si substrate is transparent. However, photons with $h\nu = 0.9$ eV are able to fulfill the electron transition between Urbach tail-state of heterogeneous near-surface Si or transitions through deep states of SiO_2/Si interface. Threshold energy for such kind of transition $\sim$0.8 eV can be estimated from photoluminescence spectra of SiO_2/Si structures [19]. It is obvious that additional photoexcitation of free holes in the near-surface Si causes the conductivity increasing and makes observation of quenching effect more difficult. On the contrary, excitation by 0.7 eV photons creates electron-hole pairs in Ge–NCs mainly. In consequence of this features the photoexcited holes are localized directly in Ge–NCs inducing the decrease of conductivity upon photoexcitation.

6 Conclusions

The optical adsorption of SiGe/Si heterostructures with SiGe nanoislands in the infrared range is determined by interband and intraband transitions involving localized states of valence band of SiGe nanoislands. The in-plane photoconductivity of $Si_{1-x}Ge_x$ nanoheterostructures, when the ground state of SiGe nanoislands is filled with holes, is caused by intraband hole transitions between quantum states of SiGe or bound-to-continuum transitions with subsequent transport of holes through wetting layer and silicon intermediate layers. Two types of interband transitions can be observed in structures where Fermi level is higher than ground state of SiGe nanoislands. The first one is spatially indirect optical transitions of electrons from the QD valence band to the conduction band Δ_2 valley of the surrounding silicon matrix in which tensile strains are present. The second type is interband electron transitions from the HH valence band of the QDs to the Δ_4 valley of the QD conduction band. The spectral range of photoconductivity was found to depend on size, Ge content and strain value near $Si_{1-x}Ge_x$/Si interface.

Two optically induced steady-states with excess and shortage of surface conductivity values as compared to the equilibrium one were observed in structures with Ge NCs grown on silicon oxide at low temperatures. The result of photoinduced changes depends on photon energy due to different type of electronic transitions observed in Ge–NC/SiO_2/Si structures. The persistent photoconductivity behavior was observed after excitation of electron-hole pairs in Si(001) substrate due to interband absorption in Si. The conductivity excess may be attributed to spatial carrier separation by macroscopic fields in the depletion layer of near-surface Si. Interband transitions in Ge create localized holes in Ge directly, leading to optically-induced spatial redistribution of trapped positive charges between SiO_2/Si interface levels and localized states of Ge–NCs, which enhances variation of electrostatic potential in underlying Si and, therefore, decay of surface conductivity under stationary photoexcitation. Observed results demonstrate that hole trapping by Ge–NCs and interface states have a significant effect on in-plane transport in the Ge–NCs/SiO_2/Si structures. Observed possibility for optically-controlled switching between different modes of system's conductivity may be helpful for design of optical memory devices.

Acknowledgments The research was supported by the program of fundamental research of the National Academy of Sciences of Ukraine "Nanostructured systems, nanomaterials, nanotechnologies" through the Project No9/07 and by the Ministry of Education and Science of Ukraine through Project NoM/34-09.

References

1. Brunner, K.: Si/Ge nanostructures. Rep. Prog. Phys. **65**, 27–72 (2002)
2. Bimberg, D., Grundmann, M., Ledentsov, N.: Quantum Dot Heterostructures. John Wiley & Sons Ltd, Chichester (1999)
3. Miesner, C., Röthig, O., Brunner, K., Abstreiter, G.: Intra-valence band photocurrent spectroscopy of self-assembled Ge dots in Si. Appl. Phys. Lett. **76**, 1027–1029 (2000)
4. Lafontaine, H., Rowell, N.L., Janz, S., Xu, D.-X.: Growth of undulating $Si_{0.5}Ge_{0.5}$ layers for photodetectors at $\lambda = 1.55$ μm. J. Appl. Phys. **86**, 1287–1291 (1999)
5. Barnham, K., Duggan, G.: A new approach to high efficiency multi-band-gap solar cells. J. Appl. Phys. **67**, 3490–3493 (1990)
6. Luque, A., Martı, A.: Increasing the efficiency of ideal solar cells by photon induced transitions at intermediate levels. Phys. Rev. Lett. **78**, 5014–5017 (1997)

7. Chen, Y., Lu, Y.F., Tang, L.J., Wu, Y.H., Cho, B.J., Xu, X. J., Dong, J. R., Song, W. D.: Annealing and oxidation of silicon oxide films prepared by plasma-enhanced chemical vapor deposition. J. Appl. Phys. **97**, 014913 (2005)
8. Schmidt, O.G., Eberl, K.: Multiple layers of self-asssembled Ge/Si islands: photoluminescence, strain fields, material interdiffusion and island formation. Phys. Rev. B **61**, 13721–13729 (2000)
9. Tiwari, S.: Volatile and non-volatile memories in silicon with nano-crystal storage. In: IEEE International Electron Devices Meeting. on 521
10. Wu, J.H., Li, P.W.: Ge nanocrystal metal-oxide-semiconductor transistors with Ge nanocrystals formed by thermal oxidation of poly-$Si_{0.88}Ge_{0.12}$. Semicond. Sci. Technol. **22**, S89 (2007)
11. Shklyaev, A.A., Ichikawa, M.: Three-dimensional Si islands on Si(001) surfaces. Phys. Rev. B **65**, 045307 (2001)
12. Shklyaev, A.A., Shibata, M., Ichikawa, M.: High-density ultrasmall epitaxial Ge islands on Si(111) surfaces with a SiO_2 coverage. Phys. Rev. B **62**, 1540–1543 (2000)
13. Zhang, L., Ye, H., Huangfu, Y.R., Zhang, C., Liu, X.: Densely packed Ge quantum dots grown on SiO2/Si substrate. Appl. Surf. Sci. **256**, 768–772 (2009)
14. Lysenko, V.S., Gomeniuk, Y.V., Kozyrev, Y.N., Rubezhanska, M.Y., Sklyar, V.K., Kondratenko, S.V., Melnichuk, Y.Y., Teichert, C.: Effect of Ge nanoislands on lateral photoconductivity of Ge–SiO_x-Si structures. Adv. Mater. Res. **276**, 179–186 (2011)
15. Nakamura, Y., Ichikawa, M., Watanabe, K., Hatsugai, Y.: Quantum fluctuation of tunneling current in individual Ge quantum dots induced by a single-electron transfer. Appl. Phys. Lett. **90**, 153104–153106 (2007)
16. Chen, Y., Lu, Y.F., Tang, L.J., Wu, Y.H., Cho, B.J., Xu, X.J., Dong, J.R., Song, W.D.: Annealing and oxidation of silicon oxide films prepared by plasma-enhanced chemical vapor deposition. J. Appl. Phys. **97**, 014913–014922 (2005)
17. Tromp, R., Rubloff, G.W., Balk, P., LeGoues, F.K., van Loenen, E.J.: High-temperature SiO_2 decomposition at the SiO_2, Si interface. Phys. Rev. Lett. **55**, 2332–2335 (1985)
18. Skorupa, W., Rebohle, L., Gebel, T., Helm, M.: Strong blue light emission from ion implanted Si/SiO_2 structures. NATO Sci. Ser. **93**, 69–78 (2003)
19. Shklyaev, A.A., Ichikawa, M.: Extremely dense arrays of germanium and silicon nanostructures. Phys. Usp. **51**, 133–161 (2008)
20. Kwak, D.W., Park, C.J., Lee, Y.H., Kim, W.S., Cho, H.Y.: Extended defect states of Ge/Si quantum dots using optical isothermal capacitance transient spectroscopy. Nanotechnology **20**, 055201 (2009)
21. Peibst, R., de Sousa, J.S., Hofmann, K.R.: Determination of the Ge-nanocrystal/SiO_2 matrix interface trap density from the small signal response of charge stored in the nanocrystals. Phys. Rev. B **82**, 195415–195426 (2010)
22. Kasper, E., Lyutovich, K.: Properties of Silicon Germanium and SiGe: Carbon. INSPEC, The Institution of Electrical Engineers, London (2000)
23. Van de Walle C.G.: Band lineups and deformation potentials in the model-solid theory. Phys. Rev. B. **39**, 1871–1883 (1989)
24. Schmidt, O.G., Eberl, K., Rau, Y.: Strain and band-edge alignment in single and multiple layers of self-assembled Ge/Si and GeSi/Si islands. Phys. Rev. B. **62**, 16715–16720 (2000)
25. Patriarche, G., Sagnes, I., Boucaud, P., Le Thanh, V., Bouchier, D., Hernandez, C., Campidelli3, Y., Bensahel, D.: Strain and composition of capped Ge/Si self-assembled quantum dots grown by chemical vapor deposition. Appl. Phys. Lett. **77**, 370–372 (2000)
26. Van de Walle, C.G., Martin, R.M.: Theoretical calculations of heterojunction discontinuities in the Si/Ge system. Phys. Rev. B. **34**, 5621–5634 (1986)
27. Van de Walle, C.G., Martin, R.M.: Theoretical study of Si/Ge interfaces. J. Vac. Sci. Technol. B. **3**, 1256–1259 (1985)
28. Van de Walle, C.G.: Band lineups and deformation potentials in the model-solid theory. Phys. Rev. B. **39**, 1871–1883 (1989)

29. Yakimov, I., Stepina, N.P., Dvurechenskii, A.V., Nikiforov, A.I., Nenashev, A.V.: Interband absorption in charged Ge/Si type-II quantum dots. Phys. Rev. B. **63**, 045312–045317 (2001)
30. Kondratenko, S.V., Nikolenko, A.S., Vakulenko, O.V., Valakh, M.Y., Yukhymchuk, V.O., Dvurechenskii, A.V., Nikiforov, A.I.: Band offsets and photocurrent spectroscopy of Si/Ge heterostructures with quantum dots. Nanotechnology **19**, 145703 (2008)
31. Ngo, C.Y., Yoon, S.F., Fan, W.J.: Effects of size and shape on electronic states of quantum dots. Phys. Rev. B. **74**, 245331–245340 (2006)
32. Yakimov, A.I., Dvurechenskii, A.V., Nikiforov, A.I., Bloshkin, A.A., Nenashev, A.V., Volodin, V.A.: Electronic states in Ge/Si quantum dots with type-II band alignment initiated by space-charge spectroscopy. Phys. Rev. B. **73**, 115333 (2006)
33. Kondratenko, S.V., Vakulenko, O.V., Kozyrev, Y.N., Rubezhanska, M.Y., Dadykin, A.A., Naumovets, A.G., Hofer, C., Teichert, C.: Photoconductivity and field-assisted photoemission in multilayer Si/Ge heterostructures with quantum dots. Ukrainian J. Phys. **55**, 381–387 (2010)
34. Kozyrev, Y.N., Kondratenko, S.V., Rubezhanska, M.Y., Sklyar, V.K.: Quantum effects in multilayer Si–Ge nanoheterostructures. Chem. Phys. Tech. Surf. **14**, 176–185 (2008)
35. Lysenko, V.S., Gomeniuk, YuV, Strelchuk, V.V., Nikolenko, A.S., Kondratenko, S.V., Kozyrev, YuN, Rubezhanska, MYu., Teichert, C.: Carrier transfer effect on transport in p-i-n structures with Ge quantum dots. Phys. Rev. B **84**, 115425 (2011)
36. Dvurechenskii, A.V., Nenashev, A.V., Yakimov, A.I.: Electronic structure of Ge/Si quantum dots. Nanotechnology **13**, 75–80 (2002)
37. Yasutake, K., Cheng, Z., Pang, S.K., Rohatgi, A.: Modeling and characterization of interface state parameters and surface recombination velocity at plasma enhanced chemical vapor deposited SiO2–Si interface. J. Appl. Phys. **75**, 2048–2054 (1994)
38. Bonfanti, M., Grilli, E., Guzzi, M., Virgilio, M., Chrastina, D., Isella, G., Neels, A.: Optical transitions in Ge/SiGe multiple quantum wells with Ge-rich barriers. Phys. Rev. B. **78**, 041407 (2008)
39. Liu, Zhi, Cheng, Buwen, Weixuan, Hu, Shaojian, Su, Li, Chuanbo, Wang, Qiming: Enhanced photoluminescence of multilayer Ge quantum dots on Si(001) substrates by increased overgrowth temperature. Nanoscale Res. Lett. **7**, 383–387 (2012)
40. Talochkin, A.B., Chistokhin, I.B., Markov, V.A.: An electron–hole spectrum of Ge/Si structures with Ge quantum dots: photoconductivity measurements. Nanotechnology. **20**, 175401 (2009)
41. Kondratenko, S.V., Golovinskiy, S.L., Nikolenko, A.S., Vakulenko, O.V.: The lateral photoconductivity of Si, Ge structures with quantum dots. Semicond. Sci. Technol. **21**, 857–859 (2006)
42. Forbes, R.G.: Ultramicroscopy. Some comments on models for field enhancement. Ultramicroscopy. **95**, 57–65 (2001)
43. Ershov, M., Liu, H.C., Buchanan, M., Wasilewski, Z.R., Ryzhii, V.: Photoconductivity nonlinearity at high excitation power in quantum well infrared photodetectors. Appl. Phys. Lett. **70**, 414–416 (1997)
44. Fromherz, T., Kruck, P., Helm, M., Bauer, G., Nützel, J.F., Abstreiter, G.: Transverse magnetic and transverse electric polarized inter-subband absorption and photoconductivity in p-type SiGe quantum wells. Appl. Phys. Lett. **68**, 3611–3613 (1996)
45. Vakulenko, O.V., Kondratenko, S.V., Nikolenko, A.S., Golovinskiy, S.L., Kozyrev, Y.N., Rubezhanska, M.Y., Vodyanitsky, A.I.: Photoconductivity spectra of Ge/Si heterostructures with Ge QDs. Nanotechnology. **18**, 185401 (2007)
46. Moss, T.S., Burrell, G.I., Ellis, B.: Semiconductor Opto-electronics. Butterworth, London (1973)
47. Macfarlane, G.G., McLean, T.P., Quarrington, J.E., Roberts, V.: Fine spectra of germanium and silicon. J. Phys. Chem. Solids **8**, 388–392 (1959)
48. Lysenko, V.S., Gomeniuk, Y.V., Kozyrev, Y.N., Rubezhanska, M.Y., Sklyar, V.K., Kondratenko, S.V., Melnichuk, Y.Y., Teichert, C.: Effect of Ge nanoislands on lateral photoconductivity of Ge–iO_x–Si structures. Adv. Mater. Res. **276**, 179–186 (2001)

49. Urbach, F.: The long-wavelength edge of photographic sensitivity and of the electronic absorption of solids. Phys. Rev. **92**, 1324 (1953)
50. Malone, B.D., Louie, S.G., Cohen, M.L.: Electronic and optical properties of body-centered-tetragonal Si and Ge. Phys. Rev. B. **81**, 115201–115205 (2010)
51. Sheinkman, M.K., Shik, A.Y.: Long-term relaxation and residual conductivity in semiconductors. Fiz. Tekh. Poluprovodn. **10**, 209–232 (1976)
52. Queisser, H.J., Theodorou, D.E.: Decay kinetics of persistent photoconductivity in semiconductors. Phys. Rev. B. **33**, 4027–4033 (1986)
53. Queisser, H.J.: Nonexponential relaxation of conductance near semiconductor interfaces. Phys. Rev. Lett **54**, 234–236 (1985)
54. Sitenko, T.N., Tyagulskii, I.P., Lyashenko, V.I., Lysenko, V.S.: Role of surface band bending in residual conductivity formation in epitaxial GaAs films. Phys. Status Solidi A **30**, 755–763 (1975)
55. Kuryliuk, V., Korotchenkov, O., Cantarero, A.: Carrier confinement in Ge/Si quantum dots grown with an intermediate ultrathin oxide layer. Phys. Rev. B. **85**, 075406–075416 (2012)
56. Shklovkii, B.I., Efros, A.L.: Electronic Properties of Doped Semiconductors. Springer, Heidelberg (1984)

SOI-Nanowire Biosensors for High-Sensitivity Protein and Gene Detection

Yuri D. Ivanov, Tatyana O. Pleshakova, Vladimir P. Popov, Olga V. Naumova, Alexander L. Aseev and Alexander I. Archakov

Abstract The development of genome and proteomic technologies is related, first of all, to the progress recently achieved in developing of high-sensitivity, rapid methods for registration and analysis of nucleic acids and proteins. One of such methods, combining high sensitivity and high response speed, is the nanowire (NW) detection technique. This technique allows one to register analyte species in real time without using labels. Nowadays, the concentration sensitivity of such systems in registering proteins and DNAs reaches the femtomolar level. In the present chapter, we consider nanowire sensor systems based on silicon nanowire field-effect transistors. We discuss the achieved potential in detecting low-copy proteins with the help of nanowire systems. We show that it is possible to use not only antibodies, but also aptamers, as probing molecules for biospecific protein detection. We also discuss the possibility of using nanowire detectors in genome studies.

Abbreviations

AFP	Alpha-fetoprotein
Anti-GST	Glutathione S-transferase antibodies
Anti-NFATc1 Ab	Anti-NFATc1 directed antibodies capable of specifically binding to DNFATc1
AMPTMS	(3-aminopropyl)trimethoxysilane (or (Trimethoxysilyl)propylamine)

Y.D. Ivanov · T.O. Pleshakova · A.I. Archakov
Orekhovich Institute of Biomedical Chemistry, Russian Academy of Medical Science, Pogodinskaya Street 10, Moscow, Russia
e-mail: yurii.ivanov@rambler.ru

V.P. Popov (✉) · O.V. Naumova · A.L. Aseev
Rzhanov Institute of Semiconductor Physics, Russian Academy of Science, Siberian Branch, Acad. Lavrent'ev Avenue, 13, 630090 Novosibirsk, Russia
e-mail: popov@isp.nsc.ru

A. Nazarov et al. (eds.), *Functional Nanomaterials and Devices for Electronics, Sensors and Energy Harvesting*, Engineering Materials,
DOI 10.1007/978-3-319-08804-4_20

APDES	3-amino-propyldiethoxysilane
APMS (or ALDTMS)	3-aldehydepropyl-trimethoxysilane (3-trimethoxysilylpropylaldehyde)
APTES	3-aminopropyl triethoxysilane
DL	Detection limit
DMSO	Dimethyl sulfoxide
D-NFATc1	DNA-binding fragment of the nuclear factor of activated T-cells
DSP	Dithiobis (succinimidyl propionate)
GPTES (or GOPTS, or GPTMS, or GPTS)	3-glycidoxypropyltrimethoxysilane
GST	Glutathione S-transferase
KP	Potassium-phosphate buffer solution
AFP	Alpha-fetoprotein
NW	Nanowire
SOI	Silicon-on-insulator
TESU	11-triethoxysilyl (undecanal)

1 Introduction

An urgent problem in proteomics is the development of new techniques for detection of low-copy proteins with concentration sensitivity limit, detection limit, $DL \leq 10^{-15}$ M [1]. Development of high-sensitive DNA biosensors operating without using an amplification procedure is necessary for an analysis of gene expression, including that in individual cells. That is why such biosensors are of importance in molecular diagnostics and in analysis of single copies of DNA molecules. Thus, further progress in genomics and proteomics is related to the development of analytical processes which should employ, as sensor elements, high-sensitive molecular detectors, or detectors capable of registering protein or DNA molecules in counting mode. Molecular detecting systems involve atomic-force detectors [2], NEMS-based mass detectors [3], nanowire (NW) detectors [4], etc. Over the last decade, development of processes for fabrication of NW biosensors has been under way since such biosensors belong to the class of detectors that use no labels and allow real-time registration of macromolecules. The operating principle of NW biosensors is based on the registration of the current modulation across NW that occurs on the adsorption of analyte molecules on NW surface. In such a biosensor, the adsorbed molecules act as a virtual gate while the nanowire itself, with Ohmic contacts provided at its ends, is used as a nanoscale field-effect transistors (FET) [5]. The high sensitivity of NW-sensor element is defined by its high surface area/volume ratio [6]. As discussed in [7], the theoretical detection limit of a NW biosensor can reach a level of one molecule per sensor element. To date, the sensitivity at the level of single particles for NW biosensors was reached for viruses [4]. As for DNA and protein molecules, the experimentally achieved detection limit reaches femtomolar in the case of DNA [8, 9] and even subfemtomolar level in the case of proteins [10].

In this chapter, processes for fabrication of NW detectors as well as for NW functionalization and sensibilization are outlined. Then, the possibility of high-sensitivity protein detection with the help of NW detectors is analyzed. Advantages of aptamers used as biospecific probes in NW-surface sensibilization are considered. The use of aptamers instead of antibodies substantially widens the range of detected proteins and improves the stability of biochips in such biosensors. Examples illustrating the use of NW biosensors in DNA registration and sequencing are given.

2 Nanowire Detectors for Biomolecules

2.1 Fabrication Methods of Nanowire Detectors

Presently, two major approaches in the technology of NW structures are used, namely, "bottom–up" [5] and "top–down" processes [11, 12]. For instance, Lieber et al. employed a "bottom–up" strategy to fabricate 20 nm sensing-element diameter NW biosensors capable of registering single influenza virus particles [4]. A high sensitivity of such biosensors to proteins has also been demonstrated: in detecting PSA in blood serum, a DL level of 10^{-13} M was achieved [13]. An advantage offered by the latter approach to the formation of NW sensors is comparative easiness in fabricating individual elements. Simultaneously, the "bottom–up" approach shows a number of disadvantageous features. For instance, this approach is rather labor-consuming one in producing element arrays and in forming low-noise contacts to individual nanowires, which is necessary for solving problems in medical proteomics.

Until now, there remains an unsolved problem of NW growth without inducing structural defects and electrically active material inclusions acting as catalysts of epitaxial growth in the CVD process [14]. For instance, Naumova et al. [15] showed that Si nanowires grown by molecular-beam epitaxy with the use of Au drops as catalysts may acquire a set of electrically active centers nonuniformly distributed along NW. Moreover, in microfluidic [16] or Langmuir-Blodgett [17] transfer of nanowires onto a planar substrate there arises a problem to control the positioning of NWs to a particular address in the chip [18].

Another method to fabricate NW structures, the "top-down" one, seems to be more technological as this method proved to be perfectly compatible with the well-established complementary metal-oxide-semiconductor (CMOS) technology, thus providing a low-cost end product [12]. In our study, an array of monocrystalline Si

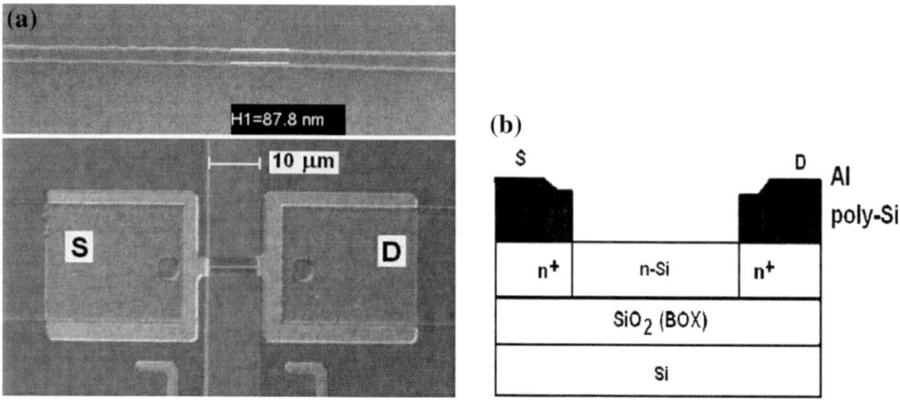

Fig. 1 **a** Image of a SOI nano-FET and the central part of the FET as viewed respectively in an electron microscope and in an optical microscope. Designations *S* and *D* denote respectively the source and the drain. Also indicated is the length of the SOI active channel: L = 10 μm. **b** Schematic of the longitudinal cross section of SOI nanowire with heavily doped S-D-poly-Si contacts and Al metallization

nanowires was fabricated using the "top–down" approach, through nanostructuring of silicon-on-insulator (SOI) layers achieved using electron-beam lithography and gas or plasma chemical etching [19]. Figure 1a shows an optical and TEM images of SOI NW sensor element of 10-μm length and 90-nm width fabricated in this way. Figure 1b shows a schematic representation of an n-type nanowire with contact drain (D) and source (S) regions involving layers of heavily doped poly-Si and aluminum (Al). A pilot biosensor based on such a sensor element contains a measuring cell whose bottom presents a nano-FET array chip built in a standard chip package.

Earlier, it was shown that the approach using the gas-phase etching instead of the ion or liquid etching allows for diminishing the density of defects in the Si/SiO_2 system. Moreover, no additional high-temperature treatments are required and, hence, a possibility of fabrication of a sensor element with high conduction ($\sim 10^{-6}$ S) at the last step of the fabrication process is provided. In the latter case, the sensor element is formed following standard CMOS process steps, when the contacts to NW and the chip periphery with control and signal processing circuitry is already formed. It was shown that a NW detector chip prepared in such a way features high sensitivity in registration of protein molecules. Interaction of bovine serum albumin (BSA) with NW surface was observed in solutions with BSA concentrations as low as 10^{-14}–10^{-15} M [19, 20].

In the future, the compatibility of the SOI-NW formation process with the conventional Si CMOS technology, the possibility of scaling down the fabrication of SOI-nano-FETs with arrays of sensor elements will allow production of low-cost NW detectors for high-sensitive clinic medical diagnostics.

2.2 *Functionalization of Si-NW Detector Surface for Protein Registration*

Functionalization of Si-NW detector sensor-element surface is necessary for ensuring a high sensitivity of the detector enabling reproducible registration of the signal from a protein [4] or a nucleic acid to be detected.

As a rule, NW-surface functionalization methods are two-step procedures involving a modification stage and a sensibilization stage.

2.3 *Modification of Si Nano-FET Sensors*

Surface modification is a chemical process favoring the formation of active groups necessary for covalent immobilization of biomolecules on a surface. Two methods suitable for nano-FET surface modification, namely, hydroxylation and silanization, can be distinguished.

The research group from Yale University was the only one to use a nano-FET surface modification based on hydroxylation [11]. This approach uses a reaction between surface groups (Si–H) and organic substances with unsaturated terminal carbon bonds (alkenes and alkanes) necessary for forming Si–C bonds. Due to such bonds formed on NW surface, the functional carbon-reagent monolayer [11, 21] capable of covalent binding of proteins was obtained. The formation of a monolayer, not a multi-layer film, is an advantage of the hydroxylation method as intended for NW surface functionalization. Besides, the NW surface thus modified is stable in aqueous solutions [22]. Disadvantages of this method are, first, the necessity to synthesize a special compound and, second, the involvement of an additional stage of unblocking active groups on the sensor-element surface.

Silanization is the most popular method to reach Si-NW functionalization. For realization of the modification scheme using silanization, the presence of native oxide layer or controllably grown oxide layer on the surface is necessary. After preliminary cleaning, the NW surface with Si–O bonds has to be immersed in a solution or in a silane-containing gaseous phase. The most popular media are silane derivative containing amino- and aldehyde-terminal groups such as APTES, APDMES, or AMPTMS (see Table 1). The choice of a particular silane derivative and silanization conditions is determined by a few factors. The thickness and quality of the silane layer on NW surface largely affects the sensitivity of the NW detector. Surface relief or involvement of several layers in the silane coating can be a consequence of excess silane concentration and/or improper incubation period duration. That is why the latter parameters must be properly chosen while optimizing silanization conditions. Quality control of silane layer can be implemented using AFM (see, e.g., [23, 24]). Formation of a monolayer is possible either using silane derivatives containing one alcoxygroup, also called silane monoderivatives [23, 25] or, alternatively, performing silanization in a gas phase [24, 25].

In silanization, the use of a silane mono-derivative with an amino-terminal group, for instance, monoalcoxide methylsilane (APDMES) (see [25]) will guarantee the formation of a monolayer coating NW surface. Yet, for a low density of active (hydroxyl) groups on Si-NW surface the use of APDMES will deteriorate the modification quality, resulting in a low density of active groups capable of forming covalent bonds with probing biomolecules.

In silanization procedure, silane derivatives yielding terminal aldehyde groups on NW surface can be used. For instance, the use of APMS (see [13]) will allow formation of thinner organic layers in comparison with APTES layers. Yet, stability of NW-surface functional layers involving aldehyde groups is rather low because the layer undergoes oxidation with atmospheric oxygen or with oxygen dissolved in the solutions. That is why nanowires functionalized in this way need to be used immediately after the functionalization procedure [21]. During silanization performed with the help of APTES, the terminal silanol groups are substituted with aminogroups capable of binding with aldehyde, carboxyl, epoxy- or succinimide cross-linker groups, the special reagents making probing molecules covalently immobilized on modified NW surface (Fig. 2).

Such a universal reactivity, and also stability of the aminosilane surface in aqueous solutions made aminosilanes popular on the whole and APTES in particular as reagents for NW surface modification. It is APTES that has been most often used for modification of surfaces in the majority of reported studies cited in the present review (see Table 1). However, during the silanization in APTES formation of silane oligomers on NW surface proved to be possible, eventually resulting in deterioration of NW-detector sensitivity. The probability of polylayer formation increases as we pass over to silanization being performed in a solution. To overcome the latter disadvantage, they use gas-phase silanization. The capabilities of this process were demonstrated in [24, 27, 28]. Results gained in our AFM studies have also proved that, during gas-phase silanization, on the surface of NW sensors an APTES monolayer of thickness 0.9–1.2 nm normally forms [24].

2.4 Sensibilization of Si-NW-Sensor Surface

To enable specific detection of biomolecular targets, sensibilization of NW surface with probing molecules covalently immobilized to target objects is used. As probing molecules in proteomics and medical diagnostics, proteins or oligonucleotides capable of specific binding to target biomolecules are employed. Probing molecules can be made covalently immobilized on NW surface with the help of standard processes. For instance, using APTES at the stage of NW surface modification, probing biomolecules can be immobilized using either glutar aldehyde [8, 12, 29–32] or carbodiimide linker [24, 26, 28, 33]. However, choosing an acceptable sensibilization strategy one must take into consideration possible

Table 1 Sensitivity of NW biosensors

Detection limit (DL), detectable species	Method of NW-surface modification[a]	Sensibilization method: probing molecules, linkers	Passivation method	Possibility of reusing	Ref. no.
Detection in buffer solutions					
8×10^{-17} M, (50 virus particles per 1 μl), influenza virus	Silanization in APMS solution	Influenza A virus and group-III adenovirus, Schiff-base formation	Passivation with ethanolamine	+	[4]
10^{-16} M, prostatic specific antigen, PSA	Silanization in APTES solution	PSA directed antibodies, APTES/GA directed antibodies, Schiff-base formation	Passivation with BSA molecules	–	[29]
10^{-16} M, bovine serum albumin, BSA	Silanization in TESU solution	BSA directed antibodies, Schiff-base formation	Passivation with ethanolamine	–	[10]
$<10^{-15}$ M, bird flu virus DNA	Silanization in APTES solution	5′-aminomodified DNA, APTES/GA directed antibodies, Schiff-base formation	Passivation with ethanolamine	–	[8]
2.5×10^{-15} M, prostatic specific antigen, PSA	Silanization in APMS solution	Monoclonal PSA directed antibodies, Schiff-base formation	Passivation with ethanilamine	+	[5]
3×10^{-15} M, antispecies antibodies	Silanization in GPTS solution	Rabbit monoclonal antibodies	Passivation with ethanolamine and gelatin	–	[23]
3.5×10^{-15} M, α-fetoprotein, AFP	Silanization in APTES vapor	AFP and APTES/EDC/NHS directed antibodies	–	–	[24]
10^{-14} M, CRP, (C-reactive protein)	Silanization in APTES solution	CRP and APTES/GA directed antibodies, Schiff-base formation	–	–	[30]
	Hydroxylation	Biotin	–	+	[11]

(continued)

Table 1 (continued)

Detection limit (DL), detectable species	Method of NW-surface modification[a]	Sensibilization method: probing molecules, linkers	Passivation method	Possibility of reusing	Ref. no.
10^{-14} M, streptavidin					
10^{-14} M, leptin, resistin	Silanization in APTES solution	Leptin or resistin directed antibodies, APTES/GA directed antibodies, Schiff-base formation	Passivation with ethanolamine	–	[12]
10^{-14} M, alpha tumor necrosis factor (ΦHO-α)	Silanization in APTES vapor	ΦHO-α and EDC/sulfoNHS directed antibodies	Passivation with ethanolamine	–	[28]
7×10^{-14} M, hepatitis B surface antigen, HBsAg	Silanization in APTES vapor	HBsAg and APTES/EDC/ NHS directed antibodies	–	–	[24]
Detection in biological liquids					
$2{,}5 \times 10^{-14}$ M, prostatic specific antigen (PSA) in blood sample	Silanization in APMS solution	PSA directed antibodies, Schiff-base formation	Passivation with ethanolamine	+	[13]
2.5 ng/ml (9×10^{-11} M), prostatic specific antigen (PSA) in blood sample	Silanization in APTES solution	PSA directed antibodies, APTES/EDC/ sulfo-NHS/ avidin	Passivation with 10 % embryonic calf serum	+	[26]
3.3×10^{-10} M, thrombin in blood sample	Silanization in APDES solution	Thrombin and APDES/succinic anhydride/EDC/ sulfo-NHS directed 5′-aminomodified aptamer	–	–	[31]
3×10^{-7} M (3 mkg/ml), C-reactive protein, (CRP), in serum sample	Silanization in APTES solution	CRP and APTES/GA directed antibodies, Schiff-base formation	–	–	[30, 31]
~10^{-15} M, (100 fg/ml), troponin T	Silanization in APTES solution	CK-MB, CK-MM, CTnT, and APTES/GA	Passivation with aminoPEG	–	[32]

(continued)

Table 1 (continued)

Detection limit (DL), detectable species	Method of NW-surface modification[a]	Sensibilization method: probing molecules, linkers	Passivation method	Possibility of reusing	Ref. no.
(CTnT), creatine kinase MM (CK-MM), and creatine kinase MB (CK-MB) in serum sample		directed antibodies, Schiff-base formation			
Detection in other media					
$<10^{-12}$ M, Leptin and resistin in culture medium for adipocites	Silanization in APTES solution	Leptin or resistin directed antibodies, APTES/GA directed antibodies, Schiff-base formation	Passivation with ethanolamine	–	[12]

[a] Substances used for modification of NW surface

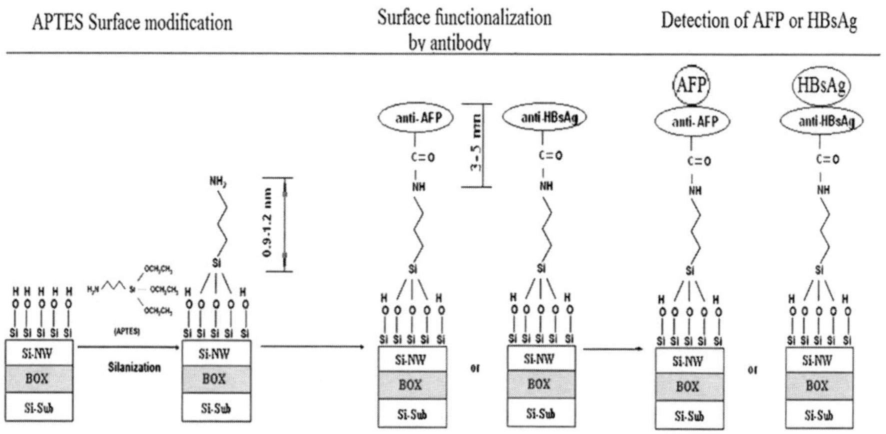

Fig. 2 Schematic illustrating the chemical modification with APTES and the functionalization of NW surface with antibodies

unwanted multilayer immobilization of biomolecules on the surface as such immobilization may seriously deteriorate the protein detection sensitivity. In [24], we showed that alpha-fetoprotein (AFP) directed antibodies and antibodies of hepatitis B surface antigen (HBsAg) can be covalently immobilized onto NW surface using a carbodimide cross-linker (Fig. 2). As an example, Fig. 3 shows an

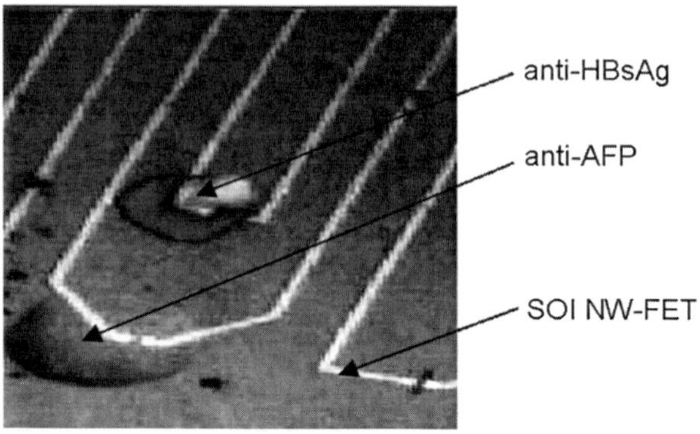

Fig. 3 An optical image of protein solution nanodroplets applied onto the surface of two nanowires (90 nm). The 3-nl volume nanodroplets contained anti-HBsAg (0.8 μM) and anti-AFP (1 μM) in KP-buffer (0.1 mM, pH 7.4) were applied onto silanized nanowires. The *arrows* point to antibody solution nanodroplets, and they also indicate the position of nanowires carrying the nanodroplets. The optical image was obtained using a BioForce Nanosciences microscope

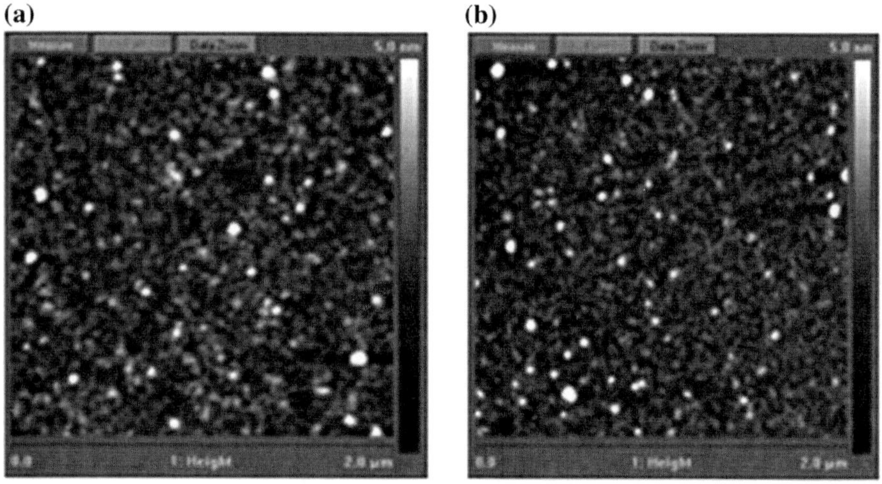

Fig. 4 AFM images of anti-HBsAg (**a**) and anti-AFP (**b**) species immobilized on a SOI surface. AFM visualization conditions: semi-contact mode, size of AFM images 2 × 2 μm^2, 5-nm is the full height of Z scale

optical image of a chip fabricated with pairwise clusterized nanowires and carrying nanodroplets that contained solutions of anti-AFP and anti-HBsAg probing molecules. The nanodroplets were applied onto NW surface by a piezorobot. Nanowires of volume $V = 3$ nl were located on individual nanowires with the spacing between them of 250 μm. In this way, we were able to form sensor regions with probes of various types. Hence, one chip with an array of sensor elements could be used to perform simultaneous detection of dissimilar target proteins.

The antibody immobilization efficiency, characterized by the immobilization density and by the layer thickness uniformity on the surface of SOI nanolayers, was evaluated using atomic force microscopy (AFM). Representative AFM images of anti-HBsAg and anti-AFP particles on a SOI wafer with heights of surface features up to 5 nm consistent with antibody sizes (see [34]) are shown in Fig. 4.

The AFM scanning data allowed us to evaluate the immobilization density of antibodies, which amounts to 1,000–3,000 molecules/100 nm^2. This density is proved to be sufficient for high-sensitivity detection of partner proteins at the level of 10^{-14}–10^{-15} M.

3 The Concentration Detection Limit of NW Biosensors

3.1 *Detection of Proteins and Oligonucleotides in Buffer Solutions*

3.1.1 Registration of Proteins

As a rule, protein adsorption leads to a linear dependence of SOI nano-FET current signal on the logarithm of concentration [35]. It should be noted that such dependence proved to be violated in some cases. For instance, Shalev et al. [37] examined the effect of BSA adsorption on the signal of an n-type NW detector with non-modified NW surface. It was shown that in the case of detection of this protein in a solution the biosensor signal exhibited a non-monotonic behavior as a function of protein concentration. This observation could be attributed to different degrees of protein oligomerization at different concentrations of proteins modifying the local distribution of charges on the surface of protein complexes in the vicinity of nanowire.

Equally, the dependence of NW-detector signal on the buffer pH value also proved to be nonlinear in some cases (see, e.g. [37]). However, it was shown that modification of NW surface with silanes improves characteristics of NW detectors and makes the dependence monotonic. For n-type silanized nanowires, an increase in conductivity with increasing the buffer pH value was observed (see [24]) while for p-type silanized nanowires the conductivity showed a decrease [7, 11].

3.1.2 Biospecific Registration of Proteins with Antibodies Used as Immobilized Probing Molecules

The NW functionalization procedures described above enable reaching a high sensitivity of NW biosensors in detection of proteins or oligonucleotides. In one of our studies, such a SOI-NW sensor was used for biospecific registration of alpha-fetoprotein (AFP) and hepatitis B surface antigen (HBsAg). The choice of the particular proteins was motivated by their diagnostic significance. A distinct stable rise of AFP concentration in blood serum is most often met in an organism suffering from hepatocellular cancer or testicle/ovary teratocarcinomas. That is why an enhanced level of AFP in blood serum can be considered as a marker of primary lever cancer [38]. HBsAg is the primary marker of hepatitis B, and in most cases it reveals itself during the incubation period [39]. In our study [24] we showed that the use of a SOI-NW detector with 90-nm wide nanowire makes it possible to achieve the registration sensitivity of HBsAg and AFP respectively at the level of 10^{-14}–10^{-15} M. The data reported in [24] showed that the addition of a solution with HBsAg and AFP molecules in KP-buffer (0.1 mM, pH 7.4) results in an increase of SOI-NW conductivity. Simultaneously, the addition of AFP molecules contained in K-phthalate buffer (0.1 mM, pH 5.2) leads to a decrease of the NW-detector current. Such an unusual biosensor response to addition of AFP in buffers with different values of pH can be attributed to the effect of solution acidity on the charge state of AFP-directed antibodies, which in turn causes the redistribution of charges in the antibody/protein complex formed on NW surface.

Table 1 gives examples illustrating the use and sensitivity limits of NW detectors obtained by various research groups. As it is seen from the table, registration of a broad class of proteins was attempted. The best sensitivity limit in protein detection in buffer solutions reached to date is 10^{-15}–10^{-16} M (see [24, 29]).

3.1.3 Biospecific Registration of Proteins with Aptamers Used as Immobilized Probing Molecules

As discussed above, since biospecific probes immobilized on NW-detector surface, as a rule in protein registration, antibodies are used. Yet, the use of antibodies is hampered by many factors, including the high production cost of antibodies, bad affine properties of antibodies resulting from their low stability, etc. Moreover, the range of available antibodies does not cover the range of proteins of interest for performing proteomic and diagnostic studies. The latter limitation can be eliminated by using aptamers as immobilized probing molecules. Aptamers are single-filament DNA- or RNA-structures obtained in vitro with the help of an SELEX (systematic evolution of ligands by exponential enrichment) process from a large ($>10^{13}$) library of randomized oligonucleotide sequences [40].

The possibility of using an aptamer as a probing molecule was demonstrated by Jolma et al. [41] with the example of detection of NFATc1 protein of coded genome contained in the 18th human chromosome. The expression level of the protein in blood increases during oncological deceases such as pancreatic cancer [42]. The NFATc1 directed aptamer was covalently immobilized onto the surface of an NW detector with the help of disuccinimide cross-linker (DSP) (see Fig. 5).

Figure 6 shows experimental data obtained in biospecific detection of NFATc1 by aptamer-sensibilized nanowires. It is seen that addition of NFATc1 solution in KP-buffer causes a reduction in the current value registered by the NW detector. Decreasing the protein concentration in the solution from 2.5×10^{-13} to 2.5×10^{-16} M, the response intensity of the detector was found to decrease.

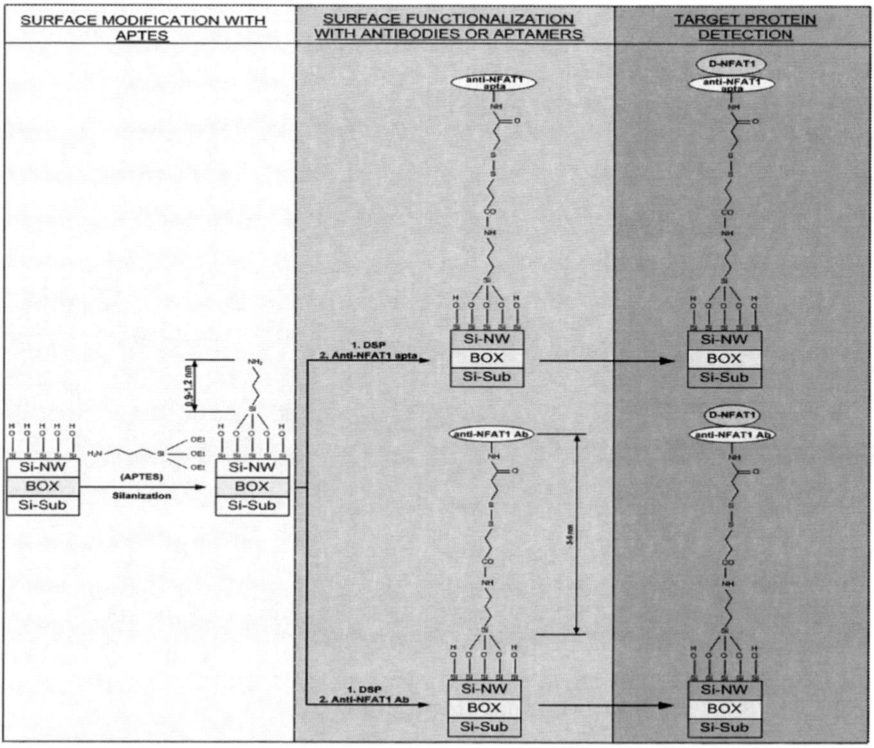

Fig. 5 Schematic illustrating the chemical modification with APTES and DSP (dithiobis succinimidyl propionate) and the cross-linker assisted functionalization of NW surface with antibodies and aptamers

The time necessary for high-sensitivity detection of biological objects in solutions with low concentrations of target species (10^{-13}–10^{-16} M) amounts to some 10 min [10, 11, 24, 26]. It should be emphasized that the experimental duration was several orders of magnitude smaller than the duration predicted by the classical model for diffusion-controlled mixing of molecules in a liquid medium [43]. As it was shown in [43], electroosmotic and electrophoretic effects due to applied voltage reduce the detection duration by several orders of magnitude as compared to the value predicted by the diffusion theory.

3.2 Protein Detection in Blood Serum

An important characteristic of biosensors defining the possibility of their use in diagnostics is the possibility of reusing the chips for such biosensors, i.e. the possibility of NW-surface regeneration after performing a test. In order to solve

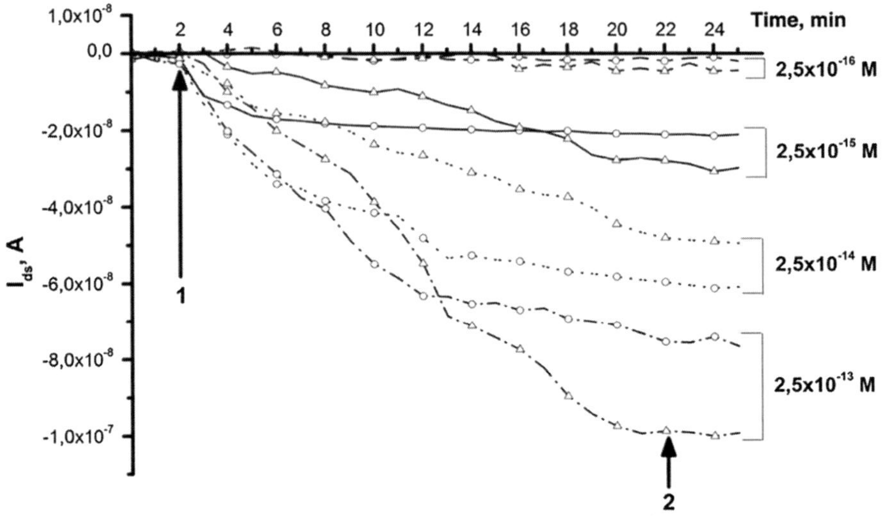

Fig. 6 ΔI_{ds} as a function of time in the biospecific detection of D-NFATc1 by NWs with immobilized antibodies or aptamers, respectively, markers (O) or (Δ). The *arrows* indicate the times at which the protein solution (1) or the cleaning buffer (2) were added to the measuring cell

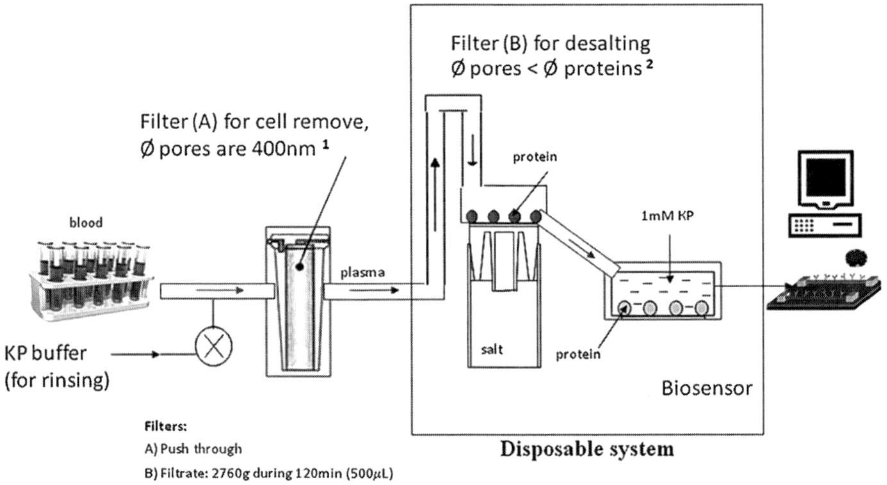

Fig. 7 Scheme of blood filtering and protein detection

this problem, proper washing buffers need to be chosen and/or reversible immobilization of probing molecules on the NW surface has to be used (Fig. 7). For instance, Lin et al. [44] used an immobilization procedure based on the reversible reaction between glutathione (GSH) and glutathione S-transferase. The bond between the proteins was found to rupture in the presence of 10 mM of GSH.

As probing molecules immobilized on NW surface, Lin et al. [44] used biotinylized GSH directed antibodies that helped to perform multiple streptavidin registration.

An original approach is the method based on the dissociation of antigen-antibody pairs using a washing buffer [13]. In [13], silane with terminal aldehyde group was used to reach NW surface modification. Surface sensibilization was realized by means of PSA directed antibodies immobilized due to the formation of the Schiff base. Blockage of the surface after the immobilization of antibodies was performed by ethanolamine. The use of such a functionalization procedure has enabled both easy destruction of anti-gene-antibody complexes in the course of buffer incubation and repeated high sensitivity registration of PSA.

Recently, a number of studies on to the detection of proteins in biological liquids have been reported (see Table 1). The detection limit in those studies was 10^{-7}–10^{-14} M. The latter values are lower than the detection limit achieved in the detection of proteins in buffer solutions. The latter circumstance most likely results from the nonspecific effect due to analyte components. A promising solution for enhancement of protein detection sensitivity by NW detectors is the use of aptamers as immobilized probing molecules [33], including modified aptamers, also called SOMAmers [45], or photoaptamers. The use of aptamers allows for harder conditions in washing the NW surface from nonspecific analyte components. Better trapping efficiency of biomacromolecules can be reached using SOMAmers because the production process of modified aptamers involves their selection in terms of low dissociation rate constant of the SOMAmer/biomacromolecule-target complex.

3.3 Detection of DNA

In recent years, there have been many attempts spent on the development of DNA biosensors allowing a quantitative real-time analysis of DNA. Zhang et al. [47] registered DNA using a Si-NW biosensor with immobilized receptor based on peptide-nucleic acid (PNA) with concentration sensitivity 10 femtomoles. A better sensitivity of 1 femtomole was achieved in DNA hybridization with oligonucleotides immobilized on the surface of a poly-Si nanowire [9]. The DNA probe was a 25-dimensional synthetic oligonucleotide. Lin et al. [8] used a polycrystalline Si-NW biosensor to detect a DNA complementary to the RNA of a highly pathogenic avian-flue (AI) virus strain. In the study, a 1-femtomole concentration sensitivity of DNA hybridization detection was achieved.

Alongside with high sensitivity, NW biosensors exhibit high specificity. For instance, it was shown that a Si-NW multi-channel biosensor allows registration of differences in hybridization between PNA probes immobilized on NW surface and DNA with various modifications in its sequence, for instance, with substitution of one, two, or three bases [46]. Note that in [47] the potential of NW detectors in early cancer diagnostics was shown. In the latter study, a Si-NW sensor with immobilized PNA probes was used to detect microRNAs in the total RNA fraction extracted out of HeLa cells. The latter process is of interest for early cancer diagnostics as normal and cancerous cells are generally assumed to contain different types and amounts of microRNA.

4 Promising Lines of Research in the Development and Application of Nanowire Detectors

4.1 Integration of Nanowire Transistors with Nanopore Matrices

The use of NW sensors is by no means restricted to the examples described above. Due to their high sensitivity, such sensors find applications in other fields. For instance, a combination of NW with a nanopore might give impetus to a new field in developing sequenators. Such a structure will allow registration of oligonucleotides at the level of single molecules through registration of a current pulse arising when a DNA molecule passes through a nanopore combined with an NW-FET channel. The latter signal emerges due to local variation of the electric potential as the DNA molecule traverses the nanopore. In [48] it was noted that, following further improvements of NW-nanopore detectors, such an approach would enable identification of DNA bases. In particular, the authors of [49, 50] suggested improvement of the spatial resolution of such a detector to be achieved by using, as a membrane, an atomically thin graphene layer [49] or F-29 protein with a built-in nanopore as a conducting FET channel between source and drain [50].

4.2 Sample Preparation Free Measurements in Strong and Diluted Solutions

Above all, further development of systems for personal genetic and immunodiagnostics will be defined by the rate of decrease of the price of such systems, and also by the rate of increase of their functionality. The decrease of price can be ensured at the expense of using NW FET arrays instead of normally used fluorescence-label registration systems, and also due to mass production of electrochemical-biochip integrated circuits, like in the case of IT-sector integrated circuits. Enhanced functionality and versatility of devices is to be achieved via the development of new sensor fabrication process, and also via introduction of novel sensor designs and analysis procedures. Thus, we can abandon awkward microfluidic sampling systems (systems for analyte dissolution and pre-conditioning) if, instead of measuring static characteristics of NW FETs, we will measure high-frequency characteristics. The latter will allow for overcoming the limits in terms of ion concentration in the solution that defines the Debye screening radius value L_D.

The concentration of ions in blood and in blood plasma reaches a value of 0.15 M, while in standard buffer solutions normally used in tests this concentration is expected to vary, depending on the analyte content of the solution, from 10 to 10^4 μM. The need for a low ion concentration in buffer solution arises because of the screening effect of ions. Yet, as it was shown by Gao et al. [46], one can overcome this negative factor due to the action of ions and measure, for instance, the biotin-streptavidin binding at a low streptavidin concentration with the help of NW detectors immediately in blood by measuring the response at a 10-MHz frequency sufficiently high for the ions in the solution not to respond to field variations in contrast to the dipole moment of complexes, which, although with a phase shift, will definitely respond to such variations, the latter respond being easily revealed by impedometry measurements. A resonant interaction can additionally enhance the detector response to the formation of antigen-antibody complexes. It is essential that a.c. measurements do not suffer from the dependence of sensor signal on the ion concentration in the solution whereas in d.c. measurements the above dependence is logarithmic. Besides, the amplitude of a.c. signal is inversely proportional to the distance from NW while in d.c. measurements the dependence is exponential, this circumstance rapidly deteriorating the detector sensitivity with decreasing the electrolyte Debye radius [51].

4.3 Sensitivity to Low-Copy Proteins in the Subfemtomole Range

A unique growth of sensitivity can be achieved due to combination of a nanowire nano-FET with a nanoelectromechanical structure (NEMS) prepared in the form of a flexible membrane occupying a non-equilibrium position in the vicinity of the field-effect transistor (NEMFET) [52]. Trapping of an analyte molecule at the membrane will cause the shift of the membrane closer to the nanowire and an exponential growth of the analyte-molecule-induced signal. Hahm et al. [7] have expressed an opinion that the sensitivity of NEMFET can be made several orders of magnitude higher than the sensitivity of any other presently known biosensors. Moreover, the detector of interest can register both charged and neutral biomolecules, without the necessity of using an additional reference electrode or additional devices, making the sensor potentially the most advantageous instrument for various low-budget applications, especially in the field of personal medicine. It is also likely that further development of the technology of layered materials (graphene, metal disulfides, etc.) existing due to van der Waals forces acting between layers will ensure fabrication of novel NEMS gates for NW FETs using the nanomechanical principle for raising detector sensitivity to biomolecules.

The analyte motion in the vicinity of nanowire as dependent on the pH value of the medium is greatly facilitated by an engineering design proposed by Kay et al. [53]. The authors of [53] have suggested using, as a transfer arm, a molecular arm formed by two segments of a single or double DNA spiral. The arm gets shortened or extended in strong dependence on the pH value of the medium in the most interesting range of pH values around pH 7. In [53], such a nanomanipulator was used for changing the distance between a semiconductor nanocrystal emitting light in the green portion of the spectrum and a red organic colorant molecule excited by the nanocrystal on FRET-induced (Fluorescent Resonant Energy Transfer) manipulator arm shortening. With the help of this device, the authors have measured the variation of pH value inside a living cell. In view of the fact that the NW detector sensitivity strongly depends on the distance between the analyte molecules and the surface of the NW detector, one can reach a high sensitivity and substantially extend the measurement range without invoking additional radiation sources or detectors, moving instead the protein closer to the surface of the NW FET sensor. The use of NEMS gates and molecular transfer arms holds promise for promoting nano-FET detectors in the subfemtomolar region ($DL < 10^{-15}$ M).

4.4 Personalized DNA Sequenators

Today, the most commercially viable semiconductor sequencing method is the product merchandized by Ion Torrent, USA (presently a department of the Life Technologies Corporation) under the Ion ProtonTM [54]. In the chip, a parallel data processing procedure is used. This procedure is implemented with an array of pH sensor cells (0.6 million cells per chip and over) prepared as deep micrometer-scale wells spaced several micrometers apart. On the bottom of the cells, gates are located. The gates are covered with a pH-sensitive dielectric layer (silicon nitride Si_3N_4 or tantalum oxide Ta_2O_5). In turn, the dielectric layer is covered with an emulsion ion-selective coating containing DNA genome fragment carriers. Incorporation of nucleotides into the growing DNA filament with the help of polymerase induces a small decrease in the pH value of the wells adjacent to the ion-sensitive transistors; the latter allows reading the nucleotide sequences in accord with the occurrence or non-occurrence of an electrically registered variation of pH on passing individual DNA-polymerase substrates (dATF, dGTF, dTTF and dCTF deoxynucleozidetriphosphates) through the cell [55].

In the Ion Proton^{TM} sequenator, an emulsion containing plastic microspheres 3 μm in diameter located on the bottom of sensor wells serves the function of DNA monoclone carriers. Expectedly, the application of micro- and nanoporous solid coatings developed at ISP SB RAS around electrochemical anodic etching and oxidation of Si, SiO_2 and Al_2O_3 to ion-sensitive layers will ensure a substantial simplification of the technology and an enhanced resolving power of future sequenators [55].

Increase in the spatial resolution of sensor elements is directly related to the increase of performance and to the reduction of the time required for identification of the pair sequence of nucleotide bases. For instance, recently developed (in 2012) Life Technologies Ion Proton System^{TM} semiconductor multi-molecular sequenator intended to deal with amplified segments formed by 100–200 base pairs of nucleotides of an initial nucleotide monomolecule and costing about 50,000 dollars in a cycle (several hours) performs an analysis of a sequence formed by 10 M base pairs. The PIITM model issued in 2013 reads up to 30 Gb per cycle due to the total number of chip cells increased from 0.6 to 660 millions. In principle, one can employ this model for performing full-genome tests. The explosive growth of the number of cells was achieved via reducing the size of pH-sensitive wells and increasing the cell density with simultaneous decrease of the Si-chip design rule. A predominant fraction of cycle duration is spent for amplification of the starting DNA molecule and for the analysis of gained data. The cost of one measurement cycle, including the cost of the chip and reagents, is within 1,000 dollars [56].

The CMOS-based platform that was developed for registration of pH signals with ion-sensitive transistors can easily be modified to allow operation with SOI NW transistors of ISP SB RAS [19, 20, 24, 36, 37] since nanostructuring and NW-FET sensibilization processes are only used at the last, low-temperature stages of the chip fabrication process while the sizes of SOI NW FET detectors, much smaller than the sizes of ion-sensitive transistors, make it possible to raise the total number of sensor cells by a factor of 10–1,000.

4.5 Requirements to an "Ideal" Biosensor for Personal Medicine

According to the opinion expressed in [57], an ideal device for personal genetic and immunodiagnostics in medicine should possess the following characteristics:

- a small sample volume (not more than 30 μl for whole blood),
- a short time sufficient for obtaining results (not longer than 5–10 min),
- simplicity,
- self-validation and self-calibration property of detector element,
- a high analysis accuracy (comparable with the accuracy being normally achieved in tests performed on laboratory equipment),
- roughly one-dollar cost of the disposable cartridge or chip for the device,
- designated purpose of chip tests.

For instance, the following combinations of target tests can be achieved on the basis of the universal platform of SOI-nano-FET sensors:

- identification of markers of somatic and infectious deceases, including hepatitis B and C;
- availability of cardiomarkers for patients with increased risk of cardiac infarction and oncomarkers;
- evaluation of hormone level for patients with endocrine-system chronic diseases;
- express identification of hospital infections in medical staff.

A key requirement for the universal platform is a good sensitivity and diagnostic specificity, which should be not lower than that achievable with conventional laboratory methods.

All the above-listed requirements can be met on the basis of FET chips. Such chips are the only artifacts produced by mankind in unprecedented amounts. Nowadays, for each Earth human there are more than 10 milliards such devices produced, with the production cost exponentially decreasing over the last 50 years. Mass production of NW FET biosensors with required performance characteristics presents no problem for advanced silicon microelectronics. The only pressing problem is integration of organic and inorganic materials in one chip and achieving an acceptably long storage life of produced biochips, and also development of their reuse methods, for instance, local-heating-based ones (see [58]).

5 Conclusions

Nanowire detectors offer considerable potential when used in genome and proteomic studies. A primary advantage offered by NW systems is their high concentration sensitivity, the registration time being as short as several minutes and the system themselves, multi-channel and portable, featuring low production and maintenance costs. The use of aptamers instead of antibodies in biospecific probes will enable a cheaper production and better performance characteristics of NW biosensors as used for registration of low-copy proteins.

The capability of NW sensors for real-time registration of nucleic acids with femtomolar sensitivity without using an amplification procedure allows performing rapid transcriptome analyses and genetic typing.

Acknowledgments The present study was supported in part by the Russian Foundation for Basic Research (Grants Nos. 12-04-31570 and 12-02-09443), by the Siberian Division of the Russian Academy of Science (Interdisciplinary Project No. 68-12-2 and State Contract No. 14.512.11.0018 concluded within the Federal Target Program "Research and development along most promising lines in modernization of Russia's scientific and technological complex for the years 2007–2013".

References

1. Archakov, A.I., Ivanov, Y.D., Lisitsa, A.V., Zgoda, V.G.: AFM fishing nanotechnology is the way to reverse the Avogadro number in proteomics. Proteomics **7**, 4–9 (2007)
2. Archakov, A.I., Ivanov, Y.D., Lisitsa, A.V., Zgoda, V.G.: Biospecific irreversible fishing coupled with atomic force microscopy for detection of extremely low-abundant proteins. Proteomics **9**, 1326–1343 (2009)
3. Naik, A.K., Hanay, M.S., Hiebert, W.K., et al.: Towards single-molecule nanomechanical mass spectrometry. Nature Nanotech **4**, 445–450 (2009)

4. Patolsky, F., Zheng, G.F., Hayden, O., et al.: Electrical detection of single viruses. Proc. Natl. Acad. Sci. U.S.A. **101**, 14017–14022 (2004)
5. Patolsky, F., Zheng, G.F., Lieber, C.M.: Fabrication of silicon nanowire devices for ultrasensitive, label-free, real-time detection of biological and chemical species. Nat. Protoc. **1**, 1711–1724 (2006)
6. Elfström, N., Juhasz, R., Sychugov, I., Engfeldt, T., Karlström, A., Linnros, J.: Surface Charge sensitivity of silicon nanowires: size dependence. Nano Lett. **7**, 2608–2612 (2007)
7. Hahm, J., Lieber, C.M.: Direct ultrasensitive electrical detecion of DNA and DNA Sequence Variations Using Nanowire Nanosensors. Nano Lett. **4**, 51–54 (2004)
8. Lin, C.-H., Hung, C.-H., Hsiao, C.-Y., et al.: Poly-silicon nanowire field–effect transistor and label-free detection of pathogenic avian influenza DNA. Biosens. Bioelectron. **24**, 3019–3024 (2009)
9. Wengaa, G., Jacquesa, E., Salaüna, A.-C., Rogela, R., et al.: Bottom-gate and Step-gate Polysilicon nanowires field effect transistors for ultrasensitive label-free biosensing application. Procedia Engineering **47**, 414–417 (2012)
10. Tian, R., Regonda, S., Gao, J., et al.: Ultrasensitive protein detection using lithographically defined Si multi-nanowire field effect transistors. Lab Chip **11**, 1952–1961 (2011)
11. Stern, E., Klemic, J.F., Routenberg, D.A., et al.: Label-free immunodetection with CMOS-compatible semiconducting nanowires. Nature **445**, 519–522 (2007)
12. Pui, T.S., Agarwal, A., Feng, Y., et al.: Ultra sensitive detection of cytokines with CMOS compartible silicon nanowire. Nanoscale **1**, 159–163 (2009)
13. Zheng, G.F., Patolsky, F., Cui, Y., et al.: Multiplexed electrical detection of cancer markers with nanowire sensor arrays. Nature Biotech **23**, 1294–1301 (2005)
14. Fan, H.J., Werner, P., Zacharias, M.: Semiconductor Nanowires: From Self-Organization to Patterned Growth. Small **2**(6), 700–717 (2006)
15. Naumova, O.V., Nastaushev Yu, V., Svitasheva, S.N., Sokolov, L.V., Zakharov, N.D., Werner, P., Gavrilova, T.A., Dultsev, F.N., Aseev, A.L.: MBE-grown Si whisker structures: morphological, optical and electrical properties. Nanotechnology **19**(5), 225708 (2008)
16. Huang, Y., et al.: Logic gates and computation from assembled nanowire building blocks. Science **294**, 1313–1317 (2001)
17. Andrea, R.T., Jiaxing, H., Yang, Peidong: Langmuir-Blodgettry Of Nanocrystals And Nanowires. Acc. Chem. Res. **41**(12), 1662–1673 (2008)
18. Lu, W., Lieber, C.M.: Nanoelectronics from the bottom up. Nature Mater **6**, 841–850 (2007)
19. Naumova, O.V., Fomin, B.I., Nasimov, D.A., et al.: SOI nanowires as sensors for charge detection. Semicond. Sci. Technol. **25**(8), 055004 (2010)
20. Naumova, O.V., Popov, V.P., Safronov, L.N., Fomin, B.I., et al.: Ultra-Thin SOI Layer Nanostructuring and Nanowire Transistor Formation for FemtoMole Electronic Biosensors. ECS Trans. **25**(10), 83–87 (2009)
21. Sudholter, E., Smet, L., Mescher, M., Uliien, D.: Organic Surface Modification of Silicon Nanowire-Based Sensor Devices, Nanowires—Implementations and Applications/Dr. Abbass Hashim (Ed.), ISBN: 978-953-307-318-7/
22. Linford, M.R., Fenter, P., Eisenberger, P.M., Chidsey, C.E.D.: Alkyl monolayers on silicon prepared from 1-alkenes and hydrogen-terminated silicon. J. Am. Chem. Soc. **117**(11), 3145–3155 (1995)
23. Carrara, S., Cavallini, A., Maruyama, Y., et al.: A new ethylene glycol-silane monolayer for highly-specific DNA detection on Silicon Chips. Surface Sci **604**, L71–L74 (2010)
24. Ivanov, YuD, Pleshakova, T.O., Kozlov, A.F., Malsagova, K.A., et al.: SOI nanowire for the high-sensitive detection of HBsAg and a-fetoprotein. Lab Chip **12**, 5104–5111 (2012)
25. Bunimovich, Y.L., Shin, Y.S., Yeo, W.S., et al.: Quantitative real-time measurements of DNA hybridization with alkylated nonoxidized silicon nanowires in electrolyte solution. J. Am. Chem. Soc. **128**, 16323–16331 (2006)
26 Stern, E., Vacic, A., Rajan, N.K., et al.: Label-free biomarker detection from whole blood. Nat. Nanotechnol. **5**, 138–142 (2010)

27. Yamada, K., Yoshii, S., Kumagai, S., Fujiwara, I., et al.: High-Density and Highly Surface Selective Adsorption of Protein–Nanoparticle Complexes by Controlling Electrostatic Interactio. Jpn. J. Appl. Phys. **45**, 4259–4264 (2006)
28. Hakim, M.M.A., Lombardinia, M., Sun, K., et al.: Thin Film Poly-crystalline Silicon Nanowire Biosensors. Nano Lett. **12**, 1868–1872 (2012)
29. Vu, C.-C., Pan, T.M., Wu, C.-S., et al.: Label-free Detection of Prostate Specific Antigen Using a Silicon Nanobelt Field-effect Transistor. Int. J. Electrochem. Sci. **7**, 4432–4442 (2012)
30. Lee, M.-H., Lee, K.-N.: Quantitative measurements of C-reactive protein using silicon nanowire arrays. Int. J. Nanomed. **3**, 117–124 (2008)
31. Lee, M.-H., Lee, D.-H., Jung, S.-W., et al.: Measurements of serum C-reactive protein levels in patients with gastric cancer and quantification using silicon nanowire arrays. Nanomed. Nanotechnol. Biol. Med. **6**, 78–83 (2010)
32. Zhang, G.-J., Chai, K.T.C., Luo, H.Z.H., et al.: Multiplexed detection of cardiac biomarkers in serum with nanowire arrays using readout ASIC. Biosens. Bioelectron. **35**, 218–223 (2012)
33. Kim, K.S., Lee, H.-S., Yang, J.-A., et al.: The fabrication, characterization and application of aptamer-functionalized Si-nanowire FET biosensors. Nanotechnology **20**, 235501 (2009)
34. Thomson, N.H.: Imaging the substructure of antibodies with tapping-mode AFM in air: the importance of a water layer on mica. J. Microsc. **217**, 193–199 (2005)
35. Carrara, S., Sacchetto, D., Doucey, M.-A. et al. Memristive-biosensors: a new detection method by using nanofabricated memristors/sensors and actuators B: Chemical, 2012. vol. 171–172. pp. 449–457
36. O.V Naumova et al., / 17 Simp" Nanostructures: Physics and Technology, Minsk, Belarus, June 22–27, 2009
37. Shalev, G., Rosenwaks, Y., Levy, I.: The interplay between pH sensitivity and label-free protein detection in immunologically modified nano-scaled field-effect transistor. Biosens. Bioelectron. **31**, 510–515 (2012)
38. Saito S., Ojima H., Ichikawa H., Hirohashi S., Kondo1 T. Molecular background of α-fetoprotein in liver cancer cells as revealed by global RNA expression analysis // Cancer Science, 2008. V. 99. P. 2402–2409
39. Jaroszewicz, J., Calle Serrano, B., Wursthorn, K., Deterding, K., et al.: Hepatitis B surface antigen (HBsAg) levels in the natural history of hepatitis B virus (HBV)-infection: A European perspective. J. Hepatology **52**, 514–522 (2010)
40. Ellington, A.D., Szostak, J.W.: In vitro selection of RNA molecules that bind specific ligands //. Nature **346**, 818–822 (1990)
41. Jolma, A., Kivioja, T., Toivonen, J., Cheng, L., et al.: Multiplexed massively parallel SELEX for characterization of human transcription factor binding specificities. Genome Res. **20**, 861–873 (2010)
42. Buchholz, M., Schatz, A., Wagner, M., Michl, P., et al.: Overexpression of c-myc in pancreatic cancer caused by ectopic activation of NFATc1 and the Ca2 +/calcineurin signaling pathway. EMBO J. **25**, 3714–3724 (2006)
43. Liu, Y., Guo, Q., Shuqiang, W., Hu, W.: Electrokinetic effects on detection time of nanowire biosensor. Appl. Phys. Lett. **100**, 153502-1–153502-4 (2012)
44. Lin, S.P., Pan, C.Y., Tseng, K.C., Lin, M.C., et al.: A reversible surface functionalized nanowire transistor to study protein–protein interactions. Nano Today **4**, 235–243 (2009)
45. http://www.somalogic.com
46. Gao, B.Z., Agarwal, A., Trigg, A.D., Singh, N., et al.: Silicon Nanowire Arrays for Label-Free Detection of DNA. Anal.Chem. **79**, 3291–3297 (2007)
47. Zhang, G.J., Chua, J.H., Chee, R.E., Agarwal, A., Wong, S.M.: Label-free direct detection of MiRNAs with silicon nanowire biosensors. Biosens. Bioelectron. **24**, 2504–2508 (2009)
48. Xie, P., Xiong, Q., Fang, Y., Qing, Q., Lieber, C.M.: Local electrical potential detection of DNA by nanowire–nanopore sensors. Nat. Nanotechnol. **7**, 119–125 (2012)
49. Banerjee, et al.: Electrochemistry at the Edge of a Single Graphene Layer in a Nanopore. ACS Nano **7**(1), 834–843 (2013)

50. Chen, Yu-Shiun, Lee, Chia-Hui, Hung, Meng-Yen, Pan, Hsu-An, Chiou, Jin-Chern, Steven Huang, G.: DNA sequencing using electrical conductance measurements of a DNA polymerase. Nat. Nanotechnol. **8**(3), 452–458 (2013)
51. Kulkarni, G.S., Zhong, Zh: Detection beyond the Debye Screening Length in a High-Frequency Nanoelectronic Biosensor. Nano Lett. **12**, 719–723 (2012)
52. Jain A., Nair P., Alam M.A. (2012) The Flexure-FET Biosensor to Break the Fundamental Sensitivity Limits of Nanobiosensors using Nonlinear ElectroMechanical. Proc. Natl. Acad. Sci. USA 109(24):9304–9308
53. Kay, E.R.: Lee Jungmin, Nocera D.G., Bawendi M.G. Conformational Control of Energy Transfer: A Mechanism for Biocompatible Nanocrystal-Based Sensors. Angew. Chem. **52**(4), 1165–1169 (2013)
54. http://www.iontorrent.com/ for Life Technologies iontorrent system
55. Parashchenko, M.A., Vandysheva, N.V., Kirienko, V.V., Filippov, N.S., et al.: Mikroelektronika **42**(1), 23–33 (2013)
56. Rothberg, J.M., et al.: An Integrated Semiconductor Device Enabling Non-Optical Genome Se-quencing. Nature **475**, 348–352 (2011)
57. Kalorama Information.: Worldwide market for in vitro diagnostic tests, 7th edn. In: 15th Anniversary Essay: The Future /IVD Technology—August 2010
58. Reddy, B., Elibol, O., Nair, P.R., Dorvel, B., Butler, F., et al.: Silicon field effect transistors as dual-use sensor-heater hybrids. Anal. Chem. **83**, 888–895 (2011)

Erratum to: Energy Harvesting Using THz Electronics

Stephen Hall, Ivona Z. Mitrovic, Naser Sedghi, Yao-Chun Shen, Yi Huang and Jason F. Ralph

Erratum to: Chapter 'Energy Harvesting Using THz Electronics' in: A. Nazarov et al. (eds.), *Functional Nanomaterials and Devices for Electronics, Sensors and Energy Harvesting*, Engineering Materials, DOI 10.1007/978-3-319-08804-4_12

The name of "Yao-chun C. Shen" should be spelled 'Yao-Chun Shen'.

Page 256, Equation 5:

$$R(V) = \frac{1}{2}\frac{I''}{I'}\Big|_{V=V_{app}} = \frac{1}{2V^2}\left(B \times \Delta E_C \times t_{ox} - \frac{1}{V}\right) \tag{5}$$

Page 257, Line: gives $C_D \sim 0.032$ fF (ignoring parasitics).

Page 258, Equation 7:

$$\varphi_n(x) = \sum_{j=1}^{M} A_{nj}\psi_j(x) = \sqrt{\frac{2}{t_{ox1} + t_{ox2}}}\sum_{j=1}^{M} A_{nj} \sin\left(\frac{\pi j x}{t_{ox1} + t_{ox2}}\right) \tag{7}$$

The online version of the original chapter can be found under DOI 10.1007/978-3-319-08804-4_12

S. Hall (✉) · I.Z. Mitrovic · N. Sedghi · Y.-C. Shen · Y. Huang · J.F. Ralph
Department of Electrical Engineering and Electronics, The University of Liverpool, Brownlow Hill, Liverpool L69 3GJ, UK
e-mail: s.hall@liverpool.ac.uk

A. Nazarov et al. (eds.), *Functional Nanomaterials and Devices for Electronics, Sensors and Energy Harvesting*, Engineering Materials,
DOI 10.1007/978-3-319-08804-4_21

MIX
Papier aus verantwortungsvollen Quellen
Paper from responsible sources
FSC® C105338

Printed by Books on Demand, Germany